PERTURBATION THEORY
FOR MATRIX EQUATIONS

STUDIES IN COMPUTATIONAL MATHEMATICS 9

Editors:

C.K. CHUI
Stanford University
Stanford, CA, USA

P. MONK
University of Delaware
Newark, DE, USA

L. WUYTACK
University of Antwerp
Antwerp, Belgium

ELSEVIER
Amsterdam – Boston – London – New York – Oxford – Paris
San Diego – San Francisco – Singapore – Sydney – Tokyo

PERTURBATION THEORY FOR MATRIX EQUATIONS

Mihail KONSTANTINOV
University of Architecture,
Civil Engineering and Geodesy
Sofia, Bulgaria

Da-Wei GU
University of Leicester
Leicester, UK

Volker MEHRMANN
Technical University of Berlin
Berlin, Germany

Petko PETKOV
Technical University of Sofia
Sofia, Bulgaria

N·H

2003

ELSEVIER
Amsterdam – Boston – London – New York – Oxford – Paris
San Diego – San Francisco – Singapore – Sydney – Tokyo

ELSEVIER SCIENCE B.V.
Sara Burgerhartstraat 25
P.O. Box 211, 1000 AE Amsterdam, The Netherlands

First edition 2003

Library of Congress Cataloging in Publication Data
A catalog record from the Library of Congress has been applied for.

British Library Cataloguing in Publication Data
A catalogue record from the British Library has been applied for.

ISBN: 0-444-51315-9
Series ISSN: 1570-579X

∞ The paper used in this publication meets the requirements of ANSI/NISO Z39.48-1992 (Permanence of Paper).
Printed in Hungary.

Preface

Matrix equations, such as Lyapunov, Sylvester, Riccati and other linear and nonlinear equations, are widely used tools in the general, stability and control theory for operator equations as well as in many application areas (systems theory, signal processing, and others). There is a huge literature on this topic that covers existence and uniqueness of solutions, numerical methods and also, more recently, perturbation analysis for special classes of such equations. Although perturbation theory is not very popular among scientists and engineers, it is essential for understanding the problems and estimating the accuracy of the computed results. Indeed, the mathematical models that are used to solve application problems are typically subject to modelling uncertainties (due to simplifications), and measurement errors in the data. Furthermore, the solution of the problem is usually carried out with numerical methods that may include approximation errors due to truncation of infinite series and/or discretization of continuous processes. In addition, the final result is contaminated by rounding errors due to the implementation of computational algorithms in finite precision arithmetic. The influence of the above uncertainties and errors on the computed result depends on the sensitivity of the problem. Thus, without a detailed perturbation analysis, it is not possible to assess the quality of the computed results.

In the last years, in a sequence of research papers, a general framework has been developed by the authors of this monograph in order to perform perturbation analysis of general matrix equations in a systematic way and it is the main goal of this monograph to present this general scheme in a concise and systematic way. Then, for several important classes of matrix equations, the framework is specialized and the perturbation results are presented. In all cases both local first order and nonlocal perturbation bounds are derived.

The general framework for perturbation analysis of matrix equations was derived in part while the authors were cooperating in the development of numerical methods in control within the European Community BRITE-EURAM III Thematic Networks Programme NICONET (contract number BRRT-CT97-5040). We thank the other partners of this network for many helpful discussions.

We also thank the Departments of Mathematics at the Technical University of Chemnitz, Technical University of Berlin, and University of Architecture, Civil Engineering and Geodesy – Sofia, the Department of Engineering at the University of Leicester, and the Department of Systems and Control at the Technical University of Sofia for providing excellent facilities to carry out this research. We also thank the DFG Research Center FZT86 "Mathematics for Key Technologies" in Berlin for the support in the final stage of preparing of this manuscript.

This book would not be accomplished without the help and understanding of our wives and children. We cordially thank them all.

Contents

Chapter 1

Introduction

This monograph is devoted to the perturbation analysis of algebraic matrix equations. In general, the perturbation analysis of a given problem is aimed at estimating the perturbation in the solution as a function of perturbations in the data, see [65, 206, 134, 135, 127] as well as [119, 16, 60] for a general treatment of this subject.

There are many reasons to look for perturbation bounds for a given problem (a *perturbation bound* is a function whose argument is the perturbation in the data and which majorizes the perturbation in the solution). Major sources of perturbations are parametric and structural uncertainties in mathematical models as well as the effects of finite precision arithmetics in the numerical simulation of the models.

Mathematical models for the description of the physical behavior of a system typically contain measurement errors, modelling errors and/or estimated parameters. When such models are treated numerically, discretization and rounding errors are introduced. Furthermore, usually a given model is applicable only for values of its parameters within certain bounds. For parameter values out of these bounds the model is not correct and the solution of the corresponding computational problem may not exist or may have no physical meaning.

Let us consider a real world example which displays these issues.

Example 1.1 To derive a mathematical model that describes the complete traffic in a realistic rail network [31], many components are needed, which include the dynamic equations for the movement of the train; the constraints for the movement, like e.g. global velocity constraints; the properties of the real network (e.g. local velocity constraints, slopes) and the complex interaction between the trains induced from the signal system and the schedule.

The most simple model for the dynamics of one train is the motion of one mass point (the center of mass of the vehicle), governed by the second order scalar differential equation $\ddot{x}(t) = f(x(t), u(t))$ with initial conditions $x(t_0) = x_0$, $\dot{x}(t_0) = \dot{x}_0$. Here $x(t)$ is the position of the point at the moment t and x_0, $\dot{x}_0$ are the initial values for the position and the velocity of the point. As the load of the vehicle changes, the mass is only determined approximately, while the center of mass may change as well. Also, due to delays and the interaction between the trains, the determination of the initial or the current position and velocity is also contaminated with errors. All in all, the model which is a high dimensional nonlinear control problem, however refined it may be, will contain a lot of simplifications and uncertainties. If one uses a numerical method to solve the resulting boundary value problem, then discretization and rounding errors occur. It is evident for everybody that uses such a train network regularly that the effect of these measurement and computational errors may be very large even if the errors themselves are small. $\diamondsuit$

This example demonstrates that it is important to know how much the solution of a problem may change when the data vary over certain admissible sets. A simple problem (which may be computationally very difficult!) is the evaluation of a scalar real continuous function φ at a given point $a \in \mathbb{R}$, i.e., the computation of $x = \varphi(a)$. Let δa and δx be perturbations in a and x with

$$\delta x = \varphi(a + \delta a) - \varphi(a).$$

Then a perturbation estimate is an inequality of the form

$$|\delta x| \leq f(|\delta a|),$$

where the perturbation bound f is a nonnegative nondecreasing function on certain interval $[0, c)$, $c > 0$, and satisfies $f(0) = 0$.

In this monograph we present basic concepts and tools in perturbation theory for the solution of computational problems in finite dimensional spaces. We, in particular, derive the following types of perturbation bounds.

– *Local* (or *asymptotic*) *bounds* . These bounds are linear or first order homogeneous functions in the data perturbations and are valid theoretically when the perturbations tend to zero. The coefficients in the linear functions are known as *condition numbers.* For particular perturbations, however, asymptotic bounds may be misleading: either because of underestimating significantly the actual perturbation they claim to estimate, or because the perturbed problem does not even have a solution. Asymptotic bounds are often obtained by using (or estimating) the Fréchet derivatives of certain mappings.

– *Linear nonlocal bounds.* Sometimes it is possible to derive linear bounds which are nonlocal and thus do not suffer from the disadvantages of the local estimates. In general it may be more difficult to get reasonable linear nonlocal estimates in comparison with the local ones.

It must be stressed that for some problems asymptotic first order bounds, both local and nonlocal, do not exist, see Chapter 2.

- *Nonlinear nonlocal bounds.* If properly defined, these bounds always estimate the true perturbation from above. Moreover, for the domain where these bounds are well defined, it is guaranteed that the perturbed problem has a solution for which the bound is valid. Nonlocal bounds are usually defined as real analytic functions in subsets of the set of admissible perturbations. The first order term of the corresponding Taylor series expansion gives an estimate from above (or is equal) to the local bound. As would be expected, nonlinear nonlocal bounds are more difficult to derive in comparison with the local bounds. A disadvantage of nonlinear bounds is that they may have smaller domain of applicability in comparison with the actual domain of perturbations for which the solution of the perturbed problem still exists. Obviously this is the price of having rigorous perturbation results.

Example 1.2 Let us illustrate the different types of bounds with a very simple but instructive example. Consider the real scalar equation $ax = b$ for $a = b = 1$. Let δa, δb be perturbations in a, b with $|\delta a| \leq \alpha < 1$, $|\delta b| \leq \beta$, and let δx be the corresponding perturbation in the solution $x = 1$. Then we have

$$\delta x = \frac{\delta b - x\delta a}{a + \delta a} = \frac{\delta b - \delta a}{1 + \delta a}, \quad |\delta a| < 1.$$

For small α, β the local estimate

$$|\delta x| \leq \alpha + \beta,$$

linear in α, β, is often used in practice, since it guarantees an accuracy of order $O(\varepsilon^2)$, $\varepsilon \to 0$, where $\varepsilon = \max\{\alpha, \beta\}$. The bound $\alpha + \beta$ may severely be violated for δa approaching -1.

A nonlocal nonlinear estimate is

$$|\delta x| \leq \frac{\alpha + \beta}{1 - \alpha}, \quad \alpha < 1.$$

This estimate is rigorous but it may be very pessimistic for δa close to 1.

For $\alpha \leq 0.5$ we may also use the nonlocal linear estimate

$$|\delta x| \leq 2(\alpha + \beta), \quad \alpha \leq 0.5.$$

This estimate is rigorous but may be pessimistic for δa close to 0.5. Note that the above nonlocal estimates "work" better for $\delta a < 0$.

Finally, there is an interesting phenomenon. For $\delta a = \delta b$ we have $\delta x = 0$ and all estimates are pessimistic.

In Figure 1.1 we give the local linear, nonlocal linear and nonlocal nonlinear bounds, respectively, for $\beta = 0$ and $0 \leq \alpha < 1$. For $\delta < 0$ the linear bound always

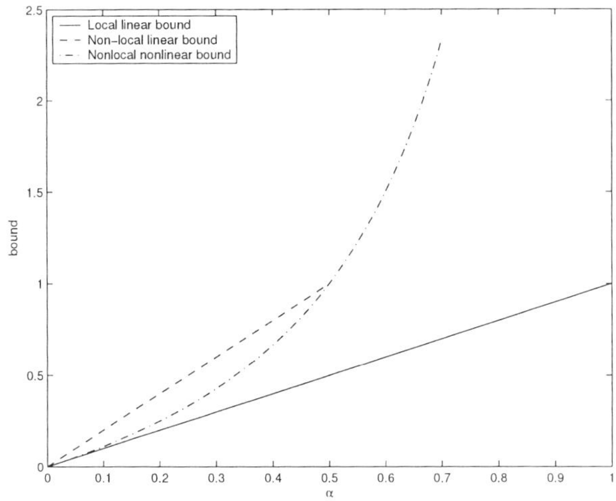

Figure 1.1: Local and nonlocal bounds

underestimates the true perturbation which in this case is equal to the nonlocal bound.
$\diamondsuit$

A very effective tool to get both local and nonlinear nonlocal perturbation bounds is the technique of *Lyapunov majorant functions* [160, 85, 135, 127], or briefly *Lyapunov majorants.* We will discuss this technique in great detail in Chapter 5 and use it for all classes of matrix equations that we consider. In order to apply this technique, the perturbed problem is first rewritten as an equivalent operator equation for the perturbation in the solution. It is then shown that the corresponding operator maps a certain compact convex set (contracting to the origin when the perturbations tend to zero) into itself. Then, according to the Schauder fixed point principle, there exists a solution of the operator equation, tending to zero together with the data perturbations. If in addition the operator is a contraction, the uniqueness of the solution to the perturbed problem is guaranteed in view of the Banach fixed point principle. Estimating the domain of the operator by Lyapunov majorants gives the desired nonlocal nonlinear perturbation bounds.

Throughout this monograph we will use the following framework for the perturbation analysis of matrix equations that was suggested in [127].

Consider a general matrix equation

$$F(A, X) = 0,$$

where F is a continuous matrix valued function, $A = (A_1, \ldots, A_r)$ is a collection of matrix parameters and X is the unknown matrix. Let X be a given solution and let the data be changed from A to $A + \delta A$. Then we obtain the perturbed equation

$$F(A + \delta A, X + \delta X) = 0,$$

where δX is the perturbation in the solution.

Two major problems then arise:

- Find conditions which guarantee that the perturbed equation has a solution $\delta X = \Xi(\delta A)$, depending continuously on δA and such that $\Xi(0) = 0$.

- Derive computable bounds for a norm $\|\delta X\|$ of δX as a function of the perturbations $\delta_i = \|\delta A_i\|$.

To solve these problems we follow a *general framework for the perturbation analysis of matrix equations* [102] that consists of the following stages.

1. *Construction of an equivalent operator equation.* This is a matrix equation

$$\delta X = \Pi(\delta A, \delta X)$$

   for δX, where $\Pi(0,0) = 0$. For this purpose the technique of Fréchet derivatives is used. Via an appropriate representation, the operator equation is then represented as an equivalent matrix equation. After this, the matrix equation is vectorized as $x = \pi(a,x)$, where $x = \mathrm{vec}(\delta X)$ and $a = (a_1, \ldots, a_r)$, $a_i = \mathrm{vec}(\delta A_i)$.

2. *Calculation of condition numbers.* The quantity $\pi(a,x)$ is represented as

$$\pi_{10}(a) + \pi_{20}(a) + \pi_2(a,x),$$

   where

$$\begin{aligned} \pi_{10}(a) &= O(\|a\|),\ a \to 0, \\ \pi_{20}(a) &= o(\|a\|),\ a \to 0, \\ \pi_2(a,x) &= o(\|a\| + \|x\|),\ \|a\| + \|x\| \to 0. \end{aligned}$$

   When only one component a_i of a is nonzero, then the quantity $\|\pi_{10}(a)\|/\|a_i\|$ is asymptotically bounded by K_i, the absolute condition number for the solution X relative to perturbations in the matrix A_i. Here K_i is the asymptotic Lipschitz constant of π_{10} in a_i (if π_{10} is not Lipschitz continuous in a_i then the condition number relative to A_i does not exist). If F is Fréchet differentiable then the condition numbers (in Frobenius norm) are the spectral norms of certain matrices depending on the Fréchet derivatives of F.

3. *Derivation of local perturbation bounds and overall measures of conditioning.* The maximum of $\|\pi_{10}(a)\|$ under the constraints $\|a_i\| \leq \delta_i$, $i = 1, \ldots, r$ is estimated to obtain local perturbation bounds and overall measures of conditioning. These are solutions to complicated optimization problems. For the solution of those problems, simple and easily–computable upper bounds are derived.

4. *Construction, analysis and solution of Lyapunov majorant equations.* Setting $\delta = (\delta_1, \ldots, \delta_r)$, a Lyapunov majorant function for the operator $\pi(a, \cdot)$ is constructed, This is a function $(\delta, \rho) \mapsto h(\delta, \rho)$ such that

$$\|\pi(a, x)\|_2 \leq h(\delta, \rho)$$

provided that $\|a_i\|_2 \leq \delta_i$ and $\|x\|_2 \leq \rho$. Under certain conditions on h and δ the majorant equation

$$\rho = h(\delta, \rho) \tag{1.1}$$

has a solution $\rho_0 = f(\delta)$, where f is continuous and $f(0) = 0$. An inclusion of the type $\delta \in \Omega$, where Ω is a certain set (possibly small but finite) then guarantees that such a solution exists.

In many cases the majorant equation (1.1) is an algebraic equation. Then there are two approaches to solve this algebraic equation. For a given δ the majorant equation is either solved numerically or (if possible) analytically to determine the smallest positive root ρ_0. If there are no positive solutions then δ is too large and the method of Lyapunov majorants does not produce nonlocal perturbation bounds. This may also indicate that the perturbed equation has no solutions δX vanishing together with δP. A second approach is to majorize $h(\delta, \rho)$ by a new Lyapunov majorant $\widehat{h}(\delta, \rho)$ for which the equation $\rho = \widehat{h}(\delta, \rho)$ has a convenient closed form solution $\widehat{\rho}_0 = \widehat{f}(\delta)$ with $\widehat{f}$ continuous and $\widehat{f}(0) = 0$. This guarantees that the initial majorant equation has a solution $\rho_0 \leq \widehat{f}(\delta)$.

5. *Topological fixed point principles and nonlocal perturbation bounds.* If we have a smallest solution $f(\delta)$ of the majorant equation (or some of its upper bounds $\widehat{f}(\delta)$), then the fixed point principles of Schauder and Banach are used to prove that the equivalent vector equation has a solution x in the central, closed ball of radius $f(\delta)$. In view of the identity $\|\delta X\|_{\mathrm{F}} = \|x\|_2$ this gives the nonlocal estimate

$$\|\delta X\|_{\mathrm{F}} \leq f(\delta) \leq \widehat{f}(\delta), \ \ \delta \in \Omega.$$

The monograph is organized as follows. In Chapters 2–4 we give the problem statement and consider general problems with explicit and implicit solutions. We present the basic concepts (regularity and conditioning in particular), related to the sensitivity of computational problems. The technique of Lyapunov majorants is presented in Chapter 5 and singular problems are briefly discussed in Chapter 6.

General concepts concerning types and properties of perturbation bounds are introduced in Chapter 7. Then in Chapter 8 and 9 perturbation bounds for general and specific Sylvester equations are derived. Using symmetry, then these bounds are extended in Chapter 10 for general Lyapunov equations and in Chapter 11 for Lyapunov equations from systems and control theory.

The perturbation analysis for general quadratic equations is presented in Chapter 12. These results are then improved for continuous-time Riccati equations that arise in the control and filtering of linear time-invariant systems in Chapter 13. For systems of coupled Riccati equations the perturbation results are presented in Chapter 14. General fractional-affine equations are studied in Chapter 15. In Chapter 16 perturbation results are given for discrete-time Riccati equations that arise in control theory as well as for a class of symmetric fractional-affine equations.

The monograph includes several appendices where the following issues are considered: elements of algebra and analysis, unitary and orthogonal decompositions, Kronecker product of matrices, fixed point principles, Sylvester, Lyapunov and Lyapunov-like operators. Finally a list of notation is given that is used throughout the monograph.

Chapter 2

Perturbation problems

2.1 Introductory remarks

The aim of perturbation analysis is to study the sensitivity of computational problems or mathematical models under perturbations. This means to estimate how the solution changes when the data of the problem are changed. In a more restricted framework the objective of perturbation analysis is to provide computable bounds for the perturbation in the solution of a given problem as a function of the perturbation in the data. At present, perturbation analysis techniques are important issues in numerical analysis and control and also in all areas of science and engineering.

In this chapter principal issues in the perturbation analysis of computational problems in finite dimensional spaces (matrix equations) are discussed, which include:

- properties of the perturbation function,
- sensitivity and conditioning,
- classification of perturbation bounds,
- properties and classification of solutions and solution sets of equations,
- construction of equivalent perturbation operators,
- Lyapunov majorants,
- application of fixed point principles,
- analysis of singular problems,
- scaling of problems and error estimates.

Examples and case studies are included and are illustrated.

2.2 Problem statement

Independently of their particular nature, most problems in science and engineering may in general be formulated in one of the following two ways: as problems with *explicit solutions*, e.g. evaluating functions defined by explicit computable expressions, and as problems with *implicit solutions*, e.g. solving equations. There are also modifications of these problems such as computing canonical forms of matrices under the action of various transformation groups. This distinction is sometimes only formal, since the above ways to formulate a problem may often, at least in theory, be transformed into each other. Also, a complicated problem may be defined by a chain of explicit and implicit subproblems.

Consider a function $\Phi : \mathcal{A} \to \mathcal{X}$, where the set $\mathcal{A}$ of data, or inputs, and the set $\mathcal{X}$ of results, or outputs, are (subsets of) normed linear spaces, which are usually finite-dimensional. The function Φ is continuous and in many problems it is differentiable. For every data $A \in \mathcal{A}$ we have the result

$$X = \Phi(A) \in \mathcal{X}.$$

A particular example for such problems is the evaluation of a scalar or vector function.

Sometimes the dependence of X on A is not functional, in the sense that there may be more than one result corresponding to a given data (for some problems this type of nonuniqueness may be inherent). In this case we may consider a set-valued function $\widehat{\Phi} : \mathcal{A} \to 2^{\mathcal{X}}$ which assigns a set $\widehat{\Phi}(A)$ of solutions to every data $A \in \mathcal{A}$. A useful approach here is to derive computable perturbation bounds which hold at least for one of the solutions of the perturbed problem.

A *computational problem* is identified with the pair (Φ, A) when we deal with a single collection of data, and with the pair $(\Phi, \mathcal{A})$ when we have a family of problems with data from the set $\mathcal{A}$.

The function Φ may also be defined implicitly via the equation

$$F(A, X) = 0, \tag{2.1}$$

where $F : \mathcal{A} \times \mathcal{X} \to \mathcal{X}$ is a given continuous function. (Sometimes F is a function from $\mathcal{A} \times \mathcal{X}$ to $\mathcal{X}_1$, where $\mathcal{X}_1$ is another linear space.) The problem here is to compute the solution X for a given A and to investigate (at least locally) the behavior of the implicitly defined function $\Phi : \mathcal{A} \to \mathcal{X}$, satisfying the identity $F(B, \Phi(B)) = 0$ in a certain neighborhood of A.

Typical examples of such problems are various classes of linear and nonlinear matrix equations in linear algebra and control theory such as the equations of Sylvester, Lyapunov, Riccati and others that are discussed in detail below.

Modifications of implicit problems are various types of matrix decompositions and canonical or condensed forms of matrices. They may be formulated by the equation

$$P(C(A,U)) = 0, \tag{2.2}$$

where U is a transformation matrix from a certain matrix group Γ, $C(A,U)$ is a condensed form of A with C being a continuous function and P is a projector. The problem here is to compute both U and $C(A,U)$, where $U = U(A)$ is implicitly defined by the data A via equation (2.2), i.e. $P(C(A,U(A))) = 0$. We stress that the dependence of U on A may not be functional, especially when condensed rather than canonical forms are considered.

We consider problems in which a subset $\mathcal{A}$ of matrix r-tuples from a linear space, interpreted as data, is transformed into the set of matrices $\mathcal{X} := \mathbb{F}^{n\times m}$, interpreted as results, where $\mathbb{F}^{n\times m}$ is the space of $n \times m$ matrices over $\mathbb{F}$ and $\mathbb{F}$ is the set of real ($\mathbb{F} = \mathbb{R}$) or complex ($\mathbb{F} = \mathbb{C}$) numbers. The spaces $\mathcal{X}$ and $\mathcal{A}$ are endowed with norms or generalized norms.

Both the data and the result of a given problem may be elements of infinite-dimensional (Hilbert or Banach) spaces. But in numerical computations one can deal only with finite dimensional spaces, and actually, in a finite precision environment, only with finite sets of rational numbers. So we typically deal with data that are collections of matrices (a collection is a set with possibly repeated elements.)

The assumption that the data A is a collection of matrices is natural, since in practice all problems depend on finite collections of input parameters. But that the result X is an element of a finite-dimensional space may seem rather restrictive having in mind problems defined via differential or other functional equations. In fact, it is not.

Consider a problem, defined by the relations $G(A,Y) = 0$ and $X = H(Y)$, where $G : \mathcal{A} \times \mathcal{Y} \to \mathcal{Y}_1$ and $H : \mathcal{Y} \to \mathcal{X}$ are given functions, and the solution Y of the equation $G(A,Y) = 0$ is an intermediate result. Here the spaces $\mathcal{Y}$ and $\mathcal{Y}_1$ may be infinite-dimensional but the final result X is a finite collection of numbers, see next example as well as Examples 2.5 and 2.6 below.

Example 2.1 In Example 1.1 the solution x is a function but actually only the values of x at a certain finite set of times are needed. $\diamondsuit$

When studying the sensitivity of a problem, identified with a certain function $\Phi : \mathcal{A} \to \mathcal{X}$, we assume that Φ has some minimal smoothness properties and, in particular, that it is continuous in a neighborhood of given data A. This issue is not trivial, since even simple nonlinear equations of type (2.1), together with smooth solutions, may have discontinuous solutions $A \mapsto \Phi(A)$.

Consider a particular problem corresponding to a given *nominal data* $A \in \mathcal{A}$. When the nominal data is changed to $A + \delta A$ we get a new problem, the so called *perturbed problem*. One of the main characteristics of a problem is its *sensitivity*. Quantitatively the sensitivity is a numerical measure of the continuity properties of the function Φ near certain nominal data or in the whole data space. Functions whose values change significantly when the argument is slightly changed, correspond to sensitive problems. But "significantly" and "slightly" are not mathematical terms and even for practical purposes they must be described by some quantitative measures.

A practically useful measure of sensitivity must be connected to the parameters of the *finite precision arithmetic*. Thus, typically the sensitivity becomes a joint property of both the problem and the finite precision arithmetic.

A sensitive problem is usually computationally difficult to solve and one must expect that the computed solution is contaminated with large errors. However, in a different computational environment the same problem may be solved more accurately. Of course, the properties of the implemented numerical algorithm are also crucial for the accuracy of the solution computed in finite precision arithmetic.

As we have discussed before, one of the main purposes of perturbation analysis is to study qualitatively and quantitatively the sensitivity of individual problems (for fixed nominal data) or of classes of problems (for data from a given set). However, for many problems the domain $\mathcal{A}$ (or the set of all A such that $\Phi(A)$ is well defined) is not known a priori, e.g., when the function Φ is defined implicitly via an equation. Here it is important to get an estimate for the *natural domain* of Φ, i.e. for the largest set on which Φ can be defined. This also gives an answer to the question whether the solution of a particular perturbed problem exists.

Perturbation analysis produces *local* (or *asymptotic*) and *nonlocal* bounds for the perturbation in the result as functions of the perturbation in the data.

Local bounds are linear or first order homogeneous functions of the perturbations in the data. They are valid asymptotically, for infinitesimal perturbations $\delta A \to 0$ in the data only. They are often obtained in a relatively simple way and are in many cases easy to compute.

An example of such a local bound is the *condition number* of a problem that is widely used throughout numerical analysis.

Definition 2.2 For a given problem $X = \Phi(A)$ the finite quantity

$$K(A) := \lim_{\alpha \to 0} \sup \left\{ \frac{\|\Phi(A + \delta A) - \Phi(A))\|}{\|\delta A\|} : \delta A \neq 0, \ \|\delta A\| \leq \alpha \right\} \tag{2.3}$$

is called the *absolute condition number* of the problem $X = \Phi(A)$.

In this definition only the dependence of K on the data A is explicitly marked assuming that the function Φ is fixed. Sometimes it is convenient to write the absolute condition number also as $K(\Phi, A)$, showing its dependence on the function Φ as well.

When the data of a problems differ widely in their magnitude, then it is often better to measure the sensitivity in terms of the *relative perturbations*

$$\rho_X := \frac{\|\delta X\|}{\|X\|}, \ \rho_A := \frac{\|\delta A\|}{\|A\|}$$

in the solution and data when $X \neq 0$ and $A \neq 0$.

Definition 2.3 The quantity

$$\kappa(A) := K(A)\frac{\|A\|}{\|X\|} = K(A)\frac{\|A\|}{\|\Phi(A)\|}$$

is called the *relative condition number* of the problem (Φ, A).

We will return to condition numbers in much more detail below.

In contrast to local bounds like condition numbers, *nonlocal* perturbation bounds are usually nonlinear functions and they are valid rigorously in a neighborhood of the nominal data. The derivation of nonlocal perturbation bounds is more involved. In addition they may be pessimistic in terms of both the size of the predicted perturbation and the domain of applicability.

In matrix problems the set $\mathcal{A}$ is a subset of a linear, finite dimensional, real or complex space $\mathcal{V}$ of dimension $\dim(\mathcal{V})$. Since $\mathcal{X} = \mathbb{F}^{n\times m}$, we also have $\dim(\mathcal{X}) = nm$. Hence $\mathcal{A}$ may be identified with a subset of $\mathbb{F}^{\dim(\mathcal{A})}$ and $\mathbb{F}^{n\times m}$ with $\mathbb{F}^{nm}$. In the following we assume that $\mathcal{A}$ is an open subset of the Cartesian product $\mathcal{V}$ of $r \geq 1$ matrix spaces $\mathcal{V}_1, \ldots, \mathcal{V}_r$,

$$\mathcal{A} \subset \mathcal{V} := \mathcal{V}_1 \times \cdots \times \mathcal{V}_r, \ \mathcal{V}_i := \mathbb{F}^{m_i \times n_i} \tag{2.4}$$

and

$$\dim(\mathcal{A}) = \dim(\mathcal{V}) = m_1 n_1 + \cdots + m_r n_r.$$

Thus, the data A is a matrix r-tuple

$$A = (A_1, \ldots, A_r) \in \mathcal{V}, \ A_i \in \mathcal{V}_i. \tag{2.5}$$

When dealing with perturbation problems we use norms and generalized norms for the corresponding matrices and/or matrix r-tuples. We denote by $\|X\| \in \mathbb{R}_+$ and $|X| = [|x_{ij}|] \in \mathbb{R}_+^{m\times n}$ the *norm* and the *absolute value* of the matrix $X = [x_{ij}] \in \mathcal{X}$, respectively. Here $\|\cdot\|$ is a unitarily invariant norm such as the spectral norm $\|\cdot\|_2$ or the Frobenius norm $\|\cdot\|_{\mathrm{F}}$. For a matrix r-tuple (2.5) we use the norm

$$\|A\| := \|A_1\| + \cdots + \|A_r\|$$

or the generalized norm

$$|\!|\!|A|\!|\!| := [\|A_1\|, \ldots, \|A_r\|]^\top \in \mathbb{R}_+^r. \tag{2.6}$$

We also use the matrix absolute value (which is also a generalized norm)

$$|A| := (|A_1|, \ldots, |A_r|) \in \mathcal{V}_+ := \mathbb{R}_+^{m_1\times n_1} \times \cdots \times \mathbb{R}_+^{m_r \times n_r}. \tag{2.7}$$

The generalized norms $|\!|\!|\cdot|\!|\!|$ and $|\cdot|$ are functions of the type $\nu : \mathcal{V} \to \mathcal{K}$. Here $\mathcal{K}$ is a nonnegative cone, defining a partial order relation $\preceq$ by $x \preceq y$ if $y - x \in \mathcal{K}$. (In our case $\preceq$ is a system of component–wise inequalities.) A generalized norm satisfies a number of relations similar to those for standard scalar norms, e.g., $\nu(A) \succeq 0$, $\nu(aA) = |a|\nu(A)$ for $a \in \mathbb{F}$ and $\nu(A+B) \preceq \nu(A) + \nu(B)$. If we define a multiplication of r-tuples $\mathcal{V} \times \mathcal{V} \to \mathcal{V}$ via

$$A \circ B := (A_1 \bullet B_1, \ldots, A_r \bullet B_r),$$

where $X \bullet Y = [x_{ij} y_{ij}]$ is the Hadamard (elementwise) product of two matrices X and Y of same size, then the generalized norm satisfies the inequality

$$\nu(A \circ B) \preceq \nu(A) \bullet \nu(B).$$

Let $\Phi : \mathcal{A} \to \mathcal{X}$ be a continuous function, which maps each r-tuple data $A \in \mathcal{A}$ into the resulting matrix $X = \Phi(A) \in \mathcal{X}$. The problem of finding X for a given A is also denoted as $A \mapsto \Phi(A)$ or $X = \Phi(A)$, while the problem of finding the set $\mathcal{X}_0$ of results for all r-tuples from the subset $\mathcal{A}_0 \subset \mathcal{A}$ is denoted briefly as $\mathcal{A}_0 \to \Phi(\mathcal{A}_0)$ or $\mathcal{X}_0 = \Phi(\mathcal{A}_0)$, where $\Phi(\mathcal{A}_0)$ is the image of $\mathcal{A}_0$ under Φ. In the latter case we have a *family* of problems $X = \Phi(A)$, parametrized by the data $A \in \mathcal{A}_0$. Since the function Φ is usually fixed, the computational problem $X = \Phi(A)$ is further identified with the data A only. Similarly, the problem $\mathcal{X}_0 = \Phi(\mathcal{A}_0)$ is identified with the set $\mathcal{A}_0$.

It is often convenient to reformulate a matrix problem into vector form in order to use standard techniques of matrix theory. For this purpose we utilize the vector representations of stacking the columns of a matrix in one vector, obtaining

$$\begin{aligned} x &:= \operatorname{vec}(X) \in \mathbb{F}^{mn} \\ a_i &:= \operatorname{vec}(A_i) \in \mathbb{F}^{m_i n_i} \\ a &:= \operatorname{vec}(A) := \left[a_1^\top, \ldots, a_r^\top\right]^\top \in \mathbb{F}^{\dim(\mathcal{A})} \end{aligned} \tag{2.8}$$

for the matrices X, A_i and the r-tuple A. In this case we use the notation $x = \varphi(a)$, where $\varphi := \operatorname{vec} \circ \Phi$.

Example 2.4 A problem with explicit solution is the evaluation of a given scalar expression $x = \varphi(a)$ for $a \in \mathcal{A}$, where $\mathcal{A} \in \mathbb{F}$ and the function $\varphi : \mathcal{A} \to \mathbb{R}$ is defined by an explicit expression in terms of arithmetic operations and elementary functions, e.g.,

$$\varphi(a) = \frac{1 + a^2}{1.000001 - \sin a}, \quad a \in \mathbb{R}.$$

Another example is the evaluation of $x = \varphi(a)$, where $\varphi(a)$ is defined by the series

$$\varphi(a) = \sum_{k-0}^{\infty} c_k (a - a_0)^k, \tag{2.9}$$

convergent for $|a - a_0| < \varepsilon$. $\diamondsuit$

Example 2.5 The initial value problem

$$\begin{aligned} y'(t) &= My(t),\ t \in \mathbb{R}, \\ y(0) &= y_0, \end{aligned}$$

where $y(t) \in \mathbb{R}^n$ and $M \in \mathbb{R}^{n\times n}$, gives rise to the problem of evaluating the vector function y at a given moment t, say $t = 1$. We have $y(1) = \exp(M)y_0$, where the matrix exponential exp is defined by the convergent power series

$$\exp(M) := \sum_{k=0}^{\infty} \frac{M^k}{k!}.$$

Thus, the solution of the problem is $x := y(1)$, corresponding to the data

$$A = (A_1, A_2) := (M, y_0) \in \mathcal{V} = \mathbb{R}^{n\times n} \times \mathbb{R}^n \simeq \mathbb{R}^{n^2+n}.$$

In this case $\varphi(a) := \exp(M)y_0 \in \mathbb{R}^n$ and $\dim(\mathcal{A}) = n^2 + n$. $\diamondsuit$

Example 2.6 As a generalization of Example 2.5, let

$$\dot{y}(t) = f(y(t), t, p),\ t \in [t_0, t_{\mathrm{f}}],\ y(t_0) = y_0,$$

be an initial value problem, where $y(t)$ is a function with values in $\mathbb{R}^n$ and p is a vector of parameters. Then the value $x := y(t_{\mathrm{f}})$ of the function y at the moment t_{f} is the solution, depending on the data $A = (p, y_0)$. This problem may be written in the form $G(A, y) = 0$, $x = H(y)$, where the function $G : \mathcal{A} \times \mathcal{Y} \to \mathcal{Y}_1$ is defined as $G(A, y)(t) := y'(t) - f(y(t), t, p)$ if $t \in (t_0, t_{\mathrm{f}}]$ and $G(A, y)(t_0) := y(t_0) - y_0$. Here $\mathcal{Y}$ and $\mathcal{Y}_1$ are the spaces of differentiable and continuous functions $[t_0, t_{\mathrm{f}}] \to \mathbb{R}^n$, respectively. Thus, we have a problem with data A and result x being elements of finite-dimensional spaces, and with an intermediate result y which is a function from an infinite-dimensional space.

When solving this initial value problem numerically, we are interested in the values of y at certain points $t_1, \ldots, t_m \in (0, t_{\mathrm{f}}]$. In this case the result is

$$X := [y(t_1), \ldots, y(t_m)] \in \mathbb{R}^{n\times m}.$$

$\diamondsuit$

Let us discuss now the formulation of problems with *implicit solution*. Many such problems are defined via matrix equations. Consider the finite dimensional linear space $\mathcal{Y} := \mathbb{F}^{p\times q}$, where usually we assume that $\mathbb{F}^{p\times q}$ is isomorphic to $\mathcal{X} = \mathbb{F}^{n\times m}$, i.e., $pq = mn$. Let $\mathcal{D} \subset \mathcal{V} \times \mathcal{X}$ be a certain domain (an open and connected set). Let the equation $F(A, X) = 0$ in $X \in \mathcal{X}$ be given, where $F : \mathcal{D} \to \mathcal{Y}$ is a continuous function. Here the problem is to find the *solution*, or the *result* X for a given A, interpreted as *data*, or as a *parameter matrix*.

Usually we are interested in a particular solution $X = \Phi(A)$ of this equation, which depends continuously on the data A, but sometimes it is also necessary to determine the *solution set* of the equation

$$\Xi(A) := \{X : F(A, X) = 0\}$$

of all solutions for a fixed value of A. Using the vectorizations (2.8) we also have $f(a, x) = 0$, where $a = \mathrm{vec}(A)$, $x = \mathrm{vec}(X)$ and $f(\cdot, \cdot) := \mathrm{vec} \circ F(\mathrm{vec}^{-1}(\cdot), \mathrm{vec}^{-1}(\cdot))$.

In this statement of the problem there are some important issues such as existence and uniqueness of the solution, which will be discussed as well.

Example 2.7 Consider the algebraic equation

$$f(a, x) := a_0 x^n + a_1 x^{n-1} + \cdots + a_{n-1} x + a_n = 0.$$

Here the vector $a := [a_0, a_1, \ldots, a_n]^\top \in \mathbb{F}^{n+1}$ is the data. Any particular solution $x \in \mathbb{F}$ is a result, and if $a_0 \neq 0$, then the collection $\{x_1, \ldots, x_n\}$ of n roots x_i is the solution set $\Xi(a)$ of this equation. ◇

Example 2.8 Consider the quadratic matrix equation

$$F(A, X) := A_1 + A_2 X + X A_3 + X A_4 X = 0,$$

where $A_1 \in \mathbb{F}^{n \times m}$, $A_2 \in \mathbb{F}^{n \times n}$, $A_3 \in \mathbb{F}^{m \times m}$, $A_4 \in \mathbb{F}^{m \times n}$ are given matrix coefficients and $X \in \mathbb{F}^{n \times m}$ is the unknown matrix. Here the data is the matrix quadruple $A = (A_1, A_2, A_3, A_4)$ and the solution set $\mathcal{A}$ has dimension $\dim(\mathcal{A}) = mn + m^2 + n^2 + mn = (m + n)^2$. ◇

The two formulations of problems – with explicit and implicit solutions, are closely related. Indeed, a problem with explicit solution $X = \Phi(A)$ may always be written as an equation, e.g. $X - \Phi(A) = 0$, and often the solution X of an equation $F(A, X) = 0$ may formally be written as an explicit expression $X = \Phi(A)$, see Example 2.5 and Example 2.9 below. We stress that the availability of an explicit formula $X = \Phi(A)$ in terms of arithmetic operations and elementary or special functions does *not* necessarily mean that it is good for a reliable computation of the solution X in finite precision arithmetic.

Example 2.9 Consider the linear system $Mx = b$, where the matrix $M \in \mathbb{F}^{n \times n}$ and the vector $b \in \mathbb{F}^n$ are given and $x \in \mathbb{F}^n$ is the solution. The elements of M and b form the vector $a := [\mathrm{vec}^\top(M), b^\top]^\top \in \mathbb{F}^{n^2+n}$ of data. This problem is formulated as an equation. If the matrix M is nonsingular, then we may write $x = \varphi(a) := M^{-1} b$, obtaining a problem with explicit solution. It is well-known that to obtain x by inverting M is usually *not* recommended when the computations are done in finite precision arithmetic. ◇

Example 2.10 Consider the problem of transforming a square $n \times n$ matrix A into Schur form $T = U^{\mathrm{H}}AU$, where the matrix T is upper triangular and the matrix U is unitary, see Appendix B. Then A is the data and the pair (T, U) is the solution. In this case the transformation matrix U is implicitly determined by the system of equations $\mathrm{Low}(U^{\mathrm{H}}AU) = 0$, $U^{\mathrm{H}}U = I_n$, where Low is the projector to the subspace of lower triangular matrices. Note, however, that even if T is the canonical form of A with respect to the similarity action of the unitary group (i.e., T is uniquely determined by A), then the transformation matrix U is not uniquely determined. $\diamondsuit$

Theoretically, problems with nonuniqueness of the solution may be treated by introducing equivalence classes of solutions and set-valued mappings. From a computational point of view, however, it is important to have a particular solution rather than the whole solution set. Accordingly, the aim of perturbation analysis in such cases is to obtain computable perturbation bounds, which are valid at least for one of the solutions of the perturbed problem.

Suppose that the nominal data A is perturbed to $A + \delta A$. As a result the solution is also perturbed from $X = \Phi(A)$ to $X + \delta X = \Phi(A + \delta A)$.

One of the most important properties of a problem $X = \Phi(A)$ is its *sensitivity* which is measured by the size of the perturbation

$$\delta X = \Phi(A + \delta A) - \Phi(A)$$

in the solution X relative to a given class of perturbations δA in the data A. Intuitively, the problem is *sensitive* if small perturbations in the data lead to large perturbations in the result. Of course, the terms "small" and "large" need to be specified.

As we have mentioned above, the sensitivity of problems is the objective of perturbation analysis. Here the perturbations in the data and in the solution are expressed in terms of norms or generalized norms, see (2.6) and (2.7). Using norms $\| \cdot \|$ or generalized norms $|\!|\!| \cdot |\!|\!|$ we study the perturbations in the matrices as a whole and do not take into account the perturbations in the individual matrix elements. To deal with such perturbations, various techniques of component-wise perturbation analysis have been developed. One of them is based on the use of matrix absolute values and generalized norms $| \cdot |$ as defined in (2.7).

In summary, we have seen that there are at least three important reasons to study the sensitivity of various problems relative to perturbations in the data from a given class.

- Perturbation analysis may give an independent and deep insight in the very nature of the problem, being therefore of independent theoretical interest. For example, perturbation analysis may provide an estimate for the distance from an object of a given set, with data $A \in \mathcal{A}_0 \subset \mathcal{A}$, to the complementary set of objects with data from $\mathcal{A}\backslash\mathcal{A}_0$, see e.g. [51] for a comprehensive study of this problem. In brief, the sensitivity of a given object (or of the corresponding computational problem) is among its most important properties.

- Perturbation bounds provide a realistic framework for most problems in mathematical modelling of objects and processes. Indeed, in practice there are inevitable measurement and other parametric and/or structural uncertainties. This means that we have to deal with a family of models rather than with a single model. In this case the perturbation bounds define a tube in the space of models, to which the characteristics of the particular model actually belong. Having a model with given parameters and estimates for their values, the only thing that we can rigorously claim is that the model will behave within the tube predicted by perturbation analysis.

- When a numerically stable algorithm [101, 233] is applied to solve a problem, then the solution, computed in finite precision arithmetic, will be close to the solution of a near problem. Having tight perturbation bounds and a knowledge about the equivalent perturbation [233] for the computed solution, it is possible to derive condition and accuracy estimates, see e.g., [181]. Without such estimates, a computational algorithm cannot be recognized as reliable from the viewpoint of modern computing standards [1].

2.3 Numerical considerations

The sensitivity is one of the important factors which determine the accuracy of the computed solution when a problem is solved by a numerical algorithm in finite precision arithmetic, e.g., in a floating-point computing environment with rounding unit eps. Without going into detail, eps is half the distance from 1 to the next larger floating point number.

In finite precision arithmetic one gets the computed solution $\widetilde{X}$ which may be, or not be, close to the exact solution $X = \Phi(A)$.

Definition 2.11 The quantities $\alpha_X := \|\widetilde{X} - X\|$ and $\rho_X := \|\widetilde{X} - X\|/\|X\|$ (if $X \neq 0$) are called the *absolute* and *relative norm-wise errors* in the computed solution.

Sometimes also the relative error $\widetilde{\rho}_X := \|\widetilde{X} - X\|/\|\widetilde{X}\|$ is used in practice, since the exact solution X usually is, and remains, unknown.

The desirable case is when the magnitude of ρ_X or $\widetilde{\rho}_X$ is of order of the rounding unit eps, but often this is not the case.

We shall not define precisely the concept of numerical algorithm. The intuitive notion of an algorithm is a sequence of arithmetic operations (and possibly of evaluations of elementary functions), which are performed with relative error of the order of the rounding unit. In addition, at each step the algorithm must produce results which are in the standard range of the finite precision arithmetic in order to avoid over- and underflows.,

In the analysis of computational errors it is convenient to introduce, following [101, 232, 233], the concept of backward error.

Definition 2.12 A collection $E = (E_1, \ldots, E_r)$, such that $\widetilde{X} = \Phi(A + E)$, is called an *equivalent perturbation* (if the equivalent perturbation is not unique, we take one with minimum norm). The norm $\|E\|$, or the generalized norm $|||E|||$ of E, is called the *absolute norm-wise backward error* of the computed solution $\widetilde{X}$. If the backward error is small in the sense that $\|E\| \leq c_1 \mathrm{eps}\|A\|$, where c_1 is a moderate constant (or a low degree polynomial in the dimension of the data vector a), then the algorithm is said to be *numerically backward stable.*

Usually, backward stability is achieved not on the whole domain $\mathcal{A}$ of Φ but on a restricted subset $\mathcal{A}_0 \subset \mathcal{A}$.

We stress that the equivalent perturbation and hence the backward error depend not only on the problem $X = \Phi(A)$ but also on the finite precision arithmetic and the numerical algorithm implemented to compute $\widetilde{X}$.

Unfortunately, the equivalent perturbation E may not exist even for very simple problems solved in finite precision arithmetic, as shown in the next example. In such cases the norm-wise backward error is formally defined as ∞.

Example 2.13 Consider the computational problem

$$x = \varphi(a) := 1 + 1/a, \ a > 0.$$

For $a > 1/\mathrm{eps}$ the computed solution is $\widetilde{x} = 1$ by the definition of the rounding unit eps. To find the equivalent perturbation e we must solve the equation $1 = \varphi(a+e)$ which yields $1/(a + e) = 0$. Hence no finite equivalent perturbation e exists and the backward error is infinite. $\diamondsuit$

Together with the concept of backward stability, the notion of forward stability is also useful.

Definition 2.14 An algorithm is *numerically forward stable* on $\mathcal{A}_0$, if for every $A \in \mathcal{A}_0$ the computed solution $\widetilde{X}$ is close to the exact solution $X = \Phi(A)$ in the sense that $\|\widetilde{X} - X\| \leq c_2 \mathrm{eps}\|X\|$, where c_2 is similar to c_1 in Definition 2.12.

However, forward stability may be achieved only if the problem $X = \Phi(A)$ is not very sensitive on the whole set $\mathcal{A}_0$. For example, an algorithm for solving the linear vector algebraic equation $Mx = b$ cannot be numerically forward stable on the data set $\mathcal{A}$, consisting of all nonsingular matrices M. The reason is that there are matrices from $\mathcal{A}$ with arbitrarily large condition numbers, for which the condition of forward numerical stability does not hold. We recall that the condition number of a nonsingular matrix A with respect to inversion is defined as $\|A\| \, \|A^{-1}\|$.

In order to deal with problems for which the backward error does not exist and/or which are very sensitive (something that no algorithm is responsible for) one may use a more general notion of numerical stability, see [101].

Definition 2.15 Suppose that the computed solution $\widetilde{X}$ is close (if not equal) to some $\widehat{X} = \Phi(A + \widehat{E})$ with $\widehat{E}$ small. If a computational algorithm produces such answers for a set of computational problems with data $A \in \mathcal{A}_0$, then it is called *numerically stable* on the set $\mathcal{A}_0$. Here the closeness is interpreted in terms of the particular finite precision arithmetic as

$$\|\widetilde{X} - \widehat{X}\| \leq c_3 \mathrm{eps} \|X\|, \ \ \|\widehat{E}\| \leq c_4 \mathrm{eps} \|A\|,$$

where c_3, c_4 are moderate constants as in Definition 2.12.

Note that $\widetilde{X}$ may not be the solution of any problem with data from $\mathcal{A}_0$ as in Example 2.13 and in this case the numerically stable algorithm cannot be backward numerically stable.

It may be shown that if an algorithm is backward or forward numerically stable, then it is also numerically stable. To show that the opposite is not true in general (i.e., that numerical stability does not necessarily imply backward or forward numerical stability) is much more subtle [101].

We note also that in the bounds of Definitions 2.11–2.15, it is implicitly assumed that the norms of the involved matrices A and X are larger than 1, since in this case the rounding errors are supposed to be large. If this is not the case, then in the expressions of the form $c_i \mathrm{eps}\|Z\|$ one should formally set $c_i = 1$ if $\|Z\| \leq 1$, where Z stands for A or X.

The result $\widetilde{X}$, produced by a numerically stable algorithm from the data A, may be far from the exact solution $X = \Phi(A)$ if the problem is very sensitive and the quantity $\widehat{X} = \Phi(A + \widehat{E})$ differs significantly from X. More details about this phenomenon are given in Section 2.5. In all cases the perturbation analysis of the computational problem is an important stage in the process of obtaining a reliable numerical solution.

There are different concepts of reliability in numerical computations. Here we consider a numerical procedure *reliable* if it provides the computed solution together with sensitivity and accuracy estimates.

2.4 Component-wise and backward analysis

In this section we briefly consider the concepts of *component–wise* perturbation analysis , see, e.g., [80, 101]. In general, this type of analysis is aimed at estimating the sensitivity of the elements x_j of the solution x to perturbations in the elements a_i of the data a, or in estimating the perturbation in the solution when the elements of the data vary in a special way, e.g., when some of them remain constant (in this section we use the vector representation $x = \varphi(a)$ of a problem $X = \Phi(A)$, where $a = \text{vec}(A)$, $x = \text{vec}(X)$ and $\varphi = \text{vec} \circ \Phi$).

This analysis is useful when the perturbations in the components of a and/or x differ significantly, since in this case the norms $\|\delta a\|$ or $\|\delta x\|$ of δa and δx are relevant measures only for largest of perturbations in a or x, while for smaller or structured perturbations the corresponding bound would be pessimistic. The technique of component-wise perturbation analysis is well developed and commonly applied to various problems in linear algebra and control theory.

Another technique of perturbation analysis is the derivation of backward perturbation bounds. The aim of this type of analysis is, given certain approximate solution $\overline{x}$ of the problem $x = \varphi(a)$, to find the minimal (in certain sense) perturbation δa in the data which satisfies $\overline{x} = \varphi(a + \delta a)$. Thus, δa is a solution of a constrained minimization problem.

When studying relative perturbations of a computational problem with data $a \neq 0$, having some components $a_i = 0$, it is appropriate to introduce a norm-like function which reflects the changes in the data in a component-wise style as shown below.

Suppose that $\mathcal{A} \subset \mathbb{F}^p$. For a vector $a = [a_1, \ldots, a_p]^\top \in \mathcal{A}$, having zero entries at prescribed positions, we denote by $\text{pat}(a) = [s_1, \ldots, s_p]^\top$ the zero-pattern vector with components

$$s_i = \begin{cases} 1 & \text{if } a_i \neq 0, \\ 0 & \text{if } a_i = 0. \end{cases}$$

Define $\mathcal{Q}(a) \subset \mathcal{A}$ as the set of all $b \in \mathcal{A}$, having the same zero-pattern as a, i.e., $\text{pat}(b) = \text{pat}(a)$. Then for $b = [b_1, \ldots, b_p]^\top \in \mathcal{Q}(a)$ we define the vector b/a as

$$b/a := (\text{diag}(a_1, \ldots, a_p))^\dagger b \in \mathcal{Q}(a)$$

with components

$$(b/a)_i = \begin{cases} b_i/a_i & \text{if } a_i \neq 0, \\ 0 & \text{if } a_i = 0. \end{cases}$$

(Here $\text{diag}(a_1, \ldots, a_p)\dagger$ denotes the Moore-Penrose pseudo-inverse [83] of $\text{diag}(a_1, \ldots, a_p)$.)

Definition 2.16 The quantity

$$\|b/a\|_\infty = \min \{\nu \geq 0 : |b_i| \leq \nu |a_i|\}$$

is known as the *component–wise relative norm of* b *with respect to* a.

Replacing the inclusion $a + \delta a \in \mathcal{A}$ by $a + \delta a \in \mathcal{Q}(a)$ in the definition of the relative condition number we obtain a new quantity

$$\widehat{\kappa}(a) = \lim_{\alpha \to 0} \sup \left\{ \frac{\|\delta x\|_\infty}{\|x\|_\infty} \frac{1}{\|\delta a/a\|_\infty} : a + \delta a \in \mathcal{Q}(a), |\delta a| \preceq \alpha |a| \right\}.$$

Definition 2.17 The quantity $\widehat{\kappa}(a)$ is known as the *mixed relative condition number* of the problem $x = \varphi(a)$.

The use of the mixed condition number often gives sharper estimates, especially when computational processes in finite precision arithmetic are considered. This is due to the fact that usually no rounding errors are introduced in the zero elements of the data vector a.

Since

$$\frac{\|b\|_\infty}{\|a\|_\infty} \le \|b/a\|_\infty$$

we have

$$\widehat{\kappa}(a) \le \kappa(a),$$

where the standard relative condition number $\kappa(a)$ (see Definition 2.3) is taken with respect to the infinity norm $\|\cdot\|_\infty$.

If the Fréchet derivative $\varphi'(a)$ exists then

$$\widehat{\kappa}(a) = \frac{\|\varphi'(a)\mathrm{diag}(a_1, \ldots, a_n)\|_\infty}{\|x\|_\infty} \le \kappa(a) = \frac{\|\varphi'(a)\|_\infty \|a\|_\infty}{\|x\|_\infty},$$

and for $|\delta a| \preceq \alpha |a|$ we have

$$\frac{\|\delta x\|_\infty}{\|x\|_\infty} \le \widehat{\kappa}(a)\alpha + o(\alpha), \ \alpha \to 0.$$

In a similar way we may define yet another component-wise condition number, see [80].

Definition 2.18 The quantity

$$\widetilde{\kappa}(a) := \lim_{\alpha \to 0} \sup \left\{ \frac{\|\delta x/x\|_\infty}{\|\delta a/a\|_\infty} \right\}$$

is called *relative component-wise condition number* of the problem $x = \varphi(a)$.

If the Fréchet derivative $\varphi'(a)$ of φ at a exists and the solution $x = [x_1, \ldots, x_q]^\top$ has no zero components, then

$$\widetilde{\kappa}(a) = \|\mathrm{diag}(1/x_1, \ldots, 1/x_q)\varphi'(a)\mathrm{diag}(a_1, \ldots, a_p)\|_\infty .$$

Various relations between the condition numbers

$$\kappa(a), \ \widehat{\kappa}(a), \ \widetilde{\kappa}(a) \tag{2.10}$$

are established in [80] for the case when $\varphi = \varphi_2 \circ \varphi_1$ is a composition of two functions φ_1 and φ_2.

Usually the domain $\mathcal{A} \subset \mathbb{F}^p$ of the function φ in the computational problem $x = \varphi(a)$ is of positive $p-$dimensional measure. Thus, the perturbations δa in a are allowed to vary in a p-dimensional volume in the definition of $\kappa(a)$, and in the definition of $\widehat{\kappa}$ and $\widetilde{\kappa}$ if all components of a are nonzero. At the same time in some practical problems the perturbations in the data are allowed to vary only in a lower dimensional subset $\mathcal{T} \subset \mathcal{A}$ (see Example 2.19 below).

In this case we may consider the new *restricted* (or *structured*) computational problem $x = \psi(a)$ for the restriction $\psi := \varphi|_{\mathcal{T}}$ of φ on $\mathcal{T}$. All three condition numbers (2.10), computed for the problem $x = \psi(a)$, are referred to as *structured* condition numbers, see [80].

Example 2.19 Consider the operation $\Phi : \mathcal{GL}(n, \mathbb{F}) \to \mathcal{GL}(n, \mathbb{F})$ of matrix inversion, $\Phi(A) := A^{-1}$. (Here $GL(n, \mathbb{F})$ denotes the group of nonsingular $n \times n$ matrices with elements in $\mathbb{F}$. Restricting ourselves to the inversion of a given class of matrices, say the class $\mathcal{T}$of nonsingular Toeplitz matrices, we get the restricted problem $A \mapsto \Phi|_{\mathcal{T}}(A)$ of inversion of Toeplitz matrices A, in which the perturbations δA are subject to the constraints $A + \delta A \in \mathcal{T}$. At the same time $\Phi'(A)(\delta A) = -A^{-1}\delta A A^{-1}$ and $\kappa(\Phi, A) = \mathrm{cond}(A)$. (Recall that a Toeplitz matrix is a matrix that is constant on every diagonal.) $\diamondsuit$

Example 2.20 Consider the matrix inversion Φ from Example 2.19 as a mapping $\mathbb{F}^{n^2} \to \mathbb{F}^{n^2}$ with $x = \varphi(a)$ and $a := \mathrm{vec}(A)$, $x := \mathrm{vec}(A^{-1})$, $\varphi := \mathrm{vec} \circ \Phi$, and set

$$y := \mathrm{vec}\left(|A^{-1}|\,|A|\,|A^{-1}|\right).$$

Then $\widehat{\kappa}(a) = \|y\|_\infty / \|x\|_\infty$ and $\widetilde{\kappa}(a) = \|y/x\|_\infty$, provided that x has no zero elements. $\diamondsuit$

Another type of component-wise perturbation bounds for the computational problem $x = \varphi(a)$ are inequalities of the form

$$|\delta x| \preceq C|\delta a|, \ |\delta a| \preceq \rho \in \mathbb{R}^r_+,$$

where $C = C(a, \rho) \in \mathbb{R}^{s\times r}_+$ is the *Lipschitz matrix* of φ in the ρ-neighborhood of a. Now the influence of the i-th element of a on the j-th element of x is measured by the (j, i)-entry c_{ji} of C. If $|\cdot|$ is the corresponding matrix absolute value, then the quantity

$$\sup_{|\delta a| \preceq \rho} \left| \frac{\partial \varphi_i}{\partial a_j}(a + \delta a) \right|$$

(if it exists) is an upper bound for c_{ij}. Similarly, if $|\cdot|$ is a generalized norm, and

$$a = \begin{bmatrix} a_1 \\ \vdots \\ a_r \end{bmatrix}, \ a_i \in \mathbb{F}^{p_i}, \ \ \varphi(a) = \begin{bmatrix} \varphi_1(a) \\ \vdots \\ \varphi_s(a) \end{bmatrix},$$

where $\varphi_j : \mathbb{F}^p \to \mathbb{F}^{q_j}$, then

$$\sup_{|\delta a| \preceq \rho} \left\| \frac{\partial \varphi_i}{\partial a_j}(a + \delta a) \right\|$$

is an upper bound for c_{ij}.

In local linear component-wise perturbation estimates the matrix $|\varphi'(a)|$ may be used instead of C. We recall that in such a case the perturbation in the data must be in the asymptotic domain of the local bound, i.e., that the neglected higher order terms are of moderate size relative to the linear (or first order homogeneous) terms that are taken into account.

Nonlocal nonlinear componentwise estimates will also be discussed, e.g.,

$$|\delta x| \preceq f(|\delta a|), \ |\delta a| \preceq \rho \in \mathbb{R}^r_+, \tag{2.11}$$

where

$$f := [f_1, \dots, f_s]^\top : [0, \rho_1] \times \cdots \times [0, \rho_r] \to \mathbb{R}^s_+$$

is a continuous vector function such that f_j is nondecreasing in each of its arguments and $f(0) = 0$. Sometimes the domain of applicability of these nonlocal bounds may be quite complicated.

2.5 Error estimates

In this section we describe a general model of error estimation for a computational process in finite precision arithmetic, based on perturbation bounds for the computational problem.

2.5.1 Forward error

Consider a problem in the form $x = \varphi(a)$, where φ is a given continuous function, or a function of class C^k, $k \geq 1$, or C^∞, x is the solution and a are the data. We assume that x and a are elements of finite dimensional spaces, say $x \in \mathbb{F}^q$, $a \in \mathbb{F}^p$. Let $K = K(a)$ be the absolute condition number of the problem at the data a.

In finite precision arithmetic with roundoff unit eps the solution of the computational problem is usually contaminated with rounding errors. In this case the actual error in the computed solution depends on three main factors, [102, 134]:

- Properties of the finite precision arithmetic (the roundoff unit eps in particular),
- properties of the computational problem (the sensitivity in particular),
- properties of the computational algorithm (the numerical stability in particular).

When implementing a numerically stable algorithm, the computed solution $\widetilde{x}$ is near to the exact solution $\widehat{x} = \varphi(a + \widehat{e})$ of a slightly perturbed problem. From a quantitative point of view this means that the following inequalities are fulfilled

$$\|\widetilde{x} - \widehat{x}\| \leq \text{eps}\, M\|x\|, \quad \|\widehat{e}\| \leq \text{eps}\, N\|a\|,$$

where the constants M, N characterise the properties of the numerical algorithm (if $\|x\|$ or $\|a\|$ is less than 1 we set $M = 1$ or $N = 1$, respectively), see also Section 2.3.

It is possible to estimate the actual absolute error $\alpha_x := \|\widetilde{x} - x\|$ and the relative error (if $x \neq 0$) $\rho_x := \alpha_x/\|x\|$ in the computed solution $\widetilde{x}$ as follows. Neglecting second and higher order terms in eps, we have

$$\begin{aligned} \alpha_x &= \|\widetilde{x} - \varphi(a)\| = \|\widetilde{x} - \widehat{x} + \widehat{x} - \varphi(a)\| \\ &\leq \|\widetilde{x} - \widehat{x}\| + \|\varphi(a + \widehat{a}) - \varphi(a)\| \\ &\leq \text{eps}\, M\|x\| + K\|\widehat{e}\| \leq \text{eps}\,(M\|x\| + KN\|a\|). \end{aligned} \tag{2.12}$$

The relative error in the computed solution is estimated by dividing both sides of (2.12) by $\|x\|$:

$$\rho_x \leq \text{eps}\left(M + NK\frac{\|a\|}{\|x\|}\right) = \text{eps}\,(M + \kappa(a)N), \tag{2.13}$$

where $\kappa(a)$ is the relative condition number of the problem.

The estimate (2.13) reveals directly the main factors which determine the accuracy of the computed solution:

- the properties of finite precision arithmetic (the rounding unit eps),
- the properties of the problem (the relative condition number κ), and
- the properties of the algorithm (the constants M and N).

If a nonlinear sensitivity estimate of type $\|\delta x\| \leq f(\|\delta a\|)$ is available, then the corresponding estimates are

$$\alpha_x \leq \text{eps}\, M\|x\| + f(\text{eps}\, N\|a\|) \tag{2.14}$$

and

$$\rho_x \leq \text{eps}\, M + \frac{f(\text{eps}\, N\|a\|)}{\|x\|}.$$

Nonlinear component–wise estimates for $|\widetilde{x} - x|$, similar to (2.14), may be derived provided a component–wise bound of type $|\delta x| \preceq f(|\delta a|)$ is available. This, however, remains an open question for many computational problems.

The above numerical considerations demonstrate the crucial role of sensitivity estimates in the floating point solution of computational problems. In fact, a solution computed in finite precision arithmetic cannot be accepted as reliable unless a bound on the actual error is known [134].

2.5.2 Backward error

Consider the problem $x = \varphi(a)$, $a \neq 0$, under the assumption that the Fréchet derivative $\varphi'(a)$ of φ at the point a exist. Let $\widetilde{x}$ be an approximate solution (e.g., a solution obtained in finite precision arithmetic) and suppose that $\widetilde{x}$ is the exact solution to a slightly perturbed problem, i.e., that

$$\widetilde{x} = \varphi(a + \delta a) \tag{2.15}$$

for some (small) δa. Then we may consider the problem to estimate the smallest perturbation δa for which (2.15) holds. This leads to the concepts of absolute and relative backward errors, corresponding to the approximate solution $\widetilde{x}$.

Definition 2.21 The quantity

$$\beta(\widetilde{x}) := \min \left\{ \frac{\|\delta a\|}{\|a\|} \right\},$$

where the minimum is taken over all δa which satisfy (2.15), is said to be the *relative backward error* of the approximate solution $\widetilde{x}$.

The relative backward error may be estimated as follows. Within first order terms we have

$$\widetilde{x} = \varphi(a) + \varphi'(a)(\delta a) + o(\|\delta\|) = x + \varphi'(a)(\delta a) + o(\|\delta\|), \ \delta \to 0$$

and

$$\delta a = (\varphi'(a))^{\dagger}(\widetilde{x} - x) + o(\|\delta\|), \ \delta \to 0.$$

Therefore, the estimate for the backward error

$$\beta(\widetilde{x}) \lesssim \frac{\|(\varphi'(a))^{\dagger}\|}{\|a\|}\|x - \widetilde{x}\|$$

is proportional to the norm of the residual $x - \widetilde{x}$.

If a is structured as in (2), then the backward error is defined via

$$\beta := \min \left\{ \beta : \frac{\|\delta a_i\|}{\|a_i\|} \leq \beta \right\},$$

where the minimum is taken in accordance with the constraints (2.15).

Let $\widetilde{x}$ be an approximate solution to the equation $f(x, a) = 0$ such that

$$f(\widetilde{x}, a + \delta a) = 0$$

for some perturbation δa. Within first order terms we have

$$f(\widetilde{x}, a) + f'_a(\widetilde{x}, a)(\delta a) = 0.$$

Thus, the minimum norm perturbation δa is obtained approximately from

$$\delta a = -(f'_a(\widetilde{x}, a))^{\dagger} f(\widetilde{x}, a),$$

which in turn leads to the estimate

$$\beta(\widetilde{x}) \lesssim \frac{\|(f'_a(\widetilde{x}, a))^{\dagger}\|}{\|a\|} \|f(\widetilde{x}, a)\|.$$

The case of an equation with structured data of type (2) is treated in a similar way.

2.6 Scaling

The *scaling* of a computational problem consists of applying transformations on the data and/or intermediate (or final) results to avoid over- and underflows, or to improve the conditioning of the original problem in order to reduce the effect of rounding in finite precision arithmetic.

It must be pointed out that the conditioning of a computational problem is usually beyond the effective control of the user, although it is a common opinion that preliminary manipulations such as scaling may improve the conditioning. That this is not exactly the case is demonstrated as follows.

Consider a scaling of the computational problem $X = \Phi(A)$, consisting in the implementation of two linear nonsingular transformations

$$B = U(A), \; Y = V(X)$$

in the input and output spaces $\mathcal{A}$ and $\mathcal{X}$, respectively. As a result we get the new computational problem

$$Y = \Psi(B), \; \Psi := V \circ \Phi \circ U^{-1}.$$

If Φ is Fréchet differentiable at A with a derivative $\Phi'(A)$, then Ψ is also Fréchet differentiable at B, and

$$\Psi'(B) = V \circ \Phi'(A) \circ U^{-1}.$$

Hence, the relative condition numbers $\kappa(\Phi, A)$ of the original problem $X = \Phi(A)$ and $\kappa(\Psi, B)$ of the transformed problem $Y = \Psi(B)$ are (if $A \neq 0$) is given by

$$\begin{aligned} \kappa(\Phi, A) &= \|\Phi'(A)\| \frac{\|A\|}{\|X\|}, \\ \kappa(\Psi, B) &= \left\|V \circ \Phi'(A) \circ U^{-1}\right\| \frac{\|U(A)\|}{\|V(X)\|}. \end{aligned}$$

The scaling procedure consists in finding transformations U, V for which $\kappa(\Psi, B)$ is minimal.

If we introduce new norms in the input space $\mathcal{A}$ and the output space $\mathcal{X}$ as

$$\|A\|_U := \|U(A)\|, \ \|X\|_V = \|V(X)\|,$$

we see that the corresponding subordinate norm of $\Phi'(A)$ is

$$\|\Phi'(A)\|_{U,V} = \left\|V \circ \Phi'(A) \circ U^{-1}\right\|.$$

Hence, the relative condition number of the transformed problem $Y = \Psi(B)$ is exactly the relative condition number of the original problem $X = \Phi(A)$ but for the norms $\|\cdot\|_U$, $\|\cdot\|_V$. Hence, scaling in this case does not decrease the condition number, but rather corresponds to finding new norms in input and output spaces for which the amplification of perturbations from input to output is minimal [76]. Whether this actually improves the numerical behaviour of a particular computational algorithm, depends on the application. We stress that the use of different norms in order to improve error estimates and stability factors is a common practice, in particular in the implementation of numerical methods for the solution of differential equations.

Of course, scaling aimed to reduce the norms of matrices and vectors in order to avoid over- and underflows and eventually to reduce the rounding errors, improves the numerical behavior of the computational procedures in finite precision arithmetic.

2.7 Notes and references

Modern numerical analysis, taking into account the effects of finite precision arithmetic, starts with the fundamental works of von Neumann, see e.g. [7, 172], and A. Turing [226]. The concept of backward error and backward stability was introduced by J. Wilkinson [232, 233], see also [234]. Stability in the sense of Definition 2.15 is first considered by W. Kahan [115, 116].

General techniques for perturbation analysis of linear control problems are considered in [142, 180, 147, 127].

General properties of the perturbation operator have been considered in [133, 134, 135].

For component-wise and backward analysis in a sense similar to that considered in Section 2.4 see [30, 80, 100]. Scaling of computational problems in the framework presented in Section 2.6 has been discussed in [76].

Often computational problems $X = \Phi(A)$ are solved decomposing Φ as $\Phi_s \circ \cdots \circ \Phi_1$. In this some of the subproblems $X_{i+1} = \Phi_{i+1}(X_i)$ may be very ill-conditioned (or even singular) even if the original problem is regular and well-conditioned. The effects of such decompositions are considered in [102].

Chapter 3

Problems with explicit solutions

3.1 Introductory remarks

Although the main purpose of this monograph is the perturbation analysis of matrix equations, in this section we present some general issues concerning the sensitivity of problems with explicit solution $X = \Phi(A)$. These results may be extended to problems with implicit solution such as matrix equations $F(A, X) = 0$. Indeed, let X be a solution of this equation, corresponding to the particular value of A. Then under some natural restrictions on F (for instance under the conditions of the implicit function theorem, see [173] or Appendix A), there exists a continuous function Φ, defined in a neighborhood $\mathcal{N}_A$ of A, such that $X = \Phi(A)$ and $F(B, \Phi(B)) = 0$ for all $B \in \mathcal{N}_A$. So the general considerations about sensitivity for problems with explicit solution (explicitly defined functions) apply also to problems with implicit solution (implicitly defined functions).

3.2 Perturbation function

Consider a problem with explicit solution $X = \Phi(A)$, where $\Phi : \mathcal{A} \to \mathcal{X}$ is a continuous function and $\mathcal{A}$ is a subset of the Cartesian product $\mathcal{V}$ of matrix spaces as in (2.4). The spaces $\mathcal{V}$ and $\mathcal{X}$ are endowed with norms and generalized norms, see e.g. (2.6) and (2.7). When relative perturbations are studied, we assume in addition that $A \neq 0$ and $X \neq 0$.

Let δA be a perturbation in the data A such that $A + \delta A \in \mathcal{A}$ and let

$$\mathcal{F} := \{(A, E) : A \in \mathcal{A},\ A + E \in \mathcal{A}\} \subset \mathcal{A} \times \mathcal{V}$$

be the set of all pairs (A, E) from $\mathcal{A} \times \mathcal{V}$ such that $A + E$ is in $\mathcal{A}$. (Observe that

$\mathcal{F}$ may not be a subset of $\mathcal{A} \times \mathcal{A}$, see Example 3.1 below.) Also, for a fixed $A \in \mathcal{A}$ let $\mathcal{E}_A \subset \mathcal{V}$ be the set of all $E \in \mathcal{V}$ such that $A + E \in \mathcal{A}$. Thus

$$\mathcal{F} = \bigcup_{A \in \mathcal{A}} \{\{A\} \times \mathcal{E}_A\}.$$

Example 3.1 Let $\mathcal{V} = \mathbb{R}$ and let $\mathcal{A}$ be the open interval $(\underline{a}, \overline{a}) \subset \mathbb{R}$. Then $\mathcal{F}$ is the open parallelogram

$$\mathcal{F} = \{(a, e) : \underline{a} < a < \overline{a},\ \underline{a} - a < e < \overline{a} - a\}.$$

$\diamondsuit$

As before, denote by $\delta X = \Psi(A, \delta A)$ the perturbation in the result X, corresponding to the perturbation $A \to A + \delta A$ in the data, where

$$\Psi(A, \delta A) := \Phi(A + \delta A) - \Phi(A), \text{ for } (A, \delta A) \in \mathcal{F}. \tag{3.1}$$

Definition 3.2 The function $\Psi : \mathcal{F} \to \mathcal{X}$ is called the *perturbation function* of the family of problems $\mathcal{A} \to \Phi(\mathcal{A})$.

In this definition we emphasize the dependence of δX on both A and δA. For a fixed $A \in \mathcal{A}$ the function $\Psi(A, \cdot) : \mathcal{E}_A \to \mathcal{X}$ is the *perturbation function* of the single problem $X \mapsto \Phi(A)$.

As thus defined, the perturbation function may not be useful in practice. Indeed, one intuitively expects that if $\Psi(A, \delta A)$ is well defined for some perturbation δA then it should remain well defined for smaller perturbations. That this may not be the case when Ψ is defined on $\mathcal{F}$, is shown in the next example.

Example 3.3 For the scalar problem

$$x = 1/a,\ a \in \mathcal{A} := \mathbb{R} \backslash \{0\}$$

we have

$$\mathcal{F} = \{(a, e) : a \neq 0,\ e \neq -a\} \subset \mathbb{R}^2.$$

For $a \neq 0$ and $\delta a = -2a$ we have $(a, a + \delta a) \in \mathcal{F}$; but for the smaller perturbation $\delta a = -a$ the solution of the perturbed problem $1/(a + \delta)$ is not defined. $\diamondsuit$

Therefore, we have to impose some additional conditions for connectivity and convexity of the domain of the perturbation function Ψ as it will be done next.

We are interested in perturbations δA for which $\Psi(A, \delta A)$ tends to zero together with δA and the expression $\Psi(A, E)$ is well defined for all $E \in \mathcal{V}$ with $\|E\| \leq \|\delta A\|$ or $\|\!|E|\!\| \preceq \|\!|\delta A|\!\|$. It must be pointed out that not every perturbation δA with $(A, \delta A) \in \mathcal{F}$ satisfies this requirement. To impose additional restrictions on δA we introduce following definition.

Definition 3.4 A pair $(A, \delta A) \in \mathcal{F}$ is called *admissible* if $A + E \in \mathcal{A}$ for all $E \in \mathcal{V}$ with $\|E\| \leq \|\delta A\|$ (or $\|E\| \preceq \|\delta A\|$).

One may also determine the subset $\mathcal{F}_{\text{adm}} \subset \mathcal{F}$ of all admissible pairs $(A, \delta A)$. Since the set $\mathcal{A}$ is open then for every $A \in \mathcal{A}$ there exists $\varepsilon = \varepsilon(A) > 0$ such that all pairs $(A, \delta A)$ with $\|\delta A\| < \varepsilon$ are admissible.

Example 3.5 For the data set from Example 3.1 the set $\mathcal{F}_{\text{adm}}$ is the open square $\mathcal{F}_{\text{adm}} = \mathcal{F}^{(1)}_{\text{adm}} \cup \mathcal{F}^{(2)}_{\text{adm}}$, where

$$\begin{aligned} \mathcal{F}^{(1)}_{\text{adm}} &= \{(a, e) : \underline{a} < a \leq (\underline{a} + \overline{a})/2,\ \underline{a} - a < e < a - \underline{a}\}, \\ \mathcal{F}^{(2)}_{\text{adm}} &= \{(a, e) : (\underline{a} + \overline{a})/2 < a < \overline{a},\ a - \overline{a} < e < \overline{a} - a\}. \end{aligned}$$

$\diamondsuit$

The function $\Psi(A, \cdot) : \mathcal{E}_A \to \mathcal{X}$ depends on the matrix parameter A. When A varies over $\mathcal{A}$, we have a family of functions $\{\Psi(A, \cdot)\}_{A \in \mathcal{A}}$, which is parametrized by $A \in \mathcal{A}$.

Let a function $\Psi : \mathcal{S} \to \mathcal{X}$ be given, where $\mathcal{S}$ is a subset of $\mathcal{A} \times \mathcal{V}$. Then the question arises whether Ψ is the perturbation function for some family of problems $\mathcal{A}_0 \to \Phi(\mathcal{A}_0)$. If this is the case, the next task is to find the function $\Phi : \mathcal{A}_0 \to \mathcal{X}$ itself, i.e., to solve the functional equation

$$\Phi(A + E) - \Phi(A) = \Psi(A, E),\ (A, E) \in \mathcal{S}$$

relative to Φ for a given Ψ. In this case the function Φ will be determined up to an arbitrary additive constant matrix from $\mathcal{X}$.

If Ψ is a perturbation function then $\Psi(A, 0) = 0$. However, not every continuous function Ψ with $\Psi(A, 0) = 0$ is the perturbation function for a family of problems.

Example 3.6 Consider the function $\psi : \mathbb{F} \times \mathbb{F} \to \mathbb{F}$, defined by $\psi(a, e) = e^2 + ae$. If $\varphi(a + e) - \varphi(a) = e^2 + ae$ for some continuous function $\varphi : \mathbb{F} \to \mathbb{F}$, then setting $a = 0$ we get $\varphi(e) = \varphi(0) + e^2$. Substituting this expression back in the functional equation for φ we obtain $ae = 0$. This is an additional restriction on a and e and hence the domain of ψ cannot be $\mathbb{F} \times \mathbb{F}$ and thus, ψ is not a perturbation function. In contrast, the function $(a, e) \mapsto e^2 + 2ae$, defined on $\mathbb{F} \times \mathbb{F}$, is the perturbation function of the family of problems $x = a^2$, $a \in \mathbb{F}$. $\diamondsuit$

Proposition 3.7 *A continuous function $\Psi : \mathcal{S} \to \mathcal{X}$ is a perturbation function if and only if for some $A^0 \in \mathcal{A}$ the equation*

$$\Psi(A^0, E + A - A^0) - \Psi(A^0, A - A^0) = \Psi(A, E)$$

holds for all $(A, E) \in \mathcal{S}$.

Since we are interested in problems with continuous mappings Φ, it is reasonable to define admissible perturbation functions according to the following definition.

Definition 3.8 The function $\Psi : \mathcal{S} \to \mathcal{X}$ is *admissible* if there exists a continuous function $\Phi : \mathcal{A} \to \mathcal{X}$ such that $\Psi(A, E) = \Phi(A + E) - \Phi(A)$.

For a given $A \in \mathcal{A}$ the domain $\mathcal{E}_A \subset \mathcal{V}$ may be quite complicated even for simple problems as the next example demonstrates.

Example 3.9 Let $\mathcal{A} = \mathcal{GL}(n, \mathbb{F})$ and $\Phi(A) := A^{-1}$. Then

$$\mathcal{E}_A = \{E \in \mathbb{F}^{n \times n} : \det(A + E) \neq 0\}.$$

The boundary $\partial\mathcal{E}_A$ of $\mathcal{E}_A$ consists of all matrices E with $\det(A + E) = 0$. It is a closed algebraic variety in $\mathbb{F}^{n \times n} \simeq \mathbb{F}^{n^2}$ of degree n and codimension 1, and has a very complicated structure. $\diamondsuit$

The above considerations show that it is reasonable to restrict the perturbations δA to certain simple subsets of $\mathcal{E}_A$ containing the origin. An example of such a set is the *generalized ball*

$$\mathcal{B}_{\rho(A)} := \{E : |\!|\!|E|\!|\!| \preceq \rho(A)\},$$

where $\rho(A)$ is a given nonnegative vector. In particular, one may choose $\rho(A) = \varepsilon|\!|\!|A|\!|\!|$, where $\varepsilon > 0$ is (usually) a small parameter.

This restriction of the problem is still rather general. It includes as particular cases many structured perturbation problems as discussed below.

Taking the generalized norm in $\mathcal{A}$ as $|a| \in \mathcal{R}_+^{\dim(\mathcal{A})}$ we obtain a perturbation problem with interval data,

$$\mathcal{B}_{\rho(A)} = [\underline{a}, \overline{a}],$$

which is the most structured type of perturbation. Indeed, here we may take $\rho(a) = (\overline{a} - \underline{a})/2$.

The other extreme (most unstructured) case is when a norm in $\mathcal{A}$ is used. This leads to the ball

$$\mathcal{B}_{\rho(A)} := \{E : \|E\| \leq \rho(A)\},$$

where $\rho(A) > 0$ is now a scalar. In particular we may choose $\rho(A) = \varepsilon\|A\|$.

The general case of *structured perturbations* is also included in our statement. It corresponds to a special choice of the data set $\mathcal{A}$ and the function Φ as follows. Consider a problem $X = \Phi(A)$ under the structured perturbation $\delta A = \Theta(\Delta) \in \mathcal{A}$, where $\Delta \in \mathcal{D}$, $\mathcal{D}$ is an open subset of a finite dimensional space with dimension less than $\dim(\mathcal{A})$ and $\Theta : \mathcal{D} \to \mathcal{A}$ is a given continuous function satisfying $\Theta(0) = 0$ (in many applications Θ is a linear function, see Example 3.10 below). Defining the function $\widetilde{\Phi} : \mathcal{D} \to \mathcal{X}$ via $\widetilde{\Phi}(\Delta) := \Phi(A + \Theta(\Delta))$, we get the structured problem $X = \widetilde{\Phi}(0)$.

Example 3.10 Consider the problem of computing the eigenvalues of the matrix $A \in \mathbb{F}^{n\times n}$ under the structured perturbations $\delta A = B\Delta C$, where $B \in \mathbb{F}^{n\times p}$, $C \in \mathbb{F}^{q\times n}$ are given matrices, $\Delta \in \mathbb{F}^{p\times q}$ and $pq < n^2$. Here $\mathcal{A} = \mathbb{F}^{n\times n}$, $\mathcal{D} = \mathbb{F}^{p\times q}$ and $\Theta(\Delta) = B\Delta C$. $\diamondsuit$

Under the action of the perturbation function the set $\mathcal{B}_{\rho(A)}$ is transformed into the set

$$\Psi(A, \mathcal{B}_{\rho(A)}) := \left\{\Psi(A, E) : E \in \mathcal{B}_{\rho(A)}\right\} \subset \mathcal{X}.$$

The aim of nonlocal perturbation analysis is to give bounds for the set $\Psi(A, \mathcal{B}_{\rho(A)})$.

Although the simple set $\mathcal{B}_{\rho(A)}$ has some nice properties such as convexity, the set $\Psi(A, \mathcal{B}_{\rho(A)})$ may be of very exotic structure as it is shown in the next example.

Example 3.11 The set of solutions of the linear algebraic interval equation

$$Mx = b;\ M \in \left[\underline{M}, \overline{M}\right] \subset \mathbb{R}^{2\times 2},\ b \in [\underline{b}, \overline{b}] \subset \mathbb{R}^2,$$

where $\left[\underline{M}, \overline{M}\right] := \{M : \underline{M} \preceq M \preceq \overline{M}\}$, may have the form of a multi-ray star. $\diamondsuit$

These considerations show that it is reasonable not to estimate the set $\Psi(A, \mathcal{B}_{\rho(A)})$ itself but rather its norm-wise radius

$$\max\left\{\|\Psi(A, E)\| : E \in \mathcal{B}_{\rho(A)}\right\} \in \mathbb{R}_+.$$

A more ambitious task is to estimate the set

$$\Psi^*(A, \mathcal{B}_{\rho(A)}) := \left\{|\Psi(A, E)| : E \in \mathcal{B}_{\rho(A)}\right\} \subset \mathbb{R}_+^{m\times n}$$

of the matrix absolute values of the perturbations δX when the perturbation δA varies over the set $\mathcal{B}_{\rho(A)}$. We will do this later for linear matrix equations.

Somewhat easier, although still quite complicated, is the problem of estimating the radius vector r^* of $\Psi^*(A, \mathcal{B}_{\rho(A)})$, i.e., the minimal vector $r^* \in \mathbb{R}_+^r$ relative to the component-wise order relation $\preceq$ such that

$$\Psi^*(A, B_{\rho(A)}) \subset \mathcal{B}_{r^*}.$$

In this statement of the problem we include the task of estimating the *continuity module* $\mu : \mathcal{A} \times \mathbb{R}_+^r \to \mathbb{R}_+$ of the function Φ at the point A, given by

$$\mu(A, \delta) := \max\{\|\Psi(A, E)\| : \|E\| \preceq \delta\}.$$

In addition to analyzing nonlocal perturbation effects, there are also perturbation techniques for studying the local behavior of the perturbation δX in the solution as a function of the perturbation δA in the data. Since in practice we always have finite perturbations, it is necessary to define local properties in some quantitative way. This may be done by using the concept of asymptotic domain, introduced below.

Suppose that we can represent the continuous perturbation function Ψ as

$$\Psi(A, \delta A) = \Psi_1(A, \delta A) + o(\|\delta A\|), \ \delta A \to 0,$$

where $o(z)/z \to 0$ for $z \to 0$ and the function $\Psi_1(A, \cdot)$ is first order homogeneous, i.e.,

$$\Psi_1(A, \lambda E) = |\lambda| \Psi_1(A, E), \ \lambda \in \mathbb{F}.$$

Then we have

$$\|\delta X\| \le \omega(|\!|\!|\delta A|\!|\!|) + o(\|\delta A\|), \ \delta A \to 0,$$

where, for $\eta \in \mathbb{R}^r_+$, the function ω is defined by

$$\omega(\eta) := \max\{\|\Psi_1(A, E)\| : |\!|\!|E|\!|\!| \preceq \eta\}.$$

In practice we cannot explicitly determine the exact maximum $\omega(\eta)$ of $\|\Psi_1(A, E)\|$ over $E \in \mathcal{B}_\eta$ (except for linear equations). For this reason we use an upper bound $\omega_1(\eta) \ge \omega(\eta)$, which is easier to compute.With such a bound we have

$$\|\delta X\| \le \omega_1(|\!|\!|\delta A|\!|\!|) + o(\|\delta A\|), \ \delta A \to 0.$$

Such bounds are usually considered in chopped form $\|\delta X\| \le \omega_1(|\!|\!|\delta A|\!|\!|)$, which is obtained by neglecting higher order terms in $\|\delta A\|$. It should be noted though, that these chopped bounds may be misleading, since actually the opposite inequality $\|\delta X\| > \omega_1(|\!|\!|\delta A|\!|\!|)$ may occur if the neglected terms are large.

In order to use such bounds without a serious underestimation of the actual quantity $\|\delta X\|$, we introduce the concept of *asymptotic domain* of the chopped bound, which is the set of data perturbations for which the quantity $\omega_1(|\!|\!|\delta A|\!|\!|)$ produced by the local bound, is N times larger than the neglected terms $o(\|\delta A\|)$. Here N is a positive constant and it is desirable that $N > 1$ but even if $N \le 1$, then the local bound may still be useful. Indeed, if $\|\delta X\|$ and $\omega_1(|\!|\!|\delta A|\!|\!|)$ are both very small, then even for $\|\delta X\| > \omega_1(|\!|\!|\delta A|\!|\!|)$ the quantity $\omega_1(|\!|\!|\delta A|\!|\!|)$ may be a good approximation for the actual perturbation $\|\delta X\|$, at least concerning its order of magnitude.

The local, or asymptotic perturbation analysis produces local (usually linear or first order homogeneous) perturbation bounds $\omega_1(|\!|\!|\delta A|\!|\!|)$ for $\|\delta X\|$ by keeping first order and neglecting higher order terms in $\|\delta A\|$. There is nothing wrong with such bounds if they are used properly. But one must always bear in mind that a "practically small" or even a "practically negligible" perturbation may not be small at all in the rigorous mathematical sense, i.e., it may be far beyond the asymptotic domain of the corresponding bound. In contrast, the nonlocal perturbation analysis gives rigorous perturbation bounds which are valid in a certain (possibly small but finite) domain of perturbations in the data.

There is a variety of viewpoints about of (chopped) local bounds. From a strict mathematical position, the use of such bounds is not appropriate unless it is guaranteed that the data perturbations are in the asymptotic domain. This, of course, requires an estimate of the neglected terms which is much more difficult than the derivation of local bounds. To estimate higher order terms means in fact to derive nonlocal perturbation bounds. Although rigorous, this viewpoint could lead to difficulties in practice, where even the derivation of local bounds may be a problem.

On the other hand, there are users of methods, who apply local estimates for a wide range of perturbations, hoping that everything is fine, or simply not suspecting that something may go wrong. Actually, we experience that many users in industry do not like condition and error estimates at all, since they require extra computational time, and since the user is not trained to interpret large sensitivity estimates.

It is difficult to determine the reliable "common sense" position, which should be a compromise between these the extreme positions of mathematically rigorous and sometimes pessimistic nonlocal bounds on one hand and easy to use chopped local local bounds on the other hand.

3.3 Regularity and linear bounds

As we have discussed, the sensitivity of (numerical) problems is measured by the size of the perturbations in the solution relative to the size of the perturbations in the data. The ratio of these quantities characterizes quantitatively the local sensitivity of the problem. In this subsection we consider the fundamental concepts of well-posedness and regularity for problems with explicit solution and the closely related issue of constructing linear perturbation bounds. Some of these results can be directly extended to problems with implicit solution.

We recall that $\mathcal{A} \subset \mathcal{V}$ is an open set and $\Phi : \mathcal{A} \to \mathcal{X}$ is a given continuous function, where $\mathcal{V}$ and $\mathcal{X}$ are finite dimensional real or complex spaces. For fixed $A \in \mathcal{A}$ the evaluation of $X = \Phi(A)$ is a problem with explicit solution.

Definition 3.12 A problem $X = \Phi(A)$, where Φ is continuous in an open neighborhood of A, is called *well-posed*. We also say that Φ is *well-posed at A*. A problem, or a function, which is not well-posed at certain A is said to be *ill-posed* at A.

The family of problems $\mathcal{A} \to \Phi(\mathcal{A})$ is *well-posed* if Φ is continuous on the set $\mathcal{A}$ (we also say that Φ is *well-posed on $\mathcal{A}$*).

The concept of well-posedness is somehow trivial for problems with explicit solution, since it simply means continuity. However, this concept is much more involved for problems with implicit solution, see Chapter 4.

It is instructive to see what ill-posedness means. The function Φ is not well-posed on $\mathcal{A}$ if it is ill-posed for at least one $A \in \mathcal{A}$. If the function Φ, defined on an open neighborhood of certain A, is ill-posed at A, then one of the following situations may happen:

– The function Φ is discontinuous at A. This means that either the limit $\lim_{B\to A}\Phi(B)$ does not exist, or it exists but is different from the value $\Phi(A)$. A classification of points of discontinuity will not be discussed here.

– The function Φ is continuous at A but for any $\varepsilon > 0$, there exists a data B with $\|A - B\| < \varepsilon$ such that Φ is discontinuous at B. This means that A is an isolated point of the set of points of continuity of Φ. There even exists a function Φ which is arbitrary times differentiable at A but discontinuous at every point $B \neq A$, see Examples 3.14 and 3.15 below.

For ill-posed problems little can be said about the quantitative dependence of the perturbations in the solution on the perturbations in the data.

Definition 3.13 A problem $X = \Phi(A)$ is said to be *regular* if it is well-posed and the ratio $\|\delta X\|/\|\delta A\|$ is uniformly bounded for $\delta A \to 0$. If a problem is not regular it is called *singular*.

Regularity means that there exist positive constants α and β such that $\|\delta X\| \leq \beta\|\delta A\|$ for all δA with $\|\delta A\| \leq \alpha$. Also, a regular problem is well-posed but, of course, a well-posed problem may not be regular.

A terminology remark. There is no unified terminology in the field of perturbation analysis. Here we have adopted terminology close to the Hadamard definition of posedness in functional analysis, see [44]. Sometimes problems, which we have just referred to as "regular" or "singular", are called in the literature "well-posed" or "ill-posed", respectively.

We will now return to condition numbers. Recall the Definition 2.2. the absolute condition number of a problem $X = \Phi(A)$. If a problem $X = \Phi(A)$, with Φ a continuous function in an open neighborhood of A, is not regular because the ratio $\|\delta X\|/\|\delta A\|$ is not bounded for $\delta A \to 0$, we set $K(A) = \infty$. Thus, for every continuous function $\Phi : \mathcal{A} \to \mathcal{X}$, relation (2.3) defines the function $K : \mathcal{A} \to [0, \infty]$ from $\mathcal{A}$ to the extended real axis $[0, \infty]$.

The absolute condition number is always well defined for well-posed problems. Indeed, consider a regular problem $X = \Phi(A)$ and set

$$K_n(A) := \sup\left\{ \frac{\|\Psi(A, E)\|}{\|E\|} : E \neq 0,\ \|E\| \leq \frac{1}{n} \right\}$$

for $n = 1, 2, \ldots$. We have $0 \leq K_{n+1}(A) \leq K_n(A) < \infty$ and hence, the sequence $\{K_n(A)\}$ is nonincreasing. Since it is also bounded from below, it is convergent to some nonnegative finite value $K(A)$.

We may determine the quantity $K(A)$ even for an arbitrary function Φ, defined in an open neighborhood of A. In this case the inequality $K(A) < \infty$ alone does not imply regularity of the problem $X = \Phi(A)$. It only guarantees that Φ is continuous *at* the point A but not necessarily in a neighborhood of A, see Example 3.14.

Example 3.14 Consider the function $\varphi : \mathbb{R} \to \mathbb{R}$ defined as

$$\varphi(a) = \begin{cases} a & \text{if} \quad a \in \mathbb{R} \backslash \mathbb{Q} \\ 0 & \text{if} \quad a \in \mathbb{Q}, \end{cases}$$

where $\mathbb{Q}$ is the set of rational numbers. We have $\varphi(0) = 0$ and for $a = 0$ the perturbation function is $\psi(0, e) = \varphi(e)$. The function φ is continuous at 0 and the function ψ is continuous at $(0, 0)$. In addition, the absolute condition number for $a = 0$ is $K(0) = 1$. However, the problem $x = \varphi(a)$ is singular at every a. Indeed, the function φ is discontinuous at every point $a \neq 0$ and the problem can not be regular at $a \neq 0$. Furthermore, the function φ is continuous *only* at $a = 0$ and hence, it is not continuous in any open interval containing 0, i.e., the problem is singular everywhere. ◇

In the next example we show that for every integer $m \geq 1$ there exists a function φ, defined on $\mathbb{R}$, which is m times differentiable at a given point and is discontinuous elsewhere.

Example 3.15 Let the function $\varphi : \mathbb{R} \to \mathbb{R}$ be defined via

$$\varphi(a) = \begin{cases} a^{m+1} & \text{if} \quad a \in \mathbb{R} \backslash \mathbb{Q} \\ 0 & \text{if} \quad a \in \mathbb{Q}. \end{cases}$$

This function is m times differentiable at 0 with $\varphi^{(k)}(0) = 0$ for $k = 0, 1, \ldots, m$. At the same time φ is discontinuous at every point $a \neq 0$. The absolute condition number at $a = 0$ is zero in this case. ◇

If the Fréchet derivative $\Phi'(A)$ (see Appendix A and [188]) of the function Φ at the point A exists, then the absolute condition number may be computed as

$$K(A) = \|\Phi'(A)\|. \tag{3.2}$$

Here the norm $\|\Phi'(A)\|$ of the linear operator $\Phi'(A) : \mathcal{A} \to \mathcal{X}$ is defined as

$$\|\Phi'(A)\| := \max \{ \|\Phi'(A)(E)\| : \|E\| = 1 \}.$$

Note that $\Phi'(A)$ depends on A as a parameter. Thus, $\Phi'(A)(E)$ is the image of $E \in \mathcal{A}$ under the action of the linear mapping $\Phi'(A) : \mathcal{A} \to \mathcal{X}$.

When we consider the vectorizations (2.8) and $\varphi = \mathrm{vec} \circ \Phi$ then the linear operator $\varphi'(a)$ is identified with the $N \times M$ Jacobi matrix

$$\varphi'(a) = \left[\frac{\partial \varphi_i(a)}{\partial a_j} \right] = \left[\frac{\partial \varphi(a)}{\partial a_1}, \ldots, \frac{\partial \varphi(a)}{\partial a_M} \right]$$

of the vector function $\varphi = [\varphi_1, \ldots, \varphi_N]^\top$, evaluated at the point $a = [a_1, \ldots, a_M]^\top$, where φ_i and a_j are the components of the function φ and the argument a, and $N = mn$, $M = m_1n_1 + \cdots + m_rn_r$.

Example 3.16 Consider the function $\Phi : \mathbb{F}^{m\times n} \to \mathbb{F}^{m\times n}$, defined by

$$\Phi(A) := AC_1A + C_2A + AC_3 + C_4,$$

where $C_1 \in \mathbb{F}^{n\times m}$, $C_2 \in \mathbb{F}^{m\times m}$, $C_3 \in \mathbb{F}^{n\times n}$, $C_4 \in \mathbb{F}^{m\times n}$ are given matrix coefficients. Then the linear operator $\Phi'(A) : \mathbb{F}^{m\times n} \to \mathbb{F}^{m\times n}$ is determined as

$$\Phi'(A)(E) = (C_2 + AC_1)E + E(C_3 + C_1A).$$

Setting $a = \mathrm{vec}(A) \in \mathbb{F}^{mn}$, $A = \mathrm{vec}^{-1}(a)$, $x = \mathrm{vec}(X) \in \mathbb{F}^{mn}$, $\varphi = \mathrm{vec} \circ \Phi$ and using the Kronecker product (Appendix C), we obtain

$$\varphi'(a) = I_n \otimes \left(C_2 + \mathrm{vec}^{-1}(a)C_1\right) + \left(C_3 + C_1\mathrm{vec}^{-1}(a)\right)^\top \otimes I_m.$$

$\diamondsuit$

It may happen that the Fréchet derivatives of Φ at some points from $\mathcal{A}$ do not exist although the problem $A \mapsto \Phi(A)$ is regular for *all* data $A \in \mathcal{A}$.

Example 3.17 The function $a \mapsto \varphi(a) := \|a\|$, $a \in \mathbb{R}^M$, is not differentiable at $a = 0$ but the corresponding problem $x = \varphi(a)$ is regular with $K(a) = 1$ for all $a \in \mathbb{R}^n$. $\diamondsuit$

For regular problems we may derive component-wise bounds as follows. Consider a problem in vector form $x = \varphi(a)$, where x has size n and a is size m. If the Fréchet derivative $\varphi'(a)$ exists and is locally bounded, then for some $0 \prec \rho \in \mathbb{R}^n_+$ we have

$$|\varphi(a + \delta a) - \varphi(a)| \preceq \mathbf{L}(a, \rho)|\delta a| \tag{3.3}$$

for all δa with $|\delta a| \preceq \rho$, where $\mathbf{L}(a, \rho) = [\mathbf{l}_{ij}(a, \rho)] \in \mathbb{R}^{n\times m}_+$ is a matrix with elements

$$\mathbf{l}_{ij}(a, \rho) := \max\left\{\left|\frac{\partial \varphi_i}{\partial a_j}(a + e)\right| : |e| \preceq \rho\right\}.$$

Note that even if the Fréchet derivative does not exist, then the bound (3.3) is still valid with

$$\mathbf{l}_{ij}(a, \rho) := \max\{|\varphi_{ij}(a + e)| : |e| \preceq \rho\},$$

where

$$\varphi_{ij}(a) := \lim_{\alpha\to 0} \sup\left\{\frac{|\varphi_i(a + ze_j) - \varphi_i(a)|}{|z|} : z \neq 0,\ |z| \leq \alpha\right\}$$

and $e_1, \ldots, e_m$ are the columns of the identity matrix I_m.

There is a deep connection between differentiability and regularity as described next. It follows from the definition of regularity that the problem $X = \Phi(A)$ is regular if and only if the function Φ is *locally Lipschitz continuous*.

Definition 3.18 *A function* Φ *is* locally Lipschitz continuous *at the point* A *if*

$$\|\delta X\| = \|\Psi(A, \delta A)\| \le L(A, \alpha)\|\delta A\|, \ \|\delta A\| \le \alpha \tag{3.4}$$

for all $\alpha \in [0, \alpha_0)$ *and some* $\alpha_0 = \alpha_0(A) > 0$. *Here*

$$L(A, \alpha) := \sup\left\{\frac{\|\Psi(A, E)\|}{\|E\|} : E \ne 0, \ \|E\| \le \alpha\right\} < \infty$$

is the Lipschitz constant *of* Φ *in the closed* α*-neighborhood of* A.

Since $L(A, \alpha) \ge 0$ is nondecreasing in $\alpha > 0$, we see that $K(A) \le L(A, \alpha)$ and

$$\lim_{\alpha \to 0} L(A, \alpha) = K(A). \tag{3.5}$$

The connection between local Lipschitz continuity and local differentiability is revealed by the theorem of Rademacher.

Theorem 3.19 *If the function* Φ *is Lipschitz in a neighborhood* $\mathcal{N}_A$ *of* A *then it is almost everywhere Fréchet differentiable in* $\mathcal{N}_A$.

Thus, differentiability implies regularity, while regularity implies differentiability almost everywhere.

It is important to observe that the bound (3.4) is linear but nonlocal. Such bounds are of special interest in perturbation theory. Note that not only may $K(A)$ be obtained from $L(A, \alpha)$ via (3.5), but also vice versa. A nonlocal bound (3.4) may be constructed using the absolute condition number K via the relation

$$L(A, \alpha) := \sup\{K(A + E) : \|E\| \le \alpha\}. \tag{3.6}$$

To utilize (3.6) for a differentiable function $\varphi = \mathrm{vec} \circ \Phi$ in $a = \mathrm{vec}(A)$ one may use the property that

$$K(a) = \left\|\left[\frac{\partial \varphi_i(a)}{\partial a_j}\right]\right\|.$$

Example 3.20 Consider the problem of computing the power $\Phi(A) := A^p$ of the square matrix $A \in \mathbb{F}^{n \times n}$, where $p > 0$ is a positive integer. Then Φ is locally Lipschitz continuous for all A, and

$$L(A, \alpha) = \frac{(\|A\| + \alpha)^p - \|A\|^p}{\alpha} = \sum_{k=0}^{p-1} \binom{p}{k} \|A\|^k \alpha^{p-1-k}.$$

$\diamondsuit$

Example 3.21 Consider the problem $\Phi(A) := A^{-1}$, defined on the set of nonsingular matrices. The function Φ is locally Lipschitz continuous. Let $E \in \mathbb{F}^{n\times n}$ be any matrix with $\|E\| < \alpha_0 := \|A^{-1}\|^{-1}$. Using the representation

$$\begin{aligned}\Psi(A,E) &= (A+E)^{-1} - A^{-1}\\ &= -A^{-1}EA^{-1} + (A^{-1}E)^2(I_n + A^{-1}E)^{-1}A^{-1},\end{aligned}$$

we obtain

$$L(A,\alpha) = \frac{1}{\alpha_0^2}\left(1 + \frac{\alpha^2}{\alpha_0 - \alpha}\right),\quad 0 \le \alpha < \alpha_0.$$

$\diamondsuit$

Regularity (or local Lipschitz continuity) is a desirable property of problems, especially when they are solved in finite precision arithmetic. However, there are problems for which the function Φ grows (locally) faster than any linear function. To deal with such problems we introduce the concept of Hölder continuity.

Definition 3.22 The problem $X = \Phi(A)$ is said to be *locally Hölder continuous* if there exist quantities $\alpha_0(A) > 0$, $\gamma(A) > 0$ and $H(A,\alpha) > 0$ such that

$$\|\delta X\| \le H(A,\alpha)\|\delta A\|^{\gamma(A)},\quad \|\delta A\| \le \alpha$$

for all $\alpha \in [0, \alpha_0(A))$. Here $H(A,\alpha)$ and $\gamma(A)$ are the *Hölder constant* and *Hölder exponent* of Φ in the closed α-neighborhood of A.

If the function Φ is Hölder continuous at A with an exponent $\gamma(A) \ge 1$, then it is in fact Lipschitz continuous as well. Indeed, for $\gamma(A) = 1$ this holds by definition. Suppose that $\gamma(A) > 1$. Then for $\|E\| \le \alpha$ we have

$$\|\Psi(A,E)\| \le H(A,\alpha)\|E\|^{\gamma(A)} \le L(A,\alpha)\|E\|$$

with $L(A,\alpha) := \alpha^{\gamma(A)-1}H(A,\alpha)$.

Functions which are Hölder continuous with $\gamma(A) < 1$ grow faster than any Lipschitz continuous function in a neighborhood of A.

Example 3.23 The scalar problem $x = |a|^{\gamma_0}$, where $0 < \gamma_0 < 1$, is Hölder continuous at $a = 0$ with constant 1 and exponent $\gamma(0) = \gamma_0$. $\diamondsuit$

Example 3.24 The problem of computing a multiple root x of an algebraic equation is Hölder continuous with exponent $1/k$, where $k > 1$ is the multiplicity of x. $\diamondsuit$

Example 3.25 The problem of computing a multiple eigenvalue λ of a square matrix, corresponding to nonlinear elementary divisors, is Hölder continuous with exponent $1/k$, where $k > 1$ is the size of the largest block with eigenvalue λ in the Jordan canonical form of the matrix. $\diamondsuit$

According to the definition of regularity, a Hölder continuous problem $X = \Phi(A)$ with constant $\gamma(A) < 1$ is singular. But there are also singular problems which are not even Hölder continuous, since the function Φ grows (locally) faster than any power function.

Example 3.26 The scalar real function $\varphi : (-1, 1) \to \mathbb{R}$, defined as

$$\varphi(a) := \begin{cases} -\mathrm{sign}(a)/\ln|a| & \text{if} \quad 0 < |a| < 1 \\ 0 & \text{if} \quad a = 0 \end{cases}$$

is not Hölder continuous at $a = 0$. Note that the inverse function $\varphi^{-1} : \mathbb{R} \to (-1, 1)$, defined via

$$\varphi^{-1}(x) = \begin{cases} \mathrm{sign}(x)/\exp(-1/|x|) & \text{if} \quad x \neq 0 \\ 0 & \text{if} \quad x = 0 \end{cases}$$

is infinitely differentiable everywhere and extremely "flat" at the origin. It is analytic (i.e., representable by its Taylor series) for $x \neq 0$ but not for $x = 0$. Indeed, all derivatives of φ^{-1} vanish at $x = 0$. Hence, the Taylor series of φ^{-1} at $x = 0$ is identically zero and is thus different from φ^{-1} on any open interval, containing 0. ◇

For a regular problem $X = \Phi(A)$ we have the *asymptotic bound*

$$\|\delta X\| \leq K(A)\|\delta A\| + o(\|\delta A\|),\ \delta A \to 0, \tag{3.7}$$

where $o(z)/z \to 0$ for $z \to 0$ and the o-term is typically of the form $O(\|\delta A\|^2)$, $\delta A \to 0$. Neglecting the o-term, it would be good if the inequality

$$\|\delta X\| \leq K(A)\|\delta A\| \tag{3.8}$$

would hold, but unfortunately it does not hold in general for nonlinear problems. First, it may not be true for large $\|\delta A\|$. Second, if it holds for some small $\|\delta A\|$, there is no indication for the size of $\|\delta A\|$ (it may well happen that (3.8) is valid only for $\delta A = 0$ or under some special assumptions on δA). At the same time a true bound is for instance the inequality

$$\|\delta X\| \leq L(A, \alpha)\|\delta A\|$$

which is rigorously valid for all δA with $\|\delta A\| \leq \alpha$. However, while $K(A)$ is usually easily computable, it is much more difficult to calculate or estimate $L(A, \alpha)$.

In the following examples we study the validity of linear bounds for small finite perturbations in scalar problems.

Example 3.27 Consider the scalar problem $x = \varphi(a)$, where $\varphi : \mathbb{R} \to \mathbb{R}$ is analytic at a. Suppose that $\varphi'(a) \neq 0$ and that $\varphi^{(k)}(a)$ is the first nonzero derivative of φ at a with $k > 1$. Then we have

$$|\delta x| = |\delta a| \left| \varphi'(a) + \frac{\varphi^{(k)}(a)}{k!}(\delta a)^{k-1} \right| + O(|\delta a|^{k+1}), \ \delta a \to 0.$$

Hence, the bound (3.8) will be valid for all δa in a certain small neighborhood if and only if k is odd and $\varphi'(a)\varphi^{(k)}(a) < 0$. This in particular implies $k \geq 3$ and thus, a is an inflection point and (3.8) is not valid generically. ◇

Example 3.28 Consider the scalar function $a \mapsto \varphi(a) = a^2$ in a neighborhood of $a = 0$. Since $\varphi'(0) = 0$, then (3.8) yields $|\delta x| = 0$. Since in fact $\delta x = (\delta a)^2$, we see that the bound (3.8) is valid *only* for $\delta a = 0$. ◇

Examples 3.27 and 3.28 show that inequality (3.8) is *generically not valid* for all δA in an open neighborhood of the origin.

To avoid this delicate situation of a bound-that-may-be-violated, we proceed as follows. We introduce the special symbol $\lesssim$ to denote an inequality within first order terms of magnitude, i.e.,

$$\alpha \lesssim \beta \tag{3.9}$$

is equivalent to

$$\alpha \leq \beta + o(\beta), \ \beta \to +0. \tag{3.10}$$

We use (3.9) instead of (3.10) because in practice we do not deal with limiting processes $\beta \to +0$, but rather with finite (although possibly small) positive quantities β. With this notation we may write

$$\|\delta X\| \lesssim K(A)\|\delta A\| \tag{3.11}$$

which is a linear local estimate, i.e., it should be used in a small neighborhood of A. We again stress that in contrast to (3.11), the estimate (3.4) is linear but nonlocal.

Linear local estimates of the form (3.11) are widely used in computational practice in the chopped form (3.8) which typically produces acceptable results. However, the bound (3.8) may severely underestimate the actual perturbation $\|\delta X\|$. Indeed, consider the simple case of a scalar real function φ, having continuous second derivative on a given interval. If the second derivative of φ is small, the bound (3.8) will give satisfactory results. But if the second derivative of φ is large, there may be a serious underestimation of the actual perturbation.

Example 3.29 Let

$$x = \varphi(a) := \frac{1}{1\,000\,001 - a}, \quad a \leq 1\,000\,000.9999999.$$

For $a = 1\,000\,000$ we have $x = 1$ and $\varphi'(a) = 1$. Therefore, if $\delta a = 0.999999$, then the estimate (3.8) gives

$$|\delta x| \leq \frac{|\delta a|}{(1\,000\,001 - a)^2} = |\delta a| = 0.999999,$$

while the actual perturbation is

$$\delta x = \frac{\delta a}{|(1\,000\,001 - a)\,(1\,000\,001 - a - \delta a)|} = \frac{\delta a}{1 - \delta a} = 999\,999,$$

i.e., we have an underestimation by a factor 10^6. Note that here $\varphi''(a) = 2$ and the error is not due to the size of the second derivative, but rather to the fact that the remainder $R(a, \delta a)$ in the Taylor formula

$$\delta x = \varphi(a + \delta a) - \varphi(a) = \varphi'(a)\delta a + R(a, \delta a)$$

is large (in fact $R(a, \delta a) = 999\,998.000001$).

This example is actually not too artificial, since the relative perturbation in a is less than 10^{-6}. ♢

Recalling the definition of relative condition number $\kappa(A)$ (see Definition 2.3), we have for the relative perturbations ρ_X and ρ_A the relationship

$$\rho_X \leq \kappa(A)\rho_A + o(\rho_A), \;\; \rho_A \to 0$$

and hence,

$$\rho_X \lesssim \kappa(A)\rho_A \tag{3.12}$$

is the linear local perturbation bound.

In the definition of the relative condition number, the assumptions $X \neq 0$ and $A \neq 0$ may be too restrictive when we study the local sensitivity of a problem and one of the following three conditions holds (i) $K(A) = \infty$, (ii) $A = 0$ or (iii) $\Phi(A) = 0$. A typical example here is the evaluation of the function $a \mapsto a^\gamma : \mathbb{R}_+ \to \mathbb{R}_+$, where $\gamma \in (0, 1)$ is a parameter, at the point $a = 0$. Here all three conditions (i), (ii), (iii) hold. In such a situation, a generalization of the relative condition number may be introduced as follows. For $A \in \mathcal{A}$ set

$$\begin{aligned}
\mathcal{E}_1(A) &:= \{E \in \mathcal{A} : K(A + E) < \infty\}, \; \mathcal{E}_2(A) := \{-A\}, \\
\mathcal{E}_3(A) &:= \{E \in \mathcal{A} : \Phi(A + E) \neq 0\}
\end{aligned}$$

and

$$\mathcal{E}(A) := \mathcal{E}_1(A) \cup \mathcal{E}_2(A) \cup \mathcal{E}_3(A).$$

Definition 3.30 The limit (finite or infinite)

$$\kappa(A) := \lim_{\alpha \to 0} \sup \left\{ \frac{K(A + E)\|A + E\|}{\|\Phi(A + E)\|} : E \in \mathcal{E}(A), \; \|E\| \leq \alpha \right\} \tag{3.13}$$

is called a *generalized relative condition number* of the problem $X = \Phi(A)$.

When none of the conditions (i), (ii), (iii) holds, then the generalized relative condition number is the standard relative condition number of the problem.

Definition 3.31 A problem $X = \Phi(A)$ is called *R-regular* if its generalized relative condition number is finite.

Example 3.32 The continuous function $a \mapsto a^\gamma : \mathbb{R}_+ \to \mathbb{R}_+$, where $\gamma \in (0,1)$, is singular at $a = 0$, since $K(0) = \infty$. At the same time the generalized relative condition number exists for all $a \geq 0$ and is equal to γ. ♢

If the data A is represented as in (2.4), then the local linear bound in terms of absolute perturbations has the form

$$\|\delta X\| \lesssim \sum_{i=1}^{r} K_{A_i}(A)\|\delta A_i\| = K(A)\|\delta A\|, \tag{3.14}$$

where

$$K(A) := [K_{A_1}(A), \ldots, K_{A_r}(A)] \in \mathbb{R}_+^{1\times r}$$

is now the *vector absolute condition number* and $K_{A_i}(A)$ is the *individual absolute condition number* with respect to A_i. If $A_i \neq 0$, then an estimate in terms of relative perturbations is straightforward,

$$\delta_X \lesssim \sum_{i=1}^{r} \kappa_{A_i}(A)\delta_{A_i} = \kappa(A)\delta_A, \tag{3.15}$$

where

$$\begin{aligned} \kappa(A) &:= \left[\frac{\kappa_{A_1}(A)\|A_1\|}{\|X\|}, \ldots, \frac{\kappa_{A_r}(A)\|A_r\|}{\|X\|}\right] \in \mathbb{R}_+^{1\times r}, \\ \delta_A &:= [\delta_{A_1}, \ldots, \delta_{A_r}]^\top \in \mathbb{R}_+^r, \ \delta_{A_i} := \frac{\|\delta A_i\|}{\|A_i\|}. \end{aligned}$$

Here $\kappa(A)$ is the *vector relative condition number* and $\kappa_{A_i}(A)$ is the *individual relative condition number* with respect to A_i.

The accuracy of condition number based perturbation bounds depends strongly on the structure of the data, as shown in the next example.

Example 3.33 Consider the problem

$$x = M_1 a_1 + M_2 a_2 = Ma, \ M := [M_1, M_2], \ a := \left[a_1^\top, a_2^\top\right]^\top,$$

where the data a_1, a_2 and the result x are finite-dimensional vectors, and M_1, M_2, M are matrices of compatible dimensions. Then

$$\delta x = M_1 \delta a_1 + M_2 \delta a_2 = M \delta a.$$

Hence, in 2-norm the bound

$$\|\delta x\| \le \|M_1\|\,\|\delta a_1\| + \|M_2\|\,\|\delta a_2\| \tag{3.16}$$

based on condition numbers is valid. We also have the bound

$$\|\delta x\| \le \|M\|\,\|\delta a\| = \|M\|\sqrt{\|\delta a_1\|^2 + \|\delta a_2\|^2}. \tag{3.17}$$

Which bound is better depends on the data. If $\delta a_2 \neq 0$ and $\|M\| > \|M_1\|$, then a simple computation shows that the bound (3.17) is better than (3.16) if and only if

$$\frac{m_1 m_2 - \sqrt{m_1^2 + m_2^2 - 1}}{1 - m_1^2} < \frac{\|\delta a_1\|}{\|\delta a_2\|} < \frac{m_1 m_2 + \sqrt{m_1^2 + m_2^2 - 1}}{1 - m_1^2}$$

where $m_i := \|M_i\|/\|M\|$. At the same time, we have a third bound [140]

$$\|\delta x\| \le \sqrt{\|M_1\|^2\|\delta a_1\|^2 + 2\|M_1^{\mathrm{H}} M_2\|\|\delta a_1\|\|\delta a_2\| + \|M_2\|^2\|\delta a_2\|^2},$$

which is always better than (or equal to) the condition number based bound (3.16) since $\|M_1^{\mathrm{H}} M_2\| \le \|M_1\|\|M_2\|$. $\diamondsuit$

Often it is preferable to have a single overall condition number for a given problem even when the data are naturally presented in the form (2.4). Suppose that the problem is regular and $\|\delta A_i\| \le \varepsilon\|A_i\|$, where $\varepsilon > 0$ is a small constant. Then

$$\rho_X \lesssim \varepsilon\kappa^*(A),$$

where

$$\kappa^*(A) := \sum_{i=1}^{r} \kappa_i(A)$$

is the *overall relative condition number* of the problem.

The overall relative condition number $\kappa^*(A)$, together with the rounding unit eps of the finite precision arithmetic, is among the main factors of determining the accuracy of the solution. The relative condition number itself may be considered as "large" or "small" only within a particular computing environment. In particular, in finite precision arithmetic we have the following heuristic concepts.

A regular problem (Φ, A) is considered as *well-conditioned* if the quantity $\kappa^*(A)$ is small, and *ill-conditioned* if it is large in the context of the finite precision arithmetic with rounding unit eps, used to solve the problem. Usually the problem is considered as very well-conditioned if $\kappa^*(A) \approx 1$ and as very ill-conditioned if $\mathrm{eps}\,\kappa^*(A) \approx 1$.

The solution of very ill-conditioned problems in finite precision arithmetic may lead to a result with no correct digits. More generally, the following heuristic *rule of thumb* is often used in practice. If $\mathrm{eps}\,\kappa^*(A) < 1$, then about (or no more than)

$$-\log_{10}(\mathrm{eps}\,\kappa^*(A)) \tag{3.18}$$

correct decimal digits may be expected in the computed solution.

Example 3.34 The relative condition number of the problem $x = \varphi(a) \neq 0$ of evaluating a differentiable scalar function φ at the point $a \neq 0$ is

$$\kappa(a) = \frac{|a||\varphi'(a)|}{|\varphi(a)|}.$$

Hence, $\kappa(a)$ may be large, and the problem may be very ill-conditioned, if $|a|$ or $|\varphi'(a)|$ is large and/or $|\varphi(a)|$ is small. If $\varphi(a) = \sin a$, then the computational problem is ill-conditioned for arguments a with large absolute values and/or close to an integer multiple of π. $\diamondsuit$

A generalization of the concept of regularity of single problems to families of problems is the following.

Definition 3.35 *A family of computational problems $\mathcal{A} \to \Phi(\mathcal{A})$ is said to be* regular *if the function $\Phi : \mathcal{A} \to \mathcal{X}$ is continuous and the quantity*

$$K(\mathcal{A}) := \sup \{K(A) : \ A \in \mathcal{A}\}$$

is finite. Here $K(\mathcal{A})$ is the absolute condition number *of the family of problems $X = \Phi(A)$, parametrized by the data $A \in \mathcal{A}$.*

Hence, a family of regular problems is regular if the set of absolute condition numbers of its members is bounded.

A relative condition number for a family of problems is defined in a similar way.

Definition 3.36 *A family of computational problems $\mathcal{A} \to \Phi(\mathcal{A})$ is said to be* R-regular *if the function $\Phi : \mathcal{A} \to \mathcal{X}$ is continuous and the quantity*

$$\kappa(\mathcal{A}) := \sup \{\kappa(A) : A \in \mathcal{A}\backslash\{0\},\ \Phi(A) \neq 0\}$$

is finite. The quantity $\kappa(\mathcal{A})$ is the relative condition number *of the family of problems $X = \Phi(A)$, parametrized by the data $A \in \mathcal{A}$.*

It is interesting to note that a family which is R-regular may have members that are not regular.

A family of problems $\mathcal{A} \to \Phi(\mathcal{A})$ with Φ being a continuous function may be neither regular nor R-regular. In turn, regularity does not imply R-regularity and vice versa as shown in the next examples.

Example 3.37 For the family of problems $a \mapsto a^m$, $a \in \mathbb{R}_+$, where $m > 0$ is a parameter, the absolute and relative condition numbers are $K(a) = m|a|^{m-1}$ and $\kappa(a) = m$, respectively. (The values of K for $a = 0$, $m < 1$, and of κ for $a = 0$, are not defined.) Thus, the family is R-regular for all $m > 0$. At the same time the problem is not regular at $a = 0$ if $m < 1$. $\diamondsuit$

Example 3.38 The family of problems $a \mapsto \sin a$, $a \in [0, 2\pi)$ is regular with absolute condition number equal to 1, but is not R-regular because its relative condition numbers are not finite at $a = 0$ and $a = \pi$. $\diamondsuit$

Example 3.39 The problem $A \mapsto A^{-1}$ of inverting a square matrix is not R-regular on the set of all invertible matrices, since the relative condition number $\text{cond}(A) = \|A\| \, \|A^{-1}\|$ of the matrix A with respect to inversion is not uniformly bounded on that set. $\diamondsuit$

Example 3.40 Let $a, x \in \mathbb{F}^n$ and $x = \varphi(a) := Ma$, where $M \in \mathbb{F}^{n\times n}$ is a given nonsingular matrix. Then $K(a) = \|M\|$ and the relative condition number is

$$\kappa(a) = \frac{\|M\| \, \|a\|}{\|Ma\|}.$$

The relative condition number with respect to the whole space $\mathbb{F}^n$ is

$$\kappa(\mathbb{F}^n) = \sup \left\{ \frac{\|M\| \, \|a\|}{\|Ma\|} : a \in \mathbb{F}^n \right\} = \text{cond}(M).$$

$\diamondsuit$

3.4 Nonlocal bounds

The linear local sensitivity bounds (3.11), (3.12) may underestimate the true size of the perturbation, since they are valid only in a (usually small) asymptotic domain which is not known in practice. But in real applications perturbations in the data are always finite, of nonzero size. Hence, without a rigorous bound on the neglected higher order o–terms in the corresponding expressions, the use of local bounds may lead to erroneous conclusions. Thus, the use of local estimates for practical purposes remains, at least theoretically, unjustified unless an additional analysis of the neglected terms is made. But to obtain bounds for the neglected nonlinear terms means to find a nonlinear perturbation bound.

Sometimes it is possible to derive linear nonlocal perturbation bounds which do not underestimate the actual perturbation in the solution. The problem of the accuracy and the domain of applicability of such bounds, however, still remains open. It must be stressed that the behavior of the true perturbation is inherently strongly nonlinear even for simplest linear equations, see Section 3.5.

The disadvantages of local estimates may be overcome by using the technique of nonlinear perturbation analysis (Section 8), which has two main purposes: first, to show that a solution of the perturbed problem exists for a given range of perturbations, and second, to find a nonlocal (and in general nonlinear) perturbation bound of the form

$$\|\delta X\| \leq p(\|\delta A\|), \ \|\delta A\| \leq \alpha, \tag{3.19}$$

where $\alpha > 0$ and $p : [0, \alpha] \to R_+$ is a nondecreasing function with $p(0) = 0$. (We use the term "perturbation bound" for both the function p and the inequality (3.19).)

For the representation (2.4) the nonlinear bound has the form

$$\|\delta X\| \leq p(\|\!|\delta A|\!\|), \ \|\delta A\| \in \Omega, \tag{3.20}$$

where $\Omega \subset \mathbb{R}_+^r$ is a closed set and p is a function of r arguments, nondecreasing in each of them, and satisfying $p(0) = 0$.

When matrix absolute values are used for A and X, we have

$$|\delta X| \preceq P(|\delta A|), \ |\delta A| \in \Gamma, \tag{3.21}$$

where $\Gamma \subset \mathcal{V}_+$ is a closed set. The elements of the matrix-valued function $P : \Gamma \to \mathbb{R}_+^{m\times n}$ have the properties of the function p in (3.20).

An important property of the nonlocal bounds is that they are valid rigorously in the corresponding domains for $\|\delta A\|$, $\|\!|\delta A|\!\|$ or $|\delta A|$, in contrast to the chopped first order local estimates, where higher order terms are neglected.

A desirable property of a perturbation bound is to be unimprovable.

Definition 3.41 *The perturbation bound (3.19) is said to be* unimprovable *relative to the set of data $\mathcal{A}$ if for any positive $\eta < 1$ there exist $A \in \mathcal{A}$ and δA with $A + \delta A \in \mathcal{A}$ and $\|\delta A\| \leq \alpha$, such that the true perturbation δX satisfies $\|\delta X\| = \eta p(\|\delta A\|)$.*

Thus, an unimprovable bound is almost or exactly reached for some data (take η close to 1). Similar definitions apply also for the bounds (3.20) and (3.21). The concept of unimprovability is close to that of almost necessity, see e.g., [135]. Of course, an unimprovable estimate may as well give pessimistic results for other data and/or data perturbations.

A detailed study of the properties of perturbation bounds is presented in Chapter 7.

3.5 Case study

Consider the scalar equation $a_1 x = a_2$, where $a_1 a_2 \neq 0$, and let δa_i be perturbations in the data a_i satisfying $\delta a_1 \neq -a_1$. This equation is reduced to an explicit problem

$$(a_1, a_2) \mapsto x = \frac{a_2}{a_1}. \tag{3.22}$$

Although being quite simple, problem (3.22) reveals some important issues in perturbation analysis. We have

$$\delta x = \frac{\delta a_2 - x\delta a_1}{a_1 + \delta a_1}$$

and

$$\frac{\delta x}{x} = \frac{\delta a_2/a_2 - \delta a_1/a_1}{1 + \delta a_1/a_1}.$$

Thus, δx exposes highly nonlinear behavior in a neighborhood of the vertical line $\delta a_1 = -a_1$ in the plane of perturbations $\delta a_1, \delta a_2$. Setting

$$\rho_x := \left|\frac{\delta x}{x}\right|, \ \rho_{a_i} := \left|\frac{\delta a_i}{a_i}\right|$$

we get

$$\rho_x \leq \frac{\rho_{a_1} + \rho_{a_2}}{1 - \rho_{a_1}}. \tag{3.23}$$

This is a nonlinear nonlocal bound with domain of applicability $0 \leq \rho_{a_1} < 1$. The estimate (3.23) is unimprovable. If we set

$$\widetilde{\rho}_x := \frac{|\delta x|}{|x + \delta x|}$$

then for $0 \leq \rho_{a_1}, \rho_{a_2} < 1$ we get another unimprovable nonlinear bound as

$$\widetilde{\rho}_x \leq \frac{\rho_{a_1} + \rho_{a_2}}{1 - \rho_{a_2}}.$$

The relative condition numbers with respect to a_1 and a_2 are both equal to 1. Thus, the local linear estimate is

$$\rho_x \lesssim \rho_{a_1} + \rho_{a_2} \tag{3.24}$$

and it underestimates the true perturbation arbitrarily for δa_1 approaching $-a_1$. Take for instance $a_1 = a_2 = 1$, $\delta a_1 = -0.999999$ and $\delta a_2 = 0$. Then the true relative perturbation is $\rho_x = 999\,999$, while the local estimate (3.24) gives $\rho_x \lesssim 1$. In addition, the linear estimate formally can be written down also for $\delta a_1 = -a_1$. Here the perturbed equation $0x = a_2 + \delta a_2$ either has no solution (if $\delta a_2 \neq -a_2$) or has infinitely many solutions (if $\delta a_2 = -a_2$). In the first case the resulting estimate $\rho_x \lesssim 1 + \rho_{a_2}$ makes no sense.

To find a linear nonlocal estimate, we suppose that ρ_{a_1} is allowed to vary only in the interval $[0, 1 - \mu]$, where $\mu < 1$ is a positive constant. Then

$$\frac{1}{1 - \rho_{a_1}} \leq \frac{1}{\mu}$$

and

$$\rho_x \leq \frac{1}{\mu}\left(\rho_{a_1} + \rho_{a_2}\right), \ 0 \leq \rho_{a_1} \leq 1 - \mu.$$

Taking $\mu = 0.5$ we obtain the linear nonlocal estimate

$$\rho_x \leq 2\left(\rho_{a_1} + \rho_{a_2}\right), \ 0 \leq \rho_{a_1} \leq 0.5.$$

3.6 Notes and references

Properties of problems with explicit solution have been considered in [134, 135]. In particular, the concept of perturbation function (Section 3.2) has been introduced therein.

Condition numbers for various types of problems have been considered in [188, 89]. Condition estimators are discussed in [38, 121, 166].

Condition numbers for complex functions which are not Fréchet differentiable but have Fréchet pseudoderivatives are analyzed in [140, 145, 137, 127].

Chapter 4

Problems with implicit solutions

4.1 Introductory remarks

The general considerations made in Chapters 2 and 3 are applicable to problems with explicit or implicit solutions. However, problems with implicit solution (e.g., solving equations) have many special features which are better considered in a specific framework.

4.2 Posedness and regularity

Consider a problem formulated in terms of the equation

$$F(A, X) = 0 \tag{4.1}$$

relative to the unknown quantity $X \in \mathcal{X} = \mathbb{F}^{n \times m}$, where $A \in \mathcal{V}$ is a parameter and the set $\mathcal{V}$ is defined by (2.4). Here $F : \mathcal{D} \to \mathbb{F}^{p \times q}$ is a given continuous function, and $\mathcal{D} \subset \mathcal{V} \times \mathcal{X}$ is a domain, i.e., an open and connected set.

Equation (4.1) gives rise to several global and local objects. One global object is the set of all pairs (A, X) for which (4.1) holds. This is a manifold $\mathcal{P} \subset \mathcal{D}$ in $\mathcal{V} \times \mathcal{X}$ of generic dimension $\dim(\mathcal{V}) + mn - pq$ (for manifolds For a fixed A we also have (locally) the set $\Xi(A) \subset \mathcal{X}$ of all solutions X of (4.1) corresponding to this particular value of A. A particular solution $X \in \Xi(A)$ may or may not depend continuously on the data. A rigorous definition of these concepts follows below.

Denote by

$$\mathcal{D}_{\mathcal{V}} := \{A : (A, X) \in \mathcal{D}\} \subset \mathcal{V} \tag{4.2}$$

the projection of $\mathcal{D}$ on $\mathcal{V}$ which in this case is also a domain. Let

$$\mathcal{P} := \{(A, X) \in \mathcal{D} : F(A, X) = 0\} \subset \mathcal{D} \tag{4.3}$$

be the set of all pairs (A, X) from $\mathcal{D}$ satisfying (4.1). Thus, $\mathcal{P}$ is a variety of generic dimension $\dim(\mathcal{P}) = \dim(\mathcal{V}) + mn - pq$ and we assume that $\mathcal{P} \neq \emptyset$. Let finally

$$\mathcal{A} := \{A : (A, X) \in \mathcal{P}\} \subset \mathcal{D}_{\mathcal{V}}$$

be the set of all A from $\mathcal{D}_{\mathcal{V}}$ for which equation (4.1) is solvable. In view of the assumption that $\mathcal{P}$ is nonempty, the set $\mathcal{A}$ is also nonempty. However, even if the set $\mathcal{D}_{\mathcal{V}}$ is easily definable, to determine the set $\mathcal{A}$ may be difficult.

The unknown matrix X is defined as an *implicit function* $A \mapsto X$ via equation (4.1), see [173] and Appendix A for the implicit function theorem.

To avoid trivial results we assume that for every $A \in \mathcal{A}$ the $\mathbb{F}^{p\times q}$-valued functions $X \mapsto F(A, X)$, defined on open subsets of $\mathcal{X}$, are not identically zero.

When real nonlinear algebraic matrix equations (polynomial or rational equations in particular) are considered, it is necessary to deal with the fact that the field $\mathbb{R}$ of real numbers is not algebraically closed, i.e., that polynomial equations with real coefficients such as $x^2 + 1 = 0$ may not have real solutions. In this case the matrix $F(A, X)$ is real, provided that X and A are also real, but we usually admit complex solutions X as well. This formally corresponds to the case when $\mathcal{V}$ is a linear space over $\mathbb{R}$ while $X \in \mathbb{C}^{n\times m}$ and $F(A, X) \in \mathbb{C}^{p\times q}$ are complex matrices. Thus, real problems are naturally imbedded in a complex environment which provides some nice algebraic and geometric properties. In particular an algebraic equation of n-th degree in such an environment has always n roots counted with multiplicity.

In what follows we denote by $(F, \mathcal{A})$ the family of problems $A \mapsto X$ defined via (4.1) when A varies over $\mathcal{A}$. For a fixed $A \in \mathcal{A}$ denote by

$$\Xi(A) := \{X \in \mathcal{X} : (A, X) \in \mathcal{D} \text{ and } F(A, X) = 0\}$$

the *solution set* of equation (4.1), which is the set of all X, satisfying (4.1). Thus, $\Xi(A) \subset \mathcal{X}$ is the $\mathcal{V}$-section of the set $\mathcal{P} \subset \mathcal{D} \subset \mathcal{V} \times \mathcal{X}$.

We can also define $\Xi(A)$ for all A from the set $\mathcal{D}_{\mathcal{V}}$ and set $\Xi(A) = \emptyset$ if $A \in \mathcal{D}_{\mathcal{V}} \backslash \mathcal{A}$. Thus, the corresponding problem may be defined via the multi-valued mapping

$$A \mapsto \Xi(A) \subset \mathcal{X}, \ \ A \in \mathcal{A}.$$

In many problems we are interested in solutions of (4.1) which are matrices of a special form (e.g. symmetric or Hermitian solutions when $m = n$). Suppose that $\mathcal{X}_0 \subset \mathcal{X}$ is a given set. Then we may define the subset $\mathcal{A}_0 \subset \mathcal{A}$ of data A which gives rise to solutions $X \in \mathcal{X}_0$,

$$\mathcal{A}_0 := \{A \in \mathcal{A} : \Xi(A) \cap \mathcal{X}_0 \neq \emptyset\}.$$

An interesting phenomenon may be observed for some equations of type (4.1), namely that the solution set does not effectively depend on A as in the next two examples.

Example 4.1 Let $F(A, X)$ be a matrix, which may be written as a product $F_1(A)F_2(X)F_3(A)$, where the matrices $F_1(A)$ and $F_3(A)$ are nonsingular for all $A \in \mathcal{D}_{\mathcal{V}}$. Then equation (4.1) is equivalent to the equation $F_2(X) = 0$ which does not depend on A. $\diamondsuit$

Example 4.2 Let $x_1, x_2 \in \mathbb{F}$ be two distinct numbers and $\varphi : \mathbb{F} \to \{x_1, x_2\}$ be a surjective (and hence discontinuous) function. Then the solutions of the equation

$$(x - x_1)(x - x_2)(x - \varphi(a)) = 0, \ a \in \mathbb{F}$$

are x_1, x_2 and therefore, the solution set $\Xi(a) = \{x_1, x_2\}$ does not change with a. $\diamondsuit$

In Example 4.2 the left-hand side of the equation is not continuous. We will not consider this case in detail. However, it was included to show that even in such cases the solution set may look nice.

To avoid trivial results we will make another assumption in order to exclude cases in which equation (4.1) does not depend effectively on the parameter A.

In what follows we assume that there exist at least two parameters $A, B \in \mathcal{A}$ such that $\Xi(A) \neq \Xi(B)$. This means that equation (4.1) depends on the parameter A effectively.

The above assumptions guarantee that, for some value of A, equation (4.1) has a solution set $\Xi(A)$ which is a nontrivial subset of $\mathcal{X}$, i.e., $\Xi(A) \neq \emptyset$ and $\Xi(A) \neq \mathcal{X}$. Moreover, the solution set depends nontrivially on A in the sense that $\Xi(A)$ changes along with A.

Instead of solving the global problem to determine $\Xi(A)$ when $\Xi(A)$ contains more than one element, often it is necessary to consider only a particular solution $X \in \Xi(A)$ of equation (4.1). We are especially interested in solutions which depend continuously on the data in a certain neighborhood of a given nominal data A from $\mathcal{A}$.

Definition 4.3 The equation (4.1) is said to be *well-posed* at A if there exist a neighborhood (not necessarily open) $\mathcal{N}_A \subset \mathcal{A}$ of A and a continuous function $\Phi : \mathcal{N}_A \to \mathcal{X}$ such that $F(B, \Phi(B)) = 0$ for all $B \in \mathcal{N}_A$.

Thus, a well-posed equation has at least one solution $X := \Phi(A)$ which depends continuously on the data in a neighborhood of A.

Even if the equation is well-posed at A it may have a solution $\widehat{X} \in \Xi(A)$ for which $(A, \widehat{X})$ is an isolated point of the set $\mathcal{P}$, defined in (4.3). In this case there is no continuous function $\widehat{\Phi}$, defined in a neighborhood of A such that $\widehat{X} = \widehat{\Phi}(A)$.

Thus, a well-posed equation may have solutions which do not depend continuously on the data on neighborhoods of a given nominal value of A. The last clarification is essential, since a function is continuous by definition at the isolated points (if any) of its domain. In fact, well-posedness just means that at least one solution (but not necessarily all solutions) depends continuously on the data. The dependence of the solutions on the data may be a subtle problem even for simple equations.

Example 4.4 Consider the scalar equation in $\mathbb{F}$

$$(x - a - 1)(|x| + |a|) = 0.$$

The solution set $\Xi(a)$ here is a single-element set $\{a+1\}$ if $a \neq 0$ and a two-element set $\{1, 0\}$ if $a = 0$. The set $\mathcal{P}$ in (4.3) consists of the straight line $\{(a, a+1) : a \in \mathbb{F}\} \subset \mathbb{F}^2$ and the isolated point $(0,0) \in \mathbb{F}^2$. Hence, the equation is well-posed at any a, since we have the solution $x = a + 1$, depending continuously on a. For $a = 0$ we also have the solution $0 \in \Xi(0)$, which does not depend continously on the data. At the same time for $a = -1$ the solution $0 \in \Xi(-1)$ depends continuously on a. $\diamondsuit$

Definition 4.5 For $A \in \mathcal{A}$ denote by $\Omega(A)$ the set of all $\mathcal{X}$-valued functions Φ, defined in a neighborhood $\mathcal{N}_A \subset \mathcal{A}$ of A and satisfying $F(B, \Phi(B)) = 0$ for all $B \in \mathcal{N}_A$. The set of continuous functions from $\Omega(A)$ is denoted by $\Omega_c(A)$. The elements of $\Omega(A)$ are called *solution functions* of equation (4.1).

We are interested mainly in continuous solution functions, although discontinuous ones may also be of theoretical and practical interest. It must be stressed that equations of type (4.1) with F continuous may have discontinuous solution functions, and an equation with F discontinuous may have continuous solution functions. In fact, if an equation has two or more solution functions (continuous or not) it has also infinitely many discontinuous solution functions, see Example 4.7. This is based on the following observation. Let the function $A \mapsto \Phi(A)$ be discontinuous and the function $(A, X) \mapsto F(A, X)$ be continuous. Then the function $A \mapsto F(A, \Phi(A))$ may be continuous (the constant zero function in particular). For example, a function $(A, X) \mapsto F_1(A, X)F_2(A, X)$ may be continuous even if the matrix valued functions $(A, X) \mapsto F_i(A, X)$ are not.

Example 4.6 If the function $\varphi : \mathbb{F} \times \mathbb{F} \to \{-1, 1\}$ is surjective then it is discontinuous at least at one point of its domain (and may as well be discontinuous everywhere). At the same time $a \to \varphi^2(a)$ is the constant function, equal to 1, which is of course continuous on $\mathbb{F} \times \mathbb{F}$. $\diamondsuit$

Example 4.7 Consider the scalar equation

$$f(a, x) := (x - \varphi_1(a))(x - \varphi_2(a)) = 0,$$

where $\varphi_1, \varphi_2 : \mathbb{F} \to \mathbb{F}$ are given distinct functions which in this case are also solution functions. Let $\mathbb{F}_1$ and $\mathbb{F}_2$ be any two nontrivial subsets of $\mathbb{F}$ such that $\mathbb{F} = \mathbb{F}_1 \cup \mathbb{F}_2$ and $\mathbb{F}_1 \cap \mathbb{F}_2 = \emptyset$. We may construct a new solution function $\varphi : \mathbb{F} \to \mathbb{F}$ by the rule $\varphi(a) = \varphi_i(a)$ if $a \in \mathbb{F}_i$. Then we have the following observations.

- The solution function φ is discontinuous. By a suitable choice of the sets $\mathbb{F}_i$ it is possible to make φ discontinous at any point $a \in \mathbb{F}$, where $\varphi_1(a) \neq \varphi_2(a)$. (Take for example $\mathbb{F}_1$ as the set of all $a \in \mathbb{F}$ with $|a| \in \mathbb{Q}$.)

- If φ_1 is continuous and φ_2 is not, then f is not continuous. At the same time we still have the continuous solution function φ_1.

$\diamondsuit$

We also need the notion of a path in the set $\mathcal{A} \times \mathcal{X}$. Let $\Phi : \mathcal{N}_A \to \mathcal{X}$ be any function (for a moment we do not suppose that the function Φ is continuous nor that it is a solution function of equation (4.1)), defined in a neighborhood $\mathcal{N}_A$ of $A \in \mathcal{A}$ and let $X := \Phi(A)$.

Definition 4.8 A *path* through the point $(A, X) \in \mathcal{A} \times \mathcal{X}$ is the *graph* of Φ, which is the set

$$\Gamma(\Phi) := \{(A, \Phi(A)) : A \in \mathcal{N}_A\} \subset \mathcal{A} \times \mathcal{X}.$$

The path $\Gamma(\Phi)$ is *continuous* at (A, X) if the function Φ is continuous at A. The path is *continuous* on $\mathcal{N}_A$ if Φ is continuous on $\mathcal{N}_A$, and *smooth* on $\mathcal{N}_A$ if Φ is Fréchet differentiable on $\mathcal{N}_A$, see Appendix A. If $\Phi \in \Omega(A)$ then the path $\Gamma(\Phi)$ is a *solution path* of equation (4.1).

Together with the concept of well-posedness of an equation for a given data, we will introduce the notion of well-posedness for a particular solution.

Definition 4.9 A solution $X \in \Xi(A)$ of $F(A, X) = 0$ is said to be *well-posed* if $X = \Phi(A)$ for some $\Phi \in \Omega_c(A)$. If the solution $X \in \Xi(A)$ is well-posed then every $\Phi \in \Omega_c(A)$ with $\Phi(A) = X$ is referred to as a *supporting function* of this solution.

In other words, the solution X is well-posed if (A, X) lies on a certain continuous solution path. According to the definitions above, this continuous path may not be locally unique.

Example 4.10 The scalar equation $x^2 - a^2 = 0$ is well-posed at every $a \in \mathbb{F}$, and even every solution is well-posed. The solution $x = 0$, corresponding to $a = 0$, lies on two continuous paths, namely $\{(a, a) : a \in \mathbb{F}\}$ and $\{(a, -a) : a \in \mathbb{F}\}$, i.e., there are two supporting functions $x = a$ and $x = -a$ of the zero solution. In the real case $\mathbb{F} = \mathbb{R}$, there are even four suporting functions $x = a$, $x = -a$, $x = |a|$ and $x = -|a|$ of the zero solution. $\diamondsuit$

A solution $X \in \Xi(A)$ (well-posed or not) may lie on a path in $\mathcal{A} \times \mathcal{X}$ which is discontinuous at (A, X) or even everywhere in a neighborhood of (A, X). Also, the solution X may lie on a path which is continuous at (A, X) but is discontinuous elsewhere, see the next example.

Example 4.11 Consider the real scalar equation

$$\left(x^2 + \varphi_0^2(a)\right)(x - \varphi_1(a))(x - \varphi_2(a)) = 0,$$

where $\varphi_1, \varphi_2 : \mathbb{R} \to \mathbb{R}$ are two distinct differentiable functions and the function $\varphi_0 : \mathbb{R} \to \mathbb{R}$ is continuous. For every $a \in \mathbb{R}$ we have the smooth solution functions φ_1 and φ_2 and hence, the equation is well-posed. Suppose now that $\varphi_0(a_0) = 0$ and that $\varphi_1(a_0)\varphi_2(a_0) \neq 0$ for some $a \in \mathbb{R}$. Then $x_0 = 0$ is an isolated solution, since the point $(a_0, 0)$ does not lie on any solution path.

We also have the solution function φ, defined by $\varphi(a) = \varphi_1(a)$ if $a \in \mathbb{Q}$ and $\varphi(a) = \varphi_2(a)$ if $a \in \mathbb{R} \backslash \mathbb{Q}$. For every $a \in \mathbb{R}$ such that $\varphi_1(a) \neq \varphi_2(a)$, there is an open interval $\mathcal{N}_a \ni a$ such that φ is discontinous at every point from $\mathcal{N}_a$. If $\varphi_1(a) = \varphi_2(a)$ for some $a \in \mathbb{R}$, then φ may be continuous or even differentiable at a, being discontinuous at every point of the pierced interval $\mathcal{N}_a \backslash \{a\}$. Indeed, we may choose $\varphi_2 = 0$ and $\varphi_1(a) = a$ or $\varphi_1(a) = a^2$. $\diamondsuit$

Example 4.12 The scalar complex equation

$$a_1 x_1 + a_2 x_2 - a_3 = 0,$$

where $a = [a_1, a_2, a_3]^\top \in \mathbb{C}^3$ and $x = [x_1, x_2]^\top \in \mathbb{C}^2$, is well-posed if and only if either $|a_1|^2 + |a_2|^2 > 0$ or $a = 0$. If $|a_1|^2 + |a_2|^2 > 0$, then there is a one-parametric family of solutions

$$x_1 = \frac{\overline{a}_1 a_3}{|a_1|^2 + |a_2|^2} + a_2 z, \ x_2 = \frac{\overline{a}_2 a_3}{|a_1|^2 + |a_2|^2} - a_1 z,$$

where $z \in \mathbb{C}$ is a parameter. More generally, the linear algebraic equation $Mx = b$ is well-posed if and only if $\text{rank}([M, b]) = \text{rank}(M)$, or equivalently, $b \in \text{Rg}(M)$. $\diamondsuit$

The next problem we study is the uniqueness of the solution.

Definition 4.13 solution $X \in \Xi(A)$ is said to be *locally unique* if there exists an open neighborhood $\mathcal{N}_X$ of X such that $\mathcal{N}_X \cap \Xi(A) = \{X\}$.

Other equivalent conditions for local uniqueness of the solution are given in the next proposition.

Proposition 4.14 *The solution $X \in \Xi(A)$ is locally unique in the sense of Definition 4.13 if and only if one of the following three equivalent conditions hold.*

- *There exists an open neighborhood* $\mathcal{N}_X$ *of* X *such that* $\mathcal{N}_X^o \cap \Xi(A) = \emptyset$, *where* $\mathcal{N}_X^o := \mathcal{N}_X \backslash \{X\}$ *is the pierced open neighborhood of* X.

- *There exists* $\varepsilon = \varepsilon(A, X) > 0$ *such that for every* $Y \in \Xi(A)$ *with* $Y \neq X$ *the inequality* $\|Y - X\| > \varepsilon$ *holds.*

- *The solution* X *is an isolated point of the set* $\Xi(A)$.

Note that for a locally unique solution $X \in \Xi(A)$ (being an isolated point of $\Xi(A)$ by necessity) the point (A, X) need not be (and generically is not) an isolated point of the set $\mathcal{P}$ in (4.3). But if (A, X) is an isolated point of $\mathcal{P}$ then the solution X is an isolated point of $\Xi(A)$, i.e., it is locally unique.

Solutions that are not locally unique may be characterised as follows.

Proposition 4.15 *A solution* $X \in \Xi(A)$ *is not locally unique if and only if it is an* accumulation point *of the set* $\Xi(A)$, *i.e., if and only if there is a sequence* $\{X_n\}_1^\infty \subset \Xi(A)$ *such that* $\lim_{n\to\infty} = X$.

Proof. Indeed, if there is such a sequence, then the solution X cannot be locally unique, since any open neighborhood of X contains some member of the sequence and hence, infinitely many such members. Suppose now that X is not locally unique. Then any open ball, centered at A and of radius $1/n$ must contain at least one solution $X_n \in \Xi(A)$ (otherwise X would be an isolated point of $\Xi(A)$). Thus, we have constructed the sequence $\{X_n\}_1^\infty \subset \Xi(A)$ which is convergent to X and hence, X is an accumulation point of the solution set $\Xi(A)$. □

In the next example we give an equation such that $\Xi(A)$ contains a solution X together with a sequence $\{X_n\}_1^\infty$ which is convergent to X.

Example 4.16 Consider the scalar equation $f(a, x) = 0$, where f is the differentiable function

$$f(a, x) = (x - a)^3 \sin\left(\frac{\pi}{x - a}\right), \quad x \neq a$$

and $f(a, a) = 0$. The solution set is

$$\Xi(a) = \{a, a \pm 1, a \pm 1/2, \ldots, a \pm 1/n, \ldots\}$$

and hence, the solution $a \in \Xi(a)$ is not locally unique. ♢

In Example 4.16 the solution set $\Xi(a)$ is countable (isomorphic to $\mathbb{N}$) with a single accumulation point a. All other points $a \pm 1/n$ of $\Xi(a)$ are isolated.

Another possibility for nonuniqueness of a solution $X \in \Xi(A)$ is when $X \in M$, where $M \subset \Xi(A)$ is a connected set. All solutions of the equation from Example 4.12 are of this type. Another example is given below.

Example 4.17 For $\alpha < \beta$ define the analytic function $h_{\alpha,\beta} : \mathbb{R} \to (-1,1)$ by

$$h_{\alpha,\beta}(x) := \begin{cases} -\exp(1/(x-\alpha)) & \text{if} \quad x < \alpha \\ 0 & \text{if} \quad \alpha \le x \le \beta \\ \exp(1/(\beta - x)) & \text{if} \quad x > \beta. \end{cases}$$

The inverse function $h_{\alpha,\beta}^{-1}$ exists and is analytic on $(-1,0) \cup (0,1)$. Consider the equation $h_{\alpha,\beta}(x) - a = 0$. It has a unique solution $x = h_{\alpha,\beta}^{-1}(a)$ for any $a \in (-1,0) \cup (0,1)$. For $a = 0$, however, any $x \in [\alpha, \beta]$ is a solution, i.e., $\Xi(0) = [\alpha, \beta]$. $\diamondsuit$

A solution which is locally unique may or may not be well-posed. Since we are mainly interested in solutions which are simultaneously well-posed and locally unique, we come to the following definition.

Definition 4.18 The solution $X \in \Xi(A)$ is said to be *proper* if it is well-posed and locally unique. The solution $X \in \Xi(A)$ is *improper* if it is not proper.

That the properties well-posedness and local uniqueness are independent is clear from the next example.

Example 4.19 The solution $a \in \Xi(a)$ from Example 4.16 is well posed but not locally unique. The solution $0 \in \Xi(0)$ from Example 4.4 is locally unique but not well-posed. $\diamondsuit$

We stress that an improper real solution of a real equation may become proper if we allow complex solutions, thus imbedding the equation in a complex environment.

Example 4.20 Consider the real scalar equation $(x-1)(x^2 + a^2) = 0$. Here the set $\mathcal{P} \subset \mathbb{R}^2$ in (4.3) consists of the straight line $\{(a,1) : a \in \mathbb{R}\} \subset \mathbb{R}^2$ and the isolated point $(0,0)$. The solution set is determined by $\Xi(a) = \{1\}$ if $a \neq 0$ and $\Xi(0) = \{1, 0\}$. Hence, the real solution $0 \in \Xi(0)$ is not proper. If we allow complex solutions, then we have $\Xi(a) = \{1, \pm \imath a\}$ and the solution 0 is proper, since it lies on the paths $\{(a, \pm \imath a) : a \in \mathbb{R}\} \subset \mathbb{C}^2$. $\diamondsuit$

Below we will also introduce the concept of properness for equations. There are two alternatives. An equation may be called proper if it has at least one proper solution, or alternatively, if all its solutions are proper. We prefer the first concept.

Definition 4.21 The equation (4.1) is said to be *proper* at $A \in \mathcal{D}_{\mathcal{V}}$ with $\mathcal{D}_{\mathcal{V}}$ as in (4.2) if there is at least one proper solution $X \in \Xi(A)$. The equation (4.1) is *improper* at $A \in \mathcal{A}$ if all solutions $X \in \Xi(A)$ are improper.

Example 4.22 The algebraic equation from Example 2.7 is proper when at least one of the coefficients $a_0, \ldots, a_{n-1}$ is nonzero.

The linear algebraic equation $Mx = b$, where $M \in \mathbb{F}^{n\times m}$ and $b \in \mathbb{F}^n$, is proper if and only if $m = n = \operatorname{rank}(M)$. $\diamondsuit$

Since in fact we may not know $\mathcal{A}$ and even $\mathcal{D}_\mathcal{V}$, we may as well say that the equation is improper at $A \in \mathcal{V}$ if either $F(A, X)$ is not defined ($A \notin \mathcal{D}_\mathcal{V}$), or $F(A, X)$ is defined but the equation has no solution (i.e., $A \in \mathcal{D}_\mathcal{V}\backslash\mathcal{A}$ and $\Xi(A) = \emptyset$), or the set $\Xi(A)$ is nonempty (i.e. $A \in \mathcal{A}$) but all its elements are improper.

Example 4.23 The real scalar equation $x^2 + a^2 = 0$ is improper at every $a \in \mathbb{R}$. Indeed, for $a \neq 0$ the equation has no solution. For $a = 0$ the solution is $x = 0$, but $(0, 0)$ is an isolated point of $\mathcal{P}$ and hence, the only solution $0 \in \Xi(0)$ is improper. $\diamondsuit$

The effects discussed in Example 4.23 are due to the fact that we deal with real solutions only. If we allow complex solutions then the equation becomes proper for all $a \in \mathbb{R}$. As a matter of fact, complex algebraic equations are proper at all $A \in \mathcal{A}$. But complex nonalgebraic equations may as well be improper for some (or all) values of the parameter A.

Example 4.24 The complex scalar (nonalgebraic) equation $|x| + |a| = 0$ is solvable only for $a = 0$ and in this case the only solution is $0 \in \Xi(0)$. In this case $\mathcal{P} \subset \mathbb{C}^2$ is reduced to the single point $(0, 0)$ and hence, the equation is improper. $\diamondsuit$

So far we have introduced a large number of concepts characterizing the properties of solution sets and of particular solutions. This variety of properties is possible, since we have considered equations of type (4.1) in which the function F is continuous (or differentiable, or analytic).

If we restrict ourselves to algebraic equations, in which the function F is algebraic in both its arguments, then the set $\mathcal{P}$ will be an algebraic manifold (or variety) and the number of (possible) properties of the solutions is substantially reduced. If in particular $\mathbb{F} = \mathbb{C}$ and the function F is algebraic (polynomial or fractional-rational in particular) then all solutions $X \in \Xi(A)$ will become well-posed.

We now come to the last and most important concept, which characterizes the dependence of solutions on parameters.

Definition 4.25 The solution $X \in \Xi(A)$ is said to be *regular* if it is proper and at least one of its supporting functions Φ is Lipschitz continuous. The solution $X \in \Xi(A)$ is *singular* if it is not regular.

Regularity means that there are constants $\alpha = \alpha(A) > 0$ and $L(A, \alpha) \geq 0$, such that

$$\|\Phi(A+E) - \Phi(A)\| \leq L(A,\alpha)\|E\|, \ \|E\| \leq \alpha.$$

Note that if X is a regular solution then some of its supporting continuous functions may not be Lipschitz continuous.

Example 4.26 The scalar equation $(x-a)(x^2-a) = 0$ is well-posed at every $a \in \mathbb{F}$. The solution $0 \in \Xi(0)$ is regular, since one of its supporting functions $a \mapsto a$ is Lipschitz continuous. At the same time this solution also has the continuous supporting function $a \mapsto \sqrt{a}$ which is not Lipschitz continuous but Hölder continuous with exponent 0.5 in a neighborhood of the origin. $\diamondsuit$

Thus, a solution may be singular for one of the following reasons:

- It is not well-posed.
- It is not locally unique.
- It is well-posed and locally unique but none of its supporting functions is Lipschitz continuous.

Example 4.27 Every root of the algebraic equation from Example 2.7 with $a_0 \neq 0$ is regular (respectively singular) if it is simple (respectively multiple).

The solution of the linear equation $Mx = b$ with $M \in \mathbb{F}^{n\times m}$ is regular if and only if $m = n = \text{rank}(M)$. $\diamondsuit$

In the following we discuss proper problems in which the spaces $\mathcal{X} = \mathbb{F}^{n\times m}$ and $\mathbb{F}^{p\times q}$ are isomorphic (i.e., $mn = pq$) and the solution $X \in \Xi(A)$ is proper and in particular locally unique. In matrix theory and applications, typical problems with unique solutions are various classes of matrix equations. In linear systems theory such problems are for example the reduction to canonical forms or the pole assignment problem for single-input systems.

The case when the solution of equation (4.1) is not locally unique is not considered in detail in the remainder of this monograph. Note that this may be the case when the number of scalar unknowns mn is larger than the number pq of scalar equations. In matrix algebra such problems include some least square problems and the computation of the eigensystem or Schur system of a matrix, see Appendix B (respectively the generalized eigensystem or the generalized Schur system of a pair of matrices), while in linear systems theory an important problem with nonunique solution is the pole assignment for linear multi-input systems. It must be stressed, however, that even a single scalar equation in many scalar unknowns may have only locally unique solutions.

Example 4.28 The scalar equation $F(A, X) := \|X - \Phi(A)\| = 0$, where $\Phi : \mathcal{A} \to \mathcal{X}$ is a continuous function, has a unique solution $X = \Phi(A)$. Observe that the function F is not differentiable at the solution. $\diamondsuit$

The considerations presented so far include many possible types of solutions of the equation (4.1) with a continuous left-hand side F. If, however, the partial Fréchet derivatives of F exist at the solution, then the continuous solution functions may have some nice properties such as uniqueness and differentiability.

Suppose that for some $A \in \mathcal{A}$ equation (4.1) has a solution $X \in \Xi(A)$ such that the partial Fréchet derivative $F_X := F_X(A, X)$ of the function F relative to X at the point (A, X) exists. We recall that F_X is a linear operator $\mathbb{F}^{n\times m} \to \mathbb{F}^{p\times q}$ such that

$$F(A, X + Y) = F(A, X) + F_X(Y) + o(\|Y\|),\ Y \to 0.$$

The partial Fréchet derivative $F_A := F_A(A, X)$ of F relative to $A = (A_1, \ldots, A_r) \in \mathcal{A}$ is the r-tuple $F_A = (F_{A_1}, \ldots, F_{A_r})$, where the partial Fréchet derivatives $F_{A_i} := F_{A_i}(A, X)$ of F relative to A_i at (A, X) are linear operators $\mathcal{V}_i \to \mathcal{X}$. If both Fréchet derivatives of F in X and A exist, then we have

$$\begin{aligned} F(A+E, X+Y) &= F(A, X) + F_X(Y) + \sum_{i=1}^{r} F_{A_i}(E_i) \\ &\quad + o(\|Y\| + \|E\|);\ E, Y \to 0. \end{aligned} \tag{4.4}$$

In the following we will use the induced norm $\|\mathcal{L}\|$ of a linear operator $\mathcal{L} : \mathcal{Y} \to \mathcal{Z}$, where $\mathcal{Y}$ and $\mathcal{Z}$ are linear spaces, defined via

$$\|\mathcal{L}\| := \max\{\|\mathcal{L}(Y)\| : Y \in \mathcal{Y},\ \|Y\| = 1\}.$$

We recall the following concepts of invertibility of linear operators.

Definition 4.29 A linear operator $\mathcal{L}$ is *right invertible* if there exists a linear operator $\mathcal{L}_r^{-1} : \mathcal{Z} \to \mathcal{Y}$ (called *right inverse* of $\mathcal{L}$), such that $\mathcal{L} \circ \mathcal{L}_r^{-1} = I_{\mathcal{Z}}$, where $I_{\mathcal{Z}}$ is the identity operator in $\mathcal{Z}$. Similarly, the operator $\mathcal{L}$ is said to be *left invertible*, if there exists a linear operator $\mathcal{L}_l^{-1} : \mathcal{Z} \to \mathcal{Y}$ (called *left inverse* of $\mathcal{L}$), such that $\mathcal{L}_l^{-1} \circ \mathcal{L} = I_{\mathcal{Y}}$. If an operator $\mathcal{L}$ is left and right invertible it is *invertible*.

For an invertible operator $\mathcal{L}$ the left inverse is equal to the right inverse and is denoted by $\mathcal{L}^{-1}$.

If the operator $\mathcal{L} : \mathcal{Y} \to \mathcal{Z}$ is invertible, then the norm of the inverse operator $\mathcal{L}^{-1} : \mathcal{Z} \to \mathcal{Y}$ is obtained from

$$\begin{aligned} \|\mathcal{L}^{-1}\| &= \max\{\|\mathcal{L}^{-1}(Z)\| : \|Z\| = 1\} = \max\left\{\frac{\|\mathcal{L}^{-1}(Z)\|}{\|Z\|} : Z \neq 0\right\} \\ &= \max\left\{\frac{\|Y\|}{\|\mathcal{L}(Y)\|} : Y \neq 0\right\} = \max\left\{\frac{1}{\|\mathcal{L}(Y)\|} : \|Y\| = 1\right\} \\ &= \frac{1}{\min\{\|\mathcal{L}(Y)\| : \|Y\| = 1\}}. \end{aligned}$$

Let the matrix spaces $\mathcal{V}_i$, $\mathbb{F}^{n\times m}$ and $\mathbb{F}^{p\times q}$ be endowed with the Frobenius norm $\|\cdot\|_F$. If we consider invertible operators $\mathcal{L}: \mathbb{F}^{n\times m} \to \mathbb{F}^{p\times q}$, then it is necessary to assume that $mn = pq =: l$. We denote by $\mathbf{Lin}(p,m,n,q,\mathbb{F})$ the space of linear matrix operators $\mathcal{M}: \mathbb{F}^{m\times n} \to \mathbb{F}^{p\times q}$ and we use the abbreviations $\mathbf{Lin}(m,n,\mathbb{F}) = \mathbf{Lin}(m,m,n,n,\mathbb{F})$ and $\mathbf{Lin}(n,\mathbb{F}) = \mathbf{Lin}(n,n,n,n,\mathbb{F})$.

Any linear operator $\mathcal{L} \in \mathbf{Lin}(p,m,n,q)$ has a *matrix representation* (or briefly a *matrix*)

$$M_{\mathcal{L}} := \mathrm{Mat}(\mathcal{L}) \in \mathbb{F}^{l\times l}$$

defined by the identity $\mathrm{vec}(\mathcal{L}(X)) = M_{\mathcal{L}}\mathrm{vec}(X)$ for all $X \in \mathbb{F}^{n\times m}$. In this case the *induced norm* of $\mathcal{L}$ is determined by

$$\begin{aligned}\|\mathcal{L}\| &= \max\{\|\mathcal{L}(X)\|_F : \|X\|_F = 1\} = \max\{\|M_{\mathcal{L}}x\|_2 : \|x\|_2 = 1\}\\ &= \|M_{\mathcal{L}}\|_2 = \sigma_{\max}(M_{\mathcal{L}}).\end{aligned}$$

Similarly, for the induced norm of an invertible operator $\mathcal{L}$ we have

$$\|\mathcal{L}^{-1}\| = \|M_{\mathcal{L}}^{-1}\|_2 = \frac{1}{\sigma_{\min}(L)}.$$

Here $\sigma_{\min}(L)$ and $\sigma_{\max}(L)$ denote the minimal and maximal singular value of the matrix L, respectively, see Appendix B.

4.3 Linear bounds

Let $X \in \Xi(A)$ be a fixed solution of equation (4.1). It follows from the implicit function theorem (Appendix A) that if the linear operator F_X is invertible at (A,X) then the solution X is proper. If in addition the partial Fréchet derivative F_A exists at the point (A,X), then the solution X is regular. Indeed, as we will show, here the solution $X \in \Xi(A)$ is well-posed and linear estimates hold.

Meanwhile it is worth mentioning that the invertibility (or even the existence) of F_X is by no means necessary for the regularity of the solution.

Example 4.30 The scalar equation $f(a,x) := (x-a)^k = 0$, where $k \geq 2$ is an integer, is proper and for any $a \in \mathbb{F}$ it has only the regular solution $x = a$. At the same time the partial derivative $f_x(a,x) = k(x-a)^{k-1}$ is zero at the solution. ♢

Thus, an algebraic equation may have regular solutions of arbitrary algebraic multiplicity. This is possible only for equations whose coefficients are not arbitrary but belong to certain close algebraic varieties. Indeed, choosing $k = 2$ in Example 4.30, we get $x^2 - 2ax + a^2 = 0$. This is an equation of type $x^2 + a_1x + a_2 = 0$, where the pair (a_1, a_2) lies on the parabola $a_1^2 - 4a_2 = 0$. Another viewpoint here is that the singular problem of finding multiple roots is regularized by a special choice of the data, see Chapter 6.

Let the parameter $A \in \mathcal{A}$ in equation (4.1) be perturbed to $A + \delta A \in \mathcal{A}$. This leads to the *perturbed equation*

$$F(A + \delta A, X + \delta X) = 0 \tag{4.5}$$

in the unknown matrix perturbation δX, which may also be written as an equation $F(A + \delta A, Z) = 0$ in the perturbed solution $Z = X + \delta X$. Suppose that (4.5) has a solution for every $\delta A \in \mathcal{N}_0$, where $\mathcal{N}_0$ is an open neighborhood of 0. Then it follows from (4.1), (4.4) and (4.5) that the solution δX of the perturbed equation (4.5) satisfies

$$\begin{aligned} \delta X &= -F_X^{-1}(F_A(\delta A)) + o(\rho) = -F_X^{-1} \circ F_A(\delta A) + o(\rho) \\ &= -F_X^{-1}\left(\sum_{i=1}^{r} F_{A_i}(\delta A_i)\right) + o(\rho) = \sum_{i=1}^{r} \mathcal{L}_i(\delta A_i) + o(\rho), \ \rho \to 0, \end{aligned} \tag{4.6}$$

where $\rho := \|\delta X\| + \|\delta A\|$ and $\mathcal{L}_i : \mathcal{V}_i \to \mathbb{F}^{n \times m}$ are linear operators, determined by

$$\mathcal{L}_i(E_i) := -F_X^{-1}(F_{A_i}(E_i)) = -F_X^{-1} \circ F_{A_i}(E_i).$$

Thus, the solution $X = \Phi(A)$ of the unperturbed equation (4.1) is regular according to Definition 4.25. Moreover, the function $\Phi : \mathcal{A} \to \mathcal{X}$ is differentiable in some open neighborhood of A, and the partial Fréchet derivatives of Φ in A_i are in fact the operators $\mathcal{L}_i$,

$$\Phi_{A_i} = \mathcal{L}_i = -F_X^{-1} \circ F_{A_i}.$$

To estimate the norm of the perturbation in the solution as a function of the norms of the perturbations in the data we may use the linear bound

$$\|\delta X\| \leq \sum_{i=1}^{r} K_{A_i} \|\delta A_i\| + o(\rho), \ \rho \to 0,$$

where

$$K_{A_i} = K_{A_i}(A, X) := \left\|F_X^{-1} \circ F_{A_i}\right\|, \ i = 1, \ldots, r$$

are the *absolute condition numbers* of equation (4.1) with respect to A_i, computed at the point (A, X). We also say that K_{A_i} is the absolute condition number of the solution X, corresponding to the data A_i.

The evaluation of K_{A_i} via the induced norms $\|\mathcal{L}_i\|$ of the linear operators $\mathcal{L}_i : \mathcal{V}_i \to \mathcal{X}$ may be a difficult task in general. Of course, we have the estimate $K_{A_i} \leq \|F_X^{-1}\| \, \|F_{A_i}\|$, which often may be quite pessimistic. However, if the Frobenius norm is used, the computation of the condition numbers (at least in theory) is straightforward. Indeed, taking the vec operator from both sides of (4.6) we have

$$\delta x = \sum_{i=1}^{r} M_{\mathcal{L}_i} \delta a_i + o(\rho), \ \rho \to 0,$$

where $x := \mathrm{vec}(X)$, $a_i := \mathrm{vec}(A_i)$ and

$$M_{\mathcal{L}_i} := -(\mathrm{Mat}(F_X))^{-1}\mathrm{Mat}\,(F_{A_i})$$

is the $mn \times m_i n_i$ matrix of the linear operator $\mathcal{L}_i$. Thus,

$$\|\delta X\|_{\mathrm{F}} = \|\delta x\|_2 \leq \sum_{i=1}^{r} \|M_{\mathcal{L}_i}\|_2\, \|\delta A_i\|_{\mathrm{F}} + o(\rho), \;\; \rho \to 0$$

and hence,

$$K_{A_i} = \|M_{\mathcal{L}_i}\|_2 \, .$$

A drawback of this approach for evaluating the conditioning of the problem may be the large size of the involved matrices. Indeed, for a moderately sized matrix equation with 100×100 matrices X and A_i, the size of the matrices of the linear matrix operators will be $10\,000 \times 10\,000$. Condition and error estimates for the solution of matrix equations without forming large matrices are proposed in [179].

The existence of the Fréchet derivatives F_X, F_A and the invertibility of F_X is sufficient but not necessary for the regularity of the solution. For special choices of the parameter A, this has been already discussed in Example 4.30.

4.4 Equivalent operator equation

Suppose that the spaces $\mathcal{X} = \mathbb{F}^{n\times m}$ and $\mathbb{F}^{p\times q}$ are isomorphic, i.e., $mn = pq = l$. Then nonlocal, nonlinear sensitivity estimates for the regular solutions of (4.1) may be obtained as follows. First we transform the perturbed equation into an equivalent operator equation for the perturbation δX. We then show that the corresponding operator has a fixed point in a set whose radius is a continuous function of $\|\delta A\|$, vanishing at $\delta A = 0$. This radius is then the desired rigorous nonlocal perturbation bound for $\|\delta X\|$. For this purpose the technique of Lyapunov majorants is used, see [85, 135]. How this approach works is shown in the remaining part of this chapter.

Consider equation (4.1) along with its perturbed version (4.5). We may apply transform the perturbed equation (4.5) into an *equivalent operator equation*

$$\delta X = \Pi(\delta A, \delta X) \tag{4.7}$$

for the perturbation δX, where $\Pi : \mathcal{N}_0^{\mathcal{A}} \times \mathcal{N}_0^{\mathcal{X}} \to \mathcal{X}$ is a continuous function and $\mathcal{N}_0^{\mathcal{A}}$, $\mathcal{N}_0^{\mathcal{X}}$ are open neighborhoods of the origins in $\mathcal{A}$ and $\mathcal{X}$, respectively. Here the operator Π satisfies $\Pi(0,0) = 0$. Using the vectorizations (2.8) we also have

$$\delta x = \pi(\delta a, \delta x), \;\; \pi := \mathrm{vec} \circ \Pi.$$

There are several ways to determine the operator Π so that equation (4.7) in δX is *locally equivalent* to the perturbed equation (4.5). The concept of local equivalence here needs some clarification. Local equivalence of two equations depending on a parameter means that they have the same solution for all parameters in a (small) neighborhood of a given nominal value of the parameter. However, we may construct formally equivalent equations for δX which cannot be used for deriving meaningful results. For example, if the solution of (4.1) is well-posed, then we have $X = \Phi(A)$, where Φ is a continuous function. So we may write formally $\delta X = \Phi(A + \delta A) - \Phi(A)$, which is an equation that is locally equivalent to the perturbed equation (4.5). But since the function Φ is usually unknown (otherwise we would have had a problem with explicit solution), this equivalent equation for δX does not lead to any concrete perturbation bounds.

A general approach to construct the operator Π is based on the representation

$$\Pi(E, Y) = Y + G \circ F(A + E, X + Y) + H \circ F(A, X), \tag{4.8}$$

where $G, H : \mathcal{X} \to \mathbb{F}^{p \times q}$ are invertible (usually linear) operators and the equality $F(A, X) = 0$ is taken into account. The idea is to make $\Pi(E, 0)$ small of order $O(\|E\|)$, and to make $\Pi(0, Y)$ small of order $o(\|Y\|)$ (or even $O(\|Y\|^2)$) for $E \to 0$ and $Y \to 0$. These requirements may be written as

$$\Pi(E, Y) = O(\|E\|) + o(\|E\| + \|Y\|);\ E, Y \to 0.$$

If the partial Fréchet derivative F_X of F in X is invertible at the solution, one may choose $G = -H := -F_X^{-1}$, which corresponds to Newton's method [173] for solving the equation. If in addition the partial Fréchet derivative F_A also exists, this scheme is applied as follows.

For every $(A, X) \in \mathcal{P}$ and $(E, Y) \in \mathcal{V} \times \mathcal{X}$, such that $(A + E, X + Y) \in \mathcal{P}$, we have the identity

$$\begin{aligned} F(A + E, X + Y) &= F(A, X) + F_A(A, X)(E) + F_X(A, X)(Y) \\ &\quad + R(A, X)(E, Y), \end{aligned} \tag{4.9}$$

where

$$\begin{aligned} R(A, X)(E, Y) &:= F(A + E, X + Y) - F(A, X) - F_A(A, X)(E) \\ &\quad - F_X(A, X)(Y). \end{aligned} \tag{4.10}$$

We stress that here $F_A(A, X)(\cdot) : \mathcal{V} \to \mathcal{X}$, $F_X(A, X)(\cdot) : \mathcal{X} \to \mathcal{X}$ are linear operators, and $R(A, X)(\cdot, \cdot)$ is a mapping from a subset of $\mathcal{V} \times \mathcal{X}$ to $\mathcal{X}$, all depending on the pair (A, X) as a parameter. To make the notation more readable we use the abbreviations

$$\begin{aligned} F_A(E) &:= F_A(A, X)(E),\ F_X(Y) := F_X(A, X)(Y), \\ R(E, Y) &:= R(A, X)(E, Y), \end{aligned}$$

omitting the dependence of the corresponding expressions on the pair (A, X). We finally note that in relations (4.9) and (4.10) it is not assumed that the equalities $F(A, X) = 0$ or $F(A + E, X + Y) = 0$ hold.

If X and δX satisfy equations (4.1) and (4.5), then it follows from (4.9) that

$$F_A(\delta A) + F_X(\delta X) + R(\delta A, \delta X) = 0$$

and

$$\delta X = -F_X^{-1} \circ F_A(\delta A) - F_X^{-1} \circ R(\delta A, \delta X).$$

Therefore,

$$\delta X = \Pi(\delta A, \delta X) := \Pi_1(\delta A) + \Pi_2(\delta A, \delta X), \tag{4.11}$$

where

$$\Pi_1(E) := -F_X^{-1} \circ F_A(E),\ \Pi_2(E, Y) := -F_X^{-1} \circ R(E, Y).$$

Thus, we have that

$$\Pi(E, Y) = Y - F_X^{-1}(F(A + E, X + Y) - F(A, X)),$$

which is the representation (4.8) with $G = -H := -F_X^{-1}$.

Consider finally the case when the partial Fréchet derivative of F in A does not exist. Then we have

$$F(A + E, X + Y) = F(A, X) + F_X(Y) + S(E, Y), \tag{4.12}$$

where

$$S(E, Y) := F(A + E, X + Y) - F(A, X) - F_X(A, X)(Y). \tag{4.13}$$

If X and δX solve (4.1) and (4.5), then it follows from (4.12) that

$$F_X(\delta X) + S(\delta A, \delta X) = 0$$

and

$$\delta X = -F_X^{-1} \circ S(\delta A, \delta X).$$

Therefore,

$$\delta X = \widehat{\Pi}(\delta A, \delta X),$$

where

$$\widehat{\Pi}(E, Y) := -F_X^{-1} \circ S(E, Y).$$

Note that again

$$\widehat{\Pi}(E, Y) = Y - F_X^{-1}(F(A + E, X + Y) - F(A, X)).$$

Hence, $\widehat{\Pi} = \Pi$ and the only difference is that the additive representation (4.11) is possible for the operator Π, while for the operator $\widehat{\Pi}$ such a representation may not be valid.

The above considerations are illustrated below for the case of a general quadratic matrix equation. Let

$$\begin{aligned} F(A,X) &:= A_1 + L(A,X) + Q(A,X) \\ &:= A_1 + \sum_{i=1}^{p} L_i(A,X) + \sum_{j=1}^{q} Q_j(A,X), \end{aligned}$$

where $L_i(A,\cdot)$ is a linear matrix operator, defined by $L_i(A,X) = B_i X C_i$, and $Q_i(A,\cdot)$ is a quadratic matrix operator, determined as $Q_i(A,X) = R_i X S_i X T_i$. Here we have $r = 1 + 2p + 3q$ and

$$A = (A_1, B_1, C_1, \ldots, B_p, C_p, R_1, S_1, T_1, \ldots, R_q, S_q, T_q) = (A_1, \ldots A_r).$$

Example 4.31 Consider the equation

$$F(A,X) := A_1 + A_2 X A_3 + X A_4 X = 0.$$

We have

$$F_X(Y) = A_2 Y A_3 + X A_4 Y + Y A_4 X$$

and, for $E = (E_1, E_2, E_3, E_4)$,

$$F_A(E) = E_1 + E_2 X A_3 + A_2 X E_3 + X E_4 X.$$

Furthermore,

$$\begin{aligned} R(E,Y) &= A_2 Y E_3 + E_2 Y A_3 + E_2 (X+Y) E_3 \\ &\quad + X E_4 Y + Y E_4 X + Y(A_4 + E_4) Y. \end{aligned}$$

Setting $\varepsilon := \|E_2\| + \|E_3\| + \|E_4\| + \|Y\|$ we obtain

$$\|R(E,Y)\| \le c\varepsilon^2 + \varepsilon^3/4, \tag{4.14}$$

where $c := \max\{\|X\|, \|A_4\|, \|A_2\|/2, \|A_3\|/2\}$. $\diamondsuit$

4.5 Linear equations

There are different techniques to deal with equation (4.1) depending on whether it is linear or nonlinear. The technique for nonlinear equations involves topological fixed point principles and, of course, it works for linear equations as well. But there is also a straightforward approach to get perturbation bounds in the case of linear equations.

Definition 4.32 *Equation (4.1) is said to be* linear *if the function $F(A,\cdot)$ is affine, or equivalently, if the function $F(A,\cdot) - F(A,0)$ is linear.*

In this section we write a linear equation as

$$A_1 + \mathcal{L}(D)(X) = 0,$$

where $\mathcal{L}(D)(\cdot) : \mathcal{X} \to \mathcal{X}$ is a linear matrix operator, depending on the parameter D, i.e.,

$$\mathcal{L}(D)(\lambda X + \mu Y) = \lambda \mathcal{L}(D)(X) + \mu \mathcal{L}(D)(Y)$$

for all $X, Y \in \mathcal{X}$ and $\lambda, \mu \in \mathbb{F}$. We note that every linear matrix operator may be represented in the form

$$\mathcal{L}(D)(X) = \sum_{i=1}^{p} B_i X C_i,$$

where

$$D := (B_1, C_1, \ldots, B_k, C_k) := (A_2, \ldots, A_r).$$

is a given matrix $2k$-tuple, see Appendix E.

Hence, we may set

$$A := (A_1, B_1, C_1, \ldots, B_p, C_p) = (A_1, A_2, \ldots, A_r)$$

with $r = 2p + 1$.

Assuming that the operator $\mathcal{L}(D)$ is invertible we see that the linear equation has a unique solution, which formally may be written as

$$X = \Phi(A) := -\mathcal{L}^{-1}(D)(A_1).$$

Let the parameters D and A be perturbed to $D + \delta D$ and $A + \delta A$. If the perturbation δD is small enough (in a sense to be discussed later), the perturbed operator $\mathcal{L}(D + \delta D)$ will be invertible and the perturbed equation

$$A_1 + \delta A_1 + \mathcal{L}(D + \delta D)(X + \delta X) = 0$$

will also have a unique solution

$$\delta X = \mathcal{L}^{-1}(D)(A_1) - \mathcal{L}^{-1}(D + \delta D)(A_1 + \delta A_1).$$

Below we present two approaches to estimate the perturbation in the solution as a function of the perturbation in the data. The first one is classical and is based on the following observation.

We note first that the set of linear operators $\mathcal{X} \to \mathcal{X}$, where $\mathcal{X}$ is a linear space over $\mathbb{F}$, is also a linear space with multiplication by scalars $\mathbb{F} \times \mathcal{X} \to \mathcal{X}$ and summation $\mathcal{X} \times \mathcal{X} \to \mathcal{X}$, defined by $(\lambda \mathcal{L})(X) = \lambda \mathcal{L}(X)$ and $(\mathcal{L} + \mathcal{M})(X) = \mathcal{L}(X) + \mathcal{M}(X)$, respectively.

Let the linear operator $\mathcal{L} : \mathcal{X} \to \mathcal{X}$ be invertible. Then, if $\mathcal{L}_1$ is a (small) perturbation to $\mathcal{L}$, the resulting operator $\mathcal{L} + \mathcal{L}_1$ will be invertible, provided that

$$\left\| \mathcal{L}^{-1} \circ \mathcal{L}_1 \right\| < 1. \tag{4.15}$$

This sufficient condition for invertibility of the perturbed operator $\mathcal{L} + \mathcal{L}_1$ is also "almost" necessary in the sense that if the inequality is replaced with equality, then the operator $\mathcal{L} + \mathcal{L}_1$ may not be invertible. Indeed, choosing $\mathcal{L}_1 = -\mathcal{L}$ we have $\|\mathcal{L}^{-1} \circ \mathcal{L}_1\| = 1$, and of course, the operator $\mathcal{L} + \mathcal{L}_1 = 0$ in this case is not invertible.

Under condition (4.15) we have

$$(\mathcal{L} + \mathcal{L}_1)^{-1} = (1_{\mathcal{X}} + \mathcal{M})^{-1} \circ \mathcal{L}^{-1}, \ \mathcal{M} := \mathcal{L}^{-1} \circ \mathcal{L}_1,$$

where $1_{\mathcal{X}}$ is the identity operator in $\mathcal{X}$. Furthermore, for every integer $\nu \geq 1$ we have

$$\begin{aligned}(\mathcal{L} + \mathcal{L}_1)^{-1} &= \left(\sum_{i=0}^{\nu}(-\mathcal{M})^i + (-\mathcal{M})^{\nu+1}(1_{\mathcal{X}} + \mathcal{M})^{-1}\right) \circ \mathcal{L}^{-1} \\ &= \left(\sum_{i=0}^{\infty}(-M)^i\right) \circ L^{-1}.\end{aligned}$$

Therefore,

$$\|(\mathcal{L} + \mathcal{L}_1)^{-1}\| \leq \frac{\|\mathcal{L}^{-1}\|}{1 - \|\mathcal{M}\|} \tag{4.16}$$

and

$$(\mathcal{L} + \mathcal{L}_1)^{-1} - \mathcal{L}^{-1} = -\mathcal{M} \circ (1_{\mathcal{X}} + \mathcal{M})^{-1} \circ \mathcal{L}^{-1}.$$

When information is available only for the norm of $\mathcal{L}_1$, one may use, instead of (4.15), the stronger condition

$$\|\mathcal{L}^{-1}\| \, \|\mathcal{L}_1\| < 1.$$

In this case the quantity $\|\mathcal{M}\|$ in (4.16) should be replaced by $\|\mathcal{L}^{-1}\| \, \|\mathcal{L}_1\|$.

The above relations may be used in the perturbation analysis of linear equations setting

$$\mathcal{L} = \mathcal{L}(D), \ \mathcal{L}_1 = \mathcal{L}_1(D, \delta D) := \mathcal{L}(D + \delta D) - \mathcal{L}(D)$$

and

$$\mathcal{M} = \mathcal{M}(D, \delta D) := \mathcal{L}^{-1}(D) \circ \mathcal{L}_1(D, \delta D)$$

as follows. We may write δX as

$$\begin{aligned}\delta X &= (\mathcal{L}^{-1}(D) - \mathcal{L}^{-1}(D + \delta D))(A_1) - \mathcal{L}^{-1}(D + \delta D)(\delta A_1) \\ &= \mathcal{M}(A, \delta A)(1_{\mathcal{X}} + \mathcal{M}(A, \delta A))^{-1}(X) - \mathcal{L}^{-1}(D + \delta D)(\delta A_1).\end{aligned}$$

Let $\alpha = [\delta_2, \ldots, \delta_r]^\top \in \mathbb{R}_+^{r-1}$ be a given vector and set

$$\beta(\alpha) := \max\{\|\mathcal{L}_1(D, G)\| : \|\!|G|\!\| \preceq \alpha\},$$

where $G := (E_2, \ldots, E_r)$ and $\|\!|G|\!\| = [\|E_2\|, \ldots, \|E_r\|]^\top$.

Since β is continuous and $\beta(0) = 0$, then α may be choosen so that

$$\beta(\alpha)\|\mathcal{L}^{-1}(D)\| < 1.$$

This yields

$$\|\mathcal{M}(D, \delta D)\| \leq \|\mathcal{L}^{-1}(D)\| \, \|\mathcal{L}_1(D, \delta D)\| \leq \beta(\alpha)\|\mathcal{L}^{-1}(D)\| < 1$$

and the operator $\mathcal{L}(D + \delta D)$ is invertible for all perturbations δD with $\|\delta D\| \preceq \alpha$. Moreover, we have

$$\|\delta X\| \leq \frac{\beta(\alpha)\|\mathcal{L}^{-1}(D)\|}{1 - \beta(\alpha)\|\mathcal{L}^{-1}(D)\|}(\|X\| + \|\mathcal{L}^{-1}(D)\| \, \|\delta A_1\|). \tag{4.17}$$

If $A_1 \neq 0$ then $X \neq 0$ and we have the following bound in relative perturbations

$$\frac{\|\delta X\|}{\|X\|} \leq \frac{\kappa(D)\varepsilon_L + \left(\|A_1\| \, \|L^{-1}(D)\|/\|X\|\right)\varepsilon_{A_1}}{1 - \kappa(D)\varepsilon_L}. \tag{4.18}$$

Here

$$\kappa(D) := \|\mathcal{L}(D)\| \, \|\mathcal{L}^{-1}(D)\|$$

is the relative condition number of the operator $\mathcal{L}(D)$ with respect to inversion, and

$$\varepsilon_L := \frac{\beta(\alpha)}{\|\mathcal{L}(D)\|}, \ \varepsilon_{A_1} := \frac{\|\delta A_1\|}{\|A_1\|}.$$

Having in mind that

$$1 \leq \frac{\|A_1\| \, \|\mathcal{L}^{-1}(D)\|}{\|X\|} \leq \kappa(D),$$

we also have the less accurate bound

$$\frac{\|\delta X\|}{\|X\|} \leq \frac{\kappa(D)}{1 - \kappa(D)\varepsilon_L}\left(\varepsilon_L + \varepsilon_{A_1}\right), \tag{4.19}$$

which is usually used in numerical linear algebra, [83].

Then using the Banach fixed point principle, see Appendix D it is possible to obtain the perturbation bounds (4.17)-(4.19) in an easier and more elegant way.

We may write the perturbed linear equation in the equivalent form

$$\delta X = \Pi(\delta A, \delta X) := -\mathcal{L}^{-1}(D)(\delta A_1 + \mathcal{L}_1(D, \delta D)(X + \delta X)). \tag{4.20}$$

Assuming that $\|\delta D\| \preceq \alpha$, we have

$$\|\Pi(\delta A, \delta X)\| \leq \|\mathcal{L}^{-1}(D)\| \, \|\delta A_1\| + \beta(\alpha)\|\mathcal{L}^{-1}(D)\|(\|X\| + \|\delta X\|).$$

If $\beta(\alpha)\|L^{-1}(D)\| < 1$ and $\|\delta X\| \leq \rho$, where

$$\rho := \frac{\beta(\alpha)\|\mathcal{L}^{-1}(D)\| \left(\|X\| + \|\mathcal{L}^{-1}(D)\|\|\delta A_1\|\right)}{1 - \beta(\alpha)\|\mathcal{L}^{-1}(D)\|},$$

then

$$\|\Pi(\delta A, \delta X)\| \leq \rho.$$

Therefore, the operator $\Pi(\delta A, \cdot)$ transforms the closed ball $\mathcal{B}_\rho$, centered at the origin and of radius ρ, into itself. Since for all $Y, Z \in \mathcal{X}$

$$\|\Pi(\delta A, Y) - \Pi(\delta A, Z)\| \leq \beta(\alpha)\|\mathcal{L}^{-1}(D)\| \, \|Y - Z\|$$

and $\beta(\alpha)\|\mathcal{L}^{-1}(D)\| < 1$, the operator $\Pi(\delta A, \cdot)$ is a contraction on $\mathcal{B}_\rho$. According to the Banach fixed point principle, there exists a unique solution of the operator equation (4.20) for δX, satisfying $\|\delta X\| \leq \rho$ which is exactly the bound (4.17).

Linear equations are considered in more detail in Chapters 8, 9, 10 and 11.

4.6 Case study

In this section we shall illustrate the concepts introduced so far for the case of a real scalar quadratic equation

$$x^2 + a_1 x + a_2 = 0, \;\; a := [a_1, a_2]^\top \in \mathbb{R}^2. \tag{4.21}$$

The computational problem defined via equation (4.21) is regular if the discriminant $d(a) := a_1^2 - 4a_2$ is nonzero, and singular at the parabola

$$\Gamma := \left\{[a_1, a_2]^\top : a_2 = a_1^2/4, \;\; a_1 \in \mathbb{R}\right\} \subset \mathbb{R}^2.$$

For $a \notin \Gamma$ the condition numbers for the root x are defined taking the partial derivatives in a_i of both sides of equation (4.21)

$$(2x + a_1)\frac{\partial x}{\partial a_1} + x = 0,$$
$$(2x + a_1)\frac{\partial x}{\partial a_2} + 1 = 0.$$

Thus, for $a_1 \neq 0$, $a_2 \neq 0$, the relative condition numbers κ_i of x relative to a_i are

$$\begin{aligned}
\kappa_1 &= \left|\frac{\partial x}{\partial a_1}\right| \frac{|a_1|}{|x|} = \left|\frac{a_1}{\sqrt{d(a)}}\right|, \\
\kappa_2 &= \left|\frac{\partial x}{\partial a_2}\right| \frac{|a_2|}{|x|} = \left|\frac{a_2}{x\sqrt{d(a)}}\right|
\end{aligned}$$

(note that here $x \neq 0$ by necessity). In Figure 4.1 we show the conditioning $\kappa_1 + \kappa_2$ of the problem as a function of a for $0 \leq a_1, \; a_2 \leq 5$.

In Figure 4.2 we show the relative changes $|\delta x / x|$ in the solution of the perturbed equation

$$(x + \delta x)^2 + (a_1 + \delta a_1)(x + \delta x) + a_2 + \delta a_2 = 0$$

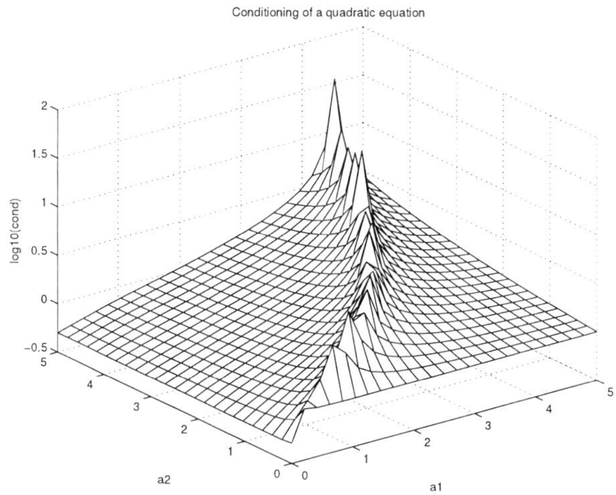

Figure 4.1: Conditioning of a quadratic equation

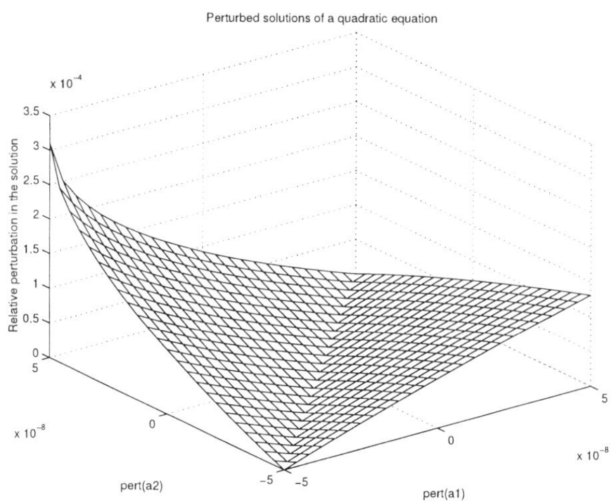

Figure 4.2: Perturbed solutions of a quadratic equation

for $a_1 = 2.0000001$, $a_2 = 1$, due to perturbations in a which satisfy $-4.998 \times 10^{-8} \leq \delta a_1, \delta a_2 \leq 4.998 \times 10^{-8}$.

In Figures 4.3 and 4.4 we show the local linear bound (based on condition numbers) and the nonlocal perturbation bounds for equation (4.21) with $a_1 = 2.0000001$, $a_2 = 1$.

In Figure 4.5 we give the distance to singularity of the quadratic equation as a function of a.

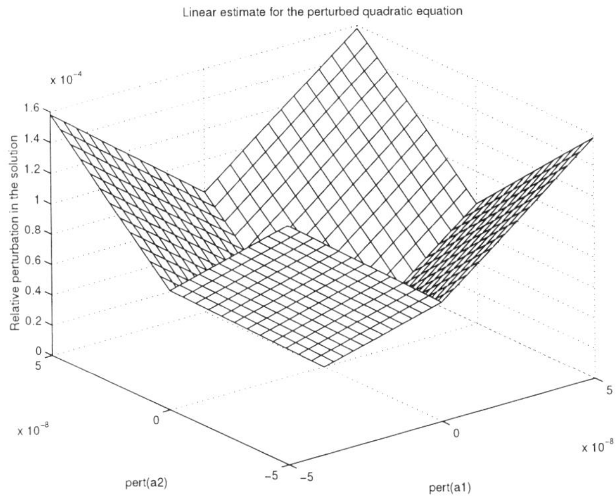

Figure 4.3: Linear estimates for a perturbed quadratic equation

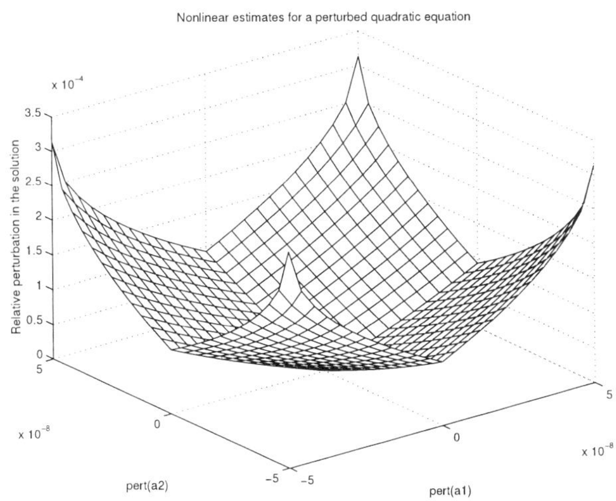

Figure 4.4: Nonlinear estimates for a perturbed quadratic equation

Finally, in Figure 4.6 we compare the magnitude of the exact perturbation $|\delta x|$ (denoted by $pert(x)$) with the linear and nonlinear perturbation bounds for $a_1 = 3$, $a_2 = 2$.

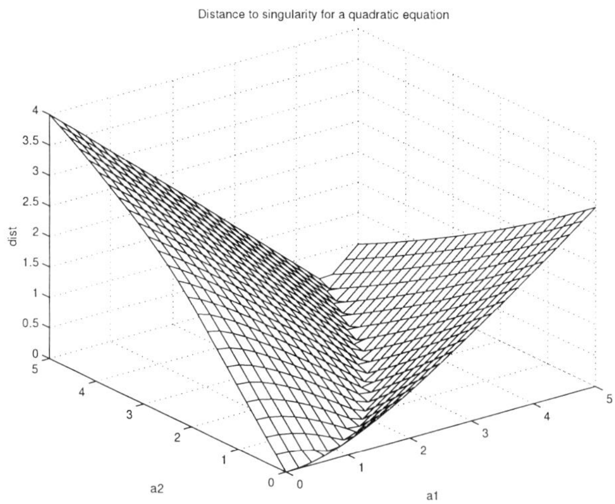

Figure 4.5: Distance to singularity for a quadratic equation

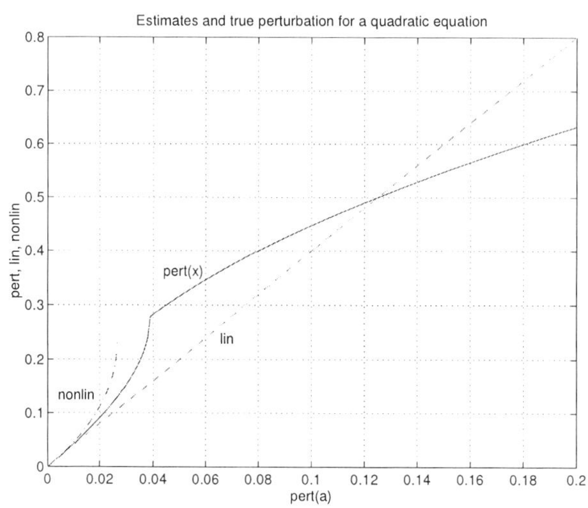

Figure 4.6: Comparison of perturbation bounds for a quadratic equation

4.7 Notes and references

General properties of problems with implicit solution are considered in [134, 135]. Perturbation bounds for various classes of equations are derived by many authors, see [74, 75, 187, 39, 149, 95, 191, 150, 66] as well as in [177, 178, 136, 140, 211, 212, 213].

A general scheme for perturbation analysis of nonlinear algebraic problems is proposed in [195].

Perturbation theory of matrix decompositions, which are special problems with implicit solution, is considered in [147], see also [28].

Chapter 5

Lyapunov majorants

5.1 Introductory remarks

In this section we describe the technique of *Lyapunov majorant functions* (or briefly, *Lyapunov majorants*) which is an important tool in the perturbation analysis of various problems in linear algebra and control theory [85, 135, 147].

A Lyapunov majorant is a nonnegative function which bounds from above the size of the equivalent operator for the perturbation in the solution. It gives rise to the so called *majorant equation* whose solution is the desired perturbation bound for the norm or generalized norm of the perturbation in the solution of the problem.

5.2 General theory

Consider the case of a nonlinear equation (4.1) together with its perturbed version (4.5) and the equivalent operator equation (4.7). Suppose that it is possible to find estimates for the size and the rate of change of the operator of Π in the form

$$\|\Pi(E, Y)\| \leq h(\eta, \rho), \tag{5.1}$$

where $\eta := \|E\|_g = [\|E_1\|, \ldots, \|E_r\|]^\top$ is the generalized norm of E, and

$$\|\Pi(E, Y) - \Pi(E, Z)\| \leq h'_\rho(\eta, \rho)\|Y - Z\| \tag{5.2}$$

for all $Y, Z \in \mathcal{X}$ with $\|Y\|, \|Z\| \leq \rho$. Here $h : G \to \mathbb{R}_+$ is a continuous function, defined in a domain $G \subset \mathbb{R}_+^r \times \mathbb{R}_+$, differentiable in its second (scalar) argument, and satisfying $h(0, 0) = 0$.

If we set

$$g(\eta, \rho) := \max\{\|\Pi(E, Y)\| : \|E\|_g \preceq \eta,\ \|Y\| \leq \rho\},$$

then the ideal case would be to choose $h(\eta, \rho) = g(\eta, \rho)$. However, the determination of $g(\eta, \rho)$ is possible only for simple linear matrix equations. In general it is only possible to find an expression $h(\eta, \rho)$ which is an upper bound for $g(\eta, \rho)$.

Note that h is a function of $r + 1$ scalar arguments. It may be more convenient to work with a function of only two arguments, setting

$$h_0(\varepsilon, \rho) := h(\varepsilon\delta^0, \rho),$$

where $\delta^0 \in \mathbb{R}^r_+$ is a given vector with positive elements.

When applying the technique of Lyapunov majorants it is convenient to introduce the concept of backward invariance of sets of nonnegative vectors as follows.

Definition 5.1 A set $\Omega \subset \mathbb{R}^r_+$ is called *backwardly invariant* if it is closed, contains a positive vector and for every $\delta \in \Omega$ the inequalities $0 \preceq \eta \preceq \delta$ yield $\eta \in \Omega$.

Setting

$$\mathcal{P}_\delta := \left\{\eta \in \mathbb{R}^r_+ : \eta \preceq \delta\right\}$$

for $\delta \in \mathbb{R}^r_+$, we see that the closed set Ω, containing a positive vector, is backwardly invariant if

$$\delta \in \Omega \Longleftrightarrow \mathcal{P}_\delta \subset \Omega.$$

For a set $M \subset \mathbb{R}^r$ denote by $\sup(M) \in \mathbb{R}^r$ and $\inf(M) \in \mathbb{R}^r$ the *supremum* and *infimum* of M, defined as $[\overline{m}_1, \ldots, \overline{m}_r]^\top$ and $[\underline{m}_1, \ldots, \underline{m}_r]^\top$, respectively, where

$$\overline{m}_i := \sup\{m_i : m \in M\}, \ \underline{m}_i := \inf\{m_i : m \in M\}$$

and $m = [m_1, \ldots, m_r]^\top$.

It is easy to verify that the following Theorem holds

Theorem 5.2 *A backwardly invariant set $\Omega \subset \mathbb{R}^r_+$ is connected, of positive measure and*

$$0 = \inf(\Omega) \in \Omega, \ \sup(\Omega) \in \Omega.$$

Note that a backwardly invariant set does not have to be convex. Actually, the structure of a backwardly invariant set may be quite complicated.

Example 5.3 The backwardly invariant subsets of $\mathbb{R}_+$ are the closed intervals $[0, a]$ with $a > 0$. In $\mathbb{R}^2_+$ a backwardly invariant set may be described as follows. Let $f : [0, a] \to \mathbb{R}_+$ be a continuous nonincreasing function with $f(0) > 0$. Then the set

$$\left\{[\delta_1, \delta_2]^\top : 0 \leq \delta_1 \leq a, \ 0 \leq \delta_2 \leq f(\delta_1)\right\}$$

is backwardly invariant. $\diamondsuit$

Next we define one of the main tools in nonlocal perturbation analysis of operator equations (matrix equations in particular) – the Lyapunov majorant functions.

Definition 5.4 A function h as in (5.1), (5.2) is called a *Lyapunov majorant function* (or briefly *Lyapunov majorant*) for the operator equation (4.1) if it satisfies the following conditions

1. The domain G admits the structure of a convex cylinder, i.e., there is a convex set $\Omega \subset \mathbb{R}_+^r$ and a continuous function $\tau : \Omega \to \mathbb{R}_+$ such that either

$$G = \{(\delta, \rho) : \rho < \tau(\delta), \delta \in \Omega\} \tag{5.3}$$

   or $G = \Omega \times \mathbb{R}_+$. In the latter case we may formally set $\tau = \infty$, reducing it to (5.3).

2. The function h is nondecreasing and strictly convex in every of its $r+1$ arguments and

$$\lim_{\rho \to \tau(\delta)} \frac{\rho}{h(\delta, \rho)} < 1 \tag{5.4}$$

   for each $\delta \in \Omega$.

3. The relations

$$h(0,0) = 0, \ h'_\rho(0,0) < 1 \tag{5.5}$$

   hold.

The importance of Lyapunov majorants may be explained as follows. If inequalities (5.1) and (5.2) hold, and for some $\rho > 0$ the relations

$$h(\delta, \rho) = \rho, \ h'_\rho(\delta, \rho) < 1 \tag{5.6}$$

are fulfilled, then the operator Π is a contraction in the ball

$$B_\rho := \{X \in \mathcal{X} : \|X\| \leq \rho\}.$$

Hence, by the Banach fixed point principle, there exists a unique solution $\delta X \in B_\rho$ to (4.1). At the same time the quantity ρ, satisfying (5.6), depends on δ, namely $\rho = f(\delta)$. Hence

$$\|\delta X\| \leq f(\|\delta A\|_g), \ \|\delta A\|_g \preceq \delta \tag{5.7}$$

is the desired nonlocal nonlinear norm-wise perturbation bound.

If only the equation $h(\delta, \rho) = \rho$ holds, then the bound (5.7) is still valid although the perturbation δX may not be unique. This will be the case e.g. in problems with nonunique solution, see [147].

Definition 5.5 The equation

$$\rho = h(\delta, \rho) \tag{5.8}$$

is referred to as a *majorant equation* for the operator equation (4.7).

One of the main problems in this approach is to determine a set $\Omega_0 \subset \mathbb{R}_+^r$ such that for $\delta \in \Omega_0$ the relations (5.6) are fulfilled for some $\rho = \rho(\delta)$. If we apply the Schauder rather than the Banach principle, then only the majorant equation $h(\delta, \rho) = \rho$ must be satisfied instead of (5.6). But if equation (4.1) admits a Lyapunov majorant, then one may *always* select a closed bounded set $\Omega_0 \subset \mathbb{R}_+^r$ with the properties listed above.

For linear matrix equations the Lyapunov majorant is an affine function in ρ,

$$h(\delta, \rho) = h_0(\delta) + h_1(\delta)\rho,$$

where the functions h_0, h_1 are nondecreasing in each argument, i.e., $h_i(\alpha) \leq h_i(\beta)$ if $\alpha \preceq \beta$, and $h_i(0) = 0$, $i = 0, 1$. In this case there exists a domain $\Omega_0 \subset \mathbb{R}_+^r$ such that $h_1(\delta) = 1$ for $\delta \in \partial\Omega_0$ and $h_1(\delta) < 1$ for $\delta \in \Omega_0^o$ (we recall that $\partial\Omega_0$ and Ω_0^o are the boundary and the interior of the set Ω_0, respectively, see Appendix A). Hence the perturbation bound here is

$$f(\delta) = \frac{h_0(\delta)}{1 - h_1(\delta)}, \quad \delta \in \Omega_0^o.$$

In the remainder of this section we consider only the case of nonlinear equations, when the Lyapunov majorant is also nonlinear.

For several important problems that we will discuss below, the Lyapunov majorant $h(\delta, \rho)$ is a polynomial in $r + 1$ variables $\delta_1, \ldots, \delta_r, \rho$ which can be written in the form

$$h(\delta, \rho) = h_0(\delta) + h_1(\delta)\rho + \cdots + h_N(\delta)\rho^N, \ N \geq 2,$$

where $h_i(\delta)$ are polynomials in δ with nonnegative coefficients, the polynomial h_N is nonzero, and

$$h_0(0) = h_1(0) = 0. \tag{5.9}$$

Hence, we have $\Omega = \mathbb{R}_+^r$, $G = \Omega \times \mathbb{R}_+$ and $\tau = \infty$. This is, for example, the case for algebraic matrix Riccati equations.

Another type of Lyapunov majorants is of the form

$$h(\delta, \rho) = \sum_{i=1}^{N} \frac{p_i(\delta, \rho)}{q_i(\delta) - r_i(\delta, \rho)},$$

where p_i, q_i, r_i are polynomials with nonnegative coefficients, such that $q_i(0) > 0$ and $r_i(0, 0) = 0$. Here $\tau(\delta)$ is the smallest among the roots of the N equations

$$q_i(\delta) - r_i(\delta, \rho) = 0, \ i = 1, \ldots, N.$$

The case when q_i are constants (and hence may be chosen as equal to 1 after an obvious scaling) is typical. Lyapunov majorants of this type arise in the non-local spectral perturbation analysis of matrices and matrix pencils with distinct eigenvalues [141].

In general, the Lyapunov majorant has a power series expansion

$$h(\delta, \rho) = \sum_{i=0}^{\infty} h_i(\delta)\rho^i, \tag{5.10}$$

where $h_i(\delta)$ are power series in δ with nonnegative coefficients, and h_0, h_1 satisfy (5.9). Here $\tau(\delta)$ is the radius of convergence of the series in the right-hand side of (5.10).

Example 5.6 Consider the scalar equation

$$\delta x = \pi(\delta a, \delta x) := \frac{\delta a(1 + \delta x)}{2 + \delta a \delta x}.$$

Setting $\delta := |\delta a|$, for $|\delta x| \leq \rho$, then we get the estimate

$$|\pi(\delta a, \delta x)| \leq h(\delta, \rho) := \frac{(1 + \rho)\delta}{2 - \rho\delta}.$$

The domain Ω coincides with $\mathbb{R}_+$. Furthermore,

$$\tau(\delta) = \begin{cases} 2/\delta, & \delta > 0, \\ \infty, & \delta = 0. \end{cases}$$

Here the coefficients of the power series (5.10) for the Lyapunov majorant are given by

$$h_0(\delta) = \frac{\delta}{2};\ h_i(\delta) = \frac{(2 + \delta)\delta^i}{2^{i+1}},\ i \geq 1.$$

$\diamondsuit$

In the technique of Lyapunov majorants a crucial role is played by the majorant equation (5.8) in the unknown quantity ρ, where $\delta \in \Omega$ is considered as a vector-parameter.

The solvability theory for equation (5.8) may be quite complex. It may have two real solutions, a double solution or no solutions at all depending on whether the parameter δ belongs to a certain bounded set $\Omega_0 \subset \Omega$, to (a part of) its boundary $\partial\Omega_0$, or is outside of Ω_0, respectively.

Although Ω is convex, the set Ω_0 may not in general be convex. The set Ω_0 has the following important property. For each $\delta_0 \in \Omega_0$ all nonnegative vectors δ with $\delta \preceq \delta_0$ also belong to Ω_0, i.e.,

$$\delta_0 \in \Omega_0 \Longleftrightarrow \{\delta : \delta \preceq \delta_0\} \subset \Omega_0.$$

Definition 5.7 *Let $S \subset \mathbb{R}_+^r$ be a bounded set. Then S is said to be* quasi-convex *if the following conditions hold:*

1. *For all $\omega \in S$ the set $[0, \omega] := \{t\omega : t \in [0, 1]\}$ is contained in S.*
2. *For all $\omega \in \partial S$ the sets $\{t\omega : t > 1\}$ and S are disjoint.*

Thus, a quasi-convex set S satisfies the convexity condition only for pairs $0, \omega \in S$. It is obvious that a convex set is also quasi-convex, but the opposite is not true in general. Intuitively, our notion of convex and quasi-convex subsets of $\mathbb{R}_+^r$ is covered by the sets S_1, $S_2 \subset \mathbb{R}_+^2$ from the example below.

Example 5.8 The set

$$S_1 := \left\{(\omega_1, \omega_2) : 0 \leq \omega_2 \leq 1 - \omega_1^2, 0 \leq \omega_1 \leq 1\right\} \subset \mathbb{R}_+^2$$

is convex, but

$$S_2 := \left\{(\omega_1, \omega_2) : 0 \leq \omega_2 \leq (1 - \omega_1)^2, 0 \leq \omega_1 \leq 1\right\} \subset \mathbb{R}_+^2$$

is only quasi-convex, see Figure 5.1. $\diamondsuit$

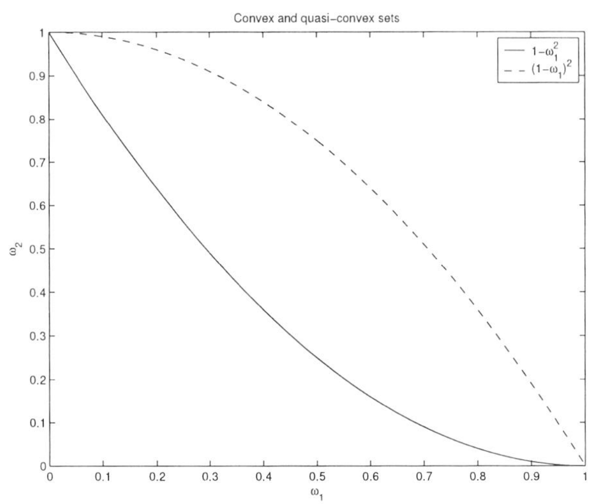

Figure 5.1: Convex and quasi-convex sets

While Condition 1. in the definition of a quasi-convex set is clear, Condition 2. needs some explanation. It is introduced in order to exclude some exotic sets, satisfying only Condition 1. Indeed, consider the cactus, obtained as the union of the set S_2 from Example 5.8 and a number of needles

$$[\omega, \mu\omega] := \{(1 + t)\omega : 0 \leq t \leq \mu - 1\},$$

where $\omega = \left(\omega_1, (1 - \omega_1)^2\right)$ is a point on the boundary ∂S_2, and $\mu > 1$. This set satisfies only Condition 1. and is, therefore, not quasi-convex.

The following characterization of quasi-convex sets is easily verified.

Theorem 5.9 *A bounded set $S \subset \mathbb{R}_+^r$ is quasi-convex if and only if the following two conditions are fulfilled:*

1. For all $\omega \in \mathbb{R}_+^r$ the set $\{t\omega : t \geq 0\}$ has exactly one common point with the boundary ∂S of S.

2. For all $\omega \in S$ and $\varepsilon > 0$ the set $B_\varepsilon(\omega) \cap S$ is of positive measure, where $B_\varepsilon(\omega) \subset \mathbb{R}^r$ is the ball centered at ω and of radius ε.

Using the concept of quasi-convexity we may formulate and prove the following theorem, which justifies the use of Lyapunov majorants in the perturbation analysis of nonlinear equations.

Theorem 5.10 *There exists a quasi-convex set $\Omega_0 \subset \Omega$, such that one and only one of the following three assertions for equation (5.8) holds:*

(i) If $\delta \in \Omega_0$, then equation (5.8) has two roots

$$\rho_1(\delta) < \rho_2(\delta)$$

in the interval $[0, \psi(\delta))$ (Figure 5.2), and we may choose

$$f(\delta) = \rho_1(\delta), \ \delta \in \Omega_0 \tag{5.11}$$

in the perturbation bound (5.7).

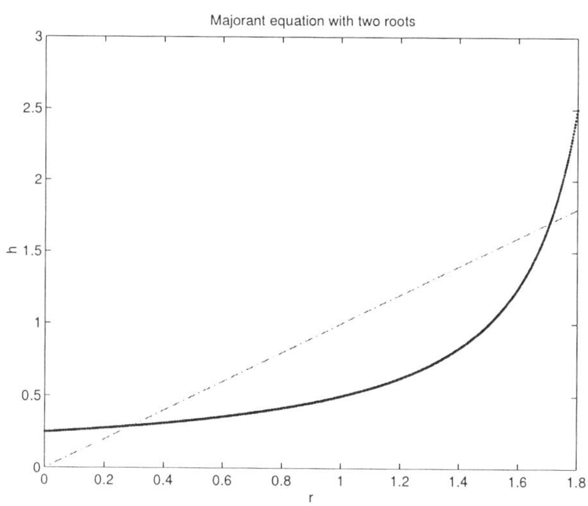

Figure 5.2: Majorant equation with two roots

(ii) For some $\delta_0 \in \partial\Omega_0$ equation (5.8) has a double root

$$\rho_1(\delta_0) = \rho_2(\delta_0)$$

in the interval $[0, \psi(\delta))$ *(Figure 5.3) and we may again choose*

$$f(\delta_0) = \rho_1(\delta_0), \ \ \delta_0 \in \partial\Omega_0 \tag{5.12}$$

in (5.7).

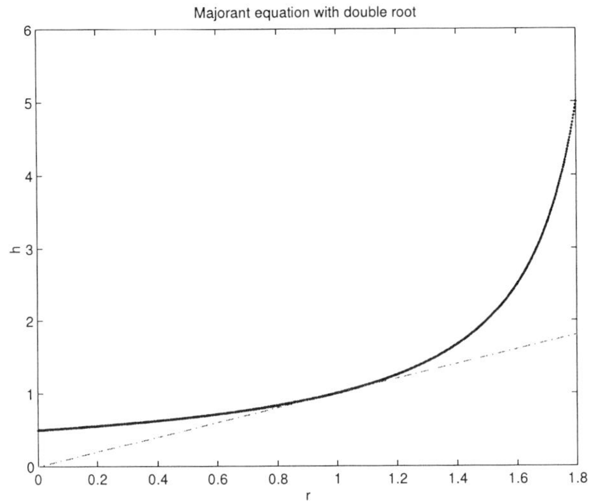

Figure 5.3: Majorant equation with double root

(iii) If $\delta \notin \Omega_0$*, then equation (5.8) has no real roots in the interval* $[0, \tau(\delta))$ *(Figure 5.4).*

Proof. Let the Lyapunov majorant have the form (5.10). We first note that since the function $h(\delta, \cdot) : [0, \tau(\delta)) \to \mathbb{R}_+$ is convex, equation (5.8) may have two different roots, one (double) root, or no roots.

For some $\alpha > 0$ and for all $\delta \in \Omega$ with $\|\delta\| < \alpha$, there exists a unique quantity $\rho_0 = \rho_0(\delta) < \tau(\delta)$ such that

$$h'_\rho(\delta, \rho_0) = 1. \tag{5.13}$$

Indeed, relation (5.5) implies that $h'_\rho(\delta, 0) = h_1(\delta) < 1$ for $\|\delta\|$ sufficiently small. At the same time, according to (5.4), we have $h'_\rho(\delta, \rho) > 1$ for ρ less than but sufficiently close to $\psi(\delta)$. Hence, (5.13) holds for some $\rho_0 \in (0, \psi(\delta))$. That this ρ_0 is unique, follows from the convexity of the function $h(\delta, \cdot) : [0, \psi(\delta)) \to \mathbb{R}_+$.

We now show that there exist two positive quantities $d_0 < d_1$ such that

$$h(\delta, \rho_0) < \rho_0, \ \|\delta\| < d_0 \tag{5.14}$$

and

$$h(\delta, \rho_0) > \rho_0, \ \|\delta\| \geq d_1. \tag{5.15}$$

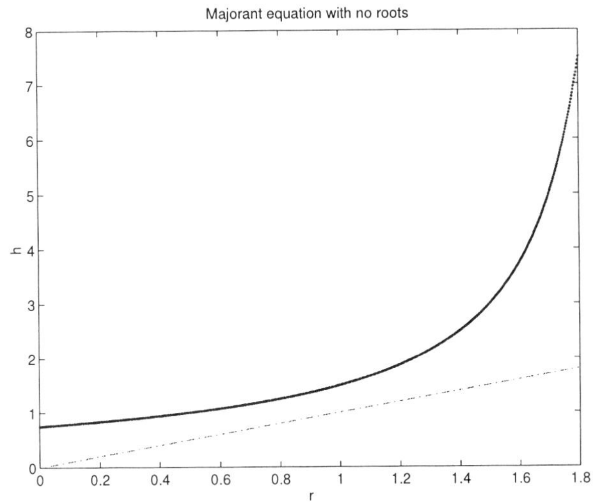

Figure 5.4: Majorant equation with no roots

Consider first the case when the sum in (5.10) is finite, i.e., $h_i(\delta) = 0$, $i > N$, for some integer $N \geq 2$. Suppose that

$$h(\delta, \rho_0) = \sum_{i=0}^{N} h_i(\delta)\rho_0^i \geq \rho_0. \tag{5.16}$$

Since

$$h'_\rho(\delta, \rho_0) = \sum_{i=1}^{N} i h_i(\delta)\rho_0^{i-1} = 1, \tag{5.17}$$

we get

$$\rho_0 = \sum_{i=1}^{N} i h_i(\delta)\rho_0^i. \tag{5.18}$$

Relations (5.16) and (5.18) yield

$$\sum_{i=0}^{N} h_i(\delta)\rho_0^i \geq \sum_{i=1}^{N} i h_i(\delta)\rho_0^i$$

and

$$h_0(\delta) \geq \sum_{i=2}^{N} (i-1) h_i(\delta)\rho_0^i.$$

Hence, for $i \geq 2$,

$$\rho_0 \leq \left(\frac{h_0(\delta)}{(i-1)h_i(\delta)} \right)^{1/i}$$

and, using (5.17), we get

$$
\begin{aligned}
1 &\leq h_1(\delta) + \sum_{i=2}^{N} i h_i(\delta) \left(\frac{h_0(\delta)}{(i-1)h_i(\delta)} \right)^{1-1/i} \\
&= h_1(\delta) + \sum_{i=2}^{N} i(i-1)^{1/i-1} (h_0(\delta))^{1-1/i} (h_i(\delta))^{1/i} \qquad (5.19) \\
&< h_1(\delta) + 2\sqrt{h_0(\delta)} \sum_{i=2}^{N} (h_i(\delta))^{1/i}.
\end{aligned}
$$

Since $h_0(0) = h_1(0) = 0$, inequality (5.19), and hence (5.16), is not valid for $\|\delta\|$ sufficiently small. This proves (5.14).

Relation (5.15) follows from the second inequality in (5.5).

Consider the implications of relations (5.14) and (5.15). We have actually divided the convex domain Ω into the disjoint union of three parts:

$$
\Omega = \Omega_1 \cup \Omega_2 \cup \Omega_3,
$$

where

$$
\begin{aligned}
\Omega_1 &:= \{\delta \in \Omega : \|\delta\| < d_0\}, \\
\Omega_2 &:= \{\delta \in \Omega : d_0 \leq \|\delta\| < d_1\}, \\
\Omega_3 &:= \{\delta \in \Omega : \|\delta\| \geq d_1\}.
\end{aligned}
$$

For $\delta \in \Omega_1$ equation (5.8) has two roots. Indeed, consider the function $\chi : G \to \mathbb{R}$, given by

$$
\chi(\delta, \rho) := h(\delta, \rho) - \rho.
$$

The function

$$
\chi(\delta, \cdot) : [0, \psi(\delta)) \to \mathbb{R}
$$

is convex and has an unique minimum at the point ρ_0. Since

$$
\begin{aligned}
\chi(\delta, 0) &= h_0(\delta) \geq 0, \\
\chi(\delta, \rho_0) &= h(\delta, \rho_0) - \rho_0 < 0, \\
\lim_{\rho \to \psi(\delta)} \chi(\delta, \rho) &> 0,
\end{aligned}
$$

it follows that $\chi(\delta, \cdot)$ has at least two roots $\rho_1 < \rho_0$ and $\rho_2 > \rho_0$ in the interval $[0, \psi(\delta))$. But the function $\chi(\delta, \cdot)$ is convex and has at most two roots, which are then exactly ρ_1 and ρ_2.

Let now $\delta \in \Omega_3$. Then $\chi(\delta, \rho_0) = h(\delta, \rho_0) - \rho_0 > 0$, i.e., the minimum of $\chi(\delta, \cdot)$ is positive and hence $\chi(\delta, \cdot)$ has no roots.

Finally consider the case $\delta \in \Omega_2$. Select two points

$$
\delta_0 := \frac{d_0}{2\|\delta\|} \delta \in \Omega_1
$$

and

$$\delta_1 := \frac{2d_1}{d_0}\delta_0 \in \Omega_3$$

such that $\|\delta_0\| = d_0/2$ and $\|\delta_1\| = d_1$, and consider the homotopy $\omega : [0,1] \times [0, \psi(\delta))$, defined by

$$\omega(t, \rho) = h(\delta_0 + t(\delta_1 - \delta_0), \rho) - \rho,$$

which connects the points $h(\delta_0, \rho) - \rho = \omega(0, \rho)$ and $h(\delta_1, \rho) - \rho = \omega(1, \rho)$. For t fixed, the function $\omega(t, \cdot) : [0, \psi(\delta)) \to \mathbb{R}$ has two, one or no roots. Let $t_0 = t_0(\delta)$ be the supremum of all $t \in [0,1]$ such that $\omega(t, \cdot)$ has two roots, and $t_1 = t_1(\delta)$ be the infimum of all $t \in [0,1]$ for which $\omega(t, \cdot)$ has no roots. Obviously $0 < t_0 \leq t_1 < 1$.

We now show that $t_0 = t_1$. Indeed, suppose that $t_0 < t_1$. Then for all $t \in (t_0, t_1)$ the function $\omega(t, \cdot)$ does not have two roots but does have at least one root, i.e., it has exactly one root and is, in particular, nonnegative. But, since $\delta_0 \neq 0$, we see that $\omega(t, \rho)$ is strictly increasing in t. Hence, for $t < \tau$ with $t, \tau \in (t_0, t_1)$ we get $\omega(\tau, \rho) > \omega(t, \rho) \geq 0$. This is a contradiction to the fact that $\omega(\tau, \rho)$ has one root. Hence the assumption $t_0 < t_1$ is false and we have $t_0 = t_1$.

Setting

$$t^* := -\frac{d_0}{2d_1 - d_0},$$

we obtain

$$\delta_0 + t(\delta_1 - \delta_0) = \left(d_1 - \frac{d_0}{2}\right) \frac{\delta}{\|\delta\|}(t - t^*).$$

Hence the set

$$\Omega_0 := \{\delta_0 + t(\delta_1 - \delta_0) : t \in [t^*, t_0),\ \delta \in \Omega\}$$

with boundary

$$\partial\Omega_0 = \{\delta_0 + t_0(\delta_1 - \delta_0) : \delta \in \Omega\}$$

has the desired properties. □

As a result of (5.11) and (5.12) we find the estimate

$$\|\delta X\| \leq f(\|\delta A\|_g),\ \|\delta A\|_g \in \mathrm{cl}(\Omega_0) := \Omega_0 \cup \partial\Omega_0. \tag{5.20}$$

We will now analyze the three cases in Theorem 5.9 in detail.

Theorem 5.11 *The function $f : \mathrm{cl}(\Omega_0) \to \mathbb{R}_+$ in (5.11) is nondecreasing in each of its r arguments and satisfies $f(0) = 0$. Moreover, in the domain Ω_0 the function f is real analytic, i.e. it may be represented by its Taylor series*

$$f(\delta) = \sum_{i=0}^{\infty} f_i(\delta),$$

where

$$f_i(\delta) = O(\|\delta\|^i),\ \delta \to 0$$

are homogeneous polynomials in δ of degree i.

Note that the boundary $\partial\Omega_0$ of the domain Ω_0 may be obtained by eliminating $\rho > 0$ from the system of equations

$$h(\delta, \rho) = \rho, \ h'_\rho(\delta, \rho) = 1.$$

Having a Lyapunov majorant, the next step is to solve the majorant equation and in particular to find its small solution ρ_0 (whenever it exists). It is highly desirable to do that finding the dependence of ρ_0 in δ in an explicit form. Of course, given a fixed δ the majorant equation can always be solved numerically and if it has two roots $0 \leq \widehat{\rho}_0 \leq \widehat{\rho}_1$ one can choose $\widehat{\rho}_0$ as a candidate for the small solution vanishing together with δ. This 'numerical' approach may or may not work. The problem is that it is not clear whether for the computed solution $\widehat{\rho}_0$ there is indeed a continuous function f with $\widehat{\rho}_0 \simeq f(\delta)$ and $f(0) = 0$ (i.e., whether a small solution exists). The next example shows that the numerical approach may be misleading.

Example 5.12 Let

$$h(\delta, \rho) = 6\delta + \frac{\delta\rho^2}{1-\rho}, \ \delta \in \mathbb{R}_+. \tag{5.21}$$

Then the majorant equation is formally equivalent to the quadratic equation

$$(1+\delta)\rho^2 - (1+6\delta)\rho + 6\delta = 0, \ \rho \neq 1.$$

For $\delta \leq (3-\sqrt{6})/6 \simeq 0.09175$ the small solution

$$\rho_0 = f(\delta) = \frac{12\delta}{1 + 6\delta + \sqrt{12\delta^2 - 12\delta + 1}}$$

is of order 6δ and indeed tends to zero with $\delta \to 0$. However, for $\delta \geq (3+\sqrt{6})/6 \simeq 0.90825$ the quadratic equation has roots which are not small because δ cannot be small. For example, if $\delta = 1$ then the roots are 1.5 and 1. Of course, the latter case should in fact be excluded from consideration since $h(\delta, \rho)$ in (5.21) is defined only for $\rho < 1$. But in practice, cases like this may cause problems. $\diamondsuit$

In many applications the expression $h(\delta, \rho)$ has the form

$$h(\delta, \rho) = g_1(\delta, \rho) + \frac{g_2(\delta, \rho)}{g(\delta) - g_3(\delta, \rho)},$$

where $g_i(\delta, \rho)$ are polynomials in ρ, $g_i(\delta, \rho) = \sum_{j=0}^{r_i} a_{i,j}(\delta)\rho^j$, $i = 1, 2, 3$. Here the coefficients $a_{i,j}(\delta)$ and $g(\delta)$ are polynomials in $\delta \in \mathbb{R}_+^k$ with nonnegative coefficients, and $g(0) > a_{3,0}(0)$. Also, we must have $a_{1,0}(0) = a_{2,0}(0) = 0$ and $a_{1,1}(0) + a_{2,1}(0) < 1$. In this case $h(\delta, \rho)$ is well defined if $\delta \in \mathbb{R}_+^k$ and $\rho < \psi(\delta)$, where $\psi(\delta)$ is the smallest positive root of the algebraic equation $g(\delta) = g_3(\delta, \rho)$.

Furthermore, the majorant equation can be reduced to an algebraic equation of degree $r := \max\{r_2, r_1 + r_3, r_3 + 1\}$ in ρ, namely

$$d(\delta, \rho) := \sum_{j=0}^{r} d_j(\delta)\rho^j = 0, \;\; \rho < \psi(\delta). \tag{5.22}$$

Note that the coefficients $d_j(\delta)$ may not be nonnegative and/or nondecreasing in δ.

Here the surface $\mathcal{S} \subset \mathbb{R}_+^k$ is defined by the equation $\Delta(\delta) = 0$, where $\Delta(\delta)$ is the discriminant of d, see [9] for the corresponding definition. In this case equation (5.22) (and hence the majorant equation) has a double nonnegative root. The discriminant of d may be constructed by different schemes (whenever appropriate we omit the dependence of d and d_j on their arguments). Let $r \geq 2$ and consider the derivative $d_\rho(\delta, \rho) = \sum_{j=0}^{r-1}(j+1)d_{j+1}\rho^j$ of d in ρ which must be zero at the double root. Multiplying d by $\rho, \ldots, \rho^{r-2}$ and d_ρ by $\rho, \ldots, \rho^{r-1}$ in view of $d = 0$ and $d_\rho = 0$, we obtain $2r - 1$ homogeneous linear equations in the quantities $1, \rho, \ldots, \rho^{2r-1}$, which can be written as a vector equation

$$T^{(r)}b^{(r)} = 0, \; T^{(r)} = \left[t_{ij}^{(r)}\right] \in \mathbb{R}_+^{(2r-1)\times(2r-1)}, \; b^{(r)} := [1, \rho, \ldots, \rho^{2r-1}]^\top \in \mathbb{R}_+^{2r-1}.$$

Here the elements $t_{ij}^{(r)}$ of T are given by

$$t_{ij}^{(r)} := \begin{cases} 0 & \text{if } 1 \leq i \leq r-1 & \text{and } j < i, \\ d_{j-i} & \text{if } 1 \leq i \leq r-1 & \text{and } i \leq j \leq r+i, \\ 0 & \text{if } 1 \leq i \leq r-1 & \text{and } j > r+i, \\ 0 & \text{if } r \leq i \leq 2r-1 & \text{and } j < i-r+1, \\ (j-i+r)d_{j-i+r} & \text{if } r \leq i \leq 2r-1 & \text{and } i-r+1 \leq j \leq i, \\ 0 & \text{if } r \leq i \leq 2r-1 & \text{and } j > i. \end{cases}$$

Example 5.13 For $r = 2$ and $r = 3$ the equations for $b^{(2)}$ and $b^{(3)}$ are

$$T^{(2)}b^{(2)} = \begin{bmatrix} d_0 & d_1 & d_2 \\ d_1 & 2d_2 & 0 \\ 0 & d_1 & 2d_2 \end{bmatrix} \begin{bmatrix} 1 \\ \rho \\ \rho^2 \end{bmatrix} = 0, \tag{5.23}$$

$$T^{(3)}b^{(3)} = \begin{bmatrix} d_0 & d_1 & d_2 & d_3 & 0 \\ 0 & d_0 & d_1 & d_2 & d_3 \\ d_1 & 2d_2 & 3d_3 & 0 & 0 \\ 0 & d_1 & 2d_2 & 3d_3 & 0 \\ 0 & 0 & d_1 & 2d_2 & 3d_3 \end{bmatrix} \begin{bmatrix} 1 \\ \rho \\ \rho^2 \\ \rho^3 \\ \rho^4 \end{bmatrix} = 0.$$

◇

The discriminant of d is $\Delta = \det(T^{(r)})$. Since $b^{(r)} \neq 0$ and having in mind that $T^{(r)}$ depends on δ, it follows that $\mathcal{S} = \left\{\delta \in \mathbb{R}_+^k : \det(T^{(r)}(\delta)) = 0\right\}$.

5.2.1 Polynomial Lyapunov majorants

In the important particular case of polynomial or pseudo polynomial [140] matrix equations $F(P, X) = 0$ the Lyapunov majorant is a polynomial in ρ,

$$h(\delta, \rho) = \sum_{j=0}^{r} a_j(\delta)\rho^j, \tag{5.24}$$

where a_i are continuous, nonnegative and nondecreasing functions of $\delta \in \mathbb{R}_+^k$ and $a_r(\delta) > 0$ for some $\delta \in \mathbb{R}_+^k$. In fact, $a_i(\delta)$ are often polynomials in δ with nonnegative coefficients. In this case the conditions $a_0(0) = 0$ and $a_1(0) < 1$ are fulfilled (in most applications we even have $a_1(0) = 0$).

Consider the majorant equation

$$\rho = \sum_{j=0}^{r} a_j(\delta)\rho^j. \tag{5.25}$$

We can always solve this equation numerically for a given $\delta \in \mathbb{R}_+^k$. Let the computed solutions be $\widehat{\rho}_0 \le \widehat{\rho}_1$. Then we can take $\widehat{\rho}_0$ as the small solution lying on a continuous path to zero. Despite of that it is still convenient to have (approximate) closed form solutions. Next we consider techniques to construct such solutions.

We denote by $\Omega_r \subset \mathbb{R}_+^k$ the set of all δ such that equation (5.25) has a small solution ρ_0, denoted as $f_r(\delta)$, where the function f_r is continuous and $f_r(0) = 0$. Upper bounds for f_r, defined for $\delta \in \widehat{\Omega}_r$, are denoted as $\widehat{f}_r$. As we shall see, Ω_1 is bounded but not closed, while for $r > 1$ the set Ω_r is compact. Obviously we have $f_{j+1}(\delta) \le f_j(\delta)$ and $\Omega_{j+1} \subset \Omega_j$, $j = 1, 2, \ldots$.

The case $r = 1$. Here the function $h(\delta, \cdot)$ is not strictly convex. Equation (5.25) has a unique solution

$$f_1(\delta) := \frac{a_0(\delta)}{1 - a_1(\delta)}, \quad \delta \in \Omega_1 \backslash \mathcal{S}_1,$$

where $\Omega_1 := \left\{\delta \in \mathbb{R}_+^k : a_1(\delta) \le 1\right\}$ and $S_1 := \left\{\delta \in \mathbb{R}_+^k : a_1(\delta) = 1\right\}$. This case arises in studying linear algebraic equations.

The case $r = 2$. Here the function $h(\delta, \cdot)$ is strictly convex for some δ. The domain for δ is

$$\Omega_2 = \left\{\delta \in \mathbb{R}_+^k : a_1(\delta) + 2\sqrt{a_0(\delta)a_2(\delta)} \le 1\right\}, \tag{5.26}$$

and the surface $\mathcal{S}_2 \subset \Omega_2$ is obtained by replacing the inequality in (5.26) by equality. For $\delta \in \Omega_2 \backslash \mathcal{S}_2$ the majorant equation has two roots, the smaller one being

$$f_2(\delta) := \frac{2a_0(\delta)}{1 - a_1(\delta) + \sqrt{(1 - a_1(\delta))^2 - 4a_0(\delta)a_2(\delta)}}.$$

For $\delta \in \mathcal{S}_2$ the majorant equation has a double root $f_2(\delta) = 2a_0(\delta)/(1 - a_1(\delta))$, $\delta \in \mathcal{S}_2$.

Similar results hold for the case when

$$h(\delta, \rho) = a_{10}(\delta) + a_{11}(\delta)\rho + \frac{a_{20} + a_{21}(\delta)\rho + a_{22}\rho^2}{g(\rho) - a_{31}(\delta)\rho}.$$

The case $r = 3$. Here the majorant equation is cubic. The surface $\mathcal{S}_3$ is obtained by $\det(T^{(3)}(\delta)) = 0$, where the matrix $T^{(3)}$ is defined by (5.23). For this case there are closed form solutions, given by the Cardano formula. But we are interested in the case when the equation has two nonnegative solutions (and hence one negative solution as well). This is the so called irreducible case when the explicit form solution is not very practical. So we shall find an approximate closed form solution.

Suppose that for a given δ such that $a_1(\delta) < 1$, equation (5.25) has two nonnegative solutions. Suppose also that $a_3(\delta) > 0$, since otherwise the majorant equation is of order less than 3.

For the small solution $\rho_0 = f_3(\delta)$ it holds that $\rho_0 \le \tau_3$, where τ_3 is the unique solution of the equation $1 = h_\rho(\delta, \rho)$, i.e., $1 = a_1 + 2a_2\tau_3 + 3a_3\tau_3^2$. Hence

$$\tau_3 = \tau_3(\delta) = \frac{1 - a_1(\delta)}{a_2(\delta) + \sqrt{a_2^2(\delta) + 3a_3(\delta)(1 - a_1(\delta))}}.$$

Furthermore for $\rho \le \tau_3$ we have

$$\rho \le a_0 + a_1\rho + (a_2 + a_3\tau_3)\rho^2. \tag{5.27}$$

The right hand side of (5.27) is again a Lyapunov majorant in the form of a second degree polynomial in ρ. So we can apply the estimates already obtained for $r = 2$ above. As a result we get the estimate

$$f_3(\delta) \le \widehat{f}_3(\delta) := \frac{2a_0(\delta)}{1 - a_1(\delta) + \sqrt{(1 - a_1(\delta))^2 - 4a_0(\delta)\widehat{a}_2(\delta)}}, \quad \delta \in \widehat{\Omega}_3, \tag{5.28}$$

where

$$\widehat{\Omega}_3 = \left\{\delta \in \mathbb{R}_+^k : a_1(\delta) + 2\sqrt{a_0(\delta)\widehat{a}_2(\delta)} \le 1\right\} \tag{5.29}$$

and $\widehat{a}_2(\delta) := a_2(\delta) + a_3(\delta)\tau_3(\delta)$.

We recall that the estimate (5.28), (5.29) is valid under the assumption that the majorant equation $\rho = a_0 + a_1\rho + a_2\rho^2 + a_3\rho^3$ *has* nonnegative solutions. And, of course, the inclusion $\delta \in \widehat{\Omega}_3$ by no means guarantees that such solutions exist (in general Ω_3 may be a proper subset of $\widehat{\Omega}_3$). Fortunately, here the existence of nonnegative solutions is easily checked by the inequality

$$\widehat{f}_3(\delta) \le \tau_3(\delta), \tag{5.30}$$

involving already computed quantities. In particular the equality in (5.30) is equivalent to $\det(T^{(3)}(\delta)) = 0$ or $\delta \in \mathcal{S}_3$. More precisely, the following result holds.

Theorem 5.14 *The following assertions are valid in case of a cubic majorant equation.*

1. *If (5.30) is fulfilled then the majorant equation has a small solution* $f_3(\delta) \le \widehat{f}_3(\delta)$. *If (5.30) is violated, then the majorant equation has no nonnegative solutions.*

2. *The equality in (5.30) describes the surface* $\mathcal{S}_3 \subset \mathbb{R}^k_+$ *on which the discriminant of the majorant equation vanishes and this equation has a double nonnegative root.*

Proof. The equality in (5.30) means that the quantity $\widehat{f}_3(\delta)$ satisfies both the majorant equation $\rho = h(\delta, \rho)$ and the equation $1 = h_\rho(\delta, \rho)$. Hence $\widehat{f}_3(\delta)$ is a double root. □

Note that inequality (5.30) is equivalent to $h(\delta, \tau_3(\delta)) \le \tau_3(\delta)$ as well as to $h_\rho(\delta, \tau_3(\delta)) \le 1$.

If $h(\delta, \widehat{f}_3(\delta)) < \widehat{f}_3(\delta)$ then we can construct better approximations by the scheme

$$\rho^{(q+1)} = \frac{\rho^{(q)} a_0(\delta)}{\rho^{(q)} - h(\delta, \rho^{(q)}) + a_0(\delta)}, \quad q = 1, 2, \ldots,$$

where $\rho^{(0)} = \widehat{f}_3(\delta)$.

Example 5.15 Consider the majorant equation $\rho = h(\delta, \rho) := \delta(1 + \rho + \rho^2 + \rho^3)$, where $\delta \ge 0$ is a scalar. Here the interval $[0, \mathcal{S}_3]$ for δ is easily obtained noting that $\mathcal{S}_3$ is the maximum of the expression $\rho/(1 + \rho + \rho^2 + \rho^3)$ in $\rho > 0$. This maximum is achieved for the positive root of the equation $2\rho^3 + \rho^2 - 1 = 0$ and is $\mathcal{S}_3 \simeq 0.27695$. We have $\tau_3(\delta) = (1 - \delta)/(\delta + \sqrt{3\delta - 2\delta^2})$ and

$$\widehat{f}_3(\delta) = \frac{2\delta}{1 - \delta + \sqrt{1 - 2\delta - \delta^2(3 + \tau_3(\delta))}}.$$

The results for the exact small solution $f_3(\delta)$ and its bound $\widehat{f}_3(\delta)$ are shown at Table 5.1. The cases when the solution does not exist are marked by double asterisk. The bound does not exist in the case marked by asterisk. We see that the bound $\widehat{f}_3(\delta)$ is good whenever applicable, i.e., for $\delta \le \mathcal{S}_3$. But it also 'works' a bit after $\mathcal{S}_3$ (for example $\delta = 0.28$) although for this value of δ the majorant equation does not have a small solution.
◇

The case $r > 3$. For $r = 4$ there is a closed form solution, which is not very suitable for practical implementation. For $r > 4$ in general there are no closed form solutions. That is why for $r > 3$ we shall construct closed form approximations for the small solution of the majorant equation as in the case $r = 3$.

Table 5.1: Solutions and bounds for a cubic majorant equation

δ	f_3	$\widehat{f}_3$
0.03000	0.03096	0.03105
0.09000	0.09999	0.10148
0.15000	0.18350	0.18968
0.21000	0.29601	0.31388
0.27000	0.52607	0.57302
$\mathcal{S}_3 \simeq 0.27695$	0.65730	0.65730
0.28000	**	0.74925
0.29000	**	*

Suppose that for a given δ such that $a_1(\delta) < 1$ and $a_r(\delta) > 0$ equation (5.25) has two nonnegative solutions. For the small solution $\rho_0 = f_r(\delta)$ we have $\rho_0 \leq \tau_r$, where $\tau_r = \tau_r(\delta)$ is the unique solution of the equation $1 = h_\rho(\delta, \rho)$,

$$1 = \sum_{j=0}^{r-1} (j+1) a_{j+1} \tau_r^j. \tag{5.31}$$

This equation has a unique solution. Indeed, $1 > a_1 = h_\rho(\delta, 0)$. On the other hand for ρ sufficiently large (take $r a_r \rho^{r-1} > 1$) we have $1 < h_\rho(\delta, \rho)$. Hence there is a solution τ_r of equation (5.31). That τ_r is unique follows from the fact that the function $h_\rho(\delta, \cdot)$ is increasing.

We have $\rho_0 \leq g(\delta, \tau_r(\delta), \rho_0) := a_0(\delta) + a_1(\delta)\rho_0 + a_2(\delta)\rho_0^2 + b(\delta, \tau_r(\delta))\rho_0^2$, where

$$b(\delta, \tau) := \sum_{j=2}^{r-1} a_{j+1}(\delta) \tau^{j-1}.$$

Here $\widehat{g}(\delta, \rho) := g(\delta, \tau_3(\delta), \rho)$ is a new Lyapunov majorant. Note that $g(\delta, \tau, \rho) \bullet h(\delta, \rho)$ for $\rho \bullet \tau$, where $\bullet$ stands for $\leq$, $=$ or $\geq$, respectively. Since for $r > 3$ there is no convenient closed form expression for τ_r we shall find an upper bound $\widehat{b}(\delta)$ for $b(\delta, \tau_r(\delta))$. It follows from (5.31) that $(j+1) a_{j+1} \tau_r^j \leq 1 - a_1$ and

$$\tau_r \leq \left(\frac{1 - a_1}{(j+1) a_{j+1}} \right)^{1/j}, \ j = 2, \ldots, r-1.$$

Hence

$$a_{j+1} \tau_r^{j-1} \leq \alpha_{j+1} := a_{j+1}^{1/j} \left(\frac{1 - a_1}{j+1} \right)^{1-1/j}, \ j = 2, \ldots, r-1$$

and

$$b(\delta, \tau_r(\delta)) \leq \widehat{b}(\delta) := \sum_{j=2}^{r-1} \alpha_{j+1}(\delta).$$

As a result we have $\rho \leq a_0(\delta) + a_1(\delta)\rho + (a_2(\delta) + \widehat{b}(\delta))\rho^2$ and

$$f_r(\delta) \leq \widehat{f}_r(\delta) := \frac{2a_0(\delta)}{1 - a_1(\delta) + \sqrt{(1 - a_1(\delta))^2 - 4a_0(\delta)(a_2(\delta) + \widehat{b}(\delta))}} \tag{5.32}$$

provided that

$$\delta \in \widehat{\Omega}_r := \left\{ \delta \in \mathbb{R}_+^k : a_1(\delta) + 2\sqrt{a_0(\delta)(a_2(\delta) + \widehat{b}(\delta))} \leq 1 \right\}. \tag{5.33}$$

Thus we have proved the following result.

Theorem 5.16 *Consider the majorant equation 5.25 for $r > 3$. If the inequality $\widehat{f}_r(\delta) \leq \tau_r(\delta)$ is fulfilled then the majorant equation has a small solution $f_r(\delta)$ for which the estimate (5.32), (5.33) holds.*

Example 5.17 The bound (5.32), (5.33) is applicable for $r = 3$ as well (in this case we shall denote the bound as $\varphi_3(\delta)$) although it will give slightly worse results than the bound (5.28), (5.29). Consider again the majorant equation from Example 5.15. Here $\varphi_3(\delta)$ is the small solution of the equation $(\delta + \alpha_3(\delta))\rho^2 - (1 - \delta)\rho + \delta = 0$. The results are shown at Table 5.2. In the case marked by an asterisk the bound φ_3 does not exist.

Table 5.2: More solutions and bounds for a cubic majorant equation

δ	f_3	φ_3
0.03000	0.03096	0.03106
0.09000	0.09999	0.10181
0.15000	0.18350	0.19190
0.21000	0.29601	0.32554
0.27000	0.52607	*

$\diamondsuit$

Example 5.18 Consider the majorant equation $\rho = h(\delta, \rho) := \delta(1 + \rho + \rho^2 + \rho^3 + \rho^4)$, where $\delta \geq 0$ is a scalar. The interval $[0, \mathcal{S}_4]$ for δ is obtained by noting that $\mathcal{S}_4$ is the maximum of $\rho/(1 + \rho + \rho^2 + \rho^3 + \rho^4)$. This maximum is achieved for the positive root of the equation $2\rho^3 + \rho^2 - 1 = 0$ and is $\mathcal{S}_3 \simeq 0.27695$. We have

$$\widehat{b}(\delta) = \alpha_3(\delta) + \alpha_4(\delta) = \delta^{1/2}\left(\frac{1 - \delta}{3}\right)^{1/2} + \delta^{1/3}\left(\frac{1 - \delta}{4}\right)^{2/3}$$

and

$$\widehat{f}_4(\delta) = \frac{2\delta}{1 - \delta + \sqrt{(1 - \delta) - 4\delta(\delta + \widehat{b}(\delta))}}.$$

The results for the small solution $f_4(\delta)$ and its bound $\widehat{f}_4(\delta)$ are shown at Table 5.3. The cases when the solution does not exist are marked by a double asterisk. The bound does not exist in the case marked by an asterisk. The bound $\widehat{f}_4(\delta)$ is satisfactory whenever applicable. We also see that the bound ceases to to exist before the critical value $\mathcal{S}_4$ for δ.

Table 5.3: Solutions and bounds for a quartic majorant equation

δ	f_4	$\widehat{f}_4$
0.02000	0.02042	0.02047
0.08000	0.08769	0.09004
0.14000	0.16831	0.18215
0.20000	0.27568	0.34254
0.26000	0.53064	*
$\mathcal{S}_4 \simeq 0.26079$	0.56774	*
0.26100	**	*

$\diamondsuit$

We conclude this subsection by justifying certain 'cheap' perturbation bounds. An interesting feature of these bounds is that while they are valid for any $r > 2$, only the first two or three terms $a_j\rho^j$ of h are taken into account explicitly. The influence of higher order terms is implicit by the requirement $\delta \in \Omega_r$.

Theorem 5.19 *Consider the majorant equation 5.25 for $r > 2$ and let $\delta \in \Omega_r \backslash \mathcal{S}_1$. Then*

$$f_r(\delta) \leq b_2(\delta) \leq b_1(\delta), \tag{5.34}$$

where

$$b_1(\delta) := \frac{2a_0}{1 - a_1(\delta)}, \quad b_2(\delta) := \frac{3a_0(\delta)}{1 - a_1(\delta) + \sqrt{(1 - a_1(\delta))^2 - 3a_0(\delta)a_2(\delta)}}. \tag{5.35}$$

Proof. We note first that the relation $\delta \in \Omega_r \subset \Omega_2$ guarantees that $a_1 + 2\sqrt{a_0 a_2} \leq 1$ and hence the quantities b_j are correctly defined by (5.35). Consider now the second estimate $f_r \leq b_1$ in (5.34). Recall that τ_r satisfies (5.31). Setting $c_l(\delta, \rho) := a_l(\delta)\rho^{j-l} + \cdots + a_r(\delta)\rho^{r-l}$, where $l = 2, 3$, we see that $a_1(\delta) + 2\tau_r(\delta)c_2(\delta, \tau_r(\delta)) \leq 1$ and hence $c_2(\delta, \tau_r(\delta)) \leq (1 - a_1(\delta))/(2\tau_r(\delta))$. On the other hand for every $\rho \leq f_r(\delta)$ we have

$$\rho \leq a_0(\delta) + a_1(\delta)\rho + c_2(\delta, \tau_r(\delta))\rho^2 \leq a_0(\delta) + a_1(\delta)\rho + (1 - a_1(\delta))\frac{\rho^2}{2\tau_r(\delta)}.$$

Since $\rho \leq \tau_r(\delta)$, we get that $\rho \leq a_0(\delta) + a_1(\delta)\rho + (1 - a_1(\delta))\rho/2$ and hence $\rho \leq b_1(\delta)$. Now the first inequality in (5.34) follows, since ρ may be chosen as $f_r(\delta)$.

Consider next the first bound $f_r \leq b_2$ in (5.34). We have $a_1(\delta) + 2a_2(\delta)\tau_r(\delta) + 3\tau_r^2(\delta)c_3(\delta, \tau_r(\delta)) \leq 1$ and hence $c_3(\delta, \tau_r(\delta)) \leq (1 - a_1(\delta) - 2a_2(\delta)\tau_r(\delta))/(3\tau_r^2(\delta))$. For every $\rho \leq f_r(\delta)$ it is fulfilled that

$$\begin{aligned} \rho &\leq a_0(\delta) + a_1(\delta)\rho + a_2(\delta)\rho^2 + c_3(\delta, \tau_r(\delta))\rho^3 \\ &\leq a_0(\delta) + a_1(\delta)\rho + a_2(\delta)\rho^2 + (1 - a_1(\delta) - 2a_2(\delta)\tau_r(\delta))\frac{\rho^3}{3\tau_r^2(\delta)}. \end{aligned}$$

Now $\rho \leq \tau_r(\delta)$ yields

$$\rho \leq a_0(\delta) + a_1(\delta)\rho + a_2(\delta)\rho^2 + (1 - a_1(\delta) - 2a_2(\delta)\rho)\rho/3$$

and $0 \leq 3a_0(\delta) - 2(1 - a_1(\delta))\rho + a_2(\delta)\rho^2$. Thus $\rho \leq b_2(\delta)$ for all $\rho \leq f_r(\delta)$.

Finally the inequality $b_2(\delta) \leq b_1(\delta)$ is verified by direct calculation. This completes the proof. □

Of course, in applying the cheap estimates (5.34) one has to check whether $\delta \in \Omega_r$. A sufficient condition for $f_r(\delta) \leq b_i(\delta)$ to be valid is $h(\delta, b_i(\delta)) \leq b_i(\delta)$.

We conclude the consideration of cheap bounds with the following remarks. For $\delta \to 0$ the small solution $f_r(\delta)$ is of asymptotic order $\alpha(\delta) + o(\|\delta\|)$, where $\alpha(\delta) := a_0(\delta)/(1 - a_1(0))$. At the same time the bound $b_2(\delta)$ is of order $\frac{3}{2}\alpha(\delta) + o(\|\delta\|)$, while $b_1(\delta)$ is of order $2\alpha(\delta) + o(\|\delta\|)$. We note finally that $b_1(\delta) = b_2(\delta)$ if and only if $\delta \in \mathcal{S}_2$, i.e., $a_1(\delta) + 2\sqrt{a_0(\delta)a_2(\delta)} = 1$.

5.2.2 Asymptotic solutions of polynomial majorant equations

Consider the problem of asymptotic expansion of the solution to the majorant equatuon. Suppose that the coefficients a_j in (5.24) can be represented as $a_j(\delta) = a_{j,0}(\delta) + a_{j,1}(\delta) + \cdots$ (the case when some of these expansions contains infinitely many terms is not excluded), where $a_{j,l}(\delta) = O(\|\delta\|^l)$, $\delta \to 0$. For example, $a_{j,l}$ can be polynomials in δ of degree l. We recall that the degree of a nonzero polynomial $\sum_{l_s \geq 0} c_{l_1 \ldots l_k}\delta_1^{l_1} \cdots \delta_k^{l_k}$ is $l := \max\{l_1 + \cdots + l_k : c_{l_1 \ldots l_k} \neq 0\}$. In particular $a_{j,0}$ are nonnegative constants. Since $h(\delta, \rho)$ is a Lyapunov majorant we have $a_{0,0} = 0$ and $a_{1,0} < 1$.

We shall represent the small solution of (5.24) as $\rho_0(\delta) = \sum_{l=1} \rho_l(\delta)$, where $\rho_l(\delta)$ is of order $\|\delta\|^l$ for $\delta \to 0$ (note that the expansion for ρ starts with the term $\rho_1(\delta) = O(\|\delta\|)$). For this purpose we shall use the technique of the *fictitious small parameter*, see [135]. Represent the coefficients a_j as $\sum_{l=0} \varepsilon^l a_{jl}$ and ρ_0 as $\rho_0 = \sum_{l=1} \varepsilon^l \rho_l$. Substitute these expressions in (5.24) and equate the coefficients of the equal powers of the parameter ε. As a result ρ_1 is immediately obtained. For the next coefficients $\rho_2, \rho_3 \ldots$, a recurrent scheme emerges as described below. Finally ε is set to 1. We have $\rho_1 = a_{0,1} + a_{1,0}\rho_1$, $\rho_2 = a_{0,2} + a_{1,0}\rho_2 + a_{1,1}\rho_1 + a_{2,0}\rho_1^2$, $\rho_3 = a_{0,3} + a_{1,0}\rho_3 + a_{1,1}\rho_2 + a_{2,1}\rho_1^2 + 2a_{2,0}\rho_1\rho_2$, etc., and $\rho_{l+1} = a_{0,l+1} + a_{1,0}\rho_{l+1} +$

$q_l(\rho_1, \ldots, \rho_l)$, $l \geq 3$, where q_l is a polynomial of degree l in $\rho_1, \ldots, \rho_l$. As a result we have

$$\rho_1 = \frac{a_{0,1}}{1 - a_{1,0}}, \ \rho_2 = \frac{a_{0,2} + a_{1,1}\rho_1 + a_{2,0}\rho_1^2}{1 - a_{1,0}}, \ \rho_3 = \frac{a_{0,3} + a_{1,1}\rho_2 + a_{2,1}\rho_1^2 + 2a_{2,0}\rho_1\rho_2}{1 - a_{1,0}}, \tag{5.36}$$

etc., and

$$\rho_{l+1} = \frac{a_{0,l+1} + q_l(\rho_1, \ldots, \rho_l)}{1 - a_{1,0}}, \ l \geq 3.$$

Example 5.20 Consider the majorant equation $\rho = \delta_1 + \delta_2 + \delta_1\rho + \rho^2 + \rho^3$, where $\delta = [\delta_1, \delta_2]^\top \in \mathbb{R}_+^2$. We have $a_{0,1} = \delta_1 + \delta_2$, $a_{1,0} = 0$, $a_{1,1} = \delta_1$, $a_{2,0} = 1$, $a_{3,0} = 1$ and the other $a_{i,j}$ are zero. Using (5.36) we obtain $\rho = (\delta_1 + \delta_2) + (2\delta_1^2 + 3\delta_1\delta_2 + \delta_2^2) + (7\delta_1^3 + 16\delta_1^2\delta_2 + 12\delta_1\delta_2^2) + O(\|\delta\|^4)$, $\delta \to 0$. ◊

Example 5.21 Consider the Lyapunov majorant $h(\delta, \rho) := \delta + \delta\rho + \rho^2 + \rho^3$, where $\delta \geq 0$ is a scalar. The maximum allowed value $\mathcal{S}_3$ for δ is easily obtained excluding δ from the equations $\rho = h(\delta, \rho)$ and $1 = h_\rho(\delta, \rho)$. As a result we obtain $\delta = 1 - 2\rho - 3\rho^2$ and $1 - 2\rho - 4\rho^2 - 2\rho^3 = 0$, which gives $\rho \simeq 0.29716$ and $\mathcal{S}_3 \simeq 0.14078$. Here the asymptotics of the small solution is

$$f_3(\delta) = \delta + 2\delta^2 + 7\delta^3 + O(\delta^4).$$

The approximate solution $\widehat{f}_3(\delta)$ is

$$\widehat{f}_3(\delta) = \delta + \frac{7}{3}\,\delta^2 + \frac{299}{36}\,\delta^3 + O(\delta^4) \simeq \delta + 2.33\,\delta^2 + 8.31\,\delta^3 + O(\delta^4).$$

The slightly worse approximation from Example 5.17 is

$$\varphi_3(\delta) = \delta + \left(2 + \frac{1}{\sqrt{3}}\right)\delta^2 + \left(\frac{20}{3} + \frac{13}{2\sqrt{3}}\right)\delta^4 + O(\delta^4) = \delta + 2.58\,\delta^2 + 7.09\,\delta^3 + O(\delta^4),$$

while the cheap bounds from (5.34) are

$$b_2(\delta) = \frac{3}{2}\,\delta + \frac{21}{8}\,\delta^2 + \frac{105}{16}\,\delta^3 + O(\delta^4) = 1.5\,\delta + 2.62\,\delta^2 + 6.56\,\delta^3 + O(\delta^4)$$

and $b_1(\delta) = 2\delta + 2\delta^2 + 2\delta^3 + O(\delta^4)$. ◊

Assume now that we have estimates of type (5.1), (5.2) for the corresponding generalized norms, i.e.

$$\begin{aligned} \|\Pi(E, Y)\|_g &\preceq h(\|E\|_g, \rho), \\ \|\Pi(E, Y) - \Pi(E, Z)\|_g &\preceq h'_\rho(\|E\|_g, \rho)\|Y - Z\|_g \end{aligned}$$

for all y, z with

$$\|Y\|_g, \|Z\|_g \preceq \rho := [\rho_1, \ldots, \rho_s]^\top \in \mathbb{R}_+^s.$$

Suppose that the function

$$h = [h_1, \dots, h_s]^\top : \mathbb{R}_+^r \times \mathbb{R}_+^s \to \mathbb{R}_+^s$$

is continuous, differentiable in its last s arguments, and $h(0,0) = 0$. We assume also that $h(\delta, \rho)$ is nondecreasing in each of its $s+r$ arguments and that the relations

$$\rho_j = o(h_j(\delta, \rho)), \ \rho_j \to \infty; \ j = 1, \dots, s$$

and

$$\mathrm{rad}(h'_\rho(0,0)) < 1$$

are fulfilled, where $\mathrm{rad}(A)$ is the spectral radius of the matrix A.

The application of the method of Lyapunov majorants here again allows to prove that there exists a closed convex domain $\Omega_0 \subset \mathbb{R}_+^r$, such that for every δA with $\|\delta A\|_g \in \Omega$, the perturbed equation (4.5) has a unique solution δX, where δX is a function of δA such that $\delta X = 0$ for $\delta A = 0$. Moreover, there exists a function

$$f = [f_1, \dots, f_s]^\top : \mathrm{cl}(\Omega_0) \to \mathbb{R}_+^s,$$

such that f_j is nondecreasing in each of its arguments, satisfies $f(0) = 0$, and

$$\|\delta X\|_g \preceq f(\|\delta A\|_g), \ \|\delta A\|_g \in \Omega_0. \tag{5.37}$$

The boundary $\partial\Omega_0$ of the domain Ω_0 is obtained by eliminating $\rho \in \mathbb{R}_+^s$ from the system of equations

$$h(\delta, \rho) = \rho, \ \det(h'_\rho(\delta, \rho) - I_s) = 0.$$

The inequality (5.20) or (5.37)) gives nonlinear nonlocal perturbation bounds for the solution in case of a regular computational problem with implicit solution.

Consider finally the case when the solution of equation (4.1) is not unique. We shall restrict ourselves to the case $q > r$, i.e., when we have more unknowns than the number of equations. Suppose that the linear operator $F_X : \mathcal{X} \to \mathcal{Y}$ is surjective, or equivalently, $\mathrm{rank}(M) = r$, where $M := \mathrm{Mat}(F_X) \in \mathbb{F}^{r \times q}$ is the matrix representation of F_X.

Let $F_X^\dagger : \mathcal{Y} \to \mathcal{X}$ be the right pseudo-inverse of F_X, such that

$$M^\dagger := \mathrm{Mat}(F_X^\dagger) = M^{\mathrm{H}}(MM^{\mathrm{H}})^{-1}.$$

Then all relations from the present section are valid with F_X^{-1} replaced by $F_X^\dagger$. In this case, however, the operator equation (4.7) is not equivalent to the original equation (4.1): all solutions of (4.7) are solutions of (4.1) but the converse may not be true. This will not cause a problem, since we usually have to find at least one solution. In addition, here it is not necessary to prove the uniqueness of the solution of the operator equation, since the original equation has no unique solution.

Thus, when using Lyapunov majorants $h(\delta, \rho)$, it is sufficient to prove that the majorant equation $\rho = h(\delta, \rho)$ has a root vanishing together with δ, without the requirement that the derivative $h'_\rho(\delta, \rho)$ (in case of a scalar h) or the spectral radius of the matrix $h'_\rho(\delta, \rho)$ (when h is a vector function) is less than 1.

5.3 Case study

Consider the real scalar quadratic equation

$$q + 2ax - sx^2 = 0, \tag{5.38}$$

where $q > 0$ and $s > 0$. The positive solution of equation (5.38) is

$$x = \frac{a+d}{s}, \ d := \sqrt{a^2 + sq}.$$

Let δq, δa and δs be perturbations in q, a and s, respectively, which preserve the form of the equation, i.e., $|\delta q| < q$ and $|\delta s| < s$. The perturbation in the solution is then

$$\delta x = \frac{s\delta a - a\delta s + s\widetilde{d} - d(s+\delta s)}{s(s+\delta s)},$$

where

$$\widetilde{d} := \sqrt{(a+\delta a)^2 + (s+\delta s)(q+\delta q)}\,.$$

Setting $c = (q, a, s)$ and $\delta s = (\delta q, \delta a, \delta s)$, the equivalent operator equation for the perturbation δx is

$$\delta x = \Pi(\delta c, \delta x) = \Pi_0(\delta s) + \Pi_1(\delta c, \delta x) + \Pi_2(\delta c, \delta x), \tag{5.39}$$

where

$$\begin{aligned} \Pi_1(\delta c) &:= (\delta q + 2x\delta a - x^2\delta s)/(2d), \\ \Pi_1(\delta c, \delta x) &:= (\delta a - x\delta s)\delta x/d, \ \Pi_2(\delta c, \delta x) := -(s+\delta s)\delta x^2/(2d). \end{aligned}$$

Hence, the Lyapunov majorant is

$$h(\delta c, \rho) = a_0(\delta c) + a_1(\delta c)\rho + a_2(\delta c)\rho^2,$$

where

$$\begin{aligned} a_0(\delta c) &:= (\delta q + 2x\delta_a + x^2\delta s)/(2d), \\ a_1(\delta c) &:= (\delta a + x\delta s)/d, \end{aligned}$$
$$a_2(\delta c) := (s + \delta s)/(2d).$$

If we take $q = a = s = 1$ and $\delta q = \varepsilon$, $\delta a = \varepsilon$ and $\delta s = -\varepsilon$, where $0 \leq \varepsilon < 1$, then after some simple computations we obtain $x = 1 + \sqrt{2}$ and

$$\delta x(\varepsilon) = \frac{\varepsilon\sqrt{2}}{1-\varepsilon}\left(1 + \sqrt{2} + \frac{1}{1+\sqrt{1+\varepsilon}}\right).$$

Furthermore,

$$a_0(\delta c) = \varepsilon(2 + 1.5\sqrt{2}),\ a_1(\delta c) = \varepsilon(1+\sqrt{2}),\ a_2(\delta c) = 0.25\sqrt{2}(1+\varepsilon).$$

As a result we have the perturbation bound

$$\delta x \leq f(\varepsilon),\ \varepsilon \leq \varepsilon_0,$$

where

$$\begin{aligned} f(\varepsilon) &:= \frac{2a_0(\delta c)}{1 - a_1(\delta c) + \sqrt{(1-a_1(\delta c))^2 - 4a_0(\delta c)a_2(\delta c)}} \\ &= \frac{\varepsilon(4+3\sqrt{2})}{1 - \varepsilon(1+\sqrt{2}) + \sqrt{1 - \varepsilon(5+4\sqrt{2})}} \end{aligned}$$

and $\varepsilon_0 = 1/(5+4\sqrt{2}) = 0.0938$ (up to four figures after the decimal point).

The expressions for the exact perturbation $\delta x(\varepsilon)$ and the bound $f(\varepsilon)$ have equal first order terms, namely

$$\begin{aligned} \delta x(\varepsilon) &= \varepsilon(2+1.5\sqrt{2}) + \varepsilon^2(2 + 1.375\sqrt{2}) + O(\varepsilon^3) \\ &= 4.1213\varepsilon + 3.9445\varepsilon^2 + O(\varepsilon^3),\ \varepsilon \to 0, \\ f(\varepsilon) &= \varepsilon(2+1.5\sqrt{2}) + \varepsilon^2(8 + 5.625\sqrt{2}) + O(\varepsilon^3) \\ &= 4.1213\varepsilon + 15.9550\varepsilon^2 + O(\varepsilon^3),\ \varepsilon \to 0. \end{aligned}$$

In Figure 5.5 we give the graphs of the functions $\delta x : [0,1) \to \mathbb{R}_+$ and $f : [0,\varepsilon_0) \to \mathbb{R}_+$.

5.4 Notes and references

The method of finite majorant equations was proposed by A.M. Lyapunov in 1893 for the analysis of series expansions of solutions of ordinary differential equations, see [162, 163]. The corresponding majorant functions are known as *Lyapunov majorants.* The use of Lyapunov majorants and fixed point principles in proving existence and uniqueness results for operator equations in nonlinear oscillation theory was proposed in [160], and, in a more general statement, in [85]. This technique had been further developed in [134, 135] for perturbation analysis of matrix problems. The combined use of Lyapunov majorants and fixed point principles in the perturbation theory for matrix equations is considered in [127], see also [147].

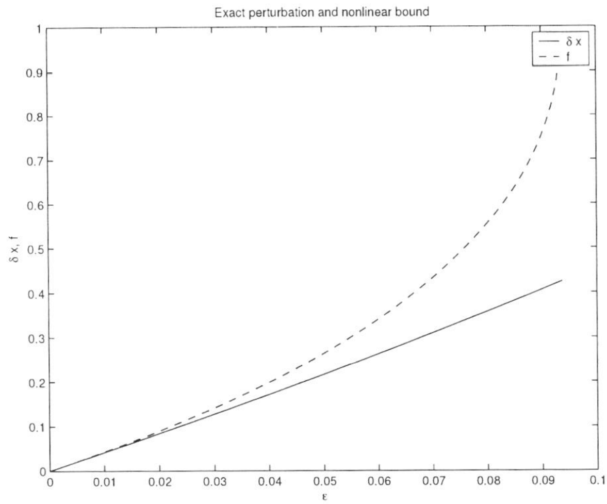

Figure 5.5: Exact perturbation and nonlinear bound

It is worth mentioning that the technique of Lyapunov majorants is in fact widely used in many problems of the general theory and perturbation theory of operator equations (including equations in abstract spaces), often without stating this explicitly.

Chapter 6

Singular problems

6.1 Introductory remarks

In this chapter we consider basic concepts for singular problems, such as the distance to singularity, and the classification and regularization of singular problems.

As discussed in Chapters 2, 3 and 4, problems with implicit solution may be regular or singular. Fortunately, regular problems are usually generic in the data space. At the same time the analysis of singular problems is important from both theoretical and practical points of view.

Singular problems are also often called *infinitely ill-conditioned*. The perturbation in the solution of a singular problem may be extremely large even if the perturbation in the data is small. A regular problem which is close to the set of singular problems may be very ill-conditioned and there is a close relation between the conditioning and the distance to singularity of a given regular problem.

Example 6.1 The problem of computing a particular solution $x = x_i$ of the algebraic equation

$$x^n + a_1 x^{n-1} + \cdots + a_{n-1} x + a_n = 0$$

is regular if x_i is a simple root, and singular if x_i is a multiple root. A solution of multiplicity k corresponds to data $a = [a_1, \ldots, a_n]^\top \in \mathbb{R}^n$, belonging to a closed $(n-k+1)$-dimensional variety in the data space $\mathbb{R}^n$. The sensitivity of multiple roots may be very high. For example, the equation

$$(x-1)^n = 0$$

has a root $x = 1$ of multiplicity n. Perturbing the constant term in the polynomial by a small quantity μ we get the equation

$$(x-1)^n - \mu = 0$$

which has roots $x+\delta x = 1 + \sqrt[n]{\mu}$. If $\mu = 10^{-n}$, then the perturbations in the roots are of magnitude 10^{-1}, i.e., they are 10^{n-1} times larger than the perturbations in the data.

One should not make an erroneous conclusion that, if an algebraic equation has only simple roots, then they are well conditioned. For instance, the famous "perfidious" polynomial equation, introduced by Wilkinson [232],

$$(x-1)(x-2)\cdots(x-20) = x^{20} - 210x^{19} + \cdots + 20! = 0$$

shows high sensitivity of the roots $x_i = i$ relative to changes in the coefficients. This is due to the very large condition number of some of the roots.$\diamondsuit$

Example 6.2 Similar to Example 6.1, the problem of computing the eigenvalues of a matrix $A \in \mathbb{F}^{n\times n}$ is regular if and only if the matrix A has only linear elementary divisors. This holds for example in the case of pairwise distinct eigenvalues. The set of singular problems (corresponding to matrices with nonlinear elementary divisors) is a closed (n^2-1)–dimensional variety in $\mathbb{F}^{n\times n}$. $\diamondsuit$

Usually, the set of singular problems $\mathcal{B}$ is a variety of co-dimension 1 in the data space $\mathcal{A}$. Thus, $\mathcal{B}$ has irreducible components, which are hyper-surfaces, described by a scalar equation $g(A) = 0$. Furthermore, although the "probability" that a particular problem is singular is zero, for a family of problems the situation changes dramatically. Even if we deal with a one-dimensional family (i.e., a curve) of problems with data $\{A_t\} \subset \mathcal{A}$, parametrized by the scalar $t \in \mathbb{R}$, most probably it will meet some hyper-surface from $\mathcal{B}$ for some t. In such a statement, singular problems seem more like a rule rather than just an exception.

An important characteristic of a regular problem is its distance to the set of singular problems. This is the norm of the smallest perturbation that transforms a regular problem into a singular one. It is intuitively clear that ill-conditioned problems are close to singularity and vice versa. The interconnection between conditioning and distance to singularity is studied in [51].

The perturbation analysis and the solution of singular problems presents a challenge in modern scientific computing. Some issues in the classification and analysis of singular problems are considered in the next subsections.

6.2 Distance to singularity

An important characteristic of a problem $X = \Phi(A)$ is its distance to the closest singular problem.

Definition 6.3 The *absolute distance* of the problem $X = \Phi(A)$ to the set $\mathcal{B} \subset \mathcal{A}$ of singular problems is the quantity

$$\Delta_{\text{abs}}(A) := \text{dist}(A, \mathcal{B}) = \inf\{\|\delta A\| : \ A + \delta A \in \mathcal{B}\}.$$

For problems with $A \neq 0$ we also define the *relative distance* $\Delta_{\text{rel}}(A)$ as

$$\Delta_{\text{rel}}(A) := \frac{\Delta_{\text{abs}}(A)}{\|A\|}.$$

Thus, a problem is singular if and only if its distance to $\mathcal{B}$ is zero.

For many problems the relative distance $\Delta_{\text{rel}}(A)$ is inversely proportional to the relative condition number κ of the problem, or to its square κ^2 [51].

Example 6.4 Let $A \in \mathbb{F}^{n\times n}$ be a nonsingular matrix and let

$$\Delta_{\text{rel}}(A) = \frac{\Delta_{\text{abs}}(A)}{\|A\|}$$

be its relative distance to the set of singular matrices, where $\Delta_{\text{abs}}(A)$ is the norm of the smallest perturbation δA, such that $A + \delta A$ is singular:

$$\Delta_{\text{abs}}(A) := \min\{\|\delta A\| : \det(A + \delta A) = 0\}.$$

Since $\Delta_{\text{abs}}(A) = 1/\|A^{-1}\|$ we have $\text{cond}(A)\Delta_{\text{rel}}(A) = 1$. ◊

Hence, the relative distance to singularity and the relative condition number (with respect to inversion) of a nonsingular matrix are reciprocal.

A problem, solved in finite precision arithmetic with roundoff unit eps, for which $\Delta_{\text{rel}}(A)$ is of order eps, is *practically singular*. Indeed, when writing the data in the computer memory, i.e., rounding A to the closest collection $\widetilde{A}$ of data with exact representation in the finite precision arithmetic, we get a problem with data $\widetilde{A}$, which may as well be singular, since $\|\widetilde{A} - A\| \approx \text{eps}\|A\|$.

6.3 Classification

In this section we classify different types of singular problems. Singular problems, corresponding to data from the set $\mathcal{B}$, are very sensitive and their numerical solutions in finite precision arithmetic may be contaminated with large errors. These errors may depend on the round-off unit eps in a highly nonlinear way, e.g., they may have magnitude of order $\text{eps}^{1/k}$, where $k > 1$. Different problems may be characterized by different values of k. But there are also problems which are so sensitive that no perturbation bound of order $\text{eps}^{1/k}$ exists.

Example 6.5 The function $\varphi : (-1, 1) \to R_+$, defined by

$$\varphi(a) = \begin{cases} 1/\log(1/|a|) & \text{if } 0 < |a| < 1 \\ 0 & \text{if } a = 0 \end{cases}$$

increases so fast in the neighborhood of the point $a = 0$, that no estimate of the type

$$|\varphi(a) - \varphi(0)| \le c\,|a|^\tau$$

exists, regardless of how small $\tau > 0$ is. ◊

We may classify singular problems according to their sensitivity, see [134], as follows. For a given problem $X = \Phi(A)$, $A \in \mathcal{A}$, with Φ at least continuous in an open neighborhood of A, and for $\alpha > 0$ small enough to ensure $A + \delta A \in \mathcal{A}$ for all δA with $\|\delta A\| \le \alpha$, set

$$\omega(A,\alpha) := \sup\{\|\delta X\| : \ \|\delta A\| \le \alpha\}.$$

Suppose that $\omega(A,\alpha)$ may be represented in the form of an asymptotic series

$$\omega(A,\alpha) = \sum_{j\ge 1} \omega_j(A,\alpha),$$

where

$$\omega_{j+1}(A,\alpha) = o(\omega_j(A,\alpha)), \ \ \alpha \to 0$$

and $o(\alpha)/\alpha \to 0$ for $\alpha \to 0$. The function ω_1 is called the *principal term* in the expansion of ω, and it determines the magnitude of ω in the neighborhood of $\alpha = 0$.

If the problem is regular then

$$\omega_1(A,\alpha) = K(A)\alpha,$$

where $K(A)$ is the absolute condition number of the problem.

Singular problems are characterised by a larger rate of increase of ω for small α, e.g.,

$$\omega_1(A,\alpha) = H\alpha^{1/k}, \ \ k > 1$$

and they may be classified by the behavior of the functions ω or ω_1.

Definition 6.6 Two problems $X = \Phi(A)$ and $Y = \Phi(B)$ are said to be *sensitivity equivalent* if there exist constants $0 < c_1 \le c_2 < \infty$, such that for some $\alpha_0 > 0$ the inequalities

$$c_1 \le \frac{\omega(A,\alpha)}{\omega(B,\alpha)} \le c_2, \ \ 0 < \alpha < \alpha_0$$

hold.

The sensitivity equivalence relation allows to divide the set $\mathcal{A}$ into pair-wise disjoint *orbits* (or *equivalence classes*) $\mathcal{A}_1, \mathcal{A}_2, \ldots$, such that two problems belong to a given orbit if and only if they are sensitivity equivalent. We suppose that the orbits are numbered such that more sensitive problems correspond to orbits with larger numbers. Thus, $\mathcal{A}_1$ may be the set of regular problems, while the union $\mathcal{A}_2 \cup \mathcal{A}_3 \cup \cdots$ of the remaining orbits is the set $\mathcal{B}$ of singular problems.

For a wide class of problems the expressions $\omega_j(A,\alpha)$ are fractional powers in α, which correspond to a function Φ that is locally Hölder continuous in a neighborhood of A. Suppose that the function Φ is locally Hölder continuous on

$\mathcal{A}$ and hence, on $\mathcal{B} \subset \mathcal{A}$. Then for every $A \in \mathcal{B}$ there exists a number $\tau \in (0,1)$ such that the quantity

$$H(A,\tau) := \lim_{\alpha \to 0} \sup \left\{ \frac{\|\delta X\|}{\|\delta A\|^\tau} : \|\delta A\| \le \alpha \right\}$$

is finite. Then we have

$$\|\delta X\| \le H(A,\tau)\|\delta A\|^\tau + o(\|\delta A\|^\tau),\ \delta A \to 0.$$

Denote by $\tau_2 < 1$ the exact upper bound of the set of all numbers τ when A varies over $\mathcal{B}$, and let $\mathcal{A}_2$ be the set of all A such that Φ is locally Hölder in a neighborhood of A with a power τ_2. Furthermore, by induction, we define powers $\tau_3,\ \tau_4, \ldots\ (\tau_2 > \tau_3 > \cdots)$ and sets $\mathcal{A}_3,\ \mathcal{A}_4, \ldots,$ such that Φ is locally Hölder continuous with a power τ_j in the neighborhood of $A \in \mathcal{A}_j$.

Setting $\tau_1 = 1$, we see that $\mathcal{A}_1$ is the set of regular problems (with the restriction $\Phi|_{\mathcal{A}_1}$ of Φ on $\mathcal{A}_1$ being Lipschitz continuous), while the set $\mathcal{B}$ of singular problems is the union of $\mathcal{A}_j$, $j \ge 2$.

Typically the orbits $\mathcal{A}_j$ are manifolds of decreasing dimensions:

$$\dim(\mathcal{A}) = \dim(\mathcal{A}_1) > \dim(\mathcal{A}_2) > \cdots.$$

If the set $\mathcal{B}$ of singular problems is defined as an algebraic variety, then τ_j are rational numbers, and often $\tau_j = 1/j$.

Example 6.7 Consider the equation

$$x^2 + a = 0$$

with data $a \in \mathbb{F}$. Here $\mathcal{A}_1 = \mathbb{F}\backslash\{0\}$ is the set of regular problems, while $\mathcal{B} = \mathcal{A}_2 = \{0\}$ is the set of singular problems. The sequence of Hölder exponents is 1 and 1/2. ◇

Example 6.8 Consider the equation

$$x^3 + a_1 x + a_2 = 0$$

with data $a = [a_1, a_2]^\top \in \mathcal{A} = \mathbb{R}^2$. Here the set $\mathcal{B}$ is the semi-cubic parabola

$$\mathcal{B} = \left\{ a : \frac{a_1^3}{27} + \frac{a_2^2}{4} = 0 \right\}.$$

It may be represented as $\mathcal{B} = \mathcal{A}_2 \cup \mathcal{A}_3$, where for $a \in \mathcal{A}_2 = \mathcal{B}\backslash\{0\}$ the equation has a double root, and for $a \in \mathcal{A}_3 := \{0\}$ it has a triple root. Thus, we have a sequence of Hölder exponents $\tau_1 = 1$, $\tau_2 = 1/2$ and $\tau_3 = 1/3$, corresponding to the manifolds $\mathcal{A}_1$, $\mathcal{A}_2$ and $\mathcal{A}_3$ with dimensions 2, 1 and 0, respectively. ◇

Example 6.9 Consider the general algebraic equation of degree n

$$f(a, x) := a_0 x^n + a_1 x^{n1} + \cdots + a_n = 0$$

with real or complex coefficients, forming the data vector $a := [a_0, a_1, \ldots, a_n]^\top \in \mathbb{F}^{n+1}$. Here we may choose $\mathcal{A}$ as the set of all a with $a_0 \neq 0$. The set $\mathcal{B}$ is defined from the condition that $f(a, x)$ and $f'_x(x, a)$ must have a common root. $\diamondsuit$

6.4 Regularization

The classification of singular problems may be effectively used in numerical analysis by applying various *regularization techniques*. These techniques are based on projecting the problem onto the nearest (regularized) problem with higher sensitivity. Sometimes regularization means an imbedding of the problem into a problem of lower sensitivity; these two approaches are dual in a certain sense.

We may regularize a problem by restricting the set of possible perturbations so that the new perturbed problem to be of higher sensitivity, since in fact the new restricted problem is less sensitive in comparison to the original one. There is no contradiction in this (strange at first glance) assertion, see Example 6.12. Indeed, the higher sensitivity of the regularized problem is exposed only to general perturbations, while relative to the restricted class of perturbations the new problem is of lower sensitivity. These phenomena are explained next.

If $A \in \mathcal{A}_j$, then Φ is locally Hölder continuous in A with an exponents τ_j, and its sensitivity increases when A approaches the boundary $\partial\mathcal{A}_j$ of $\mathcal{A}_j$. This is reflected in the increase of the coefficient $H(A, \tau_j)$ in the local estimate

$$\|\delta X\| \leq H(A, \tau_j)\|\delta A\|^{\tau_j}$$

of the perturbation in the solution. In the limiting case we have

$$\lim_{a \to \partial\mathcal{A}_j} H(A, \tau_j) = \infty.$$

Denote by $A^0 \in \mathcal{A}_{j+1}$ a point from $\mathcal{A}_{j+1}$ which is closest to A, so that $\|A^0 - A\|$ is the distance from the point A to the set $\mathcal{A}_{j+1}$, and consider the *regularized* problem

$$X = \Phi(\widehat{A}), \ \widehat{A} \in \widehat{\mathcal{A}} \subset \mathcal{A}_{j+1}. \tag{6.1}$$

The essential fact about (6.1) is that $\widehat{\mathcal{A}}$ is a neighborhood of A^0 which lies entirely in $\mathcal{A}_{j+1}$. Now the problem (6.1) is characterized with a sensitivity, determined by the power τ_j rather than τ_{j+1}, independently of the fact that $\widehat{A} \in \mathcal{A}_{j+1}$:

$$\|\delta X\| := \|\Phi(\widehat{A}) - \Phi(A^0)\| \leq \widehat{H}(A^0, \tau_j)\|\widehat{A} - A^0\|^{\tau_j},$$

where usually the quantity $\widehat{H}(A^0, \tau_j)$ is such that

$$\frac{\widehat{H}(A^0, \tau_j)}{H(A, \tau_j)} \ll 1.$$

The regularized problem (6.1) differs from the original problem $X = \Phi(A)$, since the function in (6.1) is in fact the restriction $\Phi|_{\widehat{\mathcal{A}}}$ of the original function Φ on the lower dimensional variety $\widehat{\mathcal{A}}$.

Example 6.10 Consider the quadratic equation

$$x^2 + a_1 x + a_2 = 0$$

with data $a = [a_1, a_2]^\top \in \mathbb{R}^2$ which has roots

$$x_{1,2} = \frac{-a_1 \pm \sqrt{a_1^2 - 4a_2}}{2}.$$

For a near to the parabola $\mathcal{B} \subset \mathbb{R}^2$, defined via $a_1^2 = 4a_2$, the equation becomes ill-conditioned. Indeed, we have

$$\frac{\partial x_i}{\partial a_1} = -\frac{x_i}{a_1 + 2x_i}, \ \frac{\partial x_i}{\partial a_2} = -\frac{1}{a_1 + 2x_i},$$

which gives

$$K_{i1} = \frac{|x_i|}{\sqrt{|a_1^2 - 4a_2|}}, \ K_{i2} = \frac{1}{\sqrt{|a_1^2 - 4a_2|}}.$$

Here K_{ij} is the absolute condition number of the root x_i relative to perturbations in the coefficient a_j.

For $a \in \mathcal{B}$ the equation is singular and $x_1 = x_2 = -a_1/2$. If now a_j is perturbed to $a_j + \delta a_j$, we have

$$\delta x_{1,2} = \frac{-\delta a_1 \pm \sqrt{2a_1\delta a_1 + (\delta a_1)^2 - 4\delta a_2}}{2}.$$

Hence, the perturbation may be of order

$$|\delta x_{1,2}| = \sqrt{a_1^2 + 4}\, \|\delta a\|^{1/2} + O(\|\delta a\|), \ a \to 0$$

and this is achieved for

$$\delta a_1 = \frac{a_1\|\delta a\|}{\sqrt{a_1^2 + 4}}, \ \delta a_2 = \frac{-2\|\delta a\|}{\sqrt{a_1^2 + 4}}.$$

If, however, we choose a special perturbation with

$$2a_1\delta a_1 + (\delta a_1)^2 = 4\delta a_2$$

then the perturbation in both roots is $-\delta a_1/2$ and the problem is regularized. This special perturbation in fact means that $a + \delta a \in \mathcal{B}$. It is interesting to observe that the problem is regularized also for $a_1\delta a_1 = 2\delta a_2$. In this case the perturbation in one of the roots is zero and in the other root it is $-\delta a_1$. In this second regularization the perturbed data $a + \delta a$ belongs to the straight line $\mathcal{T}$ in $\mathbb{R}^2$, parametrized as

$$\mathcal{T} := \left\{ [a_1 + t, a_1^2/4 + a_1 t/2]^\top : t \in \mathbb{R} \right\}$$

which is in fact the tangent to the parabola $\mathcal{B}$ at the point $[a_1, a_1^2/4]$. $\diamondsuit$

Example 6.10 is very instructive. In general we have the following result.

Theorem 6.11 *A singular problem $X = \Phi(A)$ with data $A \in \mathcal{B}$ is regularized if $A + \delta A$ is allowed to vary either in the variety $\widehat{\mathcal{A}} = \mathcal{B}$ or in the tangent space $\widehat{\mathcal{A}} = \mathcal{T}_\mathcal{B}(A)$ of $\mathcal{B}$ at the point A.*

The described regularization technique is applicable also to regular but ill-conditioned problems with $a \in \mathcal{A}_1$ and a large relative condition number $\kappa(A)$. In this case we may project the problem to the nearest (or some) point $A^0 \in \mathcal{A}_j$, obtaining a new family of problems $X = \Phi(A)$, $A \in \widehat{\mathcal{A}}$, which are better conditioned. Among the problems that can be solved in this way one should mention the solution of ill-conditioned linear algebraic equations, the determination of the numerical rank of a matrix, the computation of the eigenstructure of a matrix with almost multiple eigenvalues as well as the solution of some basic ill-conditioned problems in the theory of linear control systems.

Example 6.12 Consider the problem of rank determination for the matrix $A \in \mathbb{F}^{n\times n}$. Suppose that $\mathrm{rank}(A) = n$ and denote by

$$A = U\Sigma V^{\mathrm{H}}, \ \Sigma = \mathrm{diag}(\sigma_1, \ldots, \sigma_n), \quad U, V \in \mathcal{U}_n,$$

the singular value decomposition of A, see [83] and Appendix B, where $\sigma_1 \geq \sigma_2 \geq \ldots \geq \sigma_n \geq 0$ are the singular values of A. If σ_n is relatively small, say $\sigma_n \approx \mathrm{eps}\,\sigma_1$, then the computation of $\mathrm{rank}(A)$ in finite precision arithmetic with roundoff unit eps is a very ill-conditioned problem (practically a singular problem), since the rounding of the data may lead to a matrix A^*, which is not of full rank. Now let us choose a threshold $\tau > \mathrm{eps}$ and consider as zero all singular values that are less than or equal to τ. We get a new matrix

$$A^0 := U\Sigma^0 V^H, \ \Sigma^0 := \mathrm{diag}(\sigma_1, \ldots, \sigma_k, 0, \ldots, 0),$$

where the integer k is determined from $\sigma_k \geq \tau$, $\sigma_{k+1} < \tau$. The number k is the *numerical rank* of A corresponding to the threshold τ (or, briefly, the τ-numerical rank of A), and A^0 is the projection of A onto the set of matrices which are of rank k. Now the numerical rank determination in a neighborhood Ω of A^0 is a regular, and even a well-conditioned problem, if we consider the singular values of matrices $A \in \Omega$ that are less than the threshold τ, as zero. $\diamondsuit$

This regularization technique uses a *projection* of a singular problem onto the nearest (or some other close) problem of higher sensitivity. Another way to regularize an ill-posed or a singular problem is *imbedding* it into a regular problem.

Example 6.13 Consider the linear algebraic equation $Mx = b$ with $M \in \mathbb{F}^{n\times n}$ singular or close to singularity. If M is singular, then the problem is also singular but may even be ill-posed. If M is close to singularity the problem will be close to singularity or to ill-posedness. We may regularize the problem by imbedding it in the regular problem $(M + \alpha I)x_\alpha = b$, where $\alpha > 0$ is a (small) parameter. This approach is known as *Tikhonov regularization*, see [92]. $\diamondsuit$

6.5 Notes and references

Singular problems in the sense of this chapter (we recall that the problem $X = \Phi(A)$ is regular if the function Φ is continuous in a neighborhood of the data A, and singular otherwise) are called also ill-posed by some authors [51]. Here we prefer to use the classical terminology going back to H'Adamard [42, 43, 44].

The distance to singularity (sometimes called distance to ill-posedness) and related problems have been considered in [51, 52, 53, 67, 93].

The classification of singular problems from this chapter had been proposed in [133].

Regularization schemes have been proposed in [219, 218] and further developed as computational procedures in [220, 221].

Chapter 7

Perturbation bounds

7.1 Introductory remarks

In this chapter we discuss the main properties of perturbation bounds for the analysis of problems with either explicit or implicit solutions. Some important concepts are introduced and illustrated by examples.

7.2 Definitions and properties

The literature of perturbation theory is rich in various types of perturbation bounds. However, for many of them neither quantitative nor qualitative measures of exactness are discussed. Also, often the domains of applicability of some bounds are not known, or at least not stated clearly. This is particularly true for linear local perturbation bounds, based on condition numbers.

In order to compare perturbation bounds, several criteria are important. Ideally, the bound should be *rigorous*, its domain of applicability should be known and, if possible, the bound should be *sharp* or *exact* in some sense.

If the bound is too pessimistic in some cases, this should be made clear for the user.

A desirable property of a bound is to be *general* in the sense that it imposes minimum restrictions and is thus applicable to a general class of problems.

These requirements do not mean that bounds with unknown domain of applicability, as well as some heuristic (or experimentally stated) bounds are practically useless. Such bounds are of practical use, but one should be careful if a bound is not proven to be rigorous.

If the above criteria are met and the bound can also be computed numerically in a reliable way, then it should be included in software tools for solving engineering and scientific problems. We stress that without sensitivity and error estimates the corresponding software cannot be recognized as reliable. Unfortunately, some of the program systems for scientific computing do not include such estimates, and, as a result, they sometimes produce erroneous results without warning the user.

In this section we present the concepts of sharpness, exactness and attainability of perturbation bounds, which seem to be intuitively clear but nevertheless formal definitions are needed.

Let X be a solution of a regular problem with data $A = (A_1, \ldots, A_r)$, and let $X + \delta X$ be a solution, corresponding to the perturbed data $A + \delta A$. In case of an explicit problem we have $X = \Phi(A)$, where the function is locally Lipschitz continuous, and

$$\delta X = \Psi(A, \delta A) := \Phi(A + \delta A) - \Phi(A).$$

In case of an implicit problem, let X be the solution of the equation

$$F(A, X) = 0.$$

Here Φ is the supporting function, which is locally Lipschitz continuous and satisfies $F(B, \Phi(B)) = 0$ for all B from a neighborhood of the nominal data A, see Chapter 4. We set

$$\delta_X := \|\delta X\| = \|\Psi(A, \delta A)\|,$$

or $\delta_X = \delta_X(\delta A)$, denoting explicitly the dependence of the quantity δ_X only on δA for a fixed value of A.

We recall that here δA must belong to the domain $\mathcal{E}_A$, which is the set of all E such that $A + \delta A \in \mathcal{A}$ for all δA with $\|\delta A\|_g \leq \|E\|_g$, where

$$\|A\|_g := [\|A_1\|, \ldots, \|A_r\|]^\top \in \mathbb{R}^r_+.$$

Suppose that we have a perturbation bound

$$\delta_X \leq f(\|\delta A\|_g), \ \|\delta A\|_g \in \mathcal{D}, \tag{7.1}$$

where the domain $\mathcal{D} \subset \mathbb{R}^r_+$ contains a set

$$\{z \in \mathbb{R}^r : 0 \leq z_i \leq \rho_i\}$$

of positive measure ($\rho_i > 0$ for all $i = 1, \ldots, r$), and let

$$\omega(\delta) := \max\{\delta_X(\delta A) : \|\delta A\|_g \preceq \delta\} \tag{7.2}$$

be the maximal norm of the perturbations in the solution for perturbations in the data δA, varying over the generalized ball

$$\mathcal{B}_\delta = \{E : \|E\|_g \preceq \delta\}.$$

Definition 7.1 *A perturbation $\delta A = (\delta A_1, \ldots, \delta A_r)$ in the data A is called* full *if all δA_i are nonzero.*

Now we are in a position to define our first concept of exactness of a perturbation bound.

Definition 7.2 *The bound (7.1) is said to be* asymptotically sharp *if there exists* $E \in \mathbb{R}_+^r$ *such that*

$$\delta_X(\varepsilon E) = f(\varepsilon \|E\|_g) + o(\varepsilon), \ \varepsilon \to 0.$$

We note that for any $E \in \mathcal{D}$ there exists $\varepsilon_0 > 0$ such that $\varepsilon E \in \mathcal{D}$ for all $\varepsilon \in [0, \varepsilon_0]$.

Thus, asymptotical sharpness is a property that is connected to the existence of at least one infinitesimal one-parametric family of full perturbations $\{\varepsilon E\}$, $\varepsilon \to 0$, for which the bound (7.1) is asymptotically equivalent to the maximum possible perturbation (7.2) in the solution. More precisely, an asymptotically sharp bound is asymptotically equivalent to the actual perturbation for the given family of full perturbations in the sense that

$$\lim_{\varepsilon \to +0} \frac{f(\varepsilon \|E\|_g)}{\delta_X(\varepsilon E)} = 1.$$

A good perturbation bound *should be* asymptotically sharp, otherwise it may be substantially improved.

Example 7.3 For a scalar problem $x = \varphi(a)$ with φ differentiable at a, the chopped, condition number based bound is $|\delta x| \leq |\varphi'(a)| \, |\delta a|$. For $\varphi(a) = a^2$ and $a = 0$ this bound reduces to $\delta x = 0$ which is not true for all $\delta a \neq 0$. $\diamondsuit$

If we consider bounds which are asymptotically equivalent to the maximal perturbation in the solution for *all* infinitesimal perturbations, then we come to the concept of asymptotic exactness.

Definition 7.4 *The bound (7.1) is said to be* asymptotically exact *if*

$$\omega(\delta) = f(\delta) + o(\|\delta\|), \ \delta \to 0.$$

Asymptotically exact bounds are asymptotically equivalent to the maximum perturbation in the solution for all infinitesimal families $\{E\}$, $E \to 0$, of perturbations.

Of course, the most desirable property of a bound is to be exact in the sense of the following definition.

Definition 7.5 *The bound (7.1) is said to be* exact *if* $\mathcal{D} = \Omega$ *and* $f = \omega$.

Obviously, nothing more can be achieved in the norm-wise perturbation analysis than an exact bound. And, as may be expected, exact bounds are available only in rare cases.

Example 7.6 For the scalar problem $x = a^2$ the exact bound is $f(\delta) = \delta(2|a| + \delta)$. $\diamondsuit$

The only nontrivial bound in this book, that is proven to be exact, is that for the linear matrix equation $AX = C$, see Chapter 9.

Some perturbation bounds known in the literature have the property of attainability which we define as follows. Denote by $\mathcal{D}_+ \subset \mathcal{D}$ the set of all $\delta \in \mathcal{D}$ with $\delta \succ 0$.

Definition 7.7 *The bound (7.1) is said to be* attainable *if there exists a manifold* $\mathcal{M} \subset \mathcal{D}_+$ *of dimension* $\dim(\mathcal{M}) = r - 1$, *such that* $f(\delta) = \omega(\delta)$ *for* $\delta \in \mathcal{M}$.

Often attainable bounds are not even asymptotically sharp. In turn, an asymptotically exact bound may not be attainable. The next two examples of scalar linear equations illustrate these concepts.

Example 7.8 Consider the scalar equation

$$ax = c, \ a \neq 0$$

with solution $x = c/a$, and let $\delta_c := |\delta c|$, $\delta_a := |\delta a|$ and $\delta_x := |\delta x|$ be the absolute perturbations in c, a and x. For $\delta a \neq -a$ we have

$$\delta x = \frac{c + \delta c}{a + \delta a} - \frac{c}{a} = \frac{\delta c - x\delta a}{a + \delta a}.$$

Hence, the maximum absolute perturbation in x is

$$\omega(\delta) = \frac{\delta_c + |x|\delta_a}{|a| - \delta_a}$$

and the domain Ω for $\delta = [\delta_c, \delta_a]^\top \in \mathbb{R}^2_+$ is $\mathbb{R}_+ \times [0, |a|)$. Consider the following expression in δ, depending on two parameters $\alpha \geq 1$ and $\beta \geq 0$,

$$f_{\alpha,\beta}(\delta) := \frac{\alpha(\delta_c + |x|\delta_a)}{|a| - \beta\delta_a}.$$

We have five possible cases.

1. If $\alpha = 1$ and $\beta < 1$, then the inequality $\delta_x \leq f_{1,\beta}(\delta)$ may not hold and hence, $f_{1,\beta}(\delta)$ is not a bound in the strict sense.

2. If $\alpha = \beta = 1$, then the bound is exact and hence, asymptotically sharp, asymptotically exact and attainable.

3. If $\alpha = 1$ and $\beta > 1$, then the bound is asymptotically exact and hence, asymptotically sharp, but not exact and not attainable. Here $\mathcal{D} = \mathcal{R}_+ \times [0, |a|/\beta)$ is a proper subset of Ω.

4. If $\alpha > 1$ and $\beta < 1$, then the bound is not asymptotically sharp (and hence not asymptotically exact and not exact), but it is attainable. In this case it is valid in the domain $\mathcal{D} = \mathcal{R}_+ \times [0, a_0]$, where $a_0 := (\alpha - 1)|a|/(\alpha - \beta)$. The manifold $\mathcal{M}$ (see Definition 7.7) here is $\mathbb{R}_+ \times \{a_0\}$.

5. If $\alpha > 1$ and $\beta \geq 1$, then the bound has none of the pleasant properties from Definitions 7.2 – 7.5 and 7.7 but is nevertheless rigorous.

In Figure 7.1 we compare the exact quantity ω with the bound from case 4. with $|a| = 1$, $x = 1$ and $\alpha = 2$, $\beta = 0$ in the 3-dimensional space of parameters $\delta_1 = \delta_c$, $\delta_2 = \delta_a$ and f. After the intersection of the surface $\omega = (\delta_c + \delta_a)/(1 - \delta_a)$ with the plane $f = 2(\delta_c + \delta_a)$ the expression for f is not a rigorous bound. $\diamondsuit$

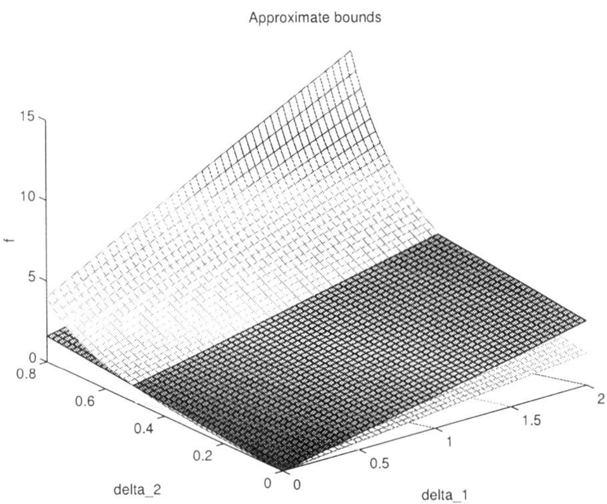

Figure 7.1: An attainable bound which is not asymptotically sharp

The next example shows that a bound may be asymptotically sharp without being asymptotically exact.

Example 7.9 Consider the equation from Example 7.8 together with the bound

$$f(\delta) := \sqrt{1 + x^2}\,\frac{\sqrt{\delta_c^2 + \delta_a^2}}{|a| - \delta_a}\,.$$

This bound is defined in the set $\mathcal{D} = \Omega$ but is not asymptotically exact. At the same time it is asymptotically sharp and attainable. Indeed, we have $f(\delta) = \omega(\delta)$ at the one-dimensional manifold $\mathcal{M}$, defined via $\delta_a = |x|\delta_c < |a|$. In Figure 7.2 we show the exact quantity ω and the bound f for $|a| = 1$ and $x = 1$. $\diamondsuit$

We will show that the concepts of asymptotical sharpness, asymptotical exactness, exactness and attainability are applicable effectively to general linear and nonlinear matrix equations (as well as to linear and nonlinear operator equations in abstract spaces) and, in particular, to the polynomial and fractional-polynomial equations that arise in control theory.

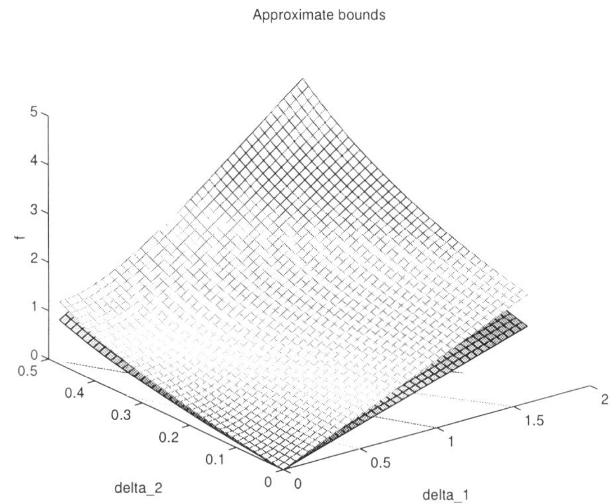

Figure 7.2: An attainable asymptotically sharp bound which is not asymptotically exact

7.3 Conservativeness of "worst case" bounds

Consider a rigorous perturbation bound

$$\delta_X \leq f(\delta),\ \delta \in \mathcal{D}$$

for the problem $X = \Phi(A)$, where $X \in \mathcal{X}$, $A \in \mathcal{A}$, $\delta_X = \|\delta X\|$,

$$\delta X = \Psi(A, \delta A) := \Phi(A + \delta A) - \Phi(A)$$

and $\|\delta A\|_g \preceq \delta$. We recall that A is a matrix collection $(A_1, \ldots, A_r)$.

Since the bound is rigorous, it is also a 'worst case' perturbation bound in the following sense. The bound is valid for all perturbations with $\|\delta A\|_g \preceq \delta$, including those for which the norm-wise perturbation δ_X in the solution is maximal. At the same time, for other perturbations, the actual perturbation δ_X may be much less than the bound $f(\delta)$ predicts (or even zero). Thus, all rigorous perturbation bounds are conservative for certain classes of particular perturbations. This is true even for exact bounds $f(\delta) = \omega(\delta)$, where

$$\omega(\delta) := \max\{\|\Phi(A + \delta A) - \Phi(A)\| : \|\delta A\|_g \preceq \delta\}$$

is the maximal perturbation in δ_X when δA varies over the set of admissible perturbations Ω.

It may happen that for a given class $\mathcal{Q}$ of perturbations δA the perturbation δX in the solution X is zero.

For an explicit problem $X = \Phi(A)$ we consider

$$\mathcal{Q} := \{E \in \mathcal{E}_A : \Psi(A, E) = 0\}.$$

For an implicit problem, defined via an equation $F(A, X) = 0$ with $F : \mathcal{A} \times \mathcal{X} \to \mathcal{Y}$, $\mathcal{Y} \simeq \mathcal{X}$, we consider

$$\mathcal{Q} := \{E \in \mathcal{D} : F(A + E, X) = 0\},$$

or, if we have already constructed the perturbation equation $\delta X = \Pi(\delta A, \delta X)$, then we consider

$$\mathcal{Q} := \{E \in \mathcal{D} : \Pi(E, 0) = 0\}.$$

In the generic case when the problem is regular and the partial Fréchet derivative $F_A(A, X)$ at (A, X) is surjective, then the set $\mathcal{Q}$ is a manifold of dimension $\dim(\mathcal{A}) - \dim(\mathcal{X})$.

Let the matrix collection A be represented as $A = (B, C)$, where B and C are in turn matrix collections. Suppose that we may rewrite the equation $F(A, X) = 0$ in the equivalent form $G(B, X) = H(C)$, where G and H are continuous functions. If B and C are perturbed to $B + \delta B$ and $C + \delta C$ we obtain the perturbed equation

$$G(B + \delta B, X + \delta X) = H(C + \delta C). \tag{7.3}$$

Suppose further that we have the perturbation bound

$$\delta_X \leq f(\beta, \gamma), \ (\beta, \gamma) \in \Omega,$$

provided that $\|\delta B\|_g \preceq \beta$, $\|\delta C\|_g \preceq \gamma$.

If the perturbations δB, δC satisfy the additional relation

$$G(B + \delta B, X) = H(C + \delta C) \tag{7.4}$$

then the perturbed equation (7.3) has a solution $\delta X = 0$ and accordingly $\delta_X = 0$. Hence, nevertheless how good the bound $f(\beta, \gamma)$ is, it may be very conservative in this particular case.

Note that relation (7.4) will be fulfilled if for example, H is the identity operator and

$$\delta C = G(B + \delta B, X) - C.$$

The most simple example here is the linear equation $BX = C$, where B is $m \times m$ and C, X are $m \times n$ matrices, respectively, with B being nonsingular, and $C \neq 0$. Assuming that $\|C^{-1}\|\|\delta B\| < 1$ and

$$\delta C = \delta B X = \delta B B^{-1} C, \tag{7.5}$$

we see that the perturbed equation

$$(B + \delta B)(X + \delta X) = C + \delta C$$

has the unique solution $\delta X = 0$ and hence, $\varepsilon_X = \|\delta X\|/\|X\| = 0$. At the same time, setting $\varepsilon_B = \|\delta B\|/\|B\|$, we have the standard perturbation bound

$$\varepsilon_X \le f(\varepsilon_B) := \frac{2\mathrm{cond}(B)\varepsilon_B}{1 - \mathrm{cond}(B)\varepsilon_B}. \tag{7.6}$$

For ε_B approaching $1/\mathrm{cond}(B)$ the bound $f(\varepsilon_B)$ becomes arbitrarily large, while the exact perturbation is zero.

The observed effect of extreme conservativeness of the perturbation bound (7.6) is not typical (or generic) and is destroyed in any neighborhood of the perturbation $(\delta C, \delta B)$. Note that the relation (7.5) defines an m^2-dimensional subspace $\mathcal{Q}$ in the $m(n+m)$-dimensional linear space of pairs (C, B). If $(\delta C, \delta B) \in \mathcal{Q}$ is such that $B + \delta B$ is close to a singular matrix, then there exists a perturbation $\overline{\delta C}$ such that $(\overline{\delta C}, \delta B) \notin \mathcal{Q}$, the quantity $\|\overline{\delta C} - \delta C\|$ is small, and the relative perturbation in the solution, corresponding to the perturbation $(\overline{\delta C}, \delta B)$, is close to the bound $f(\varepsilon_B)$.

7.4 Notes and references

In the literature there are only few studies in which the exactness of perturbation bounds is analyzed, see e.g. [135].

Chapter 8

General Sylvester equations

In this chapter we present the perturbation analysis for various types of Sylvester equations. We also derive improved first order homogeneous perturbation bounds which are applicable to large classes of nonlinear matrix equations as well.

A linear matrix equation in the form $\mathcal{L}(X) = C$, where $\mathcal{L} : \mathcal{X} \to \mathcal{Y}$ is a linear operator and $\mathcal{X}, \mathcal{Y}$ are (isomorphic) linear finite-dimensional spaces of matrices may be written as a linear vector equation $Lx = c$, where x and c are the vector representations of X and C, and L is the matrix of $\mathcal{L}$. Hence, perturbation bounds for linear matrix equations may be obtained using the perturbation theory of linear vector equations. This approach, however, neglects the specific structure of L, originating from the particular form of $\mathcal{L}$, and may lead to pessimistic bounds. As a result many of the existing perturbation estimates for particular classes of linear matrix equations may be improved and this is true for both norm-wise and component-wise bounds, which in turn may be local or nonlocal.

In this chapter we derive nonlocal nonlinear perturbation bounds for the most general type of linear matrix equations in finite-dimensional matrix spaces.

8.1 Introductory remarks

We begin the analysis with an informal introduction of some basic concepts in the theory of linear matrix equations.

Sylvester equations are linear matrix equations of the form

$$AXB + CXD + \cdots = E, \tag{8.1}$$

where $A, B, \ldots, E$ are given matrices, called *matrix coefficients*, and X is the unknown matrix, or *solution*. The matrices in (8.1) are real or complex, or may have elements from an arbitrary field. It is assumed that the sizes of all matrices are such that the matrix operations in (8.1) are correctly defined. At this stage, without significant loss of generality, the reader may assume that all matrices are real, square and of equal dimension.

The left-hand side of (8.1) defines a *linear operator* $\mathcal{L}$, namely

$$\mathcal{L}(X) = AXB + CXD + \cdots,$$

which allows to write the Sylvester equation briefly as

$$\mathcal{L}(X) = E.$$

We recall that the linearity means that

$$\mathcal{L}(\alpha X + \beta Y) = \alpha\mathcal{L}(X) + \beta\mathcal{L}(Y)$$

for all matrices X, Y and all scalars α, β.

Equation (8.1) may be written also as a linear vector equation. This may be done in many ways. Let for instance the unknown matrix X be represented by its columns x_i, i.e.,

$$X = [x_1, x_2, \ldots].$$

Then the elements of X may be stacked column-wise in a long vector

$$\mathrm{vec}(X) = \begin{bmatrix} x_1 \\ x_2 \\ \vdots \end{bmatrix}.$$

Stacking accordingly both sides of equation (8.1) (or taking the 'vec-operation' in (8.1)) we obtain the linear vector equation

$$L\mathrm{vec}(X) = \mathrm{vec}(E), \tag{8.2}$$

where

$$L = B^\top \otimes A + D^\top \otimes C + \cdots$$

is the *matrix representation or matrix* of $\mathcal{L}$ (see Appendix C for definitions and properties of the Kronecker product $\otimes$).

A natural question that may arise in connection with the vector form (8.2) of equation (8.1) is: *Why is it necessary to develop a general theory (existence, uniqueness, representation of solution, etc.) as well as perturbation theory for linear matrix equations when the corresponding theories for vector equations are well developed and understood?* There are several independent reasons and two of them are discussed below.

First of all, as we have already point out, the vectorization process may make the size of the equation (8.2) inacceptably high. It is in general better to apply methods for solving linear matrix equations of relatively modest order rather than to solve their high order vectorized formulations. Note that if $n \times n$ is the size of the coefficient matrices, then very good methods of computational complexity $O(n^3)$ exist, while the vectorized form in general leads to methods of complexity $O(n^6)$.

Moreover, perturbation theory for vector equations, when applied to (8.2) without taking into account the special structure of the matrix L, will lead to rather weak results, e.g., the corresponding perturbation bounds may be very conservative.

8.2 Motivating examples

Equations of type (8.1) arise in both mathematical theory and engineering practice. We now present some examples of such equations, associated with continuous and discrete time-invariant dynamical systems. When dealing with such systems we need the concepts of stable and convergent matrices.

Definition 8.1 *A square (real or complex) matrix A is called* stable *if its eigenvalues $\lambda_i(A)$ have negative real parts (or if the spectrum,* spect(A)*, of A lies in the open left complex half-plane $\mathbb{C}_-$). The matrix A is said to be* convergent *if its eigenvalues have absolute values less than 1 (or if its spectrum lies in the open unit disc $\mathbb{D}_1$ in the complex plane).*

Next we recall some facts about the spectra of composite matrices, see [157] and Appendix C. Let A and B be $m \times m$ and $n \times n$ matrices, respectively, and α, β be scalars. Then the eigenvalues of the matrix $\alpha I_m + \beta A$ are $\alpha + \beta\lambda_i(A)$, which may be written as

$$\mathrm{spect}(\alpha I_m + \beta A) = \{\alpha\} \oplus \beta\,\mathrm{spect}(A).$$

In turn, the eigenvalues of the matrix $L_c = I_n \otimes A + B^\top \otimes I_m$ that represents the operator

$$X \mapsto \mathcal{L}_c(X) = AX + XB$$

are $\lambda_i(A) + \lambda_k(B)$, i.e.,

$$\mathrm{spect}(\mathcal{L}_c) = \mathrm{spect}\left(I_n \otimes A + B^\top \otimes I_m\right) = \mathrm{spect}(A) \oplus \mathrm{spect}(B).$$

Finally, the eigenvalues of the matrix $L_c = B^\top \otimes A - I_{mn}$ that represents the operator

$$X \mapsto \mathcal{L}_d(X) = AXB - X$$

are $\lambda_i(A)\lambda_k(B) - 1$, i.e.,

$$\mathrm{spect}(\mathcal{L}_d) = \mathrm{spect}\left(B^\top \otimes A - I_{mn}\right) = \mathrm{spect}(A) \otimes \mathrm{spect}(B) \ominus \{1\}.$$

(For definition of operations $\oplus$, $\ominus$ and $\otimes$ with collections see Appendix A.)

Definition 8.2 *A matrix A is called* semi-stable *if its eigenvalues have nonpositive real parts and the eigenvalues with zero real part correspond to linear elementary divisors (i.e., to 1×1 blocks in the Jordan canonical form of A). The matrix A is* semi-convergent *if its eigenvalues have absolute values less than or equal to 1 and the eigenvalues with absolute value 1 correspond to linear elementary divisors.*

Consider the set of two continuous time-invariant real dynamical systems

$$\begin{aligned} x'(t) &= Ax(t),\ x(0) = x_0, \\ y'(t) &= By(t),\ y(0) = y_0, \end{aligned} \tag{8.3}$$

where $t \geq 0$ and $x(t)$ and $y(t)$ are m- and n-dimensional vectors, respectively. The states $x(t)$ and $y(t)$ are determined by

$$x(t) = \exp(At)x_0,\ y(t) = \exp(Bt)y_0,$$

where $\exp(A)$ is the *matrix exponential* of A, defined by the convergent matrix power series

$$\exp(A) = I_m + \frac{A}{1!} + \frac{A^2}{2!} + \cdots = \sum_{i=0}^{\infty} \frac{A^i}{i!}.$$

Let $Q \in \mathbb{R}^{m \times n}$ be a given matrix. An important problem in control theory and stability analysis [157] is to evaluate the integral

$$\varphi(t) := \int_0^t x^\top(s)Qy(s)\mathrm{d}s = x_0^\top \left(\int_0^t \exp(A^\top s)Q\exp(Bs)\mathrm{d}s \right) y_0,$$

as well as the improper integral

$$\lim_{t\to\infty} \varphi(t) = x_0^\top \left(\int_0^\infty \exp(A^\top s)Q\exp(Bs)\mathrm{d}s \right) y_0.$$

We have $\varphi(0) = 0$ and

$$\varphi'(t) = x^\top(t)Qy(t). \tag{8.4}$$

Consider the task of finding a Lyapunov function [158] v of the form

$$v(t) = x_0^\top P y_0 - x^\top(t)Py(t), \tag{8.5}$$

with the matrix P to be determined, and such that $v(0) = 0$ and $v'(t) = \varphi'(t)$ for $t > 0$ and all initial states x_0 and y_0. This would yield $v = \varphi$. Differentiating (8.5) in view of (8.3), we get

$$v'(t) = -(x'(t))^\top Py(t) - x^\top(t)Py'(t) = -x^\top(t)(A^\top P + PB)y(t).$$

The comparison with (8.4) shows that P must satisfy the matrix equation

$$A^\top P + PB + Q = 0. \tag{8.6}$$

This is a Sylvester equation, which is a particular case of (8.1). It has a unique solution if and only if $\lambda_i(A) + \lambda_k(B) \neq 0$ for all i, k, see Appendix C. If, in particular, both matrices A and B are stable, then there exists a unique solution P of (8.6) for every Q. Moreover, since in this case both $x(t)$ and $y(t)$ tend to zero as $t \to \infty$, we get the representation

$$P = \int_0^\infty \exp(A^\top s) Q \exp(Bs) \mathrm{d}s. \tag{8.7}$$

Consider next the matrix differential equation

$$X'(t) = AX(t) + X(t)B + C \tag{8.8}$$

with initial condition $X(0) = X_0$, where the coefficients A, B, C and the solution $X(t)$ are $m \times m$, $n \times n$, $m \times n$ and $m \times n$ matrices, respectively. The solution of (8.8) may be represented as

$$X(t) = \exp(At) X_0 \exp(Bt) + \int_0^t \exp(As) C \exp(Bs) \mathrm{d}s.$$

Equation (8.8) is autonomous (or time-invariant) in the sense that its right-hand side does not depend explicitly on t. If its right-hand side vanishes for some constant matrix, then this matrix will be a solution in the following sense.

Definition 8.3 *A constant $m \times n$ matrix R is a* steady-state solution *(or an* equilibrium state*) of the differential equation (8.8) if the substitution $R = X(t)$ annihilates the right-hand side of the equation, i.e., if R satisfies the Sylvester equation*

$$AR + RB + C = 0.$$

In this case the differential equation has a constant solution $t \mapsto R$.

If the matrix $I_n \otimes A + B^\top \otimes I_m$ is stable then we may represent the solution matrix R as

$$R = \int_0^\infty \exp(As) C \exp(Bs) \mathrm{d}s. \tag{8.9}$$

In this case the differential equation (8.8) is *globally asymptotically stable* in the sense that for every initial state X_0 the solution $X(t)$ tends to R asymptotically, i.e.,

$$\lim_{t \to \infty} X(t) = R.$$

These observations for continuous-time systems have discrete-time counterparts. Let two discrete time-invariant dynamical systems

$$\begin{aligned} x_{k+1} &= Ax_k, \\ y_{k+1} &= By_k, \end{aligned} \tag{8.10}$$

with initial states x_0, y_0 be given, where $k = 0, 1, 2, \ldots$ and x_k and y_k are m- and n-dimensional vectors. The states x_k and y_k are then given by

$$x_k = A^k x_0, \; y_k = B^k y_0.$$

Consider the problem of evaluating the power series

$$\sigma := \sum_{i=0}^{\infty} x_i^\top Q y_i = x_0^\top \left(\sum_{i=0}^{\infty} (A^\top)^i Q B^i \right) y_0 \tag{8.11}$$

for arbitrary choices of the initial states x_0, y_0 and a given matrix Q. Since $\sigma = x_0^\top S y_0$, where

$$S := \sum_{i=0}^{\infty} (A^\top)^i Q B^i, \tag{8.12}$$

we see that σ is convergent for all x_0 and y_0 if and only if the matrix series S is also convergent. If S is convergent then

$$S = Q + \sum_{i=1}^{\infty} (A^\top)^i Q B^i = Q + A^\top \left(\sum_{i=0}^{\infty} (A^\top)^i Q B^i \right) B = Q + A^\top S B.$$

Thus, we have the Sylvester equation

$$A^\top S B - S + Q = 0, \tag{8.13}$$

which is solvable for every Q if and only if $\lambda_i(A)\lambda_k(B) \neq 1$, or

$$1 \notin \operatorname{spect}(A) \otimes \operatorname{spect}(B).$$

Consider finally the matrix difference equation

$$X_{k+1} = A X_k B + C, \; k = 0, 1, 2, \ldots, \tag{8.14}$$

with initial state X_0, where the coefficient matrices A, B, C and the solution matrix X_k are $m \times m$, $n \times n$, $m \times n$ and $m \times n$, respectively. The solution of (8.14) is

$$X_k = A^k X_0 B^k + \sum_{i=0}^{k-1} A^i C B^i.$$

A constant $m \times n$ matrix T is said to be a *steady-state solution* (or an *equilibrium state*) of the difference equation (8.14) if the substitution $T = X_k$ makes its right-hand side equal to T, i.e., if T satisfies the Sylvester equation

$$T = ATB + C.$$

In this case the difference equation has a constant solution $k \mapsto T$. If the matrix $B^\top \otimes A$ is convergent, then the difference equation is globally asymptotically stable, i.e.,

$$\lim_{k \to \infty} X_k = T,$$

and the steady state solution T may be represented as

$$T = \sum_{i=0}^{\infty} A^i C B^i. \tag{8.15}$$

8.3 General linear equations

In this section we discuss general linear matrix (Sylvester) equations

$$\mathcal{L}(P)(X) = C, \tag{8.16}$$

in the unknown matrix X, where the operator

$$\mathcal{L}(P) \in \mathbf{Lin}(p, m, n, q)$$

is given by

$$\mathcal{L}(P)(X) := \sum_{k=1}^{r} P_{2k-1} X P_{2k} := \sum_{k=1}^{r} A_k X B_k$$

and $P := (P_1, \ldots, P_{2r})$. Here $P_{2k-1} = A_k$, $P_{2k} = B_k$ and $C \in \mathbb{F}^{p\times q}$ are given matrices and $mn = pq =: s$. The matrix equation (8.16) is equivalent to the vector equation

$$L(P)\mathrm{vec}(X) = \mathrm{vec}(C),$$

where

$$L(P) := \sum_{k=1}^{r} B_k^{\top} \otimes A_k \in \mathbb{F}^{s\times s} \tag{8.17}$$

is the matrix of the operator $\mathcal{L}(P)$, see [125] and Appendix E. If r is the *Sylvester index* of $\mathcal{L}(P)$, i.e., the minimum number of terms in the representation of $\mathcal{L}(P)$, see again Appendix E, then all $2r$ matrix coefficients $A_1 = P_1, \ldots, B_r = P_{2r}$ in (8.16) are nonzero. Let $P_0 := C$ and

$$D := (P_0, P) = (P_0, P_1, \ldots, P_{2r}) \in \mathbb{F}^{p\times q} \times \mathbb{F}^{p\times m} \times \mathbb{F}^{n\times q} \times \cdots \times \mathbb{F}^{p\times m} \times \mathbb{F}^{n\times q}.$$

Remark. We use two sets of notations P_j and A_k, B_k, C for the matrix coefficients in (8.16). This is done in order to keep the usual notation with coefficients A, B, C, etc., on one hand, and to have unified notations for all coefficients, on the other.

In general some of the matrices in (8.16) may be mutually dependent, for instance $A_1^{\mathrm{H}} = B_2 = A$, $B_1 = A_2 = I$, which gives rise to the Lyapunov equation

$$A^{\mathrm{H}} X + XA = C.$$

Another example is the equation

$$ABXC - XB^2 = C.$$

Since there is a large variety of such combinations, for simplicity, we assume in this chapter that the matrix coefficients A_k, B_k, C are subject to perturbations in such a way that *the matrices that are perturbed are independent.* Thus, we exclude Lyapunov equations which are considered later.

Important cases of (8.16) include the following equations.

(i) Standard matrix linear equations with possibly several right hand sides

$$AX = C. \tag{8.18}$$

(ii) Power Sylvester equations

$$\sum_{i,k=1}^{r,s} \alpha_{ik} A^i X B^k = C, \ \alpha_{ik} \in \mathbb{F}. \tag{8.19}$$

(iii) Continuous-time Sylvester equations

$$AX + XB = C. \tag{8.20}$$

(iv) Discrete-time Sylvester equations

$$AXB - \alpha X = C, \ \alpha \in \mathbb{F}. \tag{8.21}$$

(v) Linear equations in two matrix unknowns, for example

$$AX + YB = C, \tag{8.22}$$

where $A \in \mathbb{F}^{m\times m}$, $B \in \mathbb{F}^{n\times n}$ are matrix coefficients and $X \in \mathbb{F}^{m\times n}$, $Y \in \mathbb{F}^{m\times n}$ are the unknown matrices. Setting

$$U := [X,\, Y] \in \mathbb{F}^{m\times 2n} \ \text{ and } \ V := \begin{bmatrix} X \\ Y \end{bmatrix} \in \mathbb{F}^{2m\times n}$$

we may rewrite (8.22) in two equivalent forms

$$AU \begin{bmatrix} I_n \\ 0 \end{bmatrix} + U \begin{bmatrix} 0 \\ B \end{bmatrix} = C,$$

or

$$[A,\, 0]V + [0,\, I_m]VB = C.$$

(vi) General equations in several matrix unknowns $X_1, \ldots, X_\sigma$, e.g.

$$\sum_{s=1}^{\sigma} \mathcal{L}_s(X_s) := \sum_{s=1}^{\sigma}\sum_{k=1}^{r_s} A_{sk} X_s B_{sk} = C, \tag{8.23}$$

where $\mathcal{L}_s : \mathbb{F}^{m_s \times n_s} \to \mathbb{F}^{p \times q}$. Set $m := m_1 + \cdots + m_\sigma$ and $n := n_1 + \cdots + n_\sigma$. If $X = [X_{ij}] \in \mathbb{F}^{m \times n}$, where $X_{ij} \in \mathbb{F}^{m_i \times n_j}$, is a block-diagonal matrix with $X_{ss} = X_s$, then (8.23) may be written as

$$\sum_{s=1}^{\sigma} \sum_{k=1}^{r_s} \left(e_{\sigma s}^\top \otimes A_{sk}\right) X \left(e_{\sigma s} \otimes B_{sk}\right) = C,$$

where $e_{\sigma s}$ is the s-th column of I_σ.

(vii) General systems of equations in one matrix unknown, e.g.

$$\mathcal{L}_s(X) = C_s, \ s = 1, \ldots, \sigma, \tag{8.24}$$

where $L_s : \mathbb{F}^{m \times n} \to \mathbb{F}^{p_s \times q_s}$. System (8.24) may be written as a single equation in X as in (vi),

$$\sum_{s=1}^{\sigma} \sum_{k=1}^{r_s} \left(e_{\sigma s} \otimes A_{sk}\right) X \left(e_{\sigma s}^\top \otimes B_{sk}\right) = \operatorname{diag}(C_1, \ldots, C_\sigma)$$

(we use the notations from (vi) having in mind that the sizes of the involved matrices are possibly different).

(viii) General systems of equations in several matrix unknowns. These are combinations of (vi) and (vii) and are not considered in detail.

In cases (i) – (iv) we have $A \in \mathbb{F}^{m \times m}$, $B \in \mathbb{F}^{n \times n}$ and $C, X \in \mathbb{F}^{m \times n}$.

In what follows we assume that $mn = pq = s$ and that the operator $\mathcal{L}(P) \in \mathbf{Lin}(p, m, n, q, \mathbb{F})$ in (8.16) is invertible, i.e., that its matrix $L(P) \in \mathcal{R}^{s \times s}$, defined by (8.17), is nonsingular. This is equivalent to the requirement that equation (8.16) has a unique solution

$$X := \mathcal{L}^{-1}(P)(C).$$

8.4 Perturbation problem

In this section we formulate the problem of perturbation analysis for general Sylvester equations.

8.4.1 Norm-wise perturbations

Suppose that the matrices P_j in (8.16) are perturbed as

$$P_j \to P_j + \delta P_j, \ j = 0, 1, \ldots, 2r$$

and that the perturbed equation has again a unique solution (we recall that $P_0 = C$, $P_{2k-1} = A_k$ and $P_{2k} = B_k$). Then the problem is to estimate the perturbation in the solution X as a function of the perturbations δP_j in the data P_j.

We assume that the information about the perturbations δP_j is coded in the norm-wise inequalities

$$\|\delta P_j\|_{\mathrm{F}} \leq \eta_j,\ j = 0, 1, \ldots, 2r, \tag{8.25}$$

where $\eta_j \geq 0$ are given quantities. Denote by

$$\delta P := (\delta P_1, \ldots, \delta P_{2r}) = (\delta A_1, \ldots, \delta B_r)$$

and

$$\delta D := (\delta P_0, \delta P_1, \ldots, \delta P_{2r}) = (\delta C, \delta A_1, \ldots, \delta B_r)$$

the perturbations in the matrix collections P and D. Then inequalities (8.25) may be written as

$$\|\delta D\|_g \preceq \eta := [\eta_0, \eta_1, \ldots, \eta_{2r}]^\top \in \mathbb{R}_+^{2r+1},$$

where

$$\|D\|_g = [\|P_0\|_{\mathrm{F}}, \|P_1\|_{\mathrm{F}}, \ldots, \|P_{2r}\|_{\mathrm{F}}]^\top \in \mathbb{R}_+^{1+2r}.$$

If some matrix P_j is not perturbed, then we set the corresponding bound η_j to zero. However, often it it more convenient to deal only with the matrices P_j that are subject to (nonzero) perturbations. Suppose that these are the matrix coefficients

$$P_{j_1}, P_{j_2}, \ldots, P_{j_\rho},$$

where $0 \leq j_1 < \cdots < j_\rho \leq 2r$. Set

$$J := \{j_1, \ldots, j_\rho\} \subset \{0, 1, \ldots, 2r\},$$

$$E_k := P_{j_k},\ \delta_k := \eta_{j_k},\ k = 1, \ldots, \rho.$$

and

$$E := (E_1, \ldots, E_\rho),\ \delta := [\delta_1, \ldots, \delta_\rho]^\top \in \mathbb{R}_+^r.$$

The vectors η and δ are connected by the relations

$$\eta = R\delta,\ \delta = R^\top \eta,$$

where the permutation matrix

$$R = [r_{pq}] \in \mathbb{R}^{(1+2r)\times\rho},\ p = 0, 1, \ldots, 2r,\ q = 1, \ldots, \rho,$$

satisfies $R^\top R = I_\rho$, and the element r_{pq} is equal to 1 if $q = j_p$ and is zero otherwise.

Example 8.4 Consider the equation $AX + XB = C$ which corresponds to $D = (C, A, I, I, B)$. In principle it is possible to perturb all five matrix coefficients in D and we have $\|\delta P_j\|_{\mathrm{F}} \leq \eta_j$, $j = 0, 1, \ldots, 4$. If, however, only the matrices C, A, and

B are perturbed, then we have $\rho = 3$, $\eta_2 = \eta_3 = 0$, $E = (E_1, E_2, E_3) = (C, A, B)$ and $\delta_1 = \eta_0$, $\delta_2 = \eta_1$, $\delta_3 = \eta_4$. The matrix R here is

$$R = \begin{bmatrix} 1 & 0 & 0 \\ 0 & 1 & 0 \\ 0 & 0 & 0 \\ 0 & 0 & 0 \\ 0 & 0 & 1 \end{bmatrix}.$$

$\diamondsuit$

Let

$$\mathcal{P}(\delta) := \{\delta E : \|\delta E\|_g \preceq \delta\}$$

be the set of allowed perturbations in E, where

$$\|E\|_g = [\|E_1\|_{\mathrm{F}}, \ldots, \|E_\rho\|_{\mathrm{F}}]^\top \in \mathbb{R}^\rho_+.$$

The perturbed Sylvester equation is obtained by replacing E with $E + \delta E$, which results in $P \to P + \delta P$ and $D \to D + \delta D$, i.e.,

$$\mathcal{L}(P + \delta P)(Y) = C + \delta C. \tag{8.26}$$

We are interested in conditions which guarantee that the perturbed operator $\mathcal{L}(P+\delta P)$ is also invertible.

Denote by $\Omega \subset \mathbb{R}^{2\rho}_+$ the set of all δ such that the matrix

$$L(P + \delta P) = L(P) + \delta L$$

is nonsingular for every $\delta P \in \mathcal{P}_P(\delta)$, where

$$\delta L = \delta L(\delta P) := \sum_{k=1}^{r} \left(\delta B_k^\top \otimes A_k + B_k^\top \otimes \delta A_k + \delta B_k^\top \otimes \delta A_k\right).$$

In general the set Ω has a very complicated structure and we need a simpler criterion to decide how small δ must be for $L(P + \delta P)$ to be nonsingular.

The minimum singular value $\sigma_{\min}(L(P)) > 0$ of the matrix $L(P)$ may be interpreted as the distance from $\mathcal{L}(P)$ to the set of noninvertible operators.

Let Ω_l be the set of all δ, satisfying the inequality

$$l(\delta) := \sum_{k=1}^{r} (\|A_k\|_2\eta_{2k} + \|B_k\|_2\eta_{2k-1} + \eta_{2k-1}\eta_{2k}) < \sigma_{\min}(L(P))$$

(we recall that $\eta = R\delta$). The next proposition, based on the fact that

$$\|\delta L\|_2 \le l(\delta),$$

shows that the set Ω_l may indeed be used in the perturbation analysis.

Proposition 8.5 *The inclusion $\Omega_l \subset \Omega$ is valid.*

Proof. For $\delta \in \Omega_l$ the perturbed operator $\mathcal{L}(P+\delta P)$ is invertible. If $\varepsilon \preceq \delta$ then $l(\varepsilon) \le l(\delta) < \sigma_{\min}(L(P))$ and $\varepsilon \in \Omega$. Hence, $\delta \in \Omega$ and $\Omega_l \subset \Omega$ as claimed. □

For $\delta \in \Omega_l$ the perturbed equation (8.26) has a unique solution $X + \delta X$, where

$$\delta X = \delta X(\delta E) := \mathcal{L}^{-1}(P + \delta P)(C + \delta C) - \mathcal{L}^{-1}(P)(C).$$

Let

$$\delta_X = \delta_X(\delta E) := \|\delta X(\delta E)\|_{\mathrm{F}}$$

be the absolute norm-wise perturbation in the solution. For $\delta \in \Omega_l$ denote by

$$\omega(\delta) := \max\{\delta_X(\delta E) : \delta E \in \mathcal{P}(\delta)\}$$

the maximum of $\delta_X(\delta E)$ over the set $\mathcal{P}(\delta)$ of allowed perturbations δE. We note that $\omega(\delta)$ is well defined, since according to the Weierstraß theorem [96] the function $\delta_X(\cdot)$ reaches its maximum in the compact set $\mathcal{P}(\delta)$ for some $\delta E = G \in \mathcal{P}(\delta)$, i.e., $\omega(\delta) = \delta_X(G)$.

The expression $\omega(\delta)$ may be represented as $\omega_1(\delta) + \omega_2(\delta)$, where $\omega_i(\delta) = O(\|\delta\|^i)$ for $\delta \to 0$. The function $\omega_1(\cdot)$ is first order homogeneous, i.e., $\omega_1(\lambda\delta) = \lambda\omega_1(\delta)$ for $\lambda \ge 0$.

Definition 8.6 *Denote by*

$$G = G(\delta) = (G_1(\delta), \ldots, G_\rho(\delta))$$

the collection of perturbations, which produce the maximum $\omega(\delta)$ of $\delta_X(\cdot)$ according to $\omega(\delta) = \delta_X(G(\delta))$. The perturbation $G = G(\delta)$ is said to be an extremal perturbation.

Let s_k be the number of entries of E_k and set $s := s_1 + \cdots + s_\rho$. We may consider $G = G(\delta)$ as a parametrization of a s-dimensional manifold $\mathcal{S} \subset \mathbb{R}^s$:

$$\mathcal{S} := \{G(\delta) : \delta \in \Omega_l\} = \delta_X^{-1}(\omega(\delta)).$$

The manifold $\mathcal{S}$ is the pre-image of the set of maximal values of the function $\delta_X(\cdot) : \mathcal{P}_E(\delta) \to \mathbb{R}_+$ for all $\delta \in \Omega_l$. In general $\|G_{E_k}(\delta)\|_{\mathrm{F}} = \delta_k$ and $\mathcal{S}$ has several connected components.

It is usually a difficult task to construct the true bound $\omega : \Omega \to \mathcal{R}_+$. So we approach two easier problems.

- The first problem is to find a simple domain $\mathcal{D} \subset \Omega$ of positive measure in $\mathbb{R}^\rho$ such that for every $\delta \in \mathcal{D}$ and for all $\delta P \in \mathcal{P}_P(\delta)$ the perturbed operator $\mathcal{L}(P + \delta P)$ is invertible.

- The second problem is to derive a bound

$$\delta_X \leq f(\|\delta D\|_g),\ \delta D \in \mathcal{P}_D(\delta),$$

or, if we have a perturbation δD in the set $\mathcal{P}^{\mathcal{V}}(\delta) := \{\delta E : \|\delta E\|_g = \delta\} \subset \mathcal{P}(\delta)$, a bound

$$\delta_X \leq f(\delta),\ \delta \in \mathcal{D}, \tag{8.27}$$

where $f(\cdot) : \mathcal{D} \to \mathbb{R}_+$ is a continuous function, nondecreasing in each of its arguments and satisfying $f(\delta) = O(\|\delta\|)$, $\delta \to 0$. In most applications the function f is piece-wise analytic. Also, in some cases it is not differentiable at $0 \in \mathcal{D}$.

The quantity $f(\delta)$ in (8.27) is only an upper bound for the true maximal perturbation $\omega(\delta)$. One of the important tasks here is to determine how close the expressions $f(\delta)$ and $\omega(\delta)$ are, and, in particular, to decide whether $f(\delta)$ is equivalent to $\omega(\delta)$ asymptotically in the sense of the definitions from Chapter 7.

For some classes of Sylvester equations it is possible to prove that $f(\delta)$ is exactly equal to $\omega(\delta)$. Since to find $\omega(\delta)$ for a general Sylvester equation in the form (8.16) is a hopeless task, we first consider a bound $f(\delta)$ and then try to determine a class of equations for which this bound is asymptotically sharp or exact in the sense of Definitions 7.2, 7.4 .

Note that the bound (8.27) is nonlocal, since it is valid for a finite (although possibly small) domain $\mathcal{D}$ for δ. In contrast, local bounds are valid only asymptotically, i.e., for $\delta \to 0$.

In what follows we consider mainly absolute perturbation bounds, since relative bounds may be obtained from the absolute ones by simple substitution, namely

$$\varepsilon_X := \frac{\|\delta X\|_{\mathrm{F}}}{\|X\|_{\mathrm{F}}} \leq \frac{f\left(\|D_1\|_{\mathrm{F}}\varepsilon_1, \ldots, \|D_\rho\|_{\mathrm{F}}\varepsilon_\rho\right)}{\|X\|_{\mathrm{F}}},$$

where the quantities $\varepsilon_k := \delta_k/\|D_k\|_{\mathrm{F}}$ are the relative perturbations in the matrix coefficients D_k.

8.4.2 Component-wise perturbations

Another task of the perturbation analysis for matrix equations is to find component-wise perturbation bounds. In our case it is convenient to work with the vector representations

$$v_k := \mathrm{vec}(E_k),\ \delta v_k := \mathrm{vec}(\delta E_k)$$

of the matrices E_k and their perturbations δE_k.

Let

$$\Delta_k \in \mathcal{H}_k,\ k = 1, \ldots, \rho,$$

be nonzero vectors with nonnegative entries of the size of the corresponding perturbed vectors v_k, where $\mathcal{H}_k$ is one of the spaces $\mathbb{R}_+^{pq}$, $\mathbb{R}_+^{pm}$ or $\mathbb{R}_+^{nq}$. Suppose that the perturbations δv_k in v_k satisfy the component-wise inequalities $|\delta v_k| \preceq \Delta_k$. Set

$$v := (v_1, \ldots, v_\rho)$$

and

$$\Delta := (\Delta_1, \ldots, \Delta_\rho), \ |v| := (|v_1|, \ldots, |v_\rho|), \ |\delta v| := (|\delta v_1|, \ldots, |\delta v_\rho|)$$

(we recall that $|x|$ is the vector whose elements are the absolute values of the elements of the vector x). We write $v \preceq \Delta$ if $v_k \preceq \Delta_k$ for $k = 1, \ldots, \rho$.

Let

$$\Omega_\Delta \subset \mathcal{H} := \mathcal{H}_1 \times \cdots \times \mathcal{H}_\rho$$

be the set of all nonnegative collections Δ, such that the operator $\mathcal{L}(P + \delta P)$ is invertible for all δE with $|\delta v| \preceq \Delta$. Then the problem is to derive a bound for $|\delta X|$ as a function of $|\delta v|$,

$$|\mathrm{vec}(\delta X)| \preceq F(|\delta v|), \tag{8.28}$$

or, if $\delta E = \Delta$, then

$$|\mathrm{vec}(\delta X)| \preceq F(\Delta), \ \Delta \in \mathcal{D}_\Delta \subset \Omega_\Delta. \tag{8.29}$$

Here $\mathcal{D}_\Delta$ is the domain of applicability of the bound F. The function F takes values in $\mathbb{R}_+^{mn}$, satisfies $F(0) = 0$ and each of its components is a nondecreasing function of the elements of the vectors Δ_k.

Bounds of type (8.28) or (8.29) may also be local and nonlocal. The concepts of asymptotical sharpness, exactness and attainability for component-wise perturbation bounds are analogous to those for norm-wise bounds from Definitions 7.2 – 7.5 and 7.7 from Chapter 7.

8.4.3 Other perturbations

The bounds considered in the previous section may be viewed as forward in the sense that they solve the problem: given a perturbation in the data, find a bound for the perturbation in the solution. There are also other types of bounds (e.g., backward perturbation bounds [201, 101, 190]), which are connected with an approximate solution $\widehat{X}$ of equation (8.16). Note that $\widehat{X}$ may be the solution that is computed in finite precision arithmetic.

Here we may formulate the following two problems.

P1 Find a bound for $\|X - \widehat{X}\|_{\mathrm{F}}$ or $\mathrm{vec}(|X - \widehat{X}|)$.

P2 Determine the minimal perturbation δE in the data, which gives rise to the approximate solution $\widehat{X}$.

Both problems are connected. If, for instance, δE is the minimal perturbation in Problem P2, then

$$\|X - \widehat{X}\|_{\mathrm{F}} \leq f(\|\delta E\|_g)$$

and

$$\mathrm{vec}(|X - \widehat{X}|) \preceq F(|\delta v|)$$

where f and F are the bounding functions, defined in the previous section.

A direct solution of Problem P1 may be more efficient and more useful in practical computations, see for example [99]. Note that we cannot simply calculate the difference $X - \widehat{X}$, since in most cases we do not know the exact solution X.

Problem P1 is solved immediately. We rewrite (8.16) as

$$\mathcal{L}(P)(X - \widehat{X}) = \widehat{R}, \tag{8.30}$$

where $\widehat{R} := C - \mathcal{L}(P)(\widehat{X})$ is the residual, corresponding to the approximate solution $\widehat{X}$. Note that $\widehat{R}$ is an easily computable quantity. Furthermore, we have

$$\begin{aligned} \mathrm{vec}(|X - \widehat{X}|) &\preceq \left| L^{-1}(P)\mathrm{vec}(\widehat{R}) \right|, \\ \|X - \widehat{X}\|_{\mathrm{F}} &\leq \left\| L^{-1}(P)\mathrm{vec}(\widehat{R}) \right\|_2. \end{aligned}$$

Of course, for computing $X - \widehat{X}$ it is not necessary to form and invert L^{-1} but to solve equation (8.30) via an appropriate solver.

Problem P2 may be further developed as follows.

Definition 8.7 *Let a vector* $0 \prec w \in \mathbb{R}^{\rho}_{+}$ *be given. The quantity*

$$\varepsilon(\widehat{X}, w) := \min\left\{\alpha \geq 0 : \mathcal{L}(P + \delta P)(\widehat{X}) = C + \delta C,\ \|\delta E\|_g \preceq \alpha w\right\}$$

is said to be the norm-wise backward equivalent perturbation *(or* error*) of* $\widehat{X}$ *relative to* w.

We note that usually the elements w_k of w are taken as $w_k = \|\delta E_k\|_{\mathrm{F}}$ and in this case $\varepsilon(\widehat{X}, w)$ is called the *relative norm-wise error* of $\widehat{X}$. This concept was introduced and analyzed in [190] for standard linear equations $Ax = b$.

Let the collection $W = (W_1, \ldots, W_\rho) \in \mathcal{H}$ be given with $W \succeq 0$, $W_k \neq 0$, $k = 1, \ldots, \rho$.

Definition 8.8 *The quantity*

$$\varepsilon(\widehat{X}, W) := \min\left\{\alpha \geq 0 : \mathcal{L}(P + \delta P)(\widehat{X}) = C + \delta C,\ |\delta v| \preceq \varepsilon W\right\}$$

is said to be the component-wise backward equivalent perturbation *(or* error*) of* $\widehat{X}$ *relative to* W.

8.5 Local perturbation analysis

In this section we consider local perturbation bounds for general Sylvester equations. These are bounds in which only the principal term of $f(\delta)$ of order $O(\|\delta\|)$ is known. If we take only this term as a bound, it will be valid only asymptotically, i.e., for $\delta \to 0$. An application of such bounds for possibly small but nevertheless finite perturbations δ requires additional justification (e.g., an estimate of the neglected second and higher terms).

The perturbation bounds that we present are generically asymptotically sharp. They are then fully incorporated into the nonlocal, nonlinear perturbation bounds derived later. The local bounds are in general not linear but first order homogeneous functions of the vector of absolute perturbations δ. In particular, these bounds are not formulated in terms of condition numbers. The reason is that linear local bounds, based on condition numbers, may wipe out the effect of "useful" cancellations among some perturbed quantities, and thus may be more conservative than other first order homogeneous bounds.

8.5.1 Norm-wise bounds

In this section we study a slightly more general perturbation problem. Suppose that every matrix coefficient P_j is a linear combination of the matrices $E_1, \ldots, E_\rho$ such that

$$\mathrm{vec}(P_j) = \sum_{k=1}^{\rho} T_{jk} v_k, \; j = 0, 1, \ldots, 2r, \tag{8.31}$$

where $v_k := \mathrm{vec}(E_k)$. Obviously, this includes our previous statement of the perturbation problem as a particular case. Indeed, taking $T_{pq} = I$ if $q = j_p$ and $T_{pq} = 0$ otherwise, we get $P_{j_k} = E_k$.

Suppose that we have a perturbation δE in the set $\mathcal{P}^{\mathcal{V}}(\delta) := \{\delta E : \|\delta E\|_g = \delta\}$. The perturbed equation (8.26) may then be written in the equivalent form

$$\mathcal{L}(P)(\delta X) = \mathcal{M}_1(\delta E) + \mathcal{M}_2(\delta X, \delta E), \tag{8.32}$$

where $\mathcal{M}_1$ contains first order and $\mathcal{M}_2$ contains second and higher order terms in δX, δE, namely

$$\begin{aligned}\mathcal{M}_1(\delta E) &:= \delta C - \sum_{k=1}^{r} (\delta A_k X B_k + A_k X \delta B_k), \qquad (8.33)\\ \mathcal{M}_2(Z, \delta E) &:= -\sum_{k=1}^{r} (\delta A_k Z B_k + A_k Z \delta B_k + \delta A_k (X+Z) \delta B_k).\end{aligned}$$

Having in mind the dependence (8.31) of the perturbations δP_j on δE_k, it follows from (8.32) and (8.33) that

$$\begin{aligned}\operatorname{vec}[\delta X] &:= \Lambda\operatorname{vec}[\mathcal{M}_1(\delta E)] + O(\|\delta\|^2) = \sum_{k=1}^{\rho} N_k \delta v_k + O(\|\delta\|^2) \\ &= N\operatorname{vec}(\delta v) + O(\|\delta\|^2),\ \delta \to 0,\end{aligned} \tag{8.34}$$

where $\Lambda := L^{-1}$. The matrices N_k and N are determined by

$$\begin{aligned} N_k &:= \Lambda\left(T_{0k} - \sum_{i=1}^{\rho}(R_{2i-1}T_{2i-1,k} + R_{2i}T_{2i,k})\right), \\ N &:= [N_1, \ldots, N_\rho], \end{aligned} \tag{8.35}$$

where

$$\begin{aligned} R_{2i-1} &:= (B_i X)^\top \otimes I_p \in \mathbb{F}^{s\times mp}, \\ R_{2i} &:= I_q \otimes (A_i X) \in \mathbb{F}^{s\times nq}. \end{aligned} \tag{8.36}$$

Relation (8.34) makes it possible to obtain different local estimates

$$\delta_X \le \operatorname{est}(\delta) + O(\|\delta\|^2),\ \delta \to 0,$$

where $\operatorname{est}(\delta)$ is a bound for $\|N\delta v\|_2$ when $\|\delta E\|_g \preceq \delta$. To do this, one has to estimate the maximum of $\|N\delta v\|_2$ via the maximization problem

$$\|Nu\|_2 \to \max,\ u := \left[u_1^\top, \ldots, u_\rho^\top\right]^\top \in \mathbb{F}^s,\ u_k \in \mathbb{F}^{s_k}, \tag{8.37}$$

subject to the constraints $\|u_k\|_2 \le \delta_k$, $k = 1, \ldots, \rho$, or equivalently

$$\|u\|_g := [\|u_1\|_2, \ldots, \|u_\rho\|_2]^\top \preceq \delta\,. \tag{8.38}$$

Since in view of (8.38) the domain for u is compact, the maximization problem (8.37), (8.38) has a solution

$$\omega_1(\delta, N) := \|Nu^0\|_2 = \max\{\|Nu\|_2 : \|u\|_g \preceq \delta\} \tag{8.39}$$

for some u^0 with $\|u^0\|_g \preceq \delta$, which is the desired local bound. Here we write $\omega_1(\delta, N)$ for the principal asymptotic term $\omega_1(\delta)$ (which is of asymptotic order $O(\|\delta\|)$, $\delta \to 0$) of the maximal perturbation $\omega(\delta)$ in the solution in order to indicate its dependence not only on δ but on the matrix N as well.

Proposition 8.9 *The function $\omega_1(\cdot, N) : \mathbb{R}_+^\rho \to \mathbb{R}_+$ in (8.39) is first order homogeneous in the sense that for $\lambda \ge 0$ we have $\omega_1(\lambda\delta, N) = \lambda\omega_1(\delta, N)$.*

Proof. For $\lambda = 0$ the assertion of the theorem is trivial. For $\lambda > 0$ we have

$$\begin{aligned} \omega_1(\lambda\delta, N) &= \max\{\|Nu\|_2 : \|u\|_g \preceq \lambda\delta\} \\ &= \lambda\max\left\{\left\|N\frac{u}{\lambda}\right\|_2 : \left\|\frac{u}{\lambda}\right\|_g \preceq \delta\right\} = \lambda\omega_1(\delta, N). \end{aligned}$$

□

Typically the quantity $\omega_1(\delta, N)$ cannot be determined in a closed form. Since $\omega_1(\delta, N)$ is a solution of a large optimization problem (8.37), (8.38) of order s, its numerical computation may be very expensive. To overcome this problem, let us consider the following approximation of this maximization problem which is of order at most $(\rho - 1)$.

Let $\gamma_1 = 1$, $\gamma_k > 0$ for $k = 2, \ldots, \rho$, $\gamma := [\gamma_2, \ldots, \gamma_\rho]^\top \in \mathbb{R}_+^{\rho-1}$ and

$$\Upsilon := \operatorname{diag}\left(I_{s_1}, \gamma_2 I_{s_2}, \ldots, \gamma_\rho I_{s_\rho}\right) \in \mathbb{R}^{s \times s}.$$

Then

$$\|Nu\|_2 = \left\|N\Upsilon^{-1}\Upsilon u\right\|_2 \leq \left\|N\Upsilon^{-1}\right\|_2 \|\Upsilon u\|_2 = \left\|N\Upsilon^{-1}\right\|_2 \sqrt{\sum_{k=1}^{\rho} \gamma_k^2 \|u_k\|_2^2}.$$

Hence, we obtain an optimization problem of order $\rho - 1$,

$$\psi_0(\delta, N) := \min\{\psi(\gamma, \delta, N) : \gamma \succ 0\} \geq \omega_1(\delta, N), \tag{8.40}$$

where $\psi(\gamma, \delta, N) = \psi_1(\gamma, N)\psi_2(\gamma, \delta)$ and

$$\psi_1(\gamma, N) := \left\|\left[N_1, \frac{N_2}{\gamma_2}, \ldots, \frac{N_\rho}{\gamma_\rho}\right]\right\|_2, \quad \psi_2(\gamma, \delta) := \sqrt{\delta_1^2 + \sum_{k=2}^{\rho} \delta_k^2 \gamma_k^2}.$$

These considerations are justified by the following proposition.

Proposition 8.10 *The minimization problem (8.40) has a solution, i.e., there exists $\gamma^0 \succ 0$ such that $\psi_0(\delta, N) = \psi(\gamma^0, \delta, N)$.*

Proof. Denote by $\beta := \|N\|_2 \|\delta\|_2$ the value of $\psi(\gamma, \delta, N)$ for $\gamma_2 = \cdots = \gamma_\rho = 1$. Then the minimization of ψ has to be carried out only for those γ, which satisfy $\psi(\gamma, \delta, N) \leq \beta$. On the other hand for any fixed $i \in \{2, \ldots, \rho\}$ it follows that

$$\begin{aligned} \psi_1(\gamma, N) &\geq \beta_1(\gamma_i, N) := \sqrt{\frac{1}{\|L\|_2^2} + \frac{\|N_i\|_2}{\gamma_i^2}}, \\ \psi_2(\gamma, \delta) &\geq \beta_2(\gamma_i, \delta) := d\sqrt{1 + \gamma_i^2}, \end{aligned}$$

where $d := \min\{\delta_k : k = 1, \ldots, \rho\}$. Since

$$\psi(\gamma, \delta, N) \geq \beta_1(\gamma_i, N)\beta_2(\gamma_i, \delta),$$

we restrict the minimization of ψ to those γ_i for which

$$\beta_1(\gamma_i, N)\beta_2(\gamma_i, \delta) \leq \beta.$$

The last inequality is equivalent to

$$\gamma_i^4 - \left(\left(\frac{\beta^2}{d^2} - \|N_i\|_2^2\right)\|L\|_2^2 - 1\right)\gamma_i^2 + \|L\|_2^2\|N_i\|_2^2 \leq 0.$$

The corresponding algebraic fourth-degree equation has exactly two positive different roots, say $a_i < b_i$. Hence, the fourth-degree inequality for $\gamma_i > 0$ is satisfied if $a_i \leq \gamma_i \leq b_i$ and γ varies in the domain

$$\mathcal{G} := \{\gamma : a_k \leq \gamma_k \leq b_k,\ j = 1, \ldots, \rho - 1\},$$

which is compact. According to a Weierstraß theorem [96] the function $\psi(\cdot, \delta, N)$ reaches its minimum for some $\gamma^0 \in \mathcal{G}$. □

At least three simple local perturbation bounds may be derived for δ_X by solving the maximization problem (8.37), (8.38) approximately. They are functions of the perturbation vector δ and the coefficients matrix N. We combine two of these bounds in order to get a bound, which is relatively tight and asymptotically sharp in particular.

1) Applying several times the triangle inequality to (8.37), (8.38), one obtains

$$\|Nu\|_2 \leq \sum_{k=1}^{\rho} \|N_k\|_2 \|u_k\|_2 \leq \sum_{k=1}^{\rho} \|N_k\|_2 \delta_k.$$

Thus, we get the first bound

$$\delta_X \leq \text{est}_1(\delta, N) + O(\|\delta\|^2), \delta \to 0, \tag{8.41}$$

where

$$\text{est}_1(\delta, N) := \sum_{k=1}^{\rho} K_k \delta_k$$

and $K_k := \|N_k\|_2$ are the *absolute condition numbers* of the equation. This bound is linear in δ.

2) The second bound uses the result from the optimization procedure for determining $\psi_0(\delta, N) = \psi(\gamma^0, \delta, N)$ (see (8.40) and Proposition 8.10), i.e.,

$$\delta_X \leq \psi_0(\delta, N) + O(\|\delta\|^2),\ \delta \to 0.$$

We note that $\psi_0(\delta, N) \leq \psi(\gamma, \delta, N)$ for every choice of $\gamma \succ 0$. If, in particular, we take $\gamma = [1, 1, \ldots, 1]^\top$, then we obtain the second bound

$$\delta_X \leq \text{est}_2(\delta, N) + O(\|\delta\|^2),\ \delta \to 0 \tag{8.42}$$

where

$$\text{est}_2(\delta, N) := \|N\|_2 \|\delta\|_2.$$

This bound is not a linear but a first order homogeneous function in δ.

3) Using the relation

$$\begin{aligned}\|Nu\|_2^2 &= u^{\mathrm{H}}N^{\mathrm{H}}Nu = \sum_{i,k=1}^{\rho} u_i^{\mathrm{H}}N_i^{\mathrm{H}}N_k u_k \\ &\leq \sum_{i,k=1}^{\rho} \left\|N_i^{\mathrm{H}}N_k\right\|_2 \|u_i\|_2\|u_k\|_2 \leq \sum_{i,k=1}^{\rho} \left\|N_i^{\mathrm{H}}N_k\right\|_2 \delta_i\delta_k\end{aligned}$$

we have the third bound

$$\delta_X \leq \mathrm{est}_3(\delta, N) + O(\|\delta\|^2),\ \delta \to 0. \tag{8.43}$$

Here

$$\mathrm{est}_3(\delta, N) := \sqrt{\delta^\top \widehat{N}\delta}$$

and $\widehat{N} = [n_{ik}] \in \mathbb{R}_+^{\rho\times\rho}$ is a matrix with elements

$$n_{ik} := \left\|N_i^{\mathrm{H}}N_k\right\|_2;\ i,k = 1,\ldots,\rho.$$

As in case 2), the bound est_3 is a norm-like function which is first order homogeneous in δ.

We have the following relations between the bounds est_i, $i = 1, 2, 3$.

Theorem 8.11 *The quantity* $\mathrm{est}_3(\delta, N)$ *is bounded by*

$$\sqrt{n_1}\|\delta\|_2 \leq \mathrm{est}_3(\delta, N) \leq \sqrt{\|\widehat{N}\|_2}\|\delta\|_2, \tag{8.44}$$

where

$$n_1 := \min\{n_{ii} : i = 1,\ldots,\rho\}. \tag{8.45}$$

Moreover, both inequalities in (8.44) are achievable.

Proof. The right inequality is obvious. Since the matrix $\widehat{N}$ is symmetric and element-wise nonnegative, then according to the Perron-Frobenius theorem [26], its norm $\|\widehat{N}\|_2$ is an eigenvalue of $\widehat{N}$ and the corresponding eigenvector z may be taken as nonnegative. Choosing $\delta = z$, we see that the equality $\mathrm{est}(\delta, N) = \|\widehat{N}\|_2\|\delta\|_2$ is achievable.

To prove the left inequality in (8.44), suppose that the minimum in (8.45) is achieved for $i = k$, i.e., $n_1 = n_{kk}$, and consider the minimization problem

$$\nu(\delta) := \min\{x^\top \widehat{N}x\}$$

subject to the constraints

$$0 \preceq x,\ \|x\|_2 = \|\delta\|_2,$$

where $x = [x_1, \ldots, x_\rho]^\top \in \mathbb{R}^\rho$. We have $\mathrm{est}_3(\delta, N) \geq \sqrt{\nu(\delta)}$, where equality may be achieved. Furthermore,

$$x_{kk}^2 = \|\delta\|_2^2 - \sum_{i \neq k} x_i^2$$

and hence,

$$x^\top \widehat{N} x = n_{kk}\|\delta\|_2^2 + \sum_{i \neq k}(n_{ii} - n_{kk})x_i^2 + \sum_{i \neq j}\sum_{j} n_{ij}x_i x_j \geq n_{kk}\|\delta\|_2^2$$

which proves the left inequality in (8.44).

Finally, choosing $x_i = 0$ if $i \neq k$ and $x_{kk} = \|\delta\|_2$, we see that $x^\top \widehat{N} x = n_{kk}\|\delta\|_2^2$, which proves that the case of equality $\sqrt{n_1}\|\delta\|_2 = \mathrm{est}_3(\delta, N)$ in (8.44) is achievable.
□

The bound est_1, based on condition numbers, is bounded from below by est_3. Indeed, we have

$$\mathrm{est}_3(\delta, N) \leq \sqrt{\sum_{i,j=1}^{\rho} \|N_i\|_2 \|N_j\|_2 \delta_i \delta_j} = \sum_{i=1}^{\rho} \|N_i\|_2 \delta_i = \mathrm{est}_1(\delta, N).$$

Thus, linear local bounds of type est_1, based on condition numbers, are generally less sharp than first order homogeneous local bounds of type est_3 and eventually est_2. In turn, the bounds $\mathrm{est}_2(\delta, N) = \|N\|_2\|\delta\|_2$ and $\mathrm{est}_3(\delta, N) = \sqrt{\delta^\top \widehat{N} \delta}$ are alternative, i.e., which one is better depends on the particular value of δ, see Proposition 8.12 below.

In case of single perturbations, for example when all perturbations δ_k except one are equal to zero, then all three bounds est_1, est_2 and est_3 coincide. Also, in problems such as $AX = C$ with very little specified structure, estimates in terms of condition numbers produce acceptable results. For general equations of type (8.16) with strongly specified structure, however, estimates based on condition numbers may be pessimistic.

As a result of the local perturbation analysis we have the overall homogeneous local estimate

$$\delta_X \leq \mathrm{est}(\delta, N) + O(\|\delta\|^2), \ \delta \to 0, \tag{8.46}$$

where

$$\begin{aligned} \mathrm{est}(\delta, N) &:= \min\{\mathrm{est}_2(\delta, N), \mathrm{est}_3(\delta, N)\} \\ &= \min\left\{\|N\|_2\|\delta\|_2, \sqrt{\delta^\top \widehat{N} \delta}\right\}. \end{aligned} \tag{8.47}$$

Since the bound (8.46), (8.47) is obtained by taking the minimum of the quantities $\mathrm{est}_2(\delta, N)$ and $\mathrm{est}_3(\delta, N)$, the overall estimate $\mathrm{est}(\cdot, N)$ may not be differentiable for some values of δ. Also, for $\rho > 1$ the function $\mathrm{est}(\cdot, N)$ is not differentiable at $0 \in \mathbb{R}^\rho_+$. As a result we see that $\mathrm{est}(\cdot, N)$ is a piece-wise real analytic function in the domain $\mathbb{R}^\rho_+\backslash\{0\}$.

In the general case we get the following comparison between the bounds $\mathrm{est}_2(\delta, N)$ and $\mathrm{est}_3(\delta, N)$.

Proposition 8.12 *The bounds* $\mathrm{est}_2(\delta, N)$ *and* $\mathrm{est}_3(\delta, N)$ *are alternative, i.e. which one of these expressions is smaller depends on the particular choice of* δ *and* N.

Proof. We have

$$\begin{aligned}
\|N\|_2^2 &= \left\| \begin{bmatrix} N_1^{\mathrm{H}} N_1 & \dots & N_1^{\mathrm{H}} N_\rho \\ \vdots & \ddots & \vdots \\ N_\rho^{\mathrm{H}} N_1 & \dots & N_\rho^{\mathrm{H}} N_\rho \end{bmatrix} \right\|_2 \\
&\leq \left\| \begin{bmatrix} \left\|N_1^{\mathrm{H}} N_1\right\|_2 & \dots & \left\|N_1^{\mathrm{H}} N_\rho\right\|_2 \\ \vdots & \ddots & \vdots \\ \left\|N_\rho^{\mathrm{H}} N_1\right\|_2 & \dots & \left\|N_\rho^{\mathrm{H}} N_\rho\right\|_2 \end{bmatrix} \right\|_2 = \|\widehat{N}\|_2.
\end{aligned}$$

Moreover, the inequality $\|N\|_2^2 < \|\widehat{N}\|_2$ holds for some equations. Since $\widehat{N}$ is nonnegative definite and satisfies $\widehat{N} = \widehat{N}^\top \succeq 0$, then according to the Perron-Frobenius theorem [26] an eigenvector ξ of the matrix $\widehat{N}$, corresponding to its maximum eigenvalue $\|\widehat{N}\|_2$, is nonnegative, and we may choose $\delta = \xi$ (if ξ is not strictly positive we may choose δ to be strictly positive and arbitrarily close to ξ). For this choice of δ we have $\mathrm{est}_2(\delta, N) \leq \mathrm{est}_3(\delta, N)$.

To show that the inequality $\mathrm{est}_2(\delta, N) > \mathrm{est}_3(\delta, N)$ is also possible is more subtle (the inequality $\|N\|_2^2 \geq \sigma_{\min}(N_0)$ is not helpful, since a nonnegative eigenvector, corresponding to the minimum eigenvalue of $\widehat{N}$, may not exist). Assume that $\rho > 1$, since otherwise both bounds are equal to $\|N_1\|_2\delta_1$.

First we show that

$$\|N\|_2^2 > n_1 := \min\left\{\|N_i\|_2^2 : i = 1, \dots, \rho\right\}.$$

Indeed, suppose that the opposite inequality $\|N\|_2^2 \leq n_1$ holds. Since

$$\|N\|_2^2 \geq n_2 := \max\left\{\|N_i\|_2^2 : i = 1, \dots, \rho\right\},$$

this implies $\|N\|_2^2 = n_1 = n_2$. Moreover, for any fixed $k \geq 2$ we have $\|N\|_2 \geq \|[N_1, N_k]\|_2$ and hence, $\|[N_1, N_k]\|_2 = \|N_1\|_2 = \|N_k\|_2$.

Let $N_k = U\Sigma V^{\mathrm{H}}$ be the singular value decomposition of N_k. Then

$$\begin{aligned}
\|[N_1, N_k]\|_2 &= \left\| U^{\mathrm{H}}[N_1, N_k] \begin{bmatrix} I_{s_1} & 0 \\ 0 & V \end{bmatrix} \right\|_2 \\
&= \|[U^{\mathrm{H}} N_1, \Sigma]\|_2 = \|\Sigma\|_2.
\end{aligned}$$

If a is the first row of $U^{\mathrm{H}}N_1$, then

$$\begin{aligned}\|[U^{\mathrm{H}}N_1,\Sigma]\|_2 &\geq \|\mathrm{diag}(1,0)[U^{\mathrm{H}}N_1,\Sigma]\|_2 \\ &= \|[a,\sigma_1(N_k),0]\|_2 = \sqrt{\|a\|_2^2+\sigma_1^2(N_k)} = \sigma_1(\Sigma)\end{aligned}$$

and hence, $a = 0$. Therefore, $N_1 = \Lambda$ must be singular, which is a contradiction and hence, $\|N\|_2^2 > n_1$.

Let k be the index for which $n_1 = \|N_k\|_2^2$, let $\gamma > 0$ be fixed, and let

$$\varepsilon < \frac{\gamma}{\sqrt{\rho-1}}$$

be a small positive parameter. Choosing $\delta_i = \varepsilon$ for $i \neq k$ and $\delta_k = \sqrt{\gamma^2-(\rho-1)\varepsilon^2}$, we get

$$\begin{aligned}\mathrm{est}_2(\delta,N) &= \|N\|_2\gamma, \\ \mathrm{est}_3(\delta,N) &= \sqrt{n_1}\,\gamma + O(\varepsilon),\ \varepsilon\to 0.\end{aligned}$$

Using the inequality $\|N\|_2 > \sqrt{n_1}$, we see that the inequality $\mathrm{est}_2(\delta,N) > \mathrm{est}_3(\delta,N)$ holds. $\square$

For some equations we have $\|N\|_2^2 = \|\widehat{N}\|_2$. In these cases the bound $\mathrm{est}_3(\delta,N)$ is superior to $\mathrm{est}_2(\delta,N)$, i.e., $\mathrm{est}_3(\delta,N) \leq \mathrm{est}_2(\delta,N)$, for all δ. At the same time it follows from the proof of Proposition 8.12, that the opposite is impossible, i.e., for a given equation the inequality $\mathrm{est}_3(\delta,N) \leq \mathrm{est}_2(\delta,N)$ cannot be valid for all δ.

As we have mentioned above, the bound $\mathrm{est}_3(\delta,N)$ is as least as sharp as $\mathrm{est}_1(\delta,N)$ in the sense that $\mathrm{est}_3(\delta,N) \leq \mathrm{est}_1(\delta,N)$ for all δ and N. Going further, it is interesting to see how much better $\mathrm{est}_3(\delta,N)$ can be in comparison with $\mathrm{est}_1(\delta,N)$. The following result shows that the ratio $\mathrm{est}_3(\delta,N)/\mathrm{est}_1(\delta,N)$ is bounded from below by a constant, depending only on ρ (the size of the vector δ). Indeed, we have

$$\mathrm{est}_3(\delta,N) \geq \|\tau\|_2 = \sqrt{\sum_{i=1}^{\rho}\|N_i\|_2^2\delta_i^2},$$

where $\tau \in \mathbb{R}_+^{\rho}$ is a vector with components $\tau_i := \|N_i\|_2\delta_i$. Since here $\mathrm{est}_1(\delta,N) = \|\tau\|_1$, we obtain

$$\frac{\mathrm{est}_3(\delta,N)}{\mathrm{est}_1(\delta,N)} \geq \frac{\|\tau\|_2}{\|\tau\|_1} \geq \frac{1}{\sqrt{\rho}}$$

and thus,

$$\frac{1}{\sqrt{\rho}} \leq \frac{\mathrm{est}_3(\delta,N)}{\mathrm{est}_1(\delta,N)} \leq 1. \tag{8.48}$$

The left equality is reached if $N_i^{\mathrm{H}}N_j = 0$ for $i \neq j$, and $\delta_i = 1/\|N_i\|_2$, while the right one is reached if all δ_i except one are equal to zero.

It is interesting to note that when $N_i^{\mathrm{H}} N_j = 0$, $i \neq j$, then

$$\mathrm{est}_2(\delta, N) = \max\{\|N_i\|_2 : i = 1, \ldots, \rho\}\, \|\delta\|_2$$

and we may have

$$\frac{\mathrm{est}_3(\delta, N)}{\mathrm{est}_2(\delta, N)} = \frac{\min\{\|N_i\|_2 : i = 1, \ldots, \rho\}}{\max\{\|N_i\|_2 : i = 1, \ldots, \rho\}} = \frac{n_1}{n_2}.$$

This equality is reached when all elements of δ except δ_k are equal to zero, where $\|N_k\|_2 \leq \|N_i\|_2$ for $i = 1, \ldots, \rho$. Therefore, the ratio $\mathrm{est}_3(\delta, N)/\mathrm{est}_2(\delta, N)$ may become arbitrarily close to zero. A similar argument shows that the same is valid for $\mathrm{est}_1(\delta, N)/\mathrm{est}_2(\delta, N)$.

The bound (8.46), (8.47) is generically at least asymptotically sharp as shown in the next proposition.

Proposition 8.13 *Let the right singular vector u of the matrix N, corresponding to its maximum singular value $\|N\|_2$, satisfy $\|u\|_g \succ 0$. Then the bound (8.46), (8.47) is asymptotically sharp.*

Proof. Let the perturbation δE be chosen as $\mathrm{vec}(\delta E) = u$. Then

$$\|Nu\|_2 = \|N\|_2 = \mathrm{est}_2(\|u\|_g, N) \geq \mathrm{est}(\|u\|_g, N)$$

and hence, the bound (8.46), (8.47) is asymptotically sharp. □

Since the inequality $\|u\|_g \succ 0$ holds generically, Proposition 8.13 tells us that the bound (8.46), (8.47) is asymptotically sharp generically.

The bound est_1, based on condition numbers, will be asymptotically sharp if there exist $\rho - 1$ constants $\lambda_k > 0$ such that $N_k u_k = \lambda_k N_1 u_1$, where u_j is the right singular value of the matrix N_j, corresponding to its maximum singular value $\|N_j\|_2$. In this case $\lambda_k = \|N_k\|_2 / \|N_1\|_2$ and

$$\|Nu\|_2 = \|N_1\|_2 \left(1 + \sum_{k=2}^{\rho} \lambda_k\right) = \sum_{k=1}^{\rho} \|N_k\|_2.$$

The problem whether the bound (8.46), (8.47) is asymptotically exact is more difficult and will be discussed later for particular classes of Sylvester equations.

Note that chopped local estimates of the form

$$\delta_X \leq \mathrm{est}(\delta, N),$$

obtained from (8.46) by neglecting second and higher order terms in $\|\delta\|$, may underestimate the true perturbation arbitrarily.

Example 8.14 Consider the linear scalar equation $ax = c$ with $a, c > 0$. The chopped local estimate in relative perturbations is

$$\varepsilon_x := \frac{\delta_x}{|x|} \leq \varepsilon_c + \varepsilon_a; \ \varepsilon_c := \frac{|\delta c|}{|c|}, \ \varepsilon_a := \frac{|\delta a|}{|a|},$$

while for $\delta c > 0$ and $-a < \delta a < 0$ the exact relative perturbation bound is

$$\varepsilon_x = \frac{\varepsilon_c + \varepsilon_a}{1 - \varepsilon_a}.$$

With δa approaching $-a$ the chopped bound arbitrarily underestimates the exact perturbation in the solution. $\diamondsuit$

A difficulty that arises in practice is that local estimates, being valid only asymptotically (for $\delta \to 0$) are often used nonlocally, i.e., for fixed values of δ. Even when δ seems small in the sense that the norm of the relative perturbations vector is much smaller than 1, the chopped bound may become useless due to the finite escape of the perturbed solution, namely $\delta_X \to \infty$ as $\delta P \to \widetilde{P} - P$, where $\widetilde{P} \in \partial\Omega$ and $L(\widetilde{P})$ is not invertible.

To apply local estimates rigorously one must find the so called *asymptotic domain* of the bound (see [135]), for which the neglected term $O(\|\delta\|^2)$ can be bounded as $c\|\delta\|$ for some constant c. But to estimate this constant may be as difficult as to find a nonlocal perturbation bound.

8.5.2 Component-wise bounds

A local component-wise bound for δX follows immediately from (8.34) – (8.36). Recalling that $|\mathrm{vec}(\delta E_k)| \preceq \Delta_k$, we have

$$\mathrm{vec}(|\delta X|) \preceq \sum_{k=1}^{\rho} |N_k| \Delta_k + O\left(\|\Delta\|^2\right), \ \Delta \to 0 \qquad (8.49)$$

8.6 Nonlocal perturbation analysis

In this section we present a nonlocal perturbation analysis of Sylvester equations, which gives rigorous nonlocal nonlinear bounds for the perturbation in the solution δ_X as a function of the perturbations in the data δ. A nonlocal perturbation bound is defined in a certain domain $\mathcal{D}$, where it is guaranteed that the perturbed equation still has a (unique) solution.

Nonlocal bounds often have a practical drawback: their domain of applicability may be too small, and they may produce pessimistic results for some equations, overestimating considerably the true perturbed quantities. This is due to the fact that such bounds are aimed at the worst case.

Example 8.15 Consider the perturbed equation (8.26) under a special choice of the perturbation in C, namely

$$\delta C = \mathcal{L}(P + \delta P)(\mathcal{L}^{-1}(P)(C)).$$

This artificial perturbation gives $\delta_X = 0$, but of course the perturbation analysis machinery cannot recognize this and produces its worst case bounds. ◇

We emphasize again that nonlocal perturbation bounds must be at least asymptotically sharp. This means that the first order part of the corresponding nonlocal bound must be asymptotically sharp.

8.6.1 Application of the Banach principle

Nonlocal perturbation analysis of nonsingular linear matrix equations may be done by an application of the Banach fixed point principle.

Consider the operator equation

$$x = \Phi(x, \eta), \tag{8.50}$$

where $\Phi : \mathcal{X} \times \mathcal{H} \to \mathcal{X}$ is a continuous mapping. Here $\mathcal{X}$ is a normed space with norm $\|\cdot\| : \mathcal{X} \to \mathbb{R}_+$ and $\mathcal{H}$ is a normed space with generalized norm $\|\cdot\|_g : \mathcal{H} \to \mathbb{R}_+^\rho$.

Equations of this type arise naturally in perturbation analysis problems, where x is the perturbation in the solution and η is a vector, characterizing the perturbations in the data. It is assumed that $\|\eta\|_g \preceq \delta$, where $\delta \succ 0$ is a given vector. The problem is to find a domain $\mathcal{D} \subset \mathbb{R}_+^\rho$ and a bound $\|x\| \leq f(\delta)$, $\delta \in \mathcal{D}$ with $f(\delta) = O(\|\delta\|)$ for $\delta \to 0$. For this reason we shall refer to Φ as the *equivalent perturbation operator*.

The equivalent perturbation operator is constructed as follows. Consider the (linear or nonlinear) matrix equation

$$F(E, X) = 0,$$

where $F : \mathcal{H} \times \mathcal{X} \to \mathcal{Y}$ is a continuous mapping, X is the unknown matrix and $E \in \mathcal{H}$ is a collection of matrix or vector parameters. Let X be the solution, corresponding to a particular value of E. If the partial Frechét derivative

$$F_X := F_X(E, X) : \mathcal{X} \to \mathcal{Y}$$

of F in X at the point (E, X) is invertible, than the perturbed equation

$$F(E + \delta E, X + \delta X) = 0$$

may be written in the form (8.50), where $x := \delta X$, $\eta := \delta E$ and the equivalent perturbation operator is given by

$$\Phi(x, \eta) = F_X^{-1}(F_X(x) - F(E + \eta, X + x)).$$

Example 8.16 Consider the algebraic Riccati equation

$$F(E, X) := E_1 + E_2X + XE_3 + XE_4X = 0,$$

where the unknown X and the coefficients E_i are square matrices. Then

$$F_X(Z) = (E_2 + XE_4)Z + Z(E_3 + E_4X).$$

Setting $x = \delta X$ and $\eta_i = \delta E_i$ we get the equivalent perturbation operator

$$\begin{aligned}\Phi(x, \eta) &= -F_X^{-1}(\eta_1 + \eta_2X + X\eta_3 + X\eta_4X) \\ &\quad - F_X^{-1}((\eta_2 + X\eta_4)x + x((\eta_3 + \eta_4X) + x(E_4 + \eta_4)x).\end{aligned}$$

$\diamondsuit$

We may rewrite the expression for Φ as $\Phi(x, \eta) = \Phi_1(\eta) + \Phi_2(x, \eta)$, where $\Phi_1(\eta) := \Phi(0, \eta)$ and $\Phi_2(x, \eta) := \Phi(x, \eta) - \Phi(0, \eta)$. Suppose that $\Phi(\cdot, \eta) : \mathcal{X} \to \mathcal{X}$ is affine, i.e., that $\Phi_2(\cdot, \eta) : \mathcal{X} \to \mathcal{X}$ is linear, and that $\Phi(x, 0) = 0$. Setting

$$\|\Phi_2(\cdot, \eta)\| = \max\{\|\Phi_2(x, \eta)\| : \|x\| = 1\},$$

we note that the function $\eta \mapsto \|\Phi_2(\cdot, \eta)\|$ is continuous and vanishes for $\eta = 0$. Thus, for a given $\delta \in \mathbb{R}_+^\rho$ the quantities

$$\mu(\delta) := \max\{\|\Phi_2(\cdot, \eta)\| : \|\eta\|_g \preceq \delta\}$$

and

$$\theta(\delta) := \max\{\|\Phi_1(\eta)\| : \|\eta\|_g \preceq \delta\}$$

are well defined and tend to zero as $\delta \to 0$.

Let us choose δ sufficiently small so as to have $\mu(\delta) < 1$ and denote the set of such δ by $\mathcal{D} \subset \mathbb{R}_+^\rho$. Set

$$\mathcal{B}_\delta := \{x \in \mathcal{X} : \|x\| \le f(\delta)\},$$

where

$$f(\delta) := \frac{\theta(\delta)}{1 - \mu(\delta)}. \tag{8.51}$$

For $x, y \in \mathcal{B}_\delta$ we have

$$\|\Phi(x, \eta)\| \le \|\Phi_2(x, \eta)\| + \|\Phi_1(\eta)\| \le \mu(\delta)f(\delta) + \theta(\delta) = f(\delta)$$

and

$$\|\Phi(x, \eta) - \Phi(y, \eta)\| = \|\Phi_2(x - y, \eta)\| \le \mu(\delta)\|x - y\|.$$

Therefore, $\Phi(\cdot, \eta)$ is a contraction and maps the closed set $\mathcal{B}_\delta$ into itself. According to the Banach fixed point principle there exists a unique solution x of the operator equation $x = \Phi(x, \eta)$ in the ball $\mathcal{B}_\delta$, i.e.,

$$\|x\| \le f(\delta), \ \delta \in \mathcal{D}. \tag{8.52}$$

The estimate (8.52), (8.51) is nonlocal and with a proper choice of $\theta(\delta)$ and $\mu(\delta)$ it may be asymptotically sharp, asymptotically exact, or even exact in the sense of Definitions 7.2, 7.4 and 7.5.

In order to apply this approach to the perturbation analysis of linear matrix equations we first rewrite (8.26) as an operator equation

$$\delta X = \Phi(\delta X, \delta E).$$

The main problem is then to estimate properly the quantities $\theta(\delta)$ and $\mu(\delta)$ making them as small as possible using the underlying structure of the linear operator $\mathcal{L}(P)$. This is done in a unified way, obtaining a tight bound of type (8.52). We note that for problems with minimum specified structure (such as $AX = C$) very little can be done in improving the perturbation bound. For highly structured equations of type (8.16), however, taking into account the underlying structure, one can get tight nonlocal perturbation bounds.

To derive component-wise perturbation bounds, consider again the operator equation (8.50) in $x \in \mathcal{X} = \mathbb{R}^s$ under the assumptions already made for Φ. Here $x = \mathrm{vec}(X)$, $\eta = \delta E$ and $|\eta| \preceq \Delta$. In order to study component-wise perturbations we use the following generalization of the Banach fixed point principle [135], see also Appendix D.

Since Φ is affine, there exist a vector $\Theta(\Delta) \in \mathcal{R}_+^s$ and a matrix $\Psi(\Delta) \in \mathbb{R}_+^{s\times s}$, such that

$$|\Phi_1(\eta)| \preceq \Theta(\Delta)$$

and

$$|\Phi(x,\eta) - \Phi(y,\eta)| \preceq \Psi(\Delta)|x - y|$$

for all $x, y \in \mathbb{R}^s$ and η with $|\eta| \preceq \Delta$. By the continuity of Φ and the condition $\Phi(0,0) =$ it follows that both $\Theta(\Delta)$ and $\Psi(\Delta)$ tend to zero as $\Delta \to 0$. Hence, for Δ sufficiently small we have

$$\mathrm{rad}(\Psi(\Delta)) < 1. \tag{8.53}$$

Set

$$F(\Delta) := (I_s - (\Delta))^{-1}\Theta(\Delta) \tag{8.54}$$

and let $\mathcal{B}_{F(\Delta)}$ be the set of all x with $|x| \preceq F(\Delta)$. Then for $x \in \mathcal{B}_{F(\Delta)}$ we have

$$|\Phi(x,\eta)| \preceq \Psi(\Delta)F(\Delta) + \Theta(\Delta) = F(\Delta), \tag{8.55}$$

i.e., $\Phi(\mathcal{B}_{F(\Delta)}, \eta) \subset \mathcal{B}_{F(\Delta)}$. Relations (8.53) and (8.55) show that Φ is a generalized contraction on $\mathcal{B}_{F(\Delta)}$. Hence, there exists a unique solution x of the operator equation (8.50) in the rectangle B_Δ, for which

$$|x| \preceq F(\Delta),\ \Delta \in \mathcal{D}_\Delta, \tag{8.56}$$

where $\mathcal{D}_\Delta$ is the set of all $\Delta \succeq 0$ such that $\mathrm{rad}(\Psi(\Delta)) < 1$. The relations (8.56), (8.54) give the desired component-wise perturbation bound. Here the problem is again to estimate properly the vector $\Theta(\Delta)$ and the matrix $\Psi(\Delta)$.

8.6.2 Equivalent perturbation operator

The equivalent perturbation operator Φ for a general Sylvester equation of type (8.16),

$$\mathcal{L}(P)(X) = C,$$

may be written in the form

$$\Phi(\delta X, \delta E) = \Phi_1(\delta E) + \Phi_2(\delta X, \delta E),$$

where

$$\begin{aligned} \Phi_1(\delta E) &= \mathcal{L}^{-1}(P)(\delta C) + \left(1_{(p,m,n,q)} - \mathcal{L}^{-1}(P) \circ \mathcal{L}(P + \delta P)\right)(X), \\ \Phi_2(\delta X, \delta E) &= \left(1_{(p,m,n,q)} - \mathcal{L}^{-1}(P) \circ \mathcal{L}(P + \delta P)\right)(\delta X). \end{aligned}$$

In order to get tight perturbation bounds it is necessary to estimate the norm or the components of the operator

$$1_{(p,m,n,q)} - \mathcal{L}^{-1}(P) \circ \mathcal{L}(P + \delta P)$$

as accurately as possible.

8.6.3 Norm-wise bounds

In order to apply the results from Section 8.6.1 we rewrite the perturbed Sylvester equation as

$$\delta X = \Phi(\delta X, \delta E) := \Phi_1(\delta E) + \Phi_2(\delta X, \delta E), \tag{8.57}$$

where

$$\begin{aligned} \Phi_1(\delta E) &:= \mathcal{L}^{-1}(P)\left(\delta C - \sum_{k=1}^{r}(\delta A_k X B_k + A_k X \delta B_k + \delta A_k X \delta B_k)\right), \\ \Phi_2(Z, \delta E) &:= -\mathcal{L}^{-1}(P)\left(\sum_{k=1}^{r}(\delta A_k Z B_k + A_k Z \delta B_k + \delta A_k Z \delta B_k)\right). \end{aligned} \tag{8.58}$$

Taking the vec operation on both sides of (8.57), and using (8.58) and the notation of Sections 8.5 and 8.6.1, we have

$$\begin{aligned} \|\Phi_1(\delta E)\|_{\mathrm{F}} &\le \theta(\delta) := \mathrm{est}(\delta, N) + \|X\|_2 e_2(\delta), \\ \frac{\|\Phi_2(Z, \delta E)\|_{\mathrm{F}}}{\|Z\|_{\mathrm{F}}} &\le \mu(\delta) := e_1(\delta) + e_2(\delta), \; Z \ne 0, \end{aligned}$$

where the quantities $e_i(\delta)$ are given by

$$\begin{aligned} e_1(\delta) &:= \sum_{k=1}^{\rho} l_k \delta_k, \; e_2(\delta) := \|\Lambda\|_2 \sum_{k=1}^{r} \delta_{2k-1}^0 \delta_{2k}^0, \\ l_k &:= \sum_{i=1}^{\rho}\left(\left\|\Lambda\left(B_i^\top \otimes I_p\right)\right\|_2 \|T_{2i-1,k}\|_2 + \left\|\Lambda\left(I_q \otimes A_i\right)\right\|_2 \|T_{2i,k}\|_2\right). \end{aligned}$$

Thus, we have proved the following result.

Theorem 8.17 *For general Sylvester equations the following nonlocal perturbation bound holds*

$$\delta_X \le f(\delta) := \frac{\text{est}(\delta, N) + \|X\|_2 e_2(\delta)}{1 - e_1(\delta) - e_2(\delta)}, \quad \delta \in \mathcal{D}_\delta. \tag{8.59}$$

The domain of applicability of the bound (8.59) is the set

$$\mathcal{D}_\delta := \{\delta \succeq 0 : e_1(\delta) + e_2(\delta) < 1\}. \tag{8.60}$$

We note that $e_i(\delta) = O(\|\delta\|^i)$, $\delta \to 0$. Hence, the expression $f(\delta)$ may be expanded as

$$f(\delta) = \sum_{j=1}^{k} f_j(\delta) + O(\|\delta\|^{k+1}), \ \delta \to 0,$$

where $f_j(\delta) = O(\|\delta\|^j)$, $\delta \to 0$. The first three terms in this expansion are

$$\begin{aligned} f_1(\delta) &:= \text{est}(\delta, N), \\ f_2(\delta) &:= \|X\|_2 e_2(\delta) + e_1(\delta)\text{est}(\delta, N), \\ f_3(\delta) &:= \left(e_1^2(\delta) + e_2(\delta)\right) \text{est}(\delta, N). \end{aligned}$$

When the matrices in D vary independently (or, in particular, are constant), then we have $\rho = 2r + 1$ and we may set

$$E_1 := C, \ E_{2k} := A_k, \ E_{2k+1} := B_k, \ k = 1, \ldots, r.$$

Here $\delta_k = \delta_{k-1}^0$, $k = 1, \ldots, 2r + 1$, and

$$\begin{aligned} e_1(\delta) &:= \sum_{k=1}^{2r+1} l_k \delta_k, \ l_1 := \|\Lambda\|_2, \\ l_{2i} &:= \left\|\Lambda\left(B_i^\top \otimes I_p\right)\right\|_2, \ l_{2i+1} := \|\Lambda\left(I_q \otimes A_i\right)\|_2. \end{aligned}$$

If a particular matrix P_k is not perturbed, then we have $\delta_{k+1} = \delta_k^0 = 0$ in the above relations.

Since the bound (8.46) is generically asymptotically sharp, so is the bound (8.59). The problems of asymptotical exactness and exactness, however, are more subtle and, at this stage, will be illustrated using model scalar equations.

Example 8.18 Consider the scalar equation

$$ax + xb = (a + b)x = c, \ x = c/(a + b),$$

where $a + b \neq 0$ and $|\delta c| \le \delta_c$, $|\delta a| \le \delta_a$, $|\delta b| \le \delta_b$. Here $\mathcal{L}$ acts as $\mathcal{L}(x) = (a + b)x$. The domain Ω for $\delta_a \ge 0$, $\delta_b \ge 0$ is the triangle, given by

$$\delta_a + \delta_b < |a + b|$$

and the bound (8.59) becomes

$$\delta_x \le f(\delta) = \frac{\delta_c + |x|(\delta_a + \delta_b)}{|a+b| - (\delta_a + \delta_b)}.$$

At the same time for $\delta_a + \delta_b < |a+b|$ the actual perturbation is

$$\delta x = \frac{\delta c - x(\delta a + \delta b)}{a + b + \delta a + \delta b}.$$

Based on the last expression, a simple calculation shows that $\omega(\delta) = f(\delta)$ and hence, the bound (8.59) is exact. ♢

Example 8.19 Consider the scalar equation

$$axb = abx = c, \; x = c/(ab),$$

where $ab \neq 0$ and the notation of Example 8.18 is used. The domain Ω for δ_a, δ_b is the rectangle, described by the inequalities

$$\delta_a < |a|, \; \delta_b < |b|.$$

Since

$$\delta x = \frac{\delta c - x(b\delta a + a\delta b + \delta a \delta b)}{(a + \delta a)(b + \delta b)},$$

then the exact perturbation bound is

$$\omega(\delta) = \frac{\delta_c + |x|(|b|\delta_a + |a|\delta_b - \delta_a\delta_b)}{(|a| - \delta_a)(|b| - \delta_b)}. \tag{8.61}$$

At the same time the bound (8.59) reduces to

$$\delta_x \le f(\delta) = \frac{\delta_c + |x|(|b|\delta_a + |a|\delta_b + \delta_a\delta_b)}{|ab| - |b|\delta_a - |a|\delta_b - \delta_a\delta_b}$$

and is valid in the set

$$\mathcal{D} := \{[\delta_a, \delta_b]^\top \succeq 0 : |b|\delta_a + |a|\delta_b + \delta_a\delta_b < |ab|\}.$$

The principal terms of order $O(\|\delta\|)$ of $\omega(\delta)$ and $f(\delta)$ coincide and hence, the bound (8.59) in this case is asymptotically exact. However, it is not exact. The difference between $f(\delta)$ and $\omega(\delta)$ is in the sign of the quadratic term $\delta_a\delta_b$ in the numerator and denominator of both expressions. ♢

Using the results presented in Examples 8.18 and 8.19 it may be shown that only in the cases (i) and (iii) of Section 8.3 it is reasonable to expect that the bound (8.59), (8.60) are exact for some classes of equations.

It was experimentally observed that for small δ the exact perturbation behaves more as

$$\delta_X \le \frac{\mathrm{est}(\delta, N) - \|X\|_2 e_2(\delta)}{1 - e_1(\delta) + e_2(\delta)},$$

rather than as according to (8.59), see also expression (8.61) for $\omega(\delta)$ in Example 8.19.

8.6.4 Component-wise bounds

Consider the problem of deriving nonlocal component-wise perturbation bounds for Sylvester equations. Suppose first that all matrices

$$C = P_0,\ P_1 = A_1, \ldots, P_{2r} = B_r$$

in (8.16) are perturbed and the perturbations are bounded as

$$\mathrm{vec}(|\delta P_k|) \preceq \Delta_k,\ k = 0, 1, \ldots, 2r$$

(if a particular matrix P_j is not perturbed, then in the following formulas we set $\Delta_j = 0$). Then, using (8.58), we obtain

$$\begin{aligned} |\Phi_1(\delta E)| &\preceq \Theta(\Delta) := \Theta_1(\Delta) + \Theta_2(\Delta)\mathrm{vec}(|X|), \\ |\Phi_2(Z, \delta E)| &\preceq \Psi(\Delta)|Z| := (\Psi_1(\Delta) + \Theta_2(\Delta))|Z|. \end{aligned}$$

Here the vector $\Theta_1(\Delta) \succeq 0$ and the matrices $\Theta_2(\Delta) \succeq 0$, $\Psi_1(\Delta) \succeq 0$ are determined by

$$\begin{aligned} \Theta_1(\Delta) &:= |\Lambda|\Delta_0 + \sum_{k=1}^{r} \left|\Lambda\left((XB_k)^\top \otimes I_p\right)\right| \Delta_{2k-1} \\ &\quad + \sum_{k=1}^{r} |\Lambda\left(I_q \otimes (AX)\right)| \Delta_{2k}, \\ \Theta_2(\Delta) &:= \sum_{k=1}^{r} |\Lambda| \left(W_{2k}^\top \otimes W_{2k-1}\right), \\ \Psi_1(\Delta) &:= \sum_{k=1}^{r} \left|\Lambda\left(B_k^\top \otimes I_p\right)\right| (I_n \otimes W_{2k-1}) \\ &\quad + \sum_{k=1}^{r} |\Lambda\left(I_q \otimes A_k\right)| \left(W_{2k}^\top \otimes I_m\right) \\ &\preceq \sum_{k=1}^{r} |\Lambda| \left(|B_k|^\top \otimes W_{2k-1} + W_{2k}^\top \otimes |A_k|\right), \end{aligned} \tag{8.62}$$

where

$$W_{2k-1} := \mathrm{vec}^{-1}(p, m)(\Delta_{2k-1}),\ W_{2k} := \mathrm{vec}^{-1}(n, q)(\Delta_{2k}).$$

Therefore, we have proved the following theorem.

Theorem 8.20 *For general Sylvester equations the nonlocal component-wise perturbation bound is*

$$|\mathrm{vec}(\Delta X)| \preceq F(\Delta) := (I_s - \Psi_1(\Delta) - \Theta_2(\Delta))^{-1}(\Theta_1(\Delta) + \Theta_2(\Delta)|X|). \tag{8.63}$$

The domain of applicability of the bound is

$$\Delta \in \mathcal{D}_\Delta := \{\Delta \succeq 0 : \mathrm{rad}(\Psi_1(\Delta) + \Theta_2(\Delta)) < 1\}. \tag{8.64}$$

Since $\Theta_i(\Delta)$, $\Psi_i(\Delta)$ are of order $O(\|\Delta\|^i)$ for $\Delta \to 0$, we have

$$F(\Delta) = \sum_{j=1}^{k} F_j(\Delta) + O(\|\Delta\|^{k+1}),\ \Delta \to 0,$$

where $\|F_j(\Delta)\| = O(\|\Delta\|^j)$, $\Delta \to 0$. The first three terms in the expansion of $F(\Delta)$ are

$$\begin{aligned} F_1(\Delta) &= \Theta_1(\Delta), \\ F_2(\Delta) &= \Theta_2(\Delta)|X| + \Psi_1(\Delta)\Theta_1(\Delta), \\ F_3(\Delta) &= \Psi_1(\Delta)\Theta_2(\Delta)|X| + \left(\Psi_1^2(\Delta) + \Theta_2(\Delta)\right)\Theta_1(\Delta). \end{aligned}$$

Component-wise bounds may be derived also using the following approach. Let the vector equation

$$(A + B)x = b$$

be given, where the matrix A is nonsingular and $\text{rad}(|A^{-1}B|) < 1$. Then the matrix $A + B$ is also nonsingular and the following component-wise perturbation estimate for the solution x is valid:

$$|x| \preceq \left(I - |A^{-1}B|\right)^{-1} |A^{-1}b|.$$

The trick here is to exploit fully the underlying structure of A, B and b (and hence, of the products $A^{-1}Z$) and *not* to use directly the inequalities $|A^{-1}Z| \preceq |A^{-1}|\,|Z|$. The advantage of this approach may be seen for example in the inequality for Ψ_1 in (8.62) – the second bound for $\Psi_1(\Delta)$ is obtained by direct majorization $|A^{-1}Z| \preceq |A^{-1}|\,|Z|$ and is hence, worse.

8.7 Notes and references

Algebraic linear matrix equations have been intensively studied since the times of Sylvester and Kronecker [152, 215, 214], see also [196, 193, 229]. Brief historical reference may be found in [8]. In particular, the problems of existence, uniqueness and representation of the solution are solved for such equations, see e.g. [10, 12, 19, 20, 33, 36, 56, 69, 84, 94, 101, 106, 107, 153, 155, 165, 167, 189, 206, 181, 205, 224, 228, 230, 235, 236, 241]. The properties of special linear matrix transformations have also been studied [40, 41, 79, 86, 216, 223, 222]. There is a variety of techniques, algorithms and software for solving linear matrix equations [17, 13, 14, 6, 15, 18, 25, 45, 46, 50, 72, 73, 82, 88, 91, 90, 109, 164, 175, 192, 194, 202, 239, 240]. The great interest in linear matrix equations is due in a large extent to their wide application to various areas [11, 49, 48, 59, 62, 104, 105, 103, 170, 225]. Also, the perturbation theory for linear matrix equations, including the Sylvester and Lyapunov equations arising in linear control theory, has been studied

[38, 110, 68, 95, 99, 114, 113, 112, 136]. The perturbation theory for operators in abstract spaces [119] and for general linear equations [97, 98] also applies in a large scale to the perturbation analysis of linear matrix equations. Other investigations are connected with establishing bounds on the solution of linear matrix equations are given [171, 176].

Some results concerning backward perturbation analysis are given in [99, 101, 112].

Chapter 9

Specific Sylvester equations

In this chapter we present perturbation results as well as some general properties for classes of Sylvester equations that arise in linear control theory. The results are based on those from Chapter 8.

We derive bounds of type (8.59) and (8.63) for the types of equations in (i) – (iv) in Section 8.3 and we present the expressions for $\text{est}(\delta, N)$, $e_1(\delta)$ and $e_2(\delta)$ in the norm-wise case, and for $\Theta(\Delta)$ and $\Psi(\Delta)$ in the component-wise case. The following slightly different notation is used:

- $\delta_Z \geq \|Z\|_{\text{F}}$ – the norm-wise bound for Z;
- Δ – the collection (Δ_C, Δ_A) in cases (i), or $(\Delta_C, \Delta_A, \Delta_B)$ in cases (iii), (iv), respectively. The same convention is adopted for the vector δ with elements δ_C, δ_A and δ_C, δ_A, δ_B.
- $\Delta_Z \succeq \text{vec}(|Z|)$ – a vector component-wise bound for Z;
- $W_Z = \text{vec}^{-1}(\Delta_Z) \succeq |Z|$ – a matrix component-wise bound for Z,

where Z stands for A, B, C or X. To simplify the notation, we use the same letter $\mathcal{L}$ for the Sylvester operator in all cases.

The estimates presented below are valid for both real and complex equations.

9.1 Standard linear equation

The standard linear matrix equation (8.18), namely

$$AX = C, \tag{9.1}$$

with A invertible, gives rise to some of the most popular and widely used perturbation bounds (norm-wise, component-wise, structured and backward) in numerical linear algebra [83, 101]. It is instructive to see how the concepts for various types of perturbation bounds (see Chapter 7) are applied to this equation.

We consider the nontrivial case $C \neq 0$ which implies $X \neq 0$. However, the results are valid also for the case $C = 0$ with the exception of those connected to relative perturbation bounds.

Writing the perturbed equation in the operator form

$$\delta X = \Phi(\delta X, \delta E) := A^{-1}(\delta C - \delta A X) - A^{-1}\delta A \delta X, \ \delta E := (\delta C, \delta A),$$

we get the following well known a posteriori bound

$$\delta_X \leq f(\delta) := \frac{\|A^{-1}\|_2(\delta_C + \|X\|_2\delta_A)}{1 - \|A^{-1}\|_2\delta_A}, \ \delta_A < \frac{1}{\|A^{-1}\|_2}. \tag{9.2}$$

This bound is asymptotically sharp (see Definition 7.2) and it is even asymptotically exact (Definition 7.4) as shown below. We also prove that for $m > 1$ the bound (9.2) in general cannot be exact (see Definition 7.5), and the class of equations, for which it is exact, is fully described. Note that here the exact domain for δ_A is the interval $[0, 1/\|A^{-1}\|_2)$.

For equation (9.1) the bound (8.59) yields

$$\delta_X \leq \frac{\mathrm{est}(\delta_C, \delta_A, \Lambda, N_A)}{1 - \|\Lambda\|_2\delta_A}, \ \delta_A < 1/\|\Lambda\|_2, \tag{9.3}$$

where

$$\Lambda = (I_n \otimes A)^{-1} = I_n \otimes A^{-1}, \ N_A = -\Lambda(X^\top \otimes I_m).$$

In turn, the component-wise perturbation bound for equation (9.1) is obtained as follows. If

$$\mathrm{rad}\left(|A^{-1}|W_A\right) < 1,$$

then the bound (8.63) reduces to

$$|\mathrm{vec}(\delta X)| \preceq \left(I_{n^2} - I_n \otimes \left(|A^{-1}|W_A\right)\right)(|\Lambda|\Delta_C + |N_A|\Delta_A). \tag{9.4}$$

The only visible difference between the classical bound (9.2) and the bound (9.3) is in the numerator, since the denominators coincide in view of

$$\|\Lambda\|_2 = \|A^{-1}\|_2.$$

The numerator in (9.2) is

$$\|A^{-1}\|_2(\delta_C + \|X\|_2\delta_A) = \mathrm{est}_1(\delta_C, \delta_A, \Lambda, N_A).$$

On the other hand we know that

$$\mathrm{est} \leq \mathrm{est}_3 \leq \mathrm{est}_1.$$

In fact, both bounds coincide for this case. Indeed,

$$\begin{aligned} N_A &= -\Lambda(X^\top \otimes I_m) = -(I_n \otimes A^{-1})(X^\top \otimes I_m) = -X^\top \otimes A^{-1}, \\ N &= [\Lambda, N_A] = \left[I_n, -X^\top\right] \otimes A^{-1} \end{aligned}$$

and

$$\Lambda^\top N_A = -(I_n \otimes A^{-\top})(X^\top \otimes A^{-1}) = -X^\top \otimes (AA^\top)^{-1}.$$

Hence,

$$\begin{aligned} \|N_A\|_2 &= \|A^{-1}\|_2 \|X\|_2, \ \left\|\Lambda^\top N_A\right\|_2 = \|A^{-1}\|_2^2 \|X\|_2, \\ \|[\Lambda, N_A]\|_2 &= \|A^{-1}\|_2 \left\|\left[I_n, -X^\top\right]\right\|_2 = \|A^{-1}\|_2 \sqrt{1 + \|X\|_2^2} \end{aligned}$$

and

$$\mathrm{est}_3(\delta_C, \delta_A, \Lambda, N_A) = \|A^{-1}\|_2(\delta_C + \|X\|_2 \delta_A) = \mathrm{est}_1(\delta_C, \delta_A, \Lambda, N_A).$$

Consider the bound that is obtained by minimizing the expression $\psi(\gamma, \delta, N)$ in γ, see (8.40). We have that

$$\begin{aligned} \psi(\gamma, \delta_C, \delta_A, \Lambda, N_A) &= \left\|\left[\Lambda, \frac{N_A}{\gamma}\right]\right\|_2 \sqrt{\delta_C^2 + \gamma^2 \delta_A^2} \\ &= \sqrt{\delta_C^2 + \|X\|_2^2 \delta_A^2 + \delta_A^2 \gamma^2 + \frac{\|X\|_2^2 \delta_C^2}{\gamma^2}}. \end{aligned}$$

The minimum of ψ in $\gamma > 0$ is achieved for

$$\gamma^0 = \|X\|_2 \delta_C / \delta_A$$

and is equal to est_1 (we suppose that $\delta_A > 0$, since otherwise the results are trivial).

Thus, the local bounds (with the exception of est_2) coincide with the bound est. The reason is that equation (9.1) has no specific structure. After having shown that the bound $f(\delta)$ is asymptotically sharp, we prove that it is also asymptotically exact.

Proposition 9.1 *The bounds (9.2) and the (8.59) are asymptotically exact for all Sylvester equations of type (9.1).*

Proof. Let

$$\begin{aligned} X &= Q\Sigma_X R^{\mathrm{H}} = Q\mathrm{diag}(\sigma_1(X), \ldots, \sigma_k(X), 0, \ldots, 0) R^{\mathrm{H}}, \\ A &= U\Sigma_A V^{\mathrm{H}} = U\mathrm{diag}(\sigma_1(A), \ldots, \sigma_m(A)) V^{\mathrm{H}} \end{aligned}$$

be the singular value decompositions of X and A, respectively, where

$$k := \mathrm{rank}(X).$$

Let q_j, r_i and u_j, v_j be the columns of the orthogonal matrices Q, R and U, V, respectively. If we define integers k_0 and ℓ_0 via

$$\begin{aligned} k_0 &:= \min\{i : \sigma_i(A) = \sigma_m(A)\}, \\ \ell_0 &:= \max\{i : \sigma_i(X) = \sigma_1(X)\}, \end{aligned} \tag{9.5}$$

then we have

$$\|N\mathrm{vec}(\delta E)\|_2 = \left\|\mathrm{vec}^{-1}(m,n)(N\mathrm{vec}(\delta E))\right\|_{\mathrm{F}} = \left\|A^{-1}(\delta C - \delta AX)\right\|_{\mathrm{F}},$$

where

$$\mathrm{vec}(\delta E) := [\mathrm{vec}^\top(\delta C), \mathrm{vec}^\top(\delta A)]^\top.$$

If we fix the integers $i \in \{1, \ldots, \ell_0\}$ and $j \in \{k_0, \ldots, m\}$, and choose

$$\begin{aligned} \delta C &:= \delta_C \left(e_{ni}^\top \otimes u_j\right) R^{\mathrm{H}} = \delta_C u_j r_i^{\mathrm{H}}, \\ \delta A &:= -\delta_A \left(e_{mi}^\top \otimes u_j\right) Q^{\mathrm{H}} = -\delta_A u_j q_i^{\mathrm{H}}, \end{aligned}$$

where e_{ni} is the i-th column of I_n, then

$$A^{-1}u_j = \|A^{-1}\|_2 v_j, \; q_i^{\mathrm{H}} Q \Sigma_X R^{\mathrm{H}} = \sigma_1(X) r_i.$$

Since

$$\|A^{-1}\|_2 = \frac{1}{\sigma_m(A)},$$

we get

$$\begin{aligned} \|N\mathrm{vec}(\delta E)\|_2 &= \left\|A^{-1}u_j \left(\delta_A r_i^{\mathrm{H}} + \delta_A q_i^{\mathrm{H}} Q \Sigma_X R^{\mathrm{H}}\right)\right\|_{\mathrm{F}} \\ &= (\delta_C + \|X\|_2 \delta_A) \left\|A^{-1}u_j r_i^{\mathrm{H}}\right\|_{\mathrm{F}} \\ &= \|A^{-1}\|_2 (\delta_C + \|X\|_2 \delta_A) \left\|v_j r_i^{\mathrm{H}}\right\|_{\mathrm{F}} \\ &= \|A^{-1}\|_2 (\delta_C + \|X\|_2 \delta_A) = \mathrm{est}(\delta, N) \end{aligned}$$

and hence,

$$\mathrm{est}(\delta, N) \le \omega_1(\delta, N),$$

where

$$\omega_1(\delta, N) := \max\{\|\Lambda z + N_A z_A\|_2 : \|z\|_2 \le \delta_C, \; \|z_A\|_2 \le \delta_A\}.$$

On the other hand

$$\mathrm{est}(\delta, N) \ge \omega_1(\delta, N)$$

by construction. The last two inequalities yield

$$\mathrm{est}(\delta, N) = \omega_1(\delta, N)$$

which completes the proof. $\square$

Finally we will determine the class of equations of type (9.1) for which the perturbation bound is exact. To find conditions for exactness of the bound (9.2), we consider mainly the case $n = 1$ when (9.1) is a vector equation, since it is equivalent to n vector equations for the columns of X.

Setting

$$c = \begin{bmatrix} c_1 \\ \vdots \\ c_m \end{bmatrix} := U^{\mathrm{H}} C, \; y = \begin{bmatrix} y_1 \\ \vdots \\ y_m \end{bmatrix} := V^{\mathrm{H}} X,$$

where $c_i, y_i \in \mathbb{F}^{1\times n}$, we get $\Sigma_A y = c$, i.e.,

$$\sigma_i y_i = c_i, \; i = 1, \ldots, m, \tag{9.6}$$

where $\sigma_i := \sigma_i(A)$.

We look for extremal perturbations

$$c \to c + G_c, \; \Sigma_A \to \Sigma_A + G_{\Sigma_A}$$

with

$$\|G_c\|_{\mathrm{F}} \leq \delta_C, \; \|G_{\Sigma_A}\|_{\mathrm{F}} \leq \delta_A < \sigma_m$$

in the pair (Σ_A, c) for which the norm of the perturbation

$$\delta y = (\Sigma_A + G_{\Sigma_A})^{-1}(G_c - G_{\Sigma_A} y)$$

in the solution

$$y = \Sigma_A^{-1} c = V^{\mathrm{H}} X$$

is maximal, i.e.,

$$\begin{aligned} \omega(\delta) &= \max\left\{\|(\Sigma_A + \delta\Sigma)^{-1}(\delta c - \delta\Sigma y)\|_{\mathrm{F}} : \|\delta c\|_{\mathrm{F}} \leq \delta_C, \; \|\delta\Sigma\|_{\mathrm{F}} \leq \delta_A\right\} \\ &= \|(\Sigma_A + G_{\Sigma_A})^{-1}(G_c - G_{\Sigma_A} y)\|_2. \end{aligned}$$

To do this we use the notion of an acute perturbation of a nonsingular matrix A.

Definition 9.2 *A perturbation δA of A is* acute *in the norm $\|\cdot\|$ if*

$$\|\delta A\| < \frac{1}{\|A^{-1}\|}$$

and

$$\|(A + \delta A)^{-1}\| = \frac{\|A^{-1}\|}{1 - \|A^{-1}\| \, \|\delta A\|}.$$

Often it is better to estimate $\|(A + \delta A)^{-1}\|_2$ as a function of $\|\delta A\|_{\mathrm{F}}$. Then this definition must be slightly modified, since the F-norm is not an operator norm but satisfies the inequality

$$\|AB\|_{\mathrm{F}} \leq \|A\|_2 \|B\|_{\mathrm{F}}$$

which yields

$$\|(A+\delta A)^{-1}\|_2 \le \frac{\|A^{-1}\|_2}{1-\|A^{-1}\|_2\|\delta A\|_{\mathrm{F}}}.$$

Definition 9.3 *A perturbation* δA *of* $A \in \mathcal{GL}(m)$ *with*

$$\|\delta A\|_{\mathrm{F}} < \sigma_{\min}(A)$$

is said to be F-acute *if*

$$\|(A+\delta A)^{-1}\|_2 = \frac{\|A^{-1}\|_2}{1-\|A^{-1}\|_2\|\delta A\|_{\mathrm{F}}} = \frac{1}{\sigma_{\min}(A)-\|\delta A\|_{\mathrm{F}}}.$$

Given a value α with

$$0 < \alpha < 1/\|A^{-1}\|_2,$$

there are exactly $m-k+1$ different F-acute perturbations δA with

$$\|\delta A\|_{\mathrm{F}} = \alpha,$$

namely

$$\delta A = -\alpha\, u_j v_j^{\mathrm{H}},\ j = k,\ldots,m.$$

For the matrix Σ_A the F-acute perturbations are

$$\delta \mathcal{S}_A = -\alpha\, E_{ii}(m,m)$$

with $k_0 \le i \le m$. Generically $\sigma_{m-1} > \sigma_m$ and $k_0 = m$, i.e., there is only one F-acute perturbation

$$\delta A = -\alpha\, u_m v_m^{\mathrm{H}}.$$

The properties of acute perturbations strongly depend on the underlying norm. If we consider p-acute perturbations δA in the Hölder p-norm with

$$\|\delta A\|_p < \|A^{-1}\|_p^{-1},$$

for which

$$\|(A+\delta A)^{-1}\|_p = \frac{\|A^{-1}\|_p}{1-\|A^{-1}\|_p\|\delta A\|_p},$$

then, for instance, if $m > 1$, there are infinitely many 2-acute perturbations.

It follows from the inequalities $\sigma_i > 0$ and the diagonal structure of system (9.6) that $G_{\Sigma_A} \preceq 0$ and that the i-th element of G_c must have the sign of the corresponding right-hand side c_i, provided that $n = 1$. Moreover, G_{Σ_A} must be diagonal, i.e.,

$$\begin{aligned} G_{\Sigma_A} &= -\mathrm{diag}(\varepsilon_1,\ldots,\varepsilon_m),\ \varepsilon_i \ge 0, \\ G_c &= [\gamma_1\,\mathrm{sign}(c_1),\ldots,\gamma_m\,\mathrm{sign}(c_m)]^{\top},\ \gamma_i \ge 0. \end{aligned}$$

Hence,

$$\delta y_i = \pm \frac{\gamma_i + |y_i|\varepsilon_i}{\sigma_i - \varepsilon_i}.$$

The extremal perturbation is then obtained as a solution of the maximization problem

$$\sum_{i=1}^{m} \left(\frac{\gamma_i + |y_i|\varepsilon_i}{\sigma_i - \varepsilon_i} \right)^2 \to \max \tag{9.7}$$

subject to the constraints

$$\sum_{i=1}^{m} \gamma_i^2 \leq \delta_C^2, \ \sum_{i=1}^{m} \varepsilon_i^2 \leq \delta_A^2, \tag{9.8}$$

where $\delta_A < \sigma_m$.

Using particular examples, we see that in general the bound (9.2) is not exact when $m > 1$.

Example 9.4 Consider the system (9.6) with $m = 2$, $n = 1$ and $\delta_C = \delta_A = \eta$. The bound (9.2) here is

$$f(\eta, \eta) = \left(1 + \sqrt{y_1^2 + y_2^2}\right) \frac{\eta}{\sigma_2 - \eta}.$$

The maximization problem (9.7),(9.8) in γ_i, ε_i depends on five parameters σ_1, σ_2, $|y_1|$, $|y_2|$ and η, where

$$\sigma_1 > \sigma_2 > 0, \ 0 \leq \eta < \sigma_2$$

and

$$|y_1| + |y_2| > 0.$$

Depending on the relations among these parameters we have the following two cases.

First, let $(\sigma_1 = \sigma_2)$ or $(\sigma_1 > \sigma_2$ and $|y_1| \leq |y_2|)$. Then

$$\omega(\eta, \eta) = (1 + \max\{|y_1|, |y_2|\}) \frac{\eta}{\sigma_2 - \eta}.$$

In this case the extremal perturbation G_{Σ_A} in Σ_A is F-acute. The bound $f(\eta, \eta)$ is exact if and only if $(\sigma_1 \geq \sigma_2$ and $c_1 = 0)$ or $(\sigma_1 = \sigma_2$ and $c_2 = 0)$.

Second, if $\sigma_1 > \sigma_2$ and $|y_1| > |y_2|$, then the bound $f(\eta, \eta)$ is not exact. At the same time the extremal perturbation in Σ_A may not be F-acute. Indeed, the maximum norm of the perturbation δy in y for an F-acute perturbation G_{Σ_A} of Σ_A is

$$\nu_2 := (1 + |y_2|) \frac{\eta}{\sigma_2 - \eta}.$$

Suppose that

$$(1 + |y_1|)\sigma_2 > (1 + |y_2|)\sigma_1$$

and

$$\eta < \frac{(1+|y_1|)\sigma_2 - (1+|y_2|)\sigma_1}{|y_1| - |y_2|}.$$

Then, taking the perturbations in c and Σ_A as

$$\delta c = \begin{bmatrix} \eta \\ 0 \end{bmatrix}, \quad \delta\Sigma_A = \begin{bmatrix} -\eta & 0 \\ 0 & 0 \end{bmatrix}$$

we obtain that the norm of the perturbation in y is

$$\nu_1 := (1+|y_1|)\frac{\eta}{\sigma_2 - \eta} > \nu_2.$$

Hence, the extremal perturbation, for which the norm of δy is at least ν_1, cannot be F-acute. ◇

The following proposition reveals the role of F-acute perturbations in exact bounds.

Proposition 9.5 *If the bound (9.2) is exact, then every extremal perturbation G_A in A is F-acute (this is true in the general case $n \geq 1$).*

Proof. Suppose that the bound (9.2) is exact ($f(\delta) = \omega(\delta)$) but the extremal perturbation G_A in A is not acute. Then

$$\|(A+G_A)^{-1}\|_2 < \frac{1}{\sigma_m - \delta_A}$$

which yields

$$\begin{aligned} \omega(\delta) &= \left\|(A+G_A)^{-1}(G_C - G_A X)\right\|_{\mathrm{F}} \\ &\leq \left\|(A+G_A)^{-1}\right\|_2 \|G_C - G_A X\|_{\mathrm{F}} \\ &< \frac{\|G_C - G_A X\|_{\mathrm{F}}}{\sigma_m - \delta_A} \leq \frac{\delta_C + \|X\|_2 \delta_A}{\sigma_m - \delta_A} = f(\delta), \end{aligned}$$

i.e., the bound is not exact. This contradiction shows that G_A must be F-acute. □

The converse statement to Proposition 9.5, namely that an extremal perturbation may be F-acute, while the bound (9.2) is not exact, is not true as demonstrated in Example 9.4. Hence, it is important to determine the class of equations (9.1), for which the bound (9.2) is exact.

Theorem 9.6 *If $n = 1$, then the perturbation bound (9.2) is exact if and only if there exists an integer $j \in \{k, \ldots, m\}$, such that $c_i = u_i^{\mathrm{H}} C = 0$ for $i \neq j$ (or equivalently, such that $\left\|u_j^{\mathrm{H}} C\right\|_2 = \|C\|_2$), where $u_1, \ldots, u_m$ are the columns of the matrix U in the singular value decomposition $A = U\Sigma_A V^{\mathrm{H}}$ of A.*

Proof. *Necessity.* Suppose that the bound (9.2) is exact. Then according to Proposition 9.5 the extremal perturbation G_{Σ_A} in Σ_A is F-acute, i.e., there exists an integer $j \in \{k_0, \ldots, m\}$ such that

$$\delta y_i = \begin{cases} \gamma_i/\sigma_i & \text{if} \quad i \neq j, \\ (\gamma_j + |y_j|\delta_A)/(\sigma_m - \delta_A) & \text{if} \quad i = j. \end{cases} \tag{9.9}$$

Since $\sigma_i \geq \sigma_j$ for all $i \in \{1, \ldots, m\}$, the maximum of $\|\delta y\|_2$ in $\gamma_1, \ldots, \gamma_m$ is achieved for $\gamma_i = 0$ if $i \neq j$ and $\gamma_j = \delta_C$. Hence,

$$\|\delta y\|_2 = |\delta y_j| = \frac{\delta_C + |y_j|\delta_A}{\sigma_m - \delta_A}.$$

Since the bound is exact, it follows from the comparison with the right-hand side of (9.2) that $|y_j| = \|y\|_2$. Having in mind that $y_i = u_i^{\mathrm{H}} C/\sigma_i$, we see that y and hence, C has all but one element (in the j-th position) equal to zero.

Sufficiency. Let $\left\|u_j^{\mathrm{H}} C\right\|_2 = \|C\|_2$. Then the only nonzero element of $U^{\mathrm{H}} C$ and hence, of y is in the j-th position and (9.9) holds. Choosing $\gamma_i = 0$ if $i \neq j$ and $\gamma_j = \delta_C$ we get

$$\|\delta y\|_2 = |\delta y_j| = \frac{\delta_C + |y_j|\delta_A}{\sigma_m - \delta_A} = \frac{\delta_C + \|y\|_2\delta_A}{\sigma_m - \delta_A} = f(\delta),$$

i.e., the bound $f(\delta)$ is reached and is thus exact. □

In the generic case $k_0 = m$ Theorem 9.6 tells us that the bound (9.2) is exact if and only if

$$C^{\mathrm{H}} U = [0, \ldots, 0, \pm\|C\|_2]^\top.$$

If the perturbations are measured in 2-norm, then we have

$$\|\delta X\|_2 \leq \frac{\|A^{-1}\|_2(\|\delta C\|_2 + \|X\|_2\|\delta A\|_2)}{1 - \|A^{-1}\|_2\|\delta A\|_2}. \tag{9.10}$$

The bound (9.10) is asymptotically exact for all $n \geq 1$. Similarly to Theorem 9.6 we have the following result.

Proposition 9.7 *The bound (9.10) is exact for $n = 1$ if and only if*

$$\left\|C^{\mathrm{H}}[u_k, \ldots, u_m]\right\|_2 = \|C\|_2.$$

Proof. The proof follows immediately by using the 2-acute perturbation

$$\delta\Sigma_A = \mathrm{diag}(0, -\delta_2 I_{m-k+1})$$

in system (9.6). □

It follows from $AX = C$ that

$$\|C\|_2 \leq \|A\|_2\|X\|_2$$

and

$$\frac{1}{\|X\|_2} \leq \frac{\|A\|_2}{\|C\|_2}.$$

Substituting the last inequality in (9.10) yields the well known a priori relative perturbation bound

$$\varepsilon_X \leq \frac{\operatorname{cond}_2(A)\,(\varepsilon_C + \varepsilon_A)}{1 - \operatorname{cond}_2(A)\varepsilon_A}, \tag{9.11}$$

where $\varepsilon_Z := \|\delta Z\|_2/\|Z\|_2$ for $Z = C, A$ and

$$\operatorname{cond}_2(A) := \|A\|_2\|A^{-1}\|_2.$$

Unfortunately, in general the bound (9.11) is not asymptotically sharp, this is the price for deleting the a posteriori quantity $\|X\|_2$.

The asymptotically exact (and hence, asymptotically sharp) relative perturbation bound here is

$$\varepsilon_X \leq \frac{\operatorname{cond}_2(A)\,(c\varepsilon_C + \varepsilon_A)}{1 - \operatorname{cond}_2(A)\varepsilon_A}, \tag{9.12}$$

where

$$c := \frac{\|C\|_2}{\|A\|_2\|X\|_2} = \frac{\|C\|_2}{\|A\|_2\|A^{-1}C\|_2}.$$

Since

$$\|A^{-1}C\|_2 \leq \|A^{-1}\|_2\|C\|_2,$$

we have

$$\frac{1}{\operatorname{cond}_2(A)} \leq c \leq 1.$$

Thus, if $\operatorname{cond}_2(A)$ is large, c is close or equal to $1/\operatorname{cond}_2(A)$ and $\varepsilon_A/\varepsilon_C$ is small, then the a priori bound (9.11) may be arbitrarily larger than the true a posteriori bound (9.12).

Example 9.8 Let

$$A = \begin{bmatrix} 1 & 0 \\ 0 & \varepsilon \end{bmatrix}, \; C = \begin{bmatrix} 0 \\ 1 \end{bmatrix}$$

and

$$\delta A = \begin{bmatrix} 0 & 0 \\ 0 & -\varepsilon^2 \end{bmatrix}, \; \delta C = \begin{bmatrix} 0 \\ \varepsilon \end{bmatrix},$$

where $\varepsilon > 0$ is a small parameter. The exact relative perturbation in X is

$$\varepsilon_X = \frac{2\varepsilon}{1-\varepsilon}.$$

The a priori bound (9.11) here takes the form

$$\varepsilon_X \leq f_{\mathrm{ap}}(\varepsilon) := \frac{1+\varepsilon}{1-\varepsilon},$$

while the bound (9.12) is reduced to

$$\delta'_X \leq \varphi_{\rm tr}(\varepsilon) := \frac{2\varepsilon}{1-\varepsilon}$$

(and is even exact for this particular case). We see that the ratio of the two bounds

$$\frac{\varphi_{\rm ap}(\varepsilon)}{\varphi_{\rm tr}(\varepsilon)} = \frac{1+\varepsilon}{2\varepsilon}$$

tends to infinity as ε tends to zero. $\diamondsuit$

It follows from the above considerations that the bound (9.11) is asymptotically exact (for all $n \geq 1$) if and only if $c = 1$, which is equivalent to

$$\|C\|_2 = \|A\|_2 \|X\|_2 = \|A\|_2 \|A^{-1}C\|_2. \tag{9.13}$$

This condition may be reformulated as follows.

Proposition 9.9 *Set*

$$m_0 := \max\{i : \sigma_i(A) = \sigma_1(A)\}.$$

The bound (9.11) is asymptotically exact for any $n \geq 1$ if and only if one of the following alternative conditions holds:

1. *$A = \alpha Q$, where $0 \neq \alpha \in \mathbb{R}$ and Q is real orthogonal, when $m_0 = m$ in the real case, and $A = \alpha Q$, where $0 \neq \alpha \in \mathbb{C}$ and Q is unitary in the complex case;*

2. *$u_i^{\rm H} C = 0$ for $i > m_0$, when $m_0 < m$,*

Proof. 1. In the real case we have $m_0 = m$ if and only if $A = \alpha Q$, where Q is real orthogonal, i.e. $Q \in \mathcal{O}(m, \mathbb{R})$. In this case

$$X = Q^\top C / \alpha$$

and

$$\|X\|_2 = \|C\|_2 / |\alpha|.$$

The complex case is treated similarly. Since $\|A\|_2 = |\alpha|$, we have

$$\|C\|_2 = \|A\|_2 \|X\|_2.$$

2. Consider the transformed system (9.6). The condition (9.13) is equivalent to

$$\|c\|_2^2 = \|\Sigma_A\|_2^2 \|y\|_2^2$$

which in turn gives

$$\sum_{i=1}^{m_0} c_i^2 + \sigma_1^2 \sum_{i=\nu+1}^{m} \frac{c_i^2}{\sigma_i^2} = \sum_{i=1}^{m_0} c_i^2 + \sum_{i=\nu+1}^{m} c_i^2.$$

Since

$$\sigma_1 > \sigma_{\nu+1} \geq \cdots \geq \sigma_m,$$

it follows that

$$c_i = u_i^{\mathrm{H}} C = 0$$

for $i > m_0$. □

Combining Theorems 9.6 and 9.9 we also get the following necessary and sufficient condition for exactness of the bound (9.11).

Theorem 9.10 *The bound (9.11) is exact if and only if $A = \alpha Q$, where $0 \neq \alpha \in \mathbb{R}$ and $Q \in \mathcal{O}(m, \mathbb{R})$ in the real case, and $A = \alpha Q$, where $0 \neq \alpha \in \mathbb{C}$ and Q is unitary, i.e., $Q \in \mathcal{U}(m)$ in the complex case.*

At the same time the relative bound (9.12) is exact together with the absolute bound (9.10) under the weaker condition of Proposition 9.7. When A is a scalar multiple of an orthogonal or unitary matrix as in the condition of Theorem 9.10 then $k_0 = 1$ and the condition of Theorem 9.7 holds.

Example 9.11 Let the matrices A, B and C in the Sylvester equation

$$AX + XB = C$$

be $n \times n$ diagonal with diagonal elements a_i, b_i and c_i, respectively. Let

$$\alpha := \min\{|a_i + b_j| : i, j = 1, \ldots, n\} = |a_{i_0} + b_{j_0}| > 0.$$

Then the solution X is the unique diagonal $n \times n$ matrix with diagonal elements x_i. ◊

Note that the above results depend on the used norm. For Hölder p-norms with $p \neq 2$ the conditions for various types of exactness of the perturbation bounds will be different.

In Figures 9.1 and 9.2 we show the elements of the relative perturbed solutions $\delta X / \|X\|$ of 3rd order well-conditioned and ill-conditioned linear equations generated by perturbations in the elements a_{11}, a_{21} and a_{31} of the matrix A. The perturbations in the data are represented by spheres while the perturbed solutions are represented by ellipsoids.

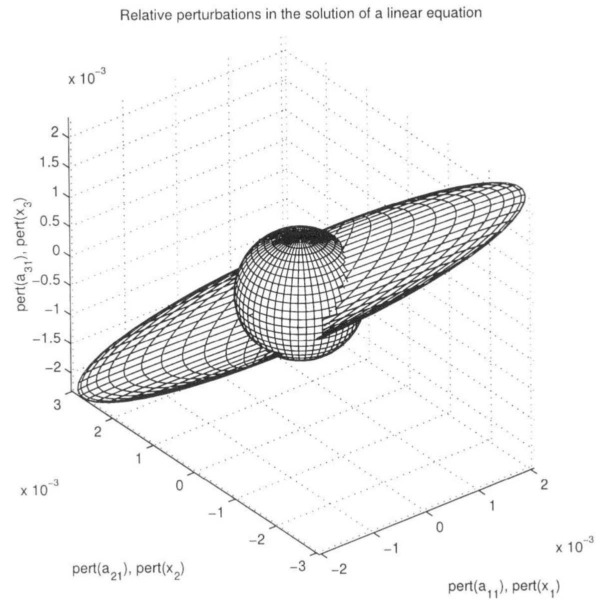

Figure 9.1: Perturbed solutions of well-conditioned linear equation

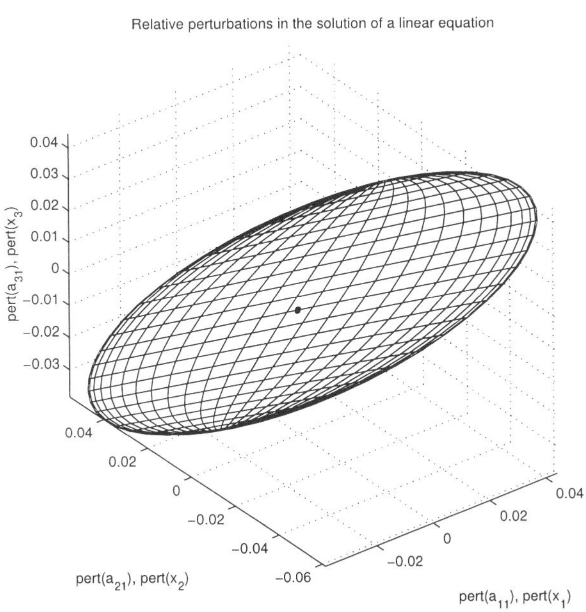

Figure 9.2: Perturbed solutions of ill-conditioned linear equation

9.2 General equations

In the case of (8.19) we have an interesting equation with (8.20) and (8.21) as particular cases. The spectrum of the linear matrix operator $\mathcal{L}$ is then

$$\operatorname{spect}(\mathcal{L}) = \left\{ \sum_{i,j=1}^{r,s} \beta_{ij}\lambda_k^i(A)\lambda_\ell^j(B);\ k \in \{1,\ldots,m\},\ \ell \in \{1,\ldots,n\} \right\},$$

where $\lambda_k(Z)$ are the eigenvalues of the matrix Z. Hence, equation (8.19) is uniquely solvable if and only if $0 \notin \operatorname{spect}(\mathcal{L})$. The bound (8.59) may be applied to this case by ordering the degrees A^i and B^j for (i,j) with $\beta_{ij} \neq 0$ as $\{P_1, P_2, \ldots\}$ and using the following result.

Proposition 9.12 *For every nonnegative integer i and*

$$\delta_A = \|\delta A\|_{\mathrm{F}}$$

the following estimate holds

$$\begin{aligned} \left\|(A+\delta A)^i - A^i\right\|_{\mathrm{F}} &\leq (\|A\|_2 + \delta_A)^i - \|A\|_2^i = \sum_{k=1}^{i} \binom{i}{k} \|A\|_2^{i-k}\delta_A^k \\ &= i\|A\|_2^{i-1}\delta_A + O(\delta_A^2),\ \delta A \to 0. \end{aligned} \quad (9.14)$$

Proof. We prove the inequality in (9.14) by induction on i. For $i = 0$ the inequality reduces to $0 \leq 1 - 1 = 0$. Suppose that it holds for $i = m \geq 1$, i.e., that

$$\alpha_m := \|(A+\delta A)^m - A^m\|_{\mathrm{F}} \leq \beta_m := (\|A\|_2 + \delta_A)^m - \|A\|_2^m.$$

For $i = m + 1$ we have

$$\begin{aligned} \alpha_{m+1} &= \left\|(A+\delta A)(A+\delta A)^m - A^{m+1}\right\|_{\mathrm{F}} \\ &= \|A((A+\delta A)^m - A^m) + \delta A(A+\delta A)^m\|_{\mathrm{F}} \\ &\leq \|A\|_2\alpha_m + \|(A+\delta A)^m\|_2\delta_A \\ &\leq \|A\|_2\beta_m + (\|A\|_2 + \delta_A)^m\delta_A = \beta_{m+1}, \end{aligned}$$

i.e., $\alpha_m \leq \beta_m$ implies $\alpha_{m+1} \leq \beta_{m+1}$ and the proof is complete. □

9.3 Continuous-time equations

The spectrum of the operator $\mathcal{L}$ in the continuous-time Sylvester equation

$$\mathcal{L}(X) := AX + XB = C \quad (9.15)$$

is

$$\begin{aligned}\operatorname{spect}(\mathcal{L}) &= \{\lambda_i(A) + \lambda_k(B) : i = 1, \ldots, m, \quad k = 1, \ldots, n\} \\ &= \operatorname{spect}(A) \oplus \operatorname{spect}(B).\end{aligned}$$

As corollaries of Theorems 8.17 and 8.20 we obtain the following results.

Corollary 9.13 *The norm-wise perturbation bound for equation (9.15) is*

$$\delta_X \leq \frac{\operatorname{est}(\delta_C, \delta_A, \delta_B, \Lambda, N_A, N_B)}{1 - \|\Lambda\|_2(\delta_A + \delta_B)}, \tag{9.16}$$

where $\Lambda = \left(B^\top \otimes I_m + I_n \otimes A\right)^{-1}$,

$$N_A = -\Lambda\left(X^\top \otimes I_m\right), \; N_B = -\Lambda(I_n \otimes X)$$

and the expression for est is given in (8.47). The bound (9.16) is valid for

$$\delta_A + \delta_B < \frac{1}{\|\Lambda\|_2} = \sigma_{\min}(\Lambda).$$

Corollary 9.14 *The component-wise bound of type (8.63) for equation (9.15) is*

$$|\delta X| \preceq (I_s - \Psi_1(\Delta))^{-1}\Theta_1(\Delta)$$

with

$$\begin{aligned}\Theta_1(\Delta) &= |\Lambda|\Delta_C + |N_A|\Delta_A + |N_B|\Delta_B, \\ \Psi_1(\Delta) &= |\Lambda(I_n \otimes W_A)| + \left|\Lambda\left(W_B^\top \otimes I_m\right)\right|\end{aligned}$$

and it is valid if $\operatorname{rad}(\Psi_1(\Delta)) < 1$.

Example 9.15 Consider the Sylvester equation

$$\mathcal{L}(X) := AX + XB = C$$

with

$$A = \begin{bmatrix} 1.5 & 0.5 \\ 0.5 & 1.5 \end{bmatrix}, \; B = \begin{bmatrix} 0.5 & -0.5 \\ -0.5 & 0.5 \end{bmatrix}, \; C = \begin{bmatrix} 1 & -2 \\ 2 & -1 \end{bmatrix}.$$

The Sylvester operator $\mathcal{L}$ is invertible and the solution is

$$X = \begin{bmatrix} 0 & -1 \\ 1 & 0 \end{bmatrix}.$$

The matrix L of $\mathcal{L}$ is

$$L = \begin{bmatrix} 2.0 & 0.5 & -0.5 & 0.0 \\ 0.5 & 2.0 & 0.0 & -0.5 \\ -0.5 & 0.0 & 2.0 & 0.5 \\ 0.0 & -0.5 & 0.5 & 2.0 \end{bmatrix}$$

and we have

$$N_C = L^{-1} = \frac{1}{12}\begin{bmatrix} 7 & -2 & 2 & -1 \\ -2 & 7 & -1 & 2 \\ 2 & -1 & 7 & -2 \\ -1 & 2 & -2 & 7 \end{bmatrix},$$

$$N_A = -L^{-1}\left(X^\top \otimes I_2\right) = \frac{1}{12}\begin{bmatrix} 2 & -1 & -7 & 2 \\ -1 & 2 & 2 & -7 \\ 7 & -2 & -2 & 1 \\ -2 & 7 & 1 & -2 \end{bmatrix},$$

$$N_B = -L^{-1}\left(I_2 \otimes X\right) = \frac{1}{12}\begin{bmatrix} 2 & 7 & 1 & 2 \\ -7 & -2 & -2 & -1 \\ 1 & 2 & 2 & 7 \\ -2 & -1 & -7 & -2 \end{bmatrix}.$$

Furthermore,

$$\|N_Y^\top N_Z\|_2 = 1, \; Y, Z \in \{C, A, B\}$$

and

$$\|N\|_2 = \|[N_C, N_A, N_B]\|_2 = \sqrt{3}.$$

Hence,

$$\mathrm{est}_2(\delta, N) = \sqrt{3}\sqrt{\delta_C^2 + \delta_A^2 + \delta_B^2} \le \mathrm{est}_1(\delta, N) = \mathrm{est}_3(\delta, N) = \delta_C + \delta_A + \delta_B$$

and the perturbation bound is

$$\delta_X \le \frac{\delta_C + \delta_A + \delta_B}{1 - \delta_A - \delta_B}.$$

$\diamondsuit$

Example 9.16 Consider the Sylvester equation

$$\mathcal{L}(X) := AX + XB = C$$

with

$$A = \begin{bmatrix} 1 & 9 \\ 0 & 1 \end{bmatrix}, \; B = \begin{bmatrix} 1 & 0 \\ 0 & 0 \end{bmatrix}, \; C = \begin{bmatrix} 2 & 9 \\ 0 & 1 \end{bmatrix}.$$

The Sylvester operator $\mathcal{L}$ is invertible and the solution is $X = I_2$. The matrix representation of $\mathcal{L}$ is

$$L = \mathrm{diag}(A + I_2, A)$$

and we have

$$N_C = L^{-1} = \frac{1}{4}\begin{bmatrix} 2 & -9 & 0 & 0 \\ 0 & 2 & 0 & 0 \\ 0 & 0 & 4 & -36 \\ 0 & 0 & 0 & 4 \end{bmatrix}$$

with

$$\nu := \|N_C\|_2 = 9.1098$$

and

$$N_A = N_B = -N_C.$$

Hence, the nonlocal perturbation bound is

$$\delta_X \leq \frac{\nu(\delta_C + \delta A + \delta_C)}{1 - \nu(\delta_A + \delta_B)}, \ \ \delta_A + \delta_B < \frac{1}{\nu}. \tag{9.17}$$

Taking the perturbations as

$$\delta C = \begin{bmatrix} 0 & 0 \\ -\varepsilon & 0 \end{bmatrix}, \ \delta A = \begin{bmatrix} 0 & 0 \\ \varepsilon & 0 \end{bmatrix}, \ \delta B = \begin{bmatrix} 0 & 0 \\ 0 & -\varepsilon \end{bmatrix},$$

where $\varepsilon > 0$ is a small parameter, a simple computation shows that the perturbed Sylvester operator is invertible if $\varepsilon < \varepsilon_0$, where

$$\varepsilon_0 = 2/(11 + \sqrt{117}) = 0.0917$$

(up to four digits) is the smaller positive root of the quadratic equation

$$\varepsilon^2 - 11\varepsilon + 1 = 0.$$

For $\varepsilon < \varepsilon_0$ the perturbation in X is determined by

$$\begin{aligned} \delta x_{11} &= \frac{18\varepsilon}{4 - 9\varepsilon}, \ \delta x_{21} = \frac{-4\varepsilon}{4 - 9\varepsilon}, \\ \delta x_{12} &= \frac{-9\varepsilon}{1 - 11\varepsilon + \varepsilon^2}, \ \delta x_{22} = \frac{\varepsilon(1 - \varepsilon)}{1 - 11\varepsilon + \varepsilon^2} \end{aligned}$$

and we have

$$\delta_X = \delta_X(\varepsilon) := \varepsilon\sqrt{\frac{340}{(4 - 9\varepsilon)^2} + \frac{19 - 2\varepsilon + \varepsilon^2}{(1 - 11\varepsilon + \varepsilon^2)^2}}, \ \ \varepsilon < 0.0917.$$

At the same time the bound (9.17) gives

$$\delta_X \leq f(\varepsilon) := \frac{27.3293\varepsilon}{1 - 18.2195\varepsilon}, \ \ \varepsilon < 0.0549.$$

◇

9.4 Discrete-time equations

The spectrum of the operator $\mathcal{L}$ of the discrete-time Sylvester equation

$$\mathcal{L}(X) := AXB - \alpha X = C \tag{9.18}$$

is

$$\begin{aligned}\operatorname{spect}(\mathcal{L}) &= \{\lambda_i(A)\lambda_k(B) - \alpha : i \in \{1,\ldots,m\},\ k \in \{1,\ldots,n\}\} \\ &= \operatorname{spect}(A) \otimes \operatorname{spect}(B) \ominus \{\alpha\}\end{aligned}$$

The application of Theorem 8.17 to equation (9.18) gives the following result.

Corollary 9.17 *The norm-wise perturbation bound for equation (9.18) is*

$$\delta_X \leq \frac{\operatorname{est}(\delta_C, \delta_A, \delta_B, \Lambda, N_A, N_B) + \|X\|_2\|\Lambda\|_2\delta_A\delta_B}{1 - l_A\delta_A - l_B\delta_B - \|\Lambda\|_2\delta_A\delta_B}, \tag{9.19}$$

where

$$\Lambda = \left(B^\top \otimes A - \alpha I_{mn}\right)^{-1}$$

and

$$\begin{aligned}N_A &= -\Lambda\left((XB)^\top \otimes I_m\right),\ N_B = -\Lambda\left(I_n \otimes (AX)\right), \\ l_A &= \left\|\Lambda\left(B^\top \otimes I_m\right)\right\|_2,\ l_B = \|\Lambda(I_n \otimes A)\|_2.\end{aligned}$$

The domain $\mathcal{D}$ for δ_A, δ_B in (9.19) is defined by the inequality

$$l_A\delta_A + l_B\delta_B + \|\Lambda\|_2\delta_A\delta_B < 1.$$

Finally, the component-wise bound (8.63) in this case is as follows.

Corollary 9.18 *The component-wise perturbation bound for equation (9.18) is*

$$|\delta X| \preceq (I_s - \Psi_1(\Delta) - \Theta_2(\Delta))^{-1}(\Theta_1(\Delta) + \Theta_2(\Delta)\operatorname{vec}(|X|)),$$

where

$$\begin{aligned}\Theta_1(\Delta) &= |\Lambda|\Delta_C + |N_A|\Delta_A + |N_B|\Delta_B,\ \Theta_2(\Delta) = |\Lambda|\left(W_B^\top \otimes W_A\right), \\ \Psi_1(\Delta) &= \left|\Lambda\left(B^\top \otimes I_m\right)\right|(I_n \otimes W_A) + |\Lambda(I_n \otimes A)|\left(W_B^\top \otimes I_m\right).\end{aligned}$$

9.5 Notes and references

Algebraic linear matrix equations have been intensively studied since the times of Sylvester and Kronecker [152, 215, 214], see also [196, 193, 229]. Brief historical reference may be found in [8]. In particular, the problems of existence, uniqueness and representation of the solution are solved for such equations, see e.g. [10, 12, 19, 20, 33, 36, 56, 69, 84, 94, 101, 106, 107, 153, 155, 165, 167, 189, 206, 181, 205, 224, 228, 230, 235, 236, 241]. The properties of special linear matrix transformations have also been studied [40, 41, 79, 86, 216, 223, 222]. There is a variety of techniques, algorithms and software for solving linear matrix equations [17, 13, 14, 6, 15, 18, 25, 45, 46, 50, 72, 73, 82, 88, 91, 90, 109, 164, 175, 192, 194,

202, 239, 240]. The great interest in linear matrix equations is due in a large extent to their wide application to various areas [11, 49, 48, 59, 62, 104, 105, 103, 170, 225].

Perturbation analysis for linear matrix equations, including the Sylvester and Lyapunov equations arising in linear control theory, has been done in [38, 110, 68, 95, 99, 114, 113, 112, 136].

Perturbation bounds for the standard vector linear equation $Ax = b$ are given in many textbooks [64, 36, 106, 122, 54, 224]. However, the problems of exactness of these bounds have benn considered here for the first time.

The perturbation theory for operators in abstract spaces [119] and for general linear equations [97, 98] also applies in a large scale to the perturbation analysis of linear matrix equations. Other investigations are connected with establishing bounds on the solution of linear matrix equations are given [171, 176].

Some results concerning backward perturbation analysis are given in [99, 101, 112]. Backward errors and conditioning for structured linear equations are considered in [97].

Chapter 10

General Lyapunov equations

10.1 Introductory remarks

In this chapter we present a complete perturbation analysis for general Lyapunov matrix equations. Local and nonlocal, norm-wise and component-wise, perturbation bounds are derived for real and complex Lyapunov equations, particular cases of which are the continuous- and discrete-time Lyapunov equations, arising in the theory of linear time-invariant systems. Results in this area have been already published in the literature for particular classes of Lyapunov equations, see e.g. [95, 125, 134].

The first order bounds are based on the standard induced norm as well as on the Lyapunov norm of Lyapunov operators. The latter norm allows to obtain tighter results for Lyapunov equations under symmetric perturbations of the constant term.

Conditions for invertibility of certain classes of Lyapunov operators are also presented.

Due to the highly specific structure of Lyapunov matrix equations, the results for complex equations cannot be deduced trivially from those for real equations. For this reason we treat real and complex equations separately.

10.2 Application to descriptor systems

Matrix Lyapunov equations arise naturally in many areas of linear systems theory. In this section we discuss the use of such equations in studying continuous and discrete time-invariant dynamic systems in descriptor form.

Consider the continuous time-invariant descriptor system

$$E\dot{x}(t) = Ax(t), \; t \in \mathbb{R}_+; \; x(0) = x_0 \in \mathbb{R}^n,$$

together with the cost functional

$$J_c(x) = \int_0^\infty x^\top(t)Cx(t)\mathrm{d}t,$$

where $x(t) \in \mathbb{R}^n$ and $E, A, C \in \mathbb{R}^{n\times n}$, $E \neq 0$, $C \geq 0$.

Suppose first that we have a regular descriptor system, i.e., that E is nonsingular and the matrix $E^{-1}A$ is stable,

$$\mathrm{spect}(E^{-1}A) \subset \mathbb{C}_-.$$

Then it follows from the Pontryagin maximum principal (see [167]) that

$$J_c(x) = x_0^\top X_c x_0,$$

where $X_c \geq 0$ is the unique solution of the Lyapunov equation

$$\mathcal{L}_c(X) + C = 0.$$

Here the continuous-time Lyapunov operator $\mathcal{L}_c \in \mathbf{Lin}(n, \mathbb{R})$ is defined as

$$\mathcal{L}_c(X) := (E^{-1}A)^\top X + XE^{-1}A.$$

If the matrix E is ill-conditioned with respect to inversion, this may cause numerical difficulties (the formation of $E^{-1}A$ should be, of course, avoided). An approach to deal with this problem is as follows. Setting

$$X := E^\top YE,$$

the descriptor Lyapunov equation in Y is

$$\mathcal{L}_c^\#(Y) + C = 0,$$

where the continuous-time descriptor Lyapunov operator $\mathcal{L}_c^\# \in \mathbf{Lin}(n, \mathbb{R})$ is defined by

$$\mathcal{L}_c^\#(Y) := A^\top YE + E^\top YA.$$

Note that the standard continuous-time Lyapunov equation

$$A^\top X + XA + C = 0$$

is a particular case of the descriptor equation for $E = I_n$.

If the matrix E is singular with

$$\mathrm{rank}(E) = r < n,$$

then let

$$E = USV^\top = U_1\Sigma V_1^\top$$

be the singular value decomposition of E, where the matrices $U = [U_1, U_2]$, $V = [V_1, V_2]$ are orthogonal, $U_1, V_1 \in \mathbb{R}^{n\times r}$, and

$$S := \mathrm{diag}(\Sigma, 0),\ \Sigma := \mathrm{diag}(\sigma_1(E), \dots, \sigma_r(E)).$$

Setting

$$y = V_1^\top x,\ z = V_2^\top x,$$

and

$$H := U^\top AV = [H_{ij}]$$

where H is a block 2×2 matrix with $H_{11} \in \mathbb{R}^{r\times r}$, we get

$$\begin{aligned} \Sigma\dot{y}(t) &= H_{11}y(t) + H_{12}z(t), \\ 0 &= H_{21}y(t) + H_{22}z(t). \end{aligned}$$

Suppose that $H_{22} \in \mathbb{R}^{(n-r)\times(n-r)}$ is nonsingular, i.e., that the descriptor system is of index 1 (this can always be achieved, see [24, 167]), and

$$x_0 \in \mathrm{Ker}\left([H_{21}, H_{22}]V^\top\right).$$

Then

$$z(t) = -H_{22}^{-1}H_{21}y(t),$$

the vector y is the state of the descriptor system

$$\Sigma\dot{y}(t) = By(t),\ t \in \mathbb{R}_+,\ y(0) = y_0 := V_1^\top x_0,$$

with

$$B := H_{11} - H_{12}H_{22}^{-1}H_{21}.$$

The cost functional takes the form

$$J_c(x) = K_c(y) := \int_0^\infty y^\top(t)Dy(t)\mathrm{d}t,$$

where

$$D := \left[I_n, -H_{21}^\top H_{22}^{-\top}\right] V^\top CV \begin{bmatrix} I_n \\ -H_{22}^{-1}H_{21} \end{bmatrix}.$$

If

$$\mathrm{spect}(\Sigma^{-1}B) \subset \mathbb{C}_-,$$

then we have

$$K_c(y) = y_0^\top Ty_0,$$

where T solves the Lyapunov equation

$$(\Sigma^{-1}B)^\top T + T\Sigma^{-1}B + D = 0,$$

i.e., the problem is reduced to the regular case.

Consider similarly the discrete-time descriptor system

$$Ex(t+1) = Ax(t),\ t \in \{0, 1, \ldots\};\ x(0) = x_0 \in \mathbb{R}^n,$$

together with the cost functional

$$J_d(x) = \sum_{t=0}^{\infty} x^\top(t) C x(t),$$

using the same notation as in the continuous-time case. Suppose first that the matrix E is nonsingular and the matrix $E^{-1}A$ is convergent, i.e.,

$$\operatorname{spect}(E^{-1}A) \subset \mathbb{D}_1.$$

Then, see [167], we have

$$J_d(x) = x_0^\top X_d x_0,$$

where X_d is the unique nonnegative definite solution of the Lyapunov equation

$$\mathcal{L}_d(X) + C = 0,$$

where the discrete-time Lyapunov operator $\mathcal{L}_d \in \mathbf{Lin}(n, \mathbb{R})$ is defined by

$$\mathcal{L}_d(X) := (E^{-1}A)^\top X E^{-1}A - X.$$

To avoid the computation of the matrix $E^{-1}A$, we set

$$X := E^\top Y E.$$

The discrete-time descriptor Lyapunov equation for Y is

$$\mathcal{L}_d^{\#}(Y) + C = 0,$$

where

$$\mathcal{L}_d^{\#}(Y) := A^\top Y A - E^\top Y E.$$

The standard discrete-time Lyapunov equation

$$A^\top X A - X + C = 0$$

is a particular case of the descriptor discrete-time Lyapunov equation, corresponding to $E = I_n$.

The case when E is singular is treated similarly as in the continuous-time case. Complex descriptor systems with $x(t) \in \mathbb{C}^n$, etc., may be studied in the same way, see [167].

10.3 Additive matrix operators

To solve the perturbation problem for general complex Lyapunov equations we need some facts about additive matrix operators. In particular, we are interested in real representations of complex additive (not necessarily linear) operators.

Consider a matrix function (or matrix operator)

$$\mathcal{F} = [f_{ij}] : \mathbb{F}^{n\times n} \to \mathbb{F}^{n\times n},$$

where $f_{ij} : \mathbb{F}^{n\times n} \to \mathbb{F}$ are scalar functions of a matrix argument. By $\mathcal{F}^\top = [f_{ji}]$ and $\mathcal{F}^{\mathrm{H}} = [\overline{f}_{ji}]$ we denote the transposed and complex conjugate transposed operators to the operator $\mathcal{F}$, respectively.

Every matrix operator

$$\mathcal{F} : \mathbb{F}^{n\times n} \to \mathbb{F}^{n\times n}$$

is equivalent to a vector function

$$f : \mathbb{F}^{n^2} \to \mathbb{F}^{n^2}$$

by setting

$$f(x) = \mathrm{vec}(\mathcal{F}(\mathrm{vec}^{-1}(x)),$$

where

$$x := \mathrm{vec}(X),\ \ X = \mathrm{vec}^{-1}(x).$$

In turn, a complex operator $\mathcal{F} : \mathbb{C}^{n\times n} \to \mathbb{C}^{n\times n}$ may be identified with the real operator

$$\mathcal{F}^{\mathbb{R}} : \mathbb{R}^{n\times n} \times \mathbb{R}^{n\times n} \to \mathbb{R}^{n\times n} \times \mathbb{R}^{n\times n},$$

that is defined as follows. Let $X = X_0 + \imath X_1$ and

$$\mathcal{F}(X) = \mathcal{F}_0(X_0, X_1) + \imath\mathcal{F}_1(X_0, X_1),$$

where X_i and $\mathcal{F}_i$ are real. Then we may set

$$\mathcal{F}^{\mathbb{R}}(X_0, X_1) := (\mathcal{F}_0(X_0, X_1), \mathcal{F}_1(X_0, X_1)).$$

We can also set

$$\mathcal{F}^{\mathbb{R}}(X_0, X_1) := \begin{bmatrix} \mathcal{F}_0(X_0, X_1) \\ \mathcal{F}_1(X_0, X_1) \end{bmatrix}.$$

In this case the co-domain of $\mathcal{F}^{\mathbb{R}}$ is $\mathbb{R}^{2n\times n}$.

If $Z = Z_0 + \imath Z_1 \in \mathbb{C}^{n\times n}$, where $Z_0, Z_1 \in \mathbb{R}^{n\times n}$, then set

$$\mathrm{vec}^{\mathbb{R}}(Z) := \begin{bmatrix} \mathrm{vec}(Z_0) \\ \mathrm{vec}(Z_1) \end{bmatrix} \in \mathbb{R}^{2n^2},\ \ Z^{\mathbb{R}} := \begin{bmatrix} Z_0 & -Z_1 \\ Z_1 & Z_0 \end{bmatrix} \in \mathbb{C}^{2n\times 2n}.$$

For

$$A, B, Z \in \mathbb{C}^{n\times n}$$

and

$$z = z_0 + \imath z_1 \in \mathbb{C}^n, \; z_0, z_1 \in \mathbb{R}^n$$

we have

$$\mathrm{vec}^{\mathbb{R}}(AZB) = (B^\top \otimes A)^{\mathbb{R}} \mathrm{vec}^{\mathbb{R}}(Z)$$

and

$$\mathrm{vec}^{\mathbb{R}}(Az) = A^{\mathbb{R}} \begin{bmatrix} z_0 \\ z_1 \end{bmatrix},$$

where

$$(B^\top \otimes A)^{\mathbb{R}} = \begin{bmatrix} B_0^\top \otimes A_0 - B_1^\top \otimes A_1 & -\left(B_1^\top \otimes A_0 + B_0^\top \otimes A_1\right) \\ B_1^\top \otimes A_0 + B_0^\top \otimes A_1 & B_0^\top \otimes A_0 - B_1^\top \otimes A_1 \end{bmatrix}.$$

Hence, if $\mathcal{L} \in \mathbf{Lin}(n, \mathbb{C})$ and

$$\mathrm{Mat}(\mathcal{L}) \in \mathbb{C}^{n^2 \times n^2}$$

is the matrix of $\mathcal{L}$, then

$$\mathrm{vec}^{\mathbb{R}}(\mathcal{L}(Z)) = \mathrm{Mat}^{\mathbb{R}}(\mathcal{L}) \mathrm{vec}^{\mathbb{R}}(Z).$$

We recall [119] the following definitions.

Definition 10.1 *An operator $\mathcal{F}$ is* additive *if*

$$\mathcal{F}(X + Y) = \mathcal{F}(X) + \mathcal{F}(Y),$$

homogeneous *if*

$$\mathcal{F}(\alpha X) = \alpha X$$

and semi-homogeneous *if*

$$\mathcal{F}(\alpha X) = \overline{\alpha} \mathcal{F}(X)$$

for all $X, Y \in \mathbb{F}^{n \times n}$ and $\alpha \in \mathbb{F}$. An operator $\mathcal{F}$ is linear *if it is additive and homogeneous,*

$$\mathcal{F}(\alpha X + \beta Y) = \alpha \mathcal{F}(X) + \beta \mathcal{F}(Y),$$

and semi-linear *if it is additive and semi-homogeneous,*

$$\mathcal{F}(\alpha X + \beta Y) = \overline{\alpha} \mathcal{F}(X) + \overline{\beta} \mathcal{F}(Y),$$

for all $X, Y \in \mathbb{F}^{n \times n}$ and $\alpha, \beta \in \mathbb{F}$.

In the real case $\mathbb{F} = \mathbb{R}$ the properties of linearity and semi-linearity coincide. Also, a complex semi-linear operator becomes linear if we consider $\mathbb{C}^{n \times n}$ as a linear space over $\mathbb{R}$ instead of $\mathbb{C}$. This is based on the observation that a linear space V over any field $\mathbb{F}$ (including $V = \mathbb{F}$) is also a linear space over any subfield $\mathbb{E}$ of $\mathbb{F}$.

Any general operator $\mathcal{L} \in \mathbf{Lin}(n, \mathbb{F})$ may be represented as

$$\mathcal{L}(X) = \sum_{k=1}^{r} A_i X B_i, \tag{10.1}$$

where $A_i, B_i \in \mathbb{F}^{n\times n}$ are given matrix coefficients and r is the Sylvester index of $\mathcal{L}$, i.e., the minimum number of terms, required in the representation of $\mathcal{L}$ as a sum of elementary linear operators $X \mapsto A_i X B_i$, see [125] and Appendix E. Similarly, a general semi-linear operator $\mathcal{M} : \mathbb{C}^{n\times n} \to \mathbb{C}^{n\times n}$ admits the representation

$$\mathcal{M}(X) = \mathcal{L}(X^{\mathrm{H}}) = \sum_{k=1}^{r} A_i X^{\mathrm{H}} B_i \tag{10.2}$$

or

$$\mathcal{M}(X) = \mathcal{N}(\overline{X}) = \sum_{k=1}^{l} C_i \overline{X} D_i, \tag{10.3}$$

where $\mathcal{L}, \mathcal{N} \in \mathbf{Lin}(n, \mathbb{C})$. The real versions of (10.2) and (10.3) are

$$\begin{aligned} \mathrm{vec}^{\mathbb{R}}(\mathcal{M}(X)) &= \mathrm{Mat}^{\mathbb{R}}(\mathcal{L}) \begin{bmatrix} P_{n^2}\mathrm{vec}(X_0) \\ -P_{n^2}\mathrm{vec}(X_1) \end{bmatrix} \\ &= \mathrm{Mat}^{\mathbb{R}}(\mathcal{L})\mathrm{diag}\left(P_{n^2}, -P_{n^2}\right)\mathrm{vec}^{\mathbb{R}}(X) \end{aligned}$$

and

$$\mathrm{vec}^{\mathbb{R}}(\mathcal{M}(X)) = \mathrm{Mat}^{\mathbb{R}}(\mathcal{N})\mathrm{diag}\left(I_{n^2}, -I_{n^2}\right)\mathrm{vec}^{\mathbb{R}}(X),$$

where

$$X = X_0 + \imath X_1 \in \mathbb{C}^{n\times n};\ X_0, X_1 \in \mathbb{R}^{n\times n}.$$

Thus, we come to the following definition.

Definition 10.2 *The* matrix representation *(or briefly, the* matrix*) of the real version $\mathcal{M}^{\mathbb{R}}$ of the semi-linear operator $\mathcal{M}$ is*

$$\begin{aligned} \mathrm{Mat}(\mathcal{M}^{\mathbb{R}}) &= \mathrm{Mat}^{\mathbb{R}}(\mathcal{L})\mathrm{diag}\left(P_{n^2}, -P_{n^2}\right) \\ &= \mathrm{Mat}^{\mathbb{R}}(\mathcal{N})\mathrm{diag}\left(I_{n^2}, -I_{n^2}\right). \end{aligned}$$

Note that a semi-linear complex operator $\mathcal{F}$ is in general not differentiable. However, its real version $\mathcal{F}^{\mathbb{R}}$ is a linear operator. We note that if $\mathcal{F}$ is a linear operator, so is $\mathcal{F}^{\top}$, while $\mathcal{F}^{\mathrm{H}}$ is semi-linear.

Taking the vec operation on both sides of the expressions (10.1) and (10.2) for a linear and a semi-linear operator we get

$$\mathrm{vec}(\mathcal{L}(X)) = L\mathrm{vec}(X)$$

and

$$\mathrm{vec}(\mathcal{M}(X)) = LP_{n^2}\mathrm{vec}(\overline{X}),$$

where

$$L := \mathrm{Mat}(\mathcal{L}) := \sum_{i=1}^{r} B_i^\top \otimes A_i \in \mathbb{F}^{n^2 \times n^2}$$

is the matrix of the linear operator $\mathcal{L}$.

We also discuss complex additive operators $\mathcal{F}$, which may be represented as sum of a linear and a semi-linear operator, i.e.,

$$\mathcal{F}(X) = \mathcal{L}_1(X) + \mathcal{L}_2(X^{\mathrm{H}}), \tag{10.4}$$

where $\mathcal{L}_1, \mathcal{L}_2 \in \mathbf{Lin}(n, \mathbb{C})$. In this case we have

$$\mathrm{vec}^{\mathbb{R}}(\mathcal{F}(X)) = \mathrm{Mat}(\mathcal{F}^{\mathbb{R}})\mathrm{vec}^{\mathbb{R}}(X),$$

where the *matrix representation* of the real version $\mathcal{F}^{\mathbb{R}}$ of the semi-linear operator $\mathcal{F}$ in (10.4) is

$$\mathrm{Mat}(\mathcal{F}^{\mathbb{R}}) := \left(\mathrm{Mat}^{\mathbb{R}}(\mathcal{L}_1) + \mathrm{Mat}^{\mathbb{R}}(\mathcal{L}_2)\mathrm{diag}\left(P_{n^2}, -P_{n^2}\right)\right) \mathrm{vec}^{\mathbb{R}}(X).$$

In the following we introduce polynomial and pseudo-polynomial operators.

Definition 10.3 *An operator* $\mathcal{F} = [f_{ij}]$ *is called* polynomial *if its elements* $f_{ij} : \mathbb{F}^{n\times n} \to \mathbb{F}$ *are polynomial functions.*

A polynomial operator $\mathcal{F}$ is globally Fréchet differentiable in the sense that for every $X_0 \in \mathbb{F}^{n\times n}$ we have

$$\mathcal{F}(X_0 + Z) = \mathcal{F}(X_0) + \mathcal{L}(Z, X_0) + \mathcal{H}(Z, X_0),$$

where $\mathcal{L}(\cdot, X_0) \in \mathbf{Lin}(n, \mathbb{F})$ and

$$\lim_{Z\to 0} \frac{\|\mathcal{H}(Z, X_0)\|}{\|Z\|} = 0.$$

In this case the linear operator $\mathcal{L}(\cdot, X_0)$ is referred to as the *Fréchet derivative* of $\mathcal{F}$ at the point X_0 and is denoted as $\mathcal{F}_X(X_0)(\cdot)$ or briefly as $\mathcal{F}_X(\cdot)$, see Appendix A.

Definition 10.4 *A complex operator* $\mathcal{F}$ *is called* pseudo-polynomial *if it may be represented as*

$$\mathcal{F}(X) = \mathcal{G}(X, X^{\mathrm{H}}), \tag{10.5}$$

where

$$\mathcal{G} : \mathbb{F}^{n\times n} \times \mathbb{F}^{n\times n} \to \mathbb{F}^{n\times n}$$

is a polynomial operator.

Pseudo-polynomial operators are not differentiable, but their real versions are real polynomial operators. If $\mathcal{F}$ is a pseudo-polynomial operator, given by (10.5), we may define the additive operator $\mathcal{F}_X(X_0)(\cdot)$ by

$$\mathcal{F}_X(X_0)(Z) := \mathcal{G}_1(X_0)(Z) + \mathcal{G}_2(X_0)(Z^{\mathrm{H}}),$$

where $\mathcal{G}_k(X_0)$ is the partial Fréchet derivative of $\mathcal{G}(X_1, X_2)$ in X_k, computed at $X_1 = X_0$, $X_2 = X_0^{\mathrm{H}}$. We have

$$F(X_0 + Z) = \mathcal{F}(X_0) + \mathcal{F}_X(X_0)(Z) + \mathcal{H}(Z, X_0),$$

where

$$\mathcal{H}(Z, X_0) = o(\|Z\|), \ Z \to 0.$$

Thus, $F_X(X_0)(\cdot)$ is an analogue of the Fréchet derivative in the case of pseudo-polynomial operators and is referred to as the *Fréchet pseudo-derivative* of $\mathcal{F}$ at the point X_0 (Appendix A). Whenever they exist, the Fréchet derivatives and pseudo-derivatives are unique.

Definition 10.5 *If $\|\cdot\|$ is a norm in $\mathbb{F}^{n\times n}$, then the* induced norm *of an operator $\mathcal{L}$ from $\mathbf{Lin}(n, \mathbb{F})$ is defined as*

$$\|\mathcal{L}\| := \max\{\|\mathcal{L}(X)\| : \|X\| = 1\}. \tag{10.6}$$

If the Frobenius norm in $\mathbb{F}^{n\times n}$ is used, then

$$\begin{aligned} \|\mathcal{L}\|_{\mathrm{F}} &:= \max\{\|\mathcal{L}(X)\|_{\mathrm{F}} : \|X\|_{\mathrm{F}} = 1\} \\ &= \max\{\|\mathrm{vec}(\mathcal{L}(X))\|_2 : \|\mathrm{vec}(X)\|_2 = 1\} \\ &= \max\{\|\mathrm{Mat}(\mathcal{L})\mathrm{vec}(X)\|_2 : \|\mathrm{vec}(X)\|_2 = 1\} \\ &= \|\mathrm{Mat}(\mathcal{L})\|_2. \end{aligned} \tag{10.7}$$

When the operator $\mathcal{M}$ is semi-linear,

$$\mathcal{M}(X) = \mathcal{L}(X^{\mathrm{H}}), \ \mathcal{L} \in \mathbf{Lin}(n, \mathbb{F}),$$

we may again define its norm via (10.6) and (10.7) and thus, the induced norm of $\mathcal{M}$ is equal to the induced norm of the underlying operator $\mathcal{L}$. However, if the complex operator $\mathcal{F}$ is only additive,

$$\mathcal{F}(X) = \mathcal{L}_1(X) + \mathcal{L}_2(X^{\mathrm{H}}), \ \mathcal{L}_1, \mathcal{L}_2 \in \mathbf{Lin}(n, \mathbb{C}), \tag{10.8}$$

then the determination of its induced norm is more subtle. Let

$$L_k = L_{k0} + \imath L_{k1} \in \mathbb{C}^{n^2\times n^2}, \ k = 1, 2,$$

be the matrix of the operator $\mathcal{L}_k$, where the matrices L_{kj} are real. Define the norm of the additive operator $\mathcal{F}$, induced by the Frobenius norm in $\mathbb{C}^{n\times n}$, via

$$\|\mathcal{F}\| := \max\{\|\mathcal{F}(X)\|_{\mathrm{F}} : \|X\|_{\mathrm{F}} \le 1\}.$$

Then

$$\|\mathcal{F}\| = \max\{\|\text{vec}(\mathcal{F}(X))\|_2 : \|\text{vec}(X)\|_2 \le 1\}.$$

Recalling that

$$\begin{aligned}\text{vec}(\mathcal{F}(X)) &= \text{vec}(\mathcal{L}_1(X)) + \text{vec}(\mathcal{L}_2(X^{\text{H}}))\\ &= L_1\text{vec}(X) + L_2P_{n^2}\text{vec}(\overline{X})\end{aligned}$$

we get

$$\|\mathcal{F}\| = \nu(L_1, L_2) := \|M(L_1, L_2)\|_2, \tag{10.9}$$

where

$$M(L_1, L_2) := \text{Mat}(\mathcal{F}^{\mathbb{R}}) = \begin{bmatrix} L_{10} + L_{20}P_{n^2} & -L_{11} + L_{21}P_{n^2} \\ L_{11} + L_{21}P_{n^2} & L_{10} - L_{20}P_{n^2} \end{bmatrix} \tag{10.10}$$

is the matrix of the real version $\mathcal{F}^{\mathbb{R}}$ of $\mathcal{F}$. Thus, we have proved the following proposition.

Proposition 10.6 *The induced norm of an additive operator $\mathcal{F}$ with a representation (10.8) is equal to the induced norm of its real representation when the underlying norm in $\mathbb{C}^{n\times n}$ is the Frobenius norm.*

Definition 10.7 *An operator $\mathcal{F} : \mathbb{F}^{n\times n} \to \mathbb{F}^{n\times n}$ is said to be* symmetric *if*

$$\mathcal{F}^H(X) = \mathcal{F}(X^H).$$

Linear symmetric matrix operators are also called Lyapunov operators.

More details about Lyapunov operators are given in [125] and Appendix F.

Every Lyapunov operator $\mathcal{F}$ may be represented as

$$\mathcal{F}(X) = \mathcal{L}(X) + \mathcal{L}^{\text{H}}(X^{\text{H}}) \tag{10.11}$$

where $\mathcal{L} \in \mathbf{Lin}(n, \mathbb{F})$. Thus, a general Lyapunov operator $\mathcal{F} \in \mathbf{Lin}(n, \mathbb{F})$ has the representation

$$\mathcal{F}(X) = \sum_{i=1}^{r_1} \left(A_iXB_i^{\text{H}} + B_iXA_i^{\text{H}}\right) + \sum_{k=1}^{r_2} \varepsilon_k C_kXC_k^{\text{H}}, \tag{10.12}$$

where $A_i, B_i, C_k \in \mathbb{F}^{n\times n}$ are given matrices and $\varepsilon_k = \pm 1$. This form seems different from (10.11) in view of (10.1), but it is not, since the symmetric monomial terms may be expressed as

$$C_kXC_k^{\text{H}} = A_kXB_k^{\text{H}} + B_kXA_k^{\text{H}},$$

where

$$A_k = \alpha_kC_k,\ B_k = \beta_kC_k,$$

and α_k, β_k are scalars from $\mathbb{F}$ with $\alpha_k\beta_k = 1/2$. However, we choose the representation (10.12), in which the symmetric terms $C_kXC_k^{\mathrm{H}}$ (if any) are grouped separately in order to reduce the number of terms in the representation of $\mathcal{L}$ as a sum of elementary linear operators. As usual, summation from 1 to 0 is considered void. Thus, $r_1 = 0$ means that there are no terms $A_iXB_i^{\mathrm{H}} + B_iXA_i^{\mathrm{H}}$, while $r_2 = 0$ means that there are no symmetric terms $C_kXC_k^{\mathrm{H}}$.

For Lyapunov operators $\mathcal{L} \in \mathbf{Lin}(n, \mathbb{F})$, in addition to the standard norm (10.7), a new symmetrized, or Lyapunov norm may be introduced, see [125] and Appendix F.

Definition 10.8 *The* symmetrized, *or* Lyapunov *norm of the operator* $\mathcal{L} \in \mathbf{Lin}(n,\mathbb{F})$ *is given by*

$$\|\mathcal{L}\|_* := \max\{\|\mathcal{L}(X)\|_{\mathrm{F}} : \|X\|_{\mathrm{F}} = 1, \ X = X^{\mathrm{H}}\}. \tag{10.13}$$

In the real case this norm may be computed via

$$\|\mathcal{L}\|_* = \|LQ\|_2, \ \mathcal{L} \in \mathbf{Lin}(n, \mathbb{R}), \tag{10.14}$$

where L is the matrix of $\mathcal{L}$ and

$$Q = [Q_{ij}]_{i,j=1}^{n,n}$$

is a specific $n^2 \times n(n+1)/2$ matrix. It is an $n \times n$ block matrix with blocks Q_{ij}, which are $n \times j$ matrices, given by:

$$\begin{aligned} Q_{ij} &:= 0 \text{ if } i > j, \\ Q_{11} &:= \begin{bmatrix} 1 \\ 0 \\ \vdots \\ 0 \end{bmatrix}, \ Q_{kk} := \begin{bmatrix} \frac{1}{\sqrt{2}} I_{k-1} & 0 \\ 0 & 1 \\ 0 & 0 \\ \vdots & \vdots \\ 0 & 0 \end{bmatrix}, \\ Q_{ij} &:= \frac{1}{\sqrt{2}} E_{ji} \text{ if } i < j. \end{aligned}$$

Here E_{ji} is an $n \times j$ matrix with a single nonzero element, equal to 1, in position (j, i). For instance, the matrices $Q = Q_n$ for $n = 2, 3, 4$ are, with $q := 1/\sqrt{2}$,

$$Q_2 = \left[\begin{array}{c|cc} 1 & 0 & 0 \\ 0 & q & 0 \\ \hline 0 & q & 0 \\ 0 & 0 & 1 \end{array}\right], Q_3 = \left[\begin{array}{c|cc|ccc} 1 & 0 & 0 & 0 & 0 & 0 \\ 0 & q & 0 & 0 & 0 & 0 \\ 0 & 0 & 0 & q & 0 & 0 \\ \hline 0 & q & 0 & 0 & 0 & 0 \\ 0 & 0 & 1 & 0 & 0 & 0 \\ 0 & 0 & 0 & 0 & q & 0 \\ \hline 0 & 0 & 0 & q & 0 & 0 \\ 0 & 0 & 0 & 0 & q & 0 \\ 0 & 0 & 0 & 0 & 0 & 1 \end{array}\right],$$

$$
Q_4 = \left[\begin{array}{c|cc|ccc|cccc}
1 & 0 & 0 & 0 & 0 & 0 & 0 & 0 & 0 & 0 \\
0 & q & 0 & 0 & 0 & 0 & 0 & 0 & 0 & 0 \\
0 & 0 & 0 & q & 0 & 0 & 0 & 0 & 0 & 0 \\
0 & 0 & 0 & 0 & 0 & 0 & q & 0 & 0 & 0 \\
\hline
0 & q & 0 & 0 & 0 & 0 & 0 & 0 & 0 & 0 \\
0 & 0 & 1 & 0 & 0 & 0 & 0 & 0 & 0 & 0 \\
0 & 0 & 0 & 0 & q & 0 & 0 & 0 & 0 & 0 \\
0 & 0 & 0 & 0 & 0 & 0 & 0 & q & 0 & 0 \\
\hline
0 & 0 & 0 & q & 0 & 0 & 0 & 0 & 0 & 0 \\
0 & 0 & 0 & 0 & q & 0 & 0 & 0 & 0 & 0 \\
0 & 0 & 0 & 0 & 0 & 1 & 0 & 0 & 0 & 0 \\
0 & 0 & 0 & 0 & 0 & 0 & 0 & 0 & q & 0 \\
\hline
0 & 0 & 0 & 0 & 0 & 0 & q & 0 & 0 & 0 \\
0 & 0 & 0 & 0 & 0 & 0 & 0 & q & 0 & 0 \\
0 & 0 & 0 & 0 & 0 & 0 & 0 & 0 & q & 0 \\
0 & 0 & 0 & 0 & 0 & 0 & 0 & 0 & 0 & 1
\end{array}\right].
$$

The determination of the symmetrized norm in the complex case is more involved. Let the matrix $L \in \mathbb{C}^{n^2 \times n^2}$ of the operator $\mathcal{L} \in \mathbf{Lin}(n, \mathbb{C})$ be represented as $L = L_0 + \imath L_1$, where $L_0, L_1 \in \mathbb{R}^{n^2 \times n^2}$. It is shown in [125] that

$$
\|\mathcal{L}\|_* = \left\| L^{\mathbb{R}} \mathrm{diag}(Q, \widehat{Q}) \right\|_2 = \left\| \begin{bmatrix} L_0 Q & -L_1 \widehat{Q} \\ L_1 Q & L_0 \widehat{Q} \end{bmatrix} \right\|_2, \quad \mathcal{L} \in \mathbf{Lin}(n, \mathbb{C}). \qquad (10.15)
$$

The $n^2 \times n(n-1)/2$ matrix $\widehat{Q}$ is obtained from Q by deleting the columns with 1's and numbered as

$$
k(k+1)/2, \ k \in \overline{1, n},
$$

and changing the sign of every second element $1/\sqrt{2}$ in each column of the reduced matrix.

The ratio

$$
\frac{\|\mathcal{L}\|_*}{\|\mathcal{L}\|} \leq 1
$$

of the symmetrized and usual norms may be arbitrarily small for some Lyapunov operators $\mathcal{L}$, when the underlying norm in $\mathbb{F}^{n \times n}$ is the Frobenius norm. Thus, the use of the symmetrized norm is preferable in order to get tighter perturbation bounds.

Definition 10.9 *An operator $\mathcal{F} : \mathbb{F}^{n \times n} \to \mathbb{F}^{n \times n}$ is called* affine *if it may be represented as*

$$
\mathcal{F}(X) = A + \mathcal{L}(X),
$$

where $A \in \mathbb{F}^{n \times n}$ and $\mathcal{L} \in \mathbf{Lin}(n, \mathbb{F})$. An affine operator $\mathcal{F}$ is called symmetric *if in the above representation $A = A^{\mathrm{H}}$ and $\mathcal{L}$ is a Lyapunov operator.*

10.4 Perturbation problem

Consider the general Lyapunov equation

$$F(X,P) := A_0 + \sum_{i=1}^{r_1} \left(A_i X B_i^{\mathrm{H}} + B_i X A_i^{\mathrm{H}}\right) + \sum_{k=1}^{r_2} \varepsilon_k C_k X C_k^{\mathrm{H}} = 0, \tag{10.16}$$

where $X \in \mathbb{F}^{n\times n}$ is the unknown matrix. The function

$$F(\cdot,P) : \mathbb{F}^{n\times n} \to \mathbb{F}^{n\times n}$$

is an affine symmetric matrix operator, depending on the parameter matrix $(1 + 2r_1 + r_2)$-tuple

$$P := (A_0; A_1, B_1, \ldots, A_{r_1}, B_{r_1}; C_1, \ldots, C_{r_2}),$$

where $A_0 = A_0^{\mathrm{H}}$. With certain abuse of notation we identify P with the set of the matrix coefficients, and write $A_j \in P$, etc.

In the real case the dependence of $F(X,\cdot)$ on the matrices $Z \in P$ is polynomial and we denote by

$$F_Z(X,P) \in \mathbf{Lin}(n,\mathbb{R}) \tag{10.17}$$

the partial Fréchet derivative of F in the corresponding matrix argument $Z = X$ or $Z \in P$, computed at the point (X,P).

In the complex case, however, $F(X,\cdot)$ is affine in A_0 and is a pseudo-polynomial operator in each $Z \in P\backslash\{A_0\}$. Hence, the partial Fréchet derivatives in $Z \in P\backslash\{A_0\}$ do not exist. In this case we use the same notation (10.17) for the partial Fréchet pseudo-derivative of F in $Z \in P\backslash\{A_0\}$, computed at the point (X,P).

Since the operator $F(\cdot,P)$ is affine, the partial Fréchet derivative

$$F_X(X_0,P)(\cdot) \in \mathbf{Lin}(n,\mathbb{F})$$

does not depend on X_0 and is in the following denoted by $F_X(\cdot)$. We assume that the operator $F_X(\cdot)$ is invertible, i.e., that its matrix $L_X := \mathrm{Mat}(F_X)$ is nonsingular. Then equation (10.16) has a unique solution X for every A_0 and, in view of $A_0 = A_0^{\mathrm{H}}$, we have $X = X^{\mathrm{H}}$.

The perturbation problem for equation (10.16) is stated as follows. Let the matrices from P be perturbed as

$$A_0 \mapsto A_0 + \delta A_0,\ A_i \mapsto A_i + \delta A_i,\ B_i \mapsto B_i + \delta B_i,\ C_k \mapsto C_k + \delta C_k,$$

where $\delta A_0 = \delta A_0^{\mathrm{H}}$. Denote by $P + \delta P$ the perturbed tuple P, in which every matrix $Z \in P$ is replaced by $Z + \delta Z$. Then the perturbed equation is

$$F(X + \delta X, P + \delta P) = 0. \tag{10.18}$$

In general, some of the matrices from P may not be perturbed and we set the corresponding perturbations to be zero. Denote by

$$\widetilde{P} := \{Z_1, Z_2, \ldots, Z_r\} \subset P$$

the set of matrices from P, which are perturbed. We also write $\widetilde{P} = (Z_1, Z_2, \ldots, Z_r)$ and, if necessary, consider $\widetilde{P}$ as an element of the linear space $(\mathbb{F}^{n\times n})^r$. For instance, given the standard continuous-time real Lyapunov equation

$$AX + XA^\top + C = 0,$$

we have

$$P = (C; A, I_n, I_n, A^\top),\ \widetilde{P} = (C, A),$$

if only perturbations in C and A are considered.

Since the operator F_X is invertible, the perturbed equation (10.18) has a unique solution $X + \delta X = (X + \delta X)^{\mathrm{H}}$ in the neighborhood of X, if the perturbation δP is sufficiently small. Moreover, in this case the elements of the real representation of δX are analytic functions of the elements of the real representations of the matrices from δP.

Denote by

$$\begin{aligned} \delta^0 &:= \left[\delta_1^0, \delta_2^0, \delta_3^0, \ldots, \delta_{2+2r_1}^0, \ldots, \delta_\nu^0\right]^\top \\ &:= \left[\delta_{A_0}, \delta_{A_1}, \delta_{B_1}, \ldots, \delta_{C_{r_1}}, \ldots, \delta_{C_{r_2}}\right]^\top \in \mathbb{R}_+^\nu \end{aligned} \tag{10.19}$$

the full norm vector of absolute perturbations $\delta_Z := \|\delta Z\|_{\mathrm{F}}$ in the data matrices, where $\nu := 1 + 2r_1 + r_2$. Let also

$$\delta := [\delta_1, \delta_2, \ldots, \delta_r]^\top := [\delta_{Z_1}, \delta_{Z_2}, \ldots, \delta_{Z_r}]^\top \in \mathbb{R}_+^r \tag{10.20}$$

be the norm vector of perturbations of the matrices δZ. Note that some of the elements of δ^0 may be zero (when the corresponding matrices are not perturbed), while all elements of δ are positive, since by assumption they are the norms of the nonzero perturbations in the matrix coefficients.

The perturbation problem is to find a bound

$$\delta_X \leq f(\delta),\ \delta \in \Omega \subset \mathbb{R}_+^r, \tag{10.21}$$

for the perturbation

$$\delta_X := \|\delta X\|_{\mathrm{F}},$$

where Ω is a given set and f is a continuous function, nondecreasing in all of its arguments and satisfying $f(0) = 0$. In the following subsections, a first order local bound

$$\delta_X \leq f_1(\delta) + O(\|\delta\|^2),\ \delta \to 0,$$

is derived, which is then incorporated in the nonlocal bound (10.21). The inclusion $\delta \in \Omega$ also guarantees that the perturbed equation (10.18) has a unique solution $X + \delta X$.

Estimates in terms of relative perturbations

$$\alpha_j = \varepsilon_{Z_j} := \frac{\|\delta Z_j\|_{\mathrm{F}}}{\|Z_j\|_{\mathrm{F}}}, \quad Z_j \in \widetilde{P},$$

for

$$\varepsilon_X := \frac{\|\delta X\|_{\mathrm{F}}}{\|X\|_{\mathrm{F}}}$$

are straightforward when $A_0 \neq 0$, and hence, $X \neq 0$. Indeed, we have

$$\varepsilon_X \leq \frac{f\left(\|Z_1\|_{\mathrm{F}}\alpha_1, \ldots, \|Z_r\|_{\mathrm{F}}\alpha_r\right)}{\|X\|_{\mathrm{F}}}.$$

In the following sections we present local and nonlocal perturbation bounds for the general Lyapunov equation (10.16).

10.5 Local perturbation analysis

10.5.1 Condition numbers

Consider the calculation of condition numbers for equation (10.16). Since $F(X,P) = 0$, the perturbed equation (10.18) may be written as

$$F(X + \delta X, P + \delta P) := F_X(\delta X) + \sum_{Z \in P} F_Z(\delta Z) + G(\delta X, \delta P) = 0, \qquad (10.22)$$

where $F_Z(\cdot)$ are the partial Fréchet derivatives (in the real case) or pseudo-derivatives (in the complex case) of $F(X, \cdot)$ in the corresponding matrix arguments $Z \in P$. In both cases $F_X(\cdot)$ is a linear symmetric operator and $F_{A_0}(\cdot)$ is the identity operator. The matrix $G(\delta X, \delta P)$ contains second and higher order terms in δX, δP.

A straightforward calculation leads to

$$\begin{aligned}
F_X(Z) &= \sum_{i=1}^{r_1} \left(A_i Z B_i^{\mathrm{H}} + B_i Z A_i^{\mathrm{H}}\right) + \sum_{k=1}^{r_2} C_k Z C_k^{\mathrm{H}}, \\
F_{A_0}(Z) &= Z, \\
F_{A_i}(Z) &= Z X B_i^{\mathrm{H}} + B_i X Z^{\mathrm{H}}, \\
F_{B_i}(Z) &= A_i X Z^{\mathrm{H}} + Z X A_i^{\mathrm{H}}, \\
F_{C_k}(Z) &= \varepsilon_k \left(Z X C_k^{\mathrm{H}} + C_k X Z^{\mathrm{H}}\right).
\end{aligned}$$

Since the operator $F_X(.)$ is invertible, we get

$$\delta X = \Phi(\delta X, \delta P) := -\sum_{Z \in \widetilde{P}} F_X^{-1} \circ F_Z(\delta Z) - F_X^{-1}(G(\delta X, \delta P)). \qquad (10.23)$$

Relation (10.23) gives

$$\delta_X \le \sum_{Z \in \widetilde{P}} K_Z \delta_Z + O(\|\delta\|^2),\ \delta \to 0, \tag{10.24}$$

where the quantities

$$K_Z := \left\| F_X^{-1} \circ F_Z \right\|,\ Z \in \widetilde{P}, \tag{10.25}$$

are the *absolute individual condition numbers* [188] of equation (10.16). Here $\|\mathcal{F}\|$ is the norm or the symmetrized norm of the corresponding linear or additive operator $\mathcal{F}$, induced by the Frobenius norm, i.e.,

$$\|\mathcal{F}\| := \max\{\|\mathcal{F}(Y)\|_{\mathrm{F}} : \|Y\|_{\mathrm{F}} = 1\},$$

see (10.7), (10.9), (10.14) and (10.15) for the corresponding explicit expressions.

If $X \neq 0$ then an estimate in terms of relative perturbations is

$$\varepsilon_X := \frac{\|\delta X\|_{\mathrm{F}}}{\|X\|_{\mathrm{F}}} \le \sum_{Z \in \widetilde{P}} k_Z \varepsilon_Z + O(\|\delta\|^2),\ \delta \to 0,$$

where the scalars

$$k_Z := K_Z \frac{\|Z\|_{\mathrm{F}}}{\|X\|_{\mathrm{F}}},\ Z \in \widetilde{P},$$

are the *relative individual condition numbers* with respect to perturbations in the matrix coefficients $Z \in \widetilde{P}$.

The calculation of the condition numbers K_Z is straightforward. First we consider the matrix representations of the partial Fréchet derivatives in X and A_0. Denote by

$$L_Z \in \mathbb{F}^{n^2 \times n^2}$$

the matrix representation of the operator $F_Z(\cdot)$, where $Z = X$ or $Z = A_0$. Noting that $(A^{\mathrm{H}})^\top = \overline{A}$ in both the real and complex case, we get

$$L_X = \sum_{i=1}^{r_1} \left(\overline{A}_i \otimes B_i + \overline{B}_i \otimes A_i\right) + \sum_{k=1}^{r_2} \varepsilon_k \overline{C}_k \otimes C_k,\ L_{A_0} = I_{n^2}$$

and

$$K_{A_0} = l := \left\|F_X^{-1}\right\|_* \le \left\|L_X^{-1}\right\|_2,$$

see (10.14) and (10.15) for the calculation of the symmetrized norm $\|\cdot\|_*$. The computation of the other matrix representations and individual condition numbers, however, is different in the real and complex case and we have to treat them separately.

Consider first the real case $\mathbb{F} = \mathbb{R}$. Here $F_Z(\cdot) \in \mathbf{Lin}(n, \mathbb{R})$ are the partial Fréchet derivatives of $F(X, \cdot)$ in all $Z \in \widetilde{P}$ at (X, P) and we denote by L_Z their matrix representations. Using (10.23), the symmetry of X and the formulae

$$\mathrm{vec}(DZE) = (E^\top \otimes D)\mathrm{vec}(Z),\ (A \otimes B)P_{n^2} = P_{n^2}(B \otimes A),$$

we obtain

$$\begin{aligned} L_{A_i} &= (B_iX) \otimes I_n + (I_n \otimes (B_iX))P_{n^2} \\ &= (I_{n^2} + P_{n^2})\left((B_iX) \otimes I_n\right), \\ L_{B_i} &= (I_{n^2} + P_{n^2})\left((A_iX) \otimes I_n\right), \\ L_{C_k} &= \varepsilon_k\,(I_{n^2} + P_{n^2})\left((C_kX) \otimes I_n\right). \end{aligned} \tag{10.26}$$

Therefore, the absolute condition numbers in the real case are

$$K_{A_0} = l, \quad K_Z = \left\|L_X^{-1}L_Z\right\|_2, \; Z \in \widetilde{P}\backslash\{A_0\}. \tag{10.27}$$

In the complex case $\mathbb{F} = \mathbb{C}$ we have

$$\begin{aligned} \mathrm{vec}\left(F_X^{-1} \circ F_{A_i}(Z)\right) &= L_X^{-1}((\overline{B}_i\overline{X}) \otimes I_n)\mathrm{vec}(Z) \\ &\quad + L_X^{-1}(I_n \otimes (B_iX))P_{n^2}\mathrm{vec}(\overline{Z}), \\ \mathrm{vec}\left(F_X^{-1} \circ F_{B_i}(Z)\right) &= L_X^{-1}((\overline{A}_i\overline{X}) \otimes I_n)\mathrm{vec}(Z) \\ &\quad + L_X^{-1}(I_n \otimes (A_iX))P_{n^2}\mathrm{vec}(\overline{Z}), \\ \mathrm{vec}\left(F_X^{-1} \circ F_{C_k}(Z)\right) &= \varepsilon_k L_X^{-1}((\overline{C}_k\overline{X}) \otimes I_n)\mathrm{vec}(Z) \\ &\quad + \varepsilon_k L_X^{-1}(I_n \otimes (C_kX))P_{n^2}\mathrm{vec}(\overline{Z}). \end{aligned} \tag{10.28}$$

Hence, we may apply relation (10.9) with

$$L_1 = L_X^{-1}((\overline{B}_i\overline{X}) \otimes I_n), \; L_2 = L_X^{-1}(I_n \otimes (B_iX)),$$

etc. Setting

$$\chi(Z) := \nu\left(L_X^{-1}(\overline{Z} \otimes I_n), L_X^{-1}(I_n \otimes Z)P_{n^2}\right) \tag{10.29}$$

(see (10.9) and (10.10)) we get

$$\begin{aligned} \left\|F_X^{-1} \circ F_{A_i}(\delta A_i)\right\|_{\mathrm{F}} &\leq \chi(B_iX)\delta_{A_i}, \\ \left\|F_X^{-1} \circ F_{B_i}(\delta B_i)\right\|_{\mathrm{F}} &\leq \chi(A_iX)\delta_{B_i}, \\ \left\|F_X^{-1} \circ F_{C_k}(\delta C_k)\right\|_{\mathrm{F}} &\leq \chi(C_kX)\delta_{C_k}, \end{aligned}$$

for $i = 1, \ldots, r_1$ and $k = 1, \ldots, r_2$. Thus, the absolute individual condition numbers relative to perturbations in A_i, B_i and C_k in the complex case are given by

$$\begin{aligned} K_{A_i} &= \chi(B_iX), \\ K_{B_i} &= \chi(A_iX), \; i = 1, \ldots, r_1, \\ K_{C_k} &= \chi(C_kX), \; k = 1, \ldots, r_2. \end{aligned}$$

A drawback of this approach is the dimension $n^2 \times n^2$ of the involved matrices. Condition and accuracy estimates, avoiding the formation and analysis of large matrices, are proposed in [179].

An overall relative condition number may be defined as follows. Since we consider $\widetilde{P}$ as an element $(Z_1, Z_2, \ldots, Z_r)$ of a linear space, we may define the product

$$\alpha\widetilde{P} = (\alpha Z_1, \alpha Z_2, \ldots, \alpha Z_r)$$

of $\widetilde{P}$, $\alpha \in \mathbb{F}$, as well as the sum

$$\widetilde{P}' + \widetilde{P}'' = (Z_1' + Z_1'', Z_2' + Z_2'', \ldots, Z_r' + Z_r'')$$

of two r-tuples $\widetilde{P}'$ and $\widetilde{P}''$. We also introduce the *generalized norm*

$$\|\widetilde{P}\|_g := [\|Z_1\|_{\mathrm{F}}, \|Z_2\|_{\mathrm{F}}, \ldots, \|Z_r\|_{\mathrm{F}}]^\top \in \mathbb{R}_+^r$$

of the r-tuple $\widetilde{P}$.

Let $\delta X = \delta X(\delta\widetilde{P})$ be the perturbation in the solution, where

$$\delta\widetilde{P} := (\delta Z_1, \delta Z_2, \ldots, \delta Z_r),$$

and let $\gamma \in \mathbb{R}^r$ be a vector with positive elements.

Definition 10.10 *The* absolute overall condition number with respect to γ *is defined as*

$$K(\gamma) := \lim_{\varepsilon\to 0} \max\left\{\|\delta X(\delta\widetilde{P})\|_{\mathrm{F}} : \|\delta\widetilde{P}\|_g \preceq \varepsilon\gamma\right\}$$

We have

$$K(\gamma) = \max\left\{\left\|\sum_{Z\in\widetilde{P}} F_X^{-1} \circ F_Z(\delta Z)\right\|_{\mathrm{F}} : \|\delta P\|_g \preceq \gamma\right\}. \tag{10.30}$$

Definition 10.11 *The* relative overall condition number with respect to γ *is*

$$\kappa(\gamma) := K(\gamma)/\|X\|_{\mathrm{F}}.$$

Definition 10.12 *If γ has a single nonzero element $\gamma_i = \|Z_i\|_{\mathrm{F}}$, then the quantities $K(\gamma)$ and $\kappa(\gamma)$ are the* individual norm-wise relative condition numbers K_{Z_i} *and k_{Z_i} relative to perturbations in the matrix $Z_i \in \widetilde{P}$. When $\gamma_j = \|Z_j\|_{\mathrm{F}}$ for all $j = 1, 2, \ldots, r$ then $K(\gamma)$ and $\kappa(\gamma)$ are the* overall relative norm-wise condition numbers *of equation (10.16).*

Unfortunately, in general there are no closed form expressions for $K(\gamma)$ and $\kappa(\gamma)$. Using the matrix expressions M_j for the operators $F_X^{-1} \circ F_{Z_j}$, $j = 1, \ldots, r$, we find that

$$K(\gamma) = \max\left\{\left\|\sum_{j=1}^r M_j z_j\right\|_2 : \|z_j\|_2 \leq \gamma_j\right\}.$$

In the next section we will derive bounds

$$K(\gamma) \leq \mathrm{est}(\gamma, M),\ M := [M_1, \ldots, M_r],$$

for $K(\gamma)$.

10.5.2 First order homogeneous bounds

In this subsection we derive local first order homogeneous perturbation bounds, which are generally better than the bounds using condition numbers.

Consider first the real case. The operator equation (10.22) for the perturbation δX may be written in vector form as

$$\mathrm{vec}(\delta X) = \sum_{Z \in P} N_Z \mathrm{vec}(\delta Z) - L_X^{-1} \mathrm{vec}(G(\delta X, \delta P)), \tag{10.31}$$

where

$$N_Z := -L_X^{-1} L_Z \in \mathbb{R}^{n^2 \times n^2}, \ Z \in P.$$

Noting that

$$\delta_X = \|\delta X\|_\mathrm{F} = \|\mathrm{vec}(\delta X)\|_2$$

and

$$\|\mathrm{vec}(\delta Z)\|_2 \le \delta_Z$$

we see that the condition number based estimate is a corollary of (10.31),

$$\delta_X \le \mathrm{est}_1(\delta, N) + O(\|\delta\|^2), \ \delta \to 0,$$

where

$$\mathrm{est}_1(\delta, N) := \sum_{Z \in \widetilde{P}} \|N_Z\|_2 \delta_Z$$

and

$$N := [N_1, N_2, \ldots, N_r] := [N_{Z_1}, N_{Z_2}, \ldots, N_{Z_r}] \in \mathbb{R}^{n^2 \times rn^2}. \tag{10.32}$$

Relation (10.31) also gives a second first order bound

$$\delta_X \le \mathrm{est}_2(\delta, N) + O(\|\delta\|^2), \ \delta \to 0, \tag{10.33}$$

where

$$\mathrm{est}_2(\delta, N) := \|N\|_2 \|\delta\|_2.$$

The bounds $\mathrm{est}_1(\delta, N)$ and $\mathrm{est}_2(\delta, N)$ are alternative, depending on the particular value of δ.

We again have a third bound, which is always less than or equal to $\mathrm{est}_1(\delta, N)$. Indeed, we have

$$\delta_X \le \mathrm{est}_3(\delta, N) + O(\|\delta\|^2), \ \delta \to 0, \tag{10.34}$$

where

$$\mathrm{est}_3(\delta, N) := \sqrt{\delta^\top N_0 \delta}.$$

Here N_0 is the $r \times r$ matrix with elements $n_{ij} := \|N_i^\top N_j\|_2$.

Since

$$\left\|N_i^\top N_j\right\|_2 \le \|N_i\|_2 \|N_j\|_2,$$

we get

$$\mathrm{est}_3(\delta, N) \leq \mathrm{est}_1(\delta, N)$$

for all perturbation vectors δ and matrices N. Hence, we have the overall estimate

$$\delta_X \leq \mathrm{est}(\delta, N) + O(\|\delta\|^2),\ \delta \to 0, \tag{10.35}$$

where

$$\mathrm{est}(\delta, N) := \min\{\mathrm{est}_2(\delta, N), \mathrm{est}_3(\delta, N)\}. \tag{10.36}$$

The local bound est in (10.35), (10.36) is a nonlinear, first order homogeneous and piece-wise real analytic function in δ.

Consider now the complex case. We have

$$\mathrm{vec}^{\mathbb{R}}(\delta X) = \sum_{Z \in \widetilde{P}} \widehat{N}_Z \mathrm{vec}^{\mathbb{R}}(\delta Z) - \mathrm{vec}^{\mathbb{R}}\left(F_X^{-1}(G(\delta X, \delta P))\right),$$

where

$$\widehat{N}_Z := -\mathrm{Mat}\left(\left(F_X^{-1} \circ F_Z\right)^{\mathbb{R}}\right) \in \mathbb{R}^{2n^2 \times 2n^2}$$

are the matrices of the real versions of the additive operators

$$-F_X^{-1} \circ F_Z,\ Z \in \widetilde{P}.$$

Using (10.10) and (10.28) we obtain

$$\widehat{N}_{A_i} = -\Psi(B_i),\ \widehat{N}_{B_i} = -\Psi(A_i),\ \widehat{N}_{C_k} = -\varepsilon_k \Psi(C_k),$$

where

$$\Psi(Z) := \begin{bmatrix} L_{10}(Z) + L_{20}(Z)P_{n^2} & -L_{11}(Z) + L_{21}(Z)P_{n^2} \\ L_{11}(Z) + L_{21}(Z)P_{n^2} & L_{10}(Z) - L_{20}(Z)P_{n^2} \end{bmatrix}$$

and

$$\begin{aligned} \begin{bmatrix} L_{10}(Z) \\ L_{11}(Z) \end{bmatrix} &= \left(L_X^{-1}\right)^{\mathbb{R}} \begin{bmatrix} \mathrm{Re}(ZX) \otimes I_n \\ -\mathrm{Im}(ZX) \otimes I_n \end{bmatrix}, \\ \begin{bmatrix} L_{20}(Z) \\ L_{21}(Z) \end{bmatrix} &= \left(L_X^{-1}\right)^{\mathbb{R}} \begin{bmatrix} I_n \otimes \mathrm{Re}(ZX) \\ I_n \otimes \mathrm{Im}(ZX) \end{bmatrix}. \end{aligned}$$

Here

$$X = X_0 + \imath X_1 \in \mathbb{C}^{n \times n}$$

is the solution to the unperturbed equation (10.16), the matrix

$$Z = Z_0 + \imath Z_1 \in \mathbb{C}^{n \times n}$$

is arbitrary,

$$X_0, X_1, Z_0, Z_1 \in \mathbb{R}^{n \times n}$$

and

$$\begin{aligned}\operatorname{Re}(ZX) &= Z_0X_0 - Z_1X_1,\\ \operatorname{Im}(ZX) &= Z_0X_1 + Z_1X_0.\end{aligned}$$

Now we can use the results for the real case, replacing the matrices $N_j \in \mathbb{R}^{n\times n}$ with $\widehat{N}_j \in \mathbb{R}^{2n^2\times 2n^2}$. The overall bound is

$$\delta_X \leq \operatorname{est}(\Delta, \widehat{N}) + O(\|\Delta\|^2),\ \Delta \to 0,$$

where

$$\widehat{N} := [\widehat{N}_1, \widehat{N}_2, \ldots, \widehat{N}_r] := \left[\widehat{N}_{Z_1}, \widehat{N}_{Z_2}, \ldots, \widehat{N}_{Z_r}\right] \in \mathbb{R}^{2n^2\times 2rn^2} \tag{10.37}$$

and in the expression for $\operatorname{est}_3(\delta, \widehat{N})$ the elements of the matrix N_0 are

$$n_{ij} = \|\widehat{N}_i^{\mathrm{H}}\widehat{N}_j\|_2.$$

10.5.3 Component-wise bounds

The local component-wise bound in the real case then follows directly from relation (10.31) as

$$|\operatorname{vec}(\delta X)| \preceq \sum_{Z\in\widetilde{\mathcal{P}}} \left|L_X^{-1}L_Z\right| |\operatorname{vec}(\delta Z)| + O(\|\delta\|^2),\ \delta \to 0.$$

To implement a component-wise bound one must have information about the perturbations in the components of the data, e.g., $|\operatorname{vec}(Z)| \preceq w_Z$, $Z \in \widetilde{P}$, where $w_Z \succeq 0$ are given vectors.

10.6 Nonlocal perturbation analysis

In this section we derive nonlinear perturbation bounds. For these we obtain a domain

$$\Omega \subset \mathbb{R}_+^r$$

and a nonlinear function

$$f : \Omega \to \mathbb{R}_+$$

such that

$$\delta_X \leq f(\delta)$$

for all $\delta \in \Omega$.

Let the tuples P and $\widetilde{P}$ be perturbed to $P + \delta P$ and $\widetilde{P} \mapsto \widetilde{P} + \delta\widetilde{P}$ and let $X + \delta X$ be the solution of the perturbed equation (10.18). In what follows we mark only the dependence on the perturbations δX and δP, recalling that X is a fixed solution of (10.16).

10.6.1 Real equations

The perturbed equation (10.18) may be rewritten in the form

$$\delta X = \Phi(\delta X, \delta P) := \Phi_0(\delta P) + \Phi_1(\delta X, \delta P), \tag{10.38}$$

where

$$\begin{aligned} \Phi_0(\delta P) &:= -F_X^{-1}(G_0(\delta P)), \\ \Phi_1(\delta X, \delta P) &:= -F_X^{-1}(G_1(\delta X, \delta P)) \end{aligned}$$

and

$$\begin{aligned} G_0(\delta P) &= \delta A_0 + R_1(X, \delta P) + R_2(X, \delta P), \\ G_1(\delta X, \delta P) &= R_1(\delta X, \delta P) + R_2(\delta X, \delta P). \end{aligned}$$

Here $R_k(\cdot, \delta P)$ are linear operators of asymptotic order k relative to δP, $\delta P \to 0$, given by

$$\begin{aligned} R_1(Z, \delta P) &:= \sum_{i=1}^{r_1} \left(\delta A_i Z B_i^\top + A_i Z \delta B_i^\top + \delta B_i Z A_i^\top + B_i Z \delta A_i^\top \right) \\ &\quad + \sum_{k=1}^{r_2} \varepsilon_k \left(\delta C_k Z C_k + C_k Z \delta C_k \right), \\ R_2(Z, \delta P) &:= \sum_{i=1}^{r_1} \left(\delta A_i Z \delta B_i^\top + \delta B_i Z \delta A_i^\top \right) + \sum_{k=1}^{r_2} \varepsilon_k \delta C_k Z \delta C_k^\top . \end{aligned}$$

If $\|Z\|_{\rm F} \le \rho$, then we have

$$\begin{aligned} \|\Phi_0(\delta P)\|_{\rm F} &\le a_0(\delta), \\ \|\Phi_1(Z, \delta P)\|_{\rm F} &\le a_1(\delta)\rho, \end{aligned}$$

where

$$\begin{aligned} a_0(\delta) &:= a_{01}(\delta) + a_{02}(\delta), \\ a_1(\delta) &:= a_{11}(\delta) + a_{12}(\delta) \end{aligned} \tag{10.39}$$

and where the quantities $a_{ik}(\delta)$ are of asymptotic order $O(\|\delta\|^k)$ for $\delta \to 0$. These are determined as follows.

In the case $i = 0$, we have

$$\begin{aligned} a_{01}(\delta) &:= \operatorname{est}(\delta, N), \\ a_{02}(\Delta) &:= \left\|F_X^{-1}\right\|_* \|X\|_2 \left(2\sum_{i=1}^{r_1} \delta_{A_i}\delta_{B_i} + \sum_{k=1}^{r_2} \delta_{C_k}^2 \right). \end{aligned} \tag{10.40}$$

and in the case $i = 1$ we get

$$\begin{aligned} a_{11}(\delta) \; &:= \; \sum_{i=1}^{r_1} \left\| L_X^{-1} \left(I_{n^2} + P_{n^2}\right) \left(B_i \otimes I_n\right) \right\|_2 \delta_{A_i} \\ &+ \sum_{i=1}^{r_1} \left\| L_X^{-1} \left(I_{n^2} + P_{n^2}\right) \left(A_i \otimes I_n\right) \right\|_2 \delta_{B_i} \\ &+ \sum_{k=1}^{r_2} \left\| L_X^{-1} \left(I_{n^2} + P_{n^2}\right) \left(C_k \otimes I_n\right) \right\|_2 \delta_{C_k}, \\ a_{12}(\delta) \; &:= \; \left\| F_X^{-1} \right\|_* \left(2 \sum_{i=1}^{r_1} \delta_{A_i} \delta_{B_i} + \sum_{k=1}^{r_2} \delta_{C_k}^2 \right). \end{aligned} \tag{10.41}$$

If

$$\|Z\|_{\mathrm{F}}, \|\widetilde{Z}\|_{\mathrm{F}} \leq \rho,$$

then a Lyapunov majorant (see [85, 135] and Chapter 5) for equation (10.38) is a function $(\delta, \rho) \mapsto h(\delta, \rho)$, defined on a subset of $\mathbb{R}_+ \times \mathcal{R}_+^r$ and satisfying the conditions

$$\|\Phi(Z, \delta P)\|_{\mathrm{F}} \leq h(\delta, \rho)$$

and

$$\|\Phi(Z, \delta P) - \Phi(\widetilde{Z}, \delta P)\|_{\mathrm{F}} \leq h'_\rho(\delta, \rho) \|Z - \widetilde{Z}\|_{\mathrm{F}}.$$

Here the Lyapunov majorant is affine in ρ and it is determined by

$$h(\delta, \rho) = a_0(\delta) + a_1(\delta)\rho.$$

In this case the fundamental majorant equation

$$h(\delta, \rho) = \rho$$

for determining the nonlocal bound $\rho = \rho(\delta)$ for δ_X gives

$$\delta_X \leq f(\delta) := \frac{a_0(\delta)}{1 - a_1(\delta)}, \quad \delta \in \Omega, \tag{10.42}$$

where

$$\Omega := \{\delta \succeq 0 : a_1(\delta) < 1\} \subset \mathbb{R}_+^r. \tag{10.43}$$

As a result of the nonlocal perturbation analysis we obtain the perturbation bound (10.42), (10.43), where the involved quantities are determined via the relations (10.39) – (10.41).

10.6.2 Complex equations

In the complex case we have again the bound (10.42), where

$$a_{01}(\delta) = \mathrm{est}(\delta, \widehat{N})),$$

the quantity $a_{02}(\delta)$ is as in the real case,

$$a_{11}(\delta) = \sum_{i=1}^{r_1} (\chi(B_i)\delta_{A_i} + \chi(A_i)\delta_{B_i}) + \sum_{k=1}^{r_2} \chi(C_k)\delta_{C_k},$$

the quantity $a_{12}(\delta)$ is again as in the real case, the matrix $\widehat{N}$ is given by (10.37) and Ξ is as in(10.29).

10.6.3 Component-wise bounds

In the derivation of nonlocal component-wise perturbation bounds for Lyapunov equations we use again the generalized Banach fixed point principle, see Appendix D.

Let the operator equation $x = \pi(x)$ be given, where $x \in \mathbb{F}^m$ and $\pi : \mathbb{F}^m \to \mathbb{F}^m$ is a continuous function. Suppose that for all $x, y \in \mathbb{F}^m$ the operator π satisfies the conditions

$$\begin{aligned} |\pi(x)| &\preceq \mu + M|x|, \\ |\pi(x) - \pi(y)| &\preceq M|x - y|, \end{aligned} \tag{10.44}$$

where $\mu \in \mathbb{R}_+^m$ and $M \in \mathbb{R}_+^{m\times m}$. If π is a generalized contraction, i.e., $\mathrm{spect}(M) \subset \mathbb{D}_1$, then there exists a unique solution $x^0 \in \mathbb{F}^m$ of the operator equation, such that

$$|x^0| \preceq (I_m - M)^{-1}\mu.$$

Suppose that we have the following component-wise bounds

$$\begin{aligned} |\delta Z| &\preceq W_Z, \\ \mathrm{vec}(|\delta Z|) &\preceq w_Z := \mathrm{vec}^{-1}(W_Z) \end{aligned}$$

for the perturbations δZ and $\mathrm{vec}(\delta Z)$ in the matrix coefficients, where $W_Z \in \mathbb{R}_+^{n\times n}$ are given matrices and $W_{A_0} = W_{A_0}^\top$. Set

$$W := (W_{Z_1}, W_{Z_2}, \ldots, W_{Z_r}).$$

Using (15.14) and (10.26), (10.28) we see that the right-hand side of the operator equation

$$x = \varphi(x, \delta P) := \mathrm{vec}(\Phi(\mathrm{vec}^{-1}(x), \delta P))$$

for $x := \mathrm{vec}(\delta X)$ satisfies the conditions (10.44) with

$$\begin{aligned} \mu(W) &:= \mu_1(W) + M_2(W)|X|, \\ M(W) &:= M_1(W) + M_2(W). \end{aligned}$$

The expressions for $\mu(W)$ and $M(W)$ are different in the real and complex case and we present them separately.

In the real case we have

$$\begin{aligned} \mu_1(W) &= \left|L_X^{-1}\right| w_{A_0} + \sum_{Z\in P} \left|L_X^{-1} L_Z\right| w_Z, \\ M_1(W) &= \sum_{i=1}^{r_1} \left|L_X^{-1}\left(I_{n^2}+P_{n^2}\right)\left(B_i\otimes I_n\right)\right|\left(I_n\otimes W_{A_i}\right) \\ &\quad + \sum_{i=1}^{r_1} \left|L_X^{-1}\left(I_{n^2}+P_{n^2}\right)\left(A_i\otimes I_n\right)\right|\left(I_n\otimes W_{B_i}\right) \\ &\quad + \sum_{k=1}^{r_2} \left|L_X^{-1}\left(I_{n^2}+P_{n^2}\right)\left(C_k\otimes I_n\right)\right|\left(I_n\otimes W_{C_k}\right), \\ M_2(W) &= \left|L_X^{-1}\right| \sum_{i=1}^{r_1}\left(W_{A_i}\otimes W_{B_i} + W_{B_i}\otimes W_{A_i}\right) \\ &\quad + \left|L_X^{-1}\right| \sum_{k=1}^{r_2} W_{C_k}\otimes W_{C_k}, \end{aligned}$$

where the matrices L_Z are determined by (10.26).

In the complex case the corresponding expressions are

$$\begin{aligned} \mu_1(W) &= \left|L_X^{-1}\right| W_{A_0} + \sum_{i=1}^{r_1}\left(Q(B_iX)W_{A_i} + Q(A_iX)W_{B_i}\right) + \sum_{k=1}^{r_2} Q(C_kX)W_{C_k}, \\ M_1(W) &= \sum_{i=1}^{r_1}\left(\left|L_X^{-1}\left(\overline{B}_i\otimes I_n\right)\right|\left(I_n\otimes W_{A_i}\right) + \left|\mathcal{L}_X^{-1}(I_n\otimes B_i)\right|\left(W_{A_i}\otimes I_n\right)\right) \\ &\quad + \sum_{i=1}^{r_1}\left(\left|L_X^{-1}\left(\overline{A}_i\otimes I_n\right)\right|\left(I_n\otimes W_{B_i}\right) + \left|\mathcal{L}_X^{-1}(I_n\otimes A_i)\right|\left(W_{B_i}\otimes I_n\right)\right) \\ &\quad + \sum_{k=1}^{r_2}\left(\left|L_X^{-1}\left(\overline{C}_k\otimes I_n\right)\right|\left(I_n\otimes W_{C_k}\right) + \left|\mathcal{L}_X^{-1}(I_n\otimes C_k)\right|\left(W_{C_k}\otimes I_n\right)\right), \end{aligned}$$

where

$$Q(Z) := \left|L_X^{-1}(\overline{X}\otimes I_n)\right| + \left|L_X^{-1}(I_n\otimes Z)P_{n^2}\right|$$

and the expression for $M_2(W)$ is as in the real case.

As a result we have the nonlocal component-wise perturbation bound

$$|\mathrm{vec}(\delta X)| \preceq \left(I_{n^2} - M(W)\right)\mu(W)$$

provided that W is small enough to ensure that

$$\mathrm{spect}(M(W)) \subset \mathbb{D}_1.$$

10.6.4 Other bounds

Let $\widetilde{X}$ be an approximate solution to the Lyapunov equation

$$F(X,P) = 0,$$

in which the Fréchet derivative

$$Z \mapsto F_X(Z) := F(X,P) - F(0,P)$$

is invertible. The matrix $\widetilde{X}$ may be, e.g., the solution, computed in finite precision arithmetic.

Denote by

$$\widetilde{R} := F(\widetilde{X},P)$$

the residual, corresponding to $\widetilde{X}$. In view of the linearity of the operator F_X we get

$$\widetilde{X} - X = F_X^{-1}(\widetilde{R}).$$

Hence

$$\|\widetilde{X} - X\|_{\mathrm{F}} \leq \left\|F_X^{-1}\right\|_* \|\widetilde{R}\|_{\mathrm{F}} \tag{10.45}$$

and

$$|\mathrm{vec}(\widetilde{X} - X)| \preceq \left|L_X^{-1}\right| \mathrm{vec}(\widetilde{R})| \tag{10.46}$$

where L_X is the matrix of the operator F_X and $\|\cdot\|_*$ is the symmetrized norm, computed via (10.14) in the real case and (10.15) in the complex case. Note that the bounds (10.45) and (10.46) are exact.

10.7 Notes and references

General Lyapunov operators have been considered in [125]. Perturbation analysis of the type presented above (for particular classes of Lyapunov equations) is given in [132].

Chapter 11

Lyapunov equations in control theory

11.1 Introductory remarks

In this chapter we use the results of the previous chapter to present a complete perturbation analysis for Lyapunov matrix equations arising in systems and control theory. Local and nonlocal norm-wise and component-wise perturbation bounds are derived for real and complex Lyapunov equations. The first order bounds are based on the standard induced norm as well as on the Lyapunov norm of Lyapunov operators. The latter norm allows to obtain tighter results for Lyapunov equations under symmetric perturbations in the constant term. Invertibility conditions for certain classes of Lyapunov operators are also presented.

separately.

11.2 General equation

The perturbation analysis of the general Lyapunov equation (10.12) is based on the norm of the inverse operator to the Lyapunov operator $\mathcal{L}$, defined by the left-hand side of the equation.

In the real case the matrix representation of $\mathcal{L}$ is

$$L = \sum_{k=1}^{r} \left(B_k^\top \otimes A_k^\top + A_k^\top \otimes B_k^\top + \varepsilon_k \left(C_k^\top \otimes C_k^\top \right) \right), \ \varepsilon_k = \pm 1.$$

Here one should recall that instead of $\|\Lambda\|_2$, where $\Lambda := L^{-1}$, we use the symmetrized norm

$$\|\Lambda\|_2^* := \max\left\{ \|\mathcal{L}(X)\|_{\mathrm{F}} : \|X\|_{\mathrm{F}} = 1, \ X = X^\top \right\} \le \|\Lambda\|_2.$$

As shown in [125], the symmetrized norm of the matrix Λ of the inverse Lyapunov operator $\mathcal{L}^{-1}$ may be obtained as

$$\|\Lambda\|_2^* = \|\Lambda Q\|_2,$$

where

$$Q := [Q_{ij}] \in \mathcal{R}^{n^2 \times n(n+1)/2}; \ i,j = 1,\ldots,n,$$

is a block upper-triangular projector ($Q^\top Q = I_{n(n+1)/2}$). The blocks $Q_{ij} \in \mathcal{R}^{n \times j}$ are defined by

$$Q_{ij} = \begin{cases} 0 & \text{if} \quad i > j, \\ [1,0,\ldots,0]^\top & \text{if} \quad i = j = 1, \\ [\mathrm{diag}(qI_{i-1}, 1), 0]^\top & \text{if} \quad i = j > 1, \\ qE_{ji}(n,j) & \text{if} \quad i < j, \end{cases}$$

where $q := 1/\sqrt{2}$.

Consider now the complex Lyapunov operator

$$\mathcal{L}_c(X) = \sum_{k=1}^{r} \left(A_k^{\mathrm{H}} X B_k + B_k^{\mathrm{H}} X A_k + \varepsilon_k \left(C_k^{\mathrm{H}} X C_k \right) \right). \tag{11.1}$$

Here the symmetrized norm of the matrix

$$\Lambda_c = L_c^{-1},$$

where

$$L_c := \sum_{k=1}^{r} \left(B_k^\top \otimes A_k^{\mathrm{H}} + A_k^\top \otimes B_k^{\mathrm{H}} + \varepsilon_k \left(C_k^\top \otimes C_k^{\mathrm{H}} \right) \right),$$

is defined as

$$\|\Lambda_c\|_2^* := \max\left\{ \|\mathcal{L}_c(X)]\|_{\mathrm{F}} : \|X\|_{\mathrm{F}} = 1, \ X = X^{\mathrm{H}} \right\} \le \|\Lambda_c\|_2$$

and may be calculated as follows. Let

$$\Lambda_c = \Lambda_0 + \imath \Lambda_1,$$

where Λ_i are real. Then

$$\|\Lambda_c\|_2^* = \left\| \begin{bmatrix} \Lambda_0 Q & -\Lambda_1 \widehat{Q} \\ \Lambda_1 Q & \Lambda_0 \widehat{Q} \end{bmatrix} \right\|_2.$$

The matrix

$$\widehat{Q} \in \mathcal{R}^{n^2 \times n(n-1)/2}$$

is obtained from Q by deleting the columns containing 1's which are numbered as $k(k+1)/2$, $k = 1, \ldots, n$, and by changing the sign of each second element q in each column of the reduced matrix. This procedure is described as follows. Let

$$R = [R_{ij}] := [\delta_{i(i+1)/2,j}] \in \mathcal{R}^{n(n+1)/2 \times n(n-1)/2},$$

where δ_{ij} is the Kronecker delta, and

$$J := \{(kn + l, k(k-1)/2 + l) : k = 1, \ldots, n-1, \, l = 1, \ldots, k\}.$$

Then

$$\widehat{Q}_{ij} = \begin{cases} (QR)_{ij} & \text{if} \quad (i,j) \notin J \\ -(QR)_{ij} & \text{if} \quad (i,j) \in J. \end{cases}$$

It must be pointed out that the ratio $\|\Lambda\|_2^*/\|\Lambda\|_2$ may be arbitrarily close to 0, i.e. the use of symmetrized norms instead of usual 2-norms for the inverse Lyapunov operators may significantly improve the perturbation bounds for both real and complex Lyapunov equations. Of course, it is also possible that $\|\Lambda\|_2^* = \|\Lambda\|_2$ and then using any of these norms gives identical results. The description of the class of Lyapunov operators for which the last equality holds is an open and probably a difficult problem, see also the discussion in [40].

11.3 Continuous-time equations

For the standard real continuous-time Lyapunov equation

$$\mathcal{L}(X) := A^\top X + XA = C \tag{11.2}$$

the spectrum of $\mathcal{L}$ is

$$\operatorname{spect}(\mathcal{L}) = \{\lambda_i(A) + \lambda_k(A) : i, k = 1, \ldots, n\}.$$

Corollary 11.1 *The norm-wise bound (8.59) for (11.2) is*

$$\delta_X \le \frac{\operatorname{est}(\delta_C, \delta_A, \Lambda, N_A)}{1 - 2\|\Lambda\|_2^* \delta_A}, \tag{11.3}$$

where

$$\begin{aligned} \Lambda &= \left(A^\top \otimes I_n + I_n \otimes A^\top\right)^{-1}, \\ N_A &= -\Lambda(I_n \otimes X + (X \otimes I_n)P_{n^2}) \\ &= -\Lambda(I_{n^2} + P_{n^2})(I_n \otimes X) \end{aligned}$$

and it is valid for

$$\delta_A < \frac{1}{2\|\Lambda\|_2^*}.$$

The next example shows that the bounds $\text{est}_2(\delta, N)$ and $\text{est}_3(\delta, N)$ are alternative for equations of type (11.2) for $n = 2$. This means that the overall expression

$$\text{est}(\delta_C, \delta_A, \Lambda, N_A) = \min\{\text{est}_2(\delta_C, \delta_A, \Lambda, N_A), \text{est}_3(\delta_C, \delta_A, \Lambda, N_A)\}$$

depends nontrivially on both bounds est_2 and est_3, being a piece-wise analytic function in $\delta \succeq 0$ for $\delta \neq 0$ and $\delta_A < 1/\left(2\|\Lambda\|_2^*\right)$.

Example 11.2 Let $n = 2$ and

$$A = \begin{bmatrix} 1.415 & 0.927 \\ 0 & 1.028 \end{bmatrix}, \quad X = \begin{bmatrix} 0.903 & 0.462 \\ 0.462 & 0.724 \end{bmatrix}.$$

Then, displaying three digits, we have

$$\|N\|_2 = 0.815 < \sqrt{\|N_0\|_2} = 0.884.$$

Hence, for δ equal to the eigenvector of the matrix N_0, corresponding to its maximum eigenvalue $\|N_0\|_2$, we have

$$\text{est}_2(\delta_C, \delta_A, \Lambda, N_A) = 0.815 < \text{est}_3(\delta_C, \delta_A, \Lambda, N_A)\} = 0.884.$$

$\diamondsuit$

Corollary 11.3 *The component-wise bound (8.63) for equation (11.2) is*

$$|\delta X| \preceq (I_{n^2} - \Psi_1(\Delta))\, \Theta_1(\Delta), \tag{11.4}$$

where

$$\begin{aligned} \Theta_1(\Delta) &= |\Lambda|\Delta_C + |N_A|\Delta_A, \\ \Psi_1(\Delta) &= |\Lambda|\left(I_n \otimes W_A^\top + W_A^\top \otimes I_n\right). \end{aligned}$$

Example 11.4 Consider the $n \times n$ continuous-time Lyapunov equation

$$\mathcal{L}(X) := A^{\mathrm{H}} X + XA = C,$$

where the matrix A is stable (i.e., $\text{Re}(\lambda_i(A)) < 0$) and $C^{\mathrm{H}} = C \leq 0$. Suppose that the matrices A and C can be simultaneously reduced to diagonal form by a unitary congruence transformation. Then we may assume that

$$A = \text{diag}(\lambda_1, \ldots, \lambda_n), \ \alpha_i := -\text{Re}(\lambda_i) > 0$$

and

$$C = -\text{diag}(\gamma_1, \ldots, \gamma_n), \ \gamma_i \geq 0.$$

The solution

$$X = \text{diag}(x_1, \ldots, x_n) \in \mathbb{R}^{n \times n}$$

is defined by

$$x_i = \gamma_i/(2\alpha_i) \geq 0.$$

Let

$$\alpha = \min\{\alpha_1, \ldots, \alpha_n\}.$$

Denote by $\mathcal{J} \subset \{1, \ldots, n\}$ the set of all indices such that $j \in \mathcal{J}$ implies $\alpha_j = \alpha$. For $j \in \mathcal{J}$ fixed, take perturbations in A and C as

$$\delta A = \delta_A E_{jj}(n), \delta C = -\delta_C E_{jj}(n).$$

Then we have $\delta X = \delta_X E_{jj}(n)$.

$\diamondsuit$

Example 11.5 Consider the Lyapunov equation

$$\mathcal{L}(X) := A^\top X + XA = C$$

in $\mathcal{R}^{2\times 2}$, where

$$A = \begin{bmatrix} 1.5 & -0.5 \\ -0.5 & 1.5 \end{bmatrix}, \ C = \begin{bmatrix} 7 & -5 \\ -5 & 7 \end{bmatrix}.$$

The solution is

$$X = \begin{bmatrix} 2 & -1 \\ -1 & 2 \end{bmatrix},$$

the matrix L of the Lyapunov operator $\mathcal{L}$ is

$$L = \frac{1}{2}\begin{bmatrix} 6 & -1 & -1 & 0 \\ -1 & 6 & 0 & -1 \\ -1 & 0 & 6 & -1 \\ 0 & -1 & -1 & 6 \end{bmatrix}$$

and we have

$$N_C = L^{-1} = \frac{1}{48}\begin{bmatrix} 17 & 3 & 3 & 1 \\ 3 & 17 & 1 & 3 \\ 3 & 1 & 17 & 3 \\ 1 & 3 & 3 & 17 \end{bmatrix}.$$

Furthermore,

$$N_A = -L^{-1}(I_4 + P_4)(I_2 \otimes X) = \frac{1}{24}\begin{bmatrix} -31 & 11 & -5 & 1 \\ 3 & -15 & -15 & 3 \\ 3 & -15 & -15 & 3 \\ 1 & -5 & 11 & -31 \end{bmatrix}.$$

Thus, we have

$$\|N_C\|_2 = 0.5, \ \|N_A\|_2 = 1.5, \ \|[N_C, N_A]\|_2 = 1.5275, \ \|N_C^\top N_A\|_2 = 0.5.$$

Consider the perturbations

$$\delta A = -\frac{\varepsilon}{2}\begin{bmatrix} 1 & 1 \\ 1 & 1 \end{bmatrix}, \ \delta C = 2\varepsilon \begin{bmatrix} 1 & 1 \\ 1 & 1 \end{bmatrix}$$

with $\|\delta A\|_{\mathrm{F}} = \varepsilon$, $\|\delta C\|_{\mathrm{F}} = 4\varepsilon$, where $0 \le \varepsilon < 1$. The perturbation in X is then

$$\delta X = \frac{2\varepsilon}{1-\varepsilon}\begin{bmatrix} 1 & 1 \\ 1 & 1 \end{bmatrix}, \ \|X\|_{\mathrm{F}} = \frac{2(1+2\varepsilon)}{1-\varepsilon}.$$

At the same time the first order bounds for δ_X are

$$\mathrm{est}_1 = 3.5000\varepsilon, \ \mathrm{est}_2 = 3.4157\varepsilon, \ \mathrm{est}_3 = 2.3452\varepsilon.$$

Hence, the norm-wise bound is

$$\delta_X \le \frac{2.3452\varepsilon}{1-\varepsilon}.$$

◇

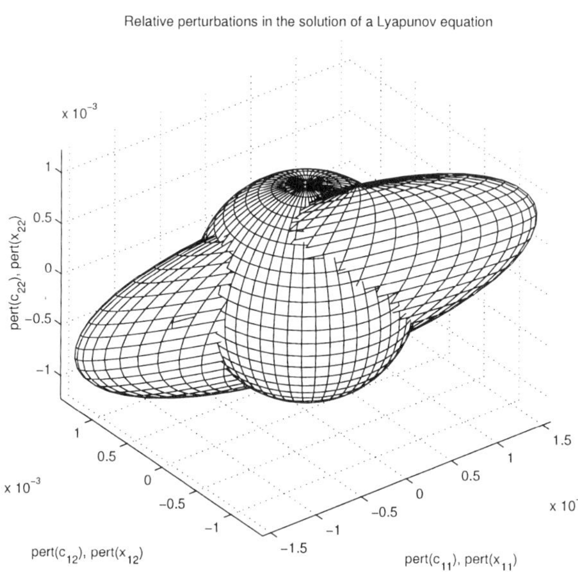

Figure 11.1: Perturbed solutions of well-conditioned Lyapunov equation

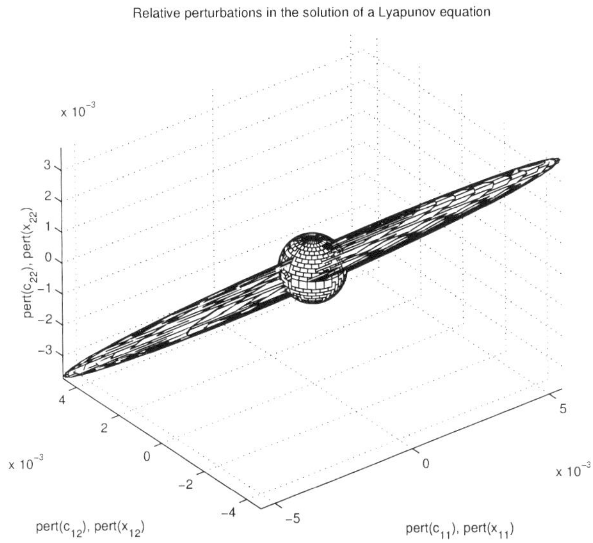

Figure 11.2: Perturbed solutions of ill-conditioned Lyapunov equation

In Figures 11.1 and 11.2 we show the relative changes in the elements of the solutions $\delta X/\|X\|$ of 2nd order well-conditioned and ill-conditioned Lyapunov equations generated by perturbations in the elements C_{11}, C_{12} and C_{22} of the matrix C. The perturbations in the data are represented by spheres while the perturbed solutions are represented by ellipsoids.

For the complex continuous-time Lyapunov equation

$$\mathcal{L}_c(X) := A^{\mathrm{H}}X + XA = C, \tag{11.5}$$

with $C = C^{\mathrm{H}}$, the spectrum of $\mathcal{L}_c$ is

$$\operatorname{spect}(\mathcal{L}_c) = \left\{\lambda_i(A) + \overline{\lambda}_k(A) : i, k = 1, \ldots, n\right\}.$$

The presence of the term δA^{H} in the perturbed equation makes it more difficult to get tight (and in particular asymptotically exact) perturbation bounds. Here we use an approach based on the real version

$$\mathcal{L}^{\mathbb{R}} : \mathbb{R}^{2n^2} \to \mathbb{R}^{2n^2}$$

of the linear operator

$$\mathcal{L} : \mathbb{C}^{n\times n} \simeq \mathbb{C}^{n^2} \to \mathbb{C}^{n\times n} \simeq \mathbb{C}^{n^2}.$$

Let the operator $\mathcal{N} : \mathbb{C}^n \to \mathbb{C}^m$ be defined via

$$\mathcal{N}(u) := Ru + S\overline{u}, \ u \in \mathbb{C}^n,$$

where $R, S \in \mathbb{C}^{m\times n}$ are given matrices. We identify $\mathcal{N}$ with the ordered pair

$$(R, S) \in \mathbb{C}^{m\times n} \times \mathbb{C}^{m\times n}$$

and write also $\mathcal{N} = \mathcal{N}(R, S)$.

For $\lambda \in \mathbb{C}$ set $\lambda\mathcal{N}(R,S) = \mathcal{N}(\lambda R, \lambda S)$. Also, if $\mathcal{N}_i = \mathcal{N}(R_i, S_i)$, $i = 1, 2$, are two operators of this type, set

$$\mathcal{N}_1 + \mathcal{N}_2 = \mathcal{N}(R_1 + R_2, S_1 + S_2).$$

Hence, the set of the operators $\mathcal{N}$ is a linear space (in fact it is isomorphic to $\mathbb{C}^{2mn}$) and we equip this space with the norm

$$\nu(\mathcal{N}) = \nu(R,S) := \max\left\{\|\mathcal{N}(u)\|_2 : u \in \mathbb{C}^n,\ \|u\|_2 \le 1\right\}. \tag{11.6}$$

This concept needs justification, since the operator $\mathcal{N}$ is additive ($\mathcal{N}(u+v) = \mathcal{N}(u) + \mathcal{N}(v)$) but not homogeneous ($\mathcal{N}(\lambda u) \neq \lambda\mathcal{N}(u)$ for $\lambda \in \mathbb{C}\backslash\mathbb{R}$) and hence it is not linear over $\mathbb{C}$ if $S \neq 0$.

Proposition 11.6 *The function $\nu : \mathbb{C}^{m\times n} \times \mathbb{C}^{m\times n} \to \mathbb{R}_+$, defined by (11.6), is a norm.*

Proof. We show that

$$\begin{aligned} \nu(\mathcal{N}) &= 0 \ \text{ if and only if } \ \mathcal{N} = 0, \text{ i.e., } R = S = 0, \\ \nu(\lambda\mathcal{N}) &= |\lambda|\nu(\mathcal{N}),\ \lambda \in \mathbb{C}, \\ \nu(\mathcal{N}_1 + \mathcal{N}_2) &\le \nu(\mathcal{N}_1) + \nu(\mathcal{N}_2), \end{aligned} \tag{11.7}$$

i.e. that ν has indeed the properties of a norm. We have

$$\begin{aligned} \nu(R,S) &= \max\left\{\|Ru + S\overline{u}\|_2 : u \in \mathbb{C}^n,\ \|u\|_2 \le 1\right\} \\ &\le \max\left\{\|Ru + Sv\|_2 : u, v \in \mathbb{C}^n;\ \|u\|_2, \|v\|_2 \le 1\right\} \\ &\le \max\left\{\left\|[R,S]\begin{bmatrix} u \\ v \end{bmatrix}\right\|_2 : \begin{bmatrix} u \\ v \end{bmatrix} \in \mathbb{C}^{2n}, \left\|\begin{bmatrix} u \\ v \end{bmatrix}\right\|_2 \le \sqrt{2}\right\} \\ &= \sqrt{2}\,\|[R,S]\|_2. \end{aligned}$$

Similarly, we get

$$\begin{aligned} \nu(R,S) &= \max\left\{\|Ru + S\overline{u}\|_2 : u \in \mathbb{C}^n,\ \|u\|_2 \le 1\right\} \\ &\ge \max\left\{\|Ru + S\overline{u}\|_2 : u \in \mathbb{R}^n,\ \|u\|_2 \le 1\right\} \\ &= \max\left\{\|(R+S)u\|_2 : u \in \mathbb{R}^n,\ \|u\|_2 \le 1\right\} \\ &\ge \max\left\{\|\mathrm{Re}(R+S)\|_2, \|\mathrm{Im}(R+S)\|_2\right\} \end{aligned}$$

and

$$\begin{aligned}\nu(R,S) &= \max\{\|Ru+S\overline{u}\|_2 : u\in\mathbb{C}^n,\ \|u\|_2\le 1\}\\ &\ge \max\{\|Ru+S\overline{u}\|_2 : u\in(\imath\mathbb{R})^n,\ \|u\|_2\le 1\}\\ &= \max\{\|(R-S)u\|_2 : u\in(\imath\mathbb{R})^n,\ \|u\|_2\le 1\}\\ &\ge \max\{\|\mathrm{Re}(R-S)\|_2, \|\mathrm{Im}(R-S)\|_2\}.\end{aligned}$$

Hence, we have

$$\begin{aligned}&\max\{\|\mathrm{Re}(R+S)\|_2, \|\mathrm{Im}(R+S)\|_2, \|\mathrm{Re}(R-S)\|_2, \|\mathrm{Im}(R-S)\|_2\}\\ &\le \nu(R,S)\le \sqrt{2}\,\|[R,S]\|_2. \qquad (11.8)\end{aligned}$$

That $\mathcal{N}=0$ (i.e., $R=S=0$) implies $\nu(0,0)=0$ is obvious. If now $\nu(R,S)=0$ then the left inequality in (10.44) gives $R+S=0$ and $R-S=0$, which yields $R=S=0$. Thus, the first condition in (11.7) is fulfilled. The second and the third relation in (11.7) follow by inspection. □

Although the operator $\mathcal{N}$ is not linear (it is not even differentiable, together with the map $z\mapsto\overline{z}$), it becomes linear if we consider $\mathbb{C}^n$ and $\mathbb{C}^m$ as linear spaces of complex vectors with $\mathbb{R}$ as the field of scalars.

To compute the norm $\nu(\mathcal{N})$, however, it is more convenient to use the real version

$$\mathcal{N}^{\mathbb{R}}:\mathbb{R}^{2n}\longrightarrow\mathbb{R}^{2m}$$

of $\mathcal{N}$ over $\mathbb{R}$, which is a linear operator.

Let

$$R=R_0+\imath R_1,\ \ S=S_0+\imath S_1$$

and

$$u=u_0+\imath u_1,$$

where R_i, S_i and u_i are real. Then the real version of $u\in\mathbb{C}^n$ is

$$u^{\mathbb{R}}=\begin{bmatrix}u_0\\ u_1\end{bmatrix}\in\mathbb{R}^{2n}.$$

Setting similarly

$$\mathcal{N}(u)=w_0+\imath w_1$$

we have

$$\mathcal{N}^{\mathbb{R}}(u^{\mathbb{R}})=\begin{bmatrix}w_0\\ w_1\end{bmatrix}=\mathrm{Mat}(R,S)\begin{bmatrix}u_0\\ u_1\end{bmatrix},$$

where

$$\mathrm{Mat}(R,S):=\mathrm{Mat}(\mathcal{N}^{\mathbb{R}})=\begin{bmatrix}R_0+S_0 & S_1-R_1\\ R_1+S_1 & R_0-S_0\end{bmatrix}$$

is the matrix representation of $\mathcal{N}^{\mathbb{R}}$. Since

$$\|\mathcal{N}(u)\|_2^2 = \|\mathcal{N}^{\mathbb{R}}[u^{\mathbb{R}}]\|_2 = \|w_0\|_2^2 + \|w_1\|_2^2,$$

we get

$$\nu(\mathcal{N}) = \nu(R, S) = \|\mathrm{Mat}(R, S)\|_2.$$

The column-wise vector operator form of the perturbed complex Lyapunov equation is

$$\begin{aligned} \mathrm{vec}(\delta X) &= \Lambda_c \left(\mathrm{vec}(C) - \left(X^\top \otimes I_n\right) P_{n^2}\mathrm{vec}(\overline{\delta A}) - (I_n \otimes X)\mathrm{vec}(\delta A)\right) \\ &\quad - \Lambda_c \mathrm{vec}(\delta A^{\mathrm{H}} \delta X + \delta X \delta A), \end{aligned}$$

where

$$\Lambda_c := L_c^{-1} = \left(I_n \otimes A^{\mathrm{H}} + A^\top \otimes I_n\right)^{-1} = \Lambda_0 + \imath \Lambda_1 \in \mathbb{C}^{n \times n}$$

and $\Lambda_i \in \mathbb{R}^{n \times n}$. Hence, the norm-wise perturbation bound for this case is

$$\delta_X \leq \frac{\mathrm{est}\left(\delta_C, \delta_A, \Lambda^{\mathbb{R}}, N_A^{\mathbb{R}}\right)}{1 - 2\|\Lambda_c\|_2^* \delta_A}, \quad \delta_A < \frac{1}{2\|\Lambda_c\|_2^*},$$

where

$$\begin{aligned} \Lambda^{\mathbb{R}} &= \begin{bmatrix} \Lambda_0 & -\Lambda_1 \\ \Lambda_1 & \Lambda_0 \end{bmatrix}, \\ N_A^{\mathbb{R}} &= -\mathrm{Mat}\left(\Lambda_c(I_n \otimes X), \Lambda_c \left(X^\top \otimes I_n\right) P_{n^2}\right). \end{aligned}$$

Corollary 11.7 *The component-wise bound for the complex Lyapunov equation (11.5) has the form (11.4) as for the real Lyapunov equation (11.2) but with the vector Θ_1 and the matrix Ψ_1 given by*

$$\begin{aligned} \Theta_1(\Delta) &= |\Lambda_c|\Delta_C + \left(|\Lambda_c (I_n \otimes X)| + \left|\Lambda_c \left(X^\top \otimes I_n\right) P_{n^2}\right|\right) \Delta_A, \\ \Psi_1(\Delta) &= |\Lambda_c| \left(I_n \otimes W_A^\top + W_A^\top \otimes I_n\right)\right). \end{aligned}$$

11.4 Continuous-time equations in descriptor form

In general, there is no simple expression for the spectrum $\mathrm{spect}(\mathcal{L})$ of the Lyapunov operator $\mathcal{L}$ in the continuous-time descriptor Lyapunov equation

$$\mathcal{L}(X) := A^\top X B + B^\top X A = C \tag{11.9}$$

Hence, $\mathrm{spect}(\mathcal{L})$ may be computed, if necessary, as $\mathrm{spect}(L)$, where

$$L := A^\top \otimes B^\top + B^\top \otimes A^\top$$

is the matrix of $\mathcal{L}$.

In the following we give three equivalent tests for invertibility of $\mathcal{L}$, based on $n \times n$ matrices. Two of them involve the inversion of an $n \times n$ matrix (or the solution of n algebraic linear vector equations of n-th order) and a spectral analysis of another $n \times n$ matrix, while the third one is based on the spectral analysis of an $n \times n$ regular matrix pencil.

Let us first recall some facts about matrix pencils. Two matrices A and B of the same size determine a *matrix pencil*

$$\mathrm{Pen}(A, B) := \{\beta A - \alpha B : \alpha, \beta \in \mathbb{C}\}.$$

Definition 11.8 *A matrix pencil* $\mathrm{Pen}(A, B)$ *is called* regular *if* A *and* B *are square matrices and* $\det(\beta A - \alpha B)$ *is not identically zero.*

Example 11.9 Let $A, B \in \mathbb{C}^{2\times 2}$. Then the pencil $\mathrm{Pen}(A, B)$ is regular if and only if

$$|\det(A)| + |\det(B)| + |a_{12}b_{21} + a_{21}b_{12} - a_{11}b_{22} - a_{22}b_{11}| > 0.$$

$\diamondsuit$

It seems natural to determine the eigenvalues of a regular pencil $\mathrm{Pen}(A, B)$ as the nonzero pairs (α, β) for which the matrix $\beta A - \alpha B$ is singular. This definition, however, may cause problems, since such pairs are not uniquely determined. Indeed, if (α, β) is such a pair, then any pair $(\tau\alpha, \tau\beta)$ with $\tau \neq 0$ should also be considered as an eigenvalue. To avoid this nonuniqueness we must not distinguish such pairs and consider them as equivalent. This is done according to the following definition.

Definition 11.10 *The pairs* (α, β) *and* (α', β') *are said to be* equivalent *(denoted as* $(\alpha, \beta) \equiv (\alpha', \beta')$*) if* $\alpha\beta' = \alpha'\beta$.

It is easy to show that the pairs (α, β) and (α', β') are equivalent if and only if there is a nonzero complex number τ such that $(\alpha', \beta') = (\tau\alpha, \tau\beta)$.

The set $\mathbb{C}^2$ of all ordered pairs (α, β) may be divided into disjoint subsets, called *equivalence classes* or *orbits*, such that two pairs belong to the same class if and only if they are equivalent. Thus, each pair $(\alpha, \beta) \in \mathbb{C}^2$ gives rise to the orbit

$$[\alpha, \beta] := \{(\tau\alpha, \tau\beta) : \tau \in \mathbb{C}\backslash\{0\}\} \subset \mathbb{C}^2.$$

The set of all equivalence classes in this case is the projective plane $\mathbb{P}^1(\mathbb{C})$.

Definition 11.11 *An element* $\gamma = \gamma(A, B) \in \mathbb{P}^1(\mathbb{C})$ *is called* eigenvalue *of the regular pencil* $\mathrm{Pen}(A, B)$ *if the matrix* $\beta A - \alpha B$ *is singular for some (and hence for every) member* (α, β) *of* γ.

If $A, B \in \mathbb{C}^{n\times n}$ and the pencil $\mathrm{Pen}(A, B)$ is regular , then there are (at most) n eigenvalues

$$\begin{aligned} \gamma_i(A,B) &= [\alpha_i(A,B), \beta_i(A,B)] \\ &= \{(\tau\alpha_i(A,B), \tau\beta_i(A,B)) : \tau \neq 0\},\ i = 1,\ldots,n, \end{aligned}$$

of $\mathrm{Pen}(A, B)$. We note that if e.g., B is nonsingular then the eigenvalues of $B^{-1}A$ are

$$\lambda_i(B^{-1}A) = \frac{\alpha_i(A,B)}{\beta_i(A,B)} \tag{11.10}$$

for any

$$\gamma_i(A,B) = [\alpha_i(A,B), \beta_i(A,B)].$$

Now we are in position to formulate necessary and sufficient conditions for invertibility of $\mathcal{L}$ in terms of $n \times n$ matrices only.

Proposition 11.12 *The operator $\mathcal{L}$ in (11.9) is invertible if and only if at least one of the following three equivalent conditions holds.*

(i) *The matrix A is nonsingular and*

$$\lambda_i(A^{-1}B) + \lambda_k(A^{-1}B) \neq 0;\ i,k = 1,\ldots,n.$$

(ii) *The matrix B is nonsingular and*

$$\lambda_i(B^{-1}A) + \lambda_k(B^{-1}A) \neq 0;\ i,k = 1,\ldots,n.$$

(iii)

$$\alpha_i(A,B)\beta_k(A,B) + \alpha_k(A,B)\beta_i(A,B) \neq 0;\ i,k = 1,\ldots,n.$$

Proof. (i) Suppose that $\mathcal{L}$ is invertible but A is singular. Then there exists $U \in \mathcal{O}(n)$ such that $AU = [A_1, 0]$, where $A_1 \in \mathcal{R}^{n\times k}$ and $k = \mathrm{rank}(A) < n$. We have

$$U^\top \mathcal{L}(X) U = \begin{bmatrix} A_1XBU \\ 0 \end{bmatrix} + \left[U^\top B^\top X A_1, 0\right] = \begin{bmatrix} \times & \times \\ \times & 0_{(n-k)\times(n-k)} \end{bmatrix},$$

where $\times$ denotes unspecified matrix block. Due to the zero bottom right block of $U^\top \mathcal{L}(X)U$ for any $X \in \mathbb{R}^{n\times n}$, we see that the operator $\mathcal{L}$ cannot be surjective and hence, not invertible. This shows that A is nonsingular.

Furthermore, let

$$L := B^\top \otimes A^\top + A^\top \otimes B^\top$$

be the matrix of $\mathcal{L}$, which is invertible together with $\mathcal{L}$. Set

$$L_1 := L(A\otimes A)^{-\top} = (A^{-1}B)^\top \otimes I_n + I_n \otimes (A^{-1}B)^\top.$$

The eigenvalues of L_1 are $\lambda_i(A^{-1}B) + \lambda_k(A^{-1}B)$; $i, k = 1, \ldots, n$, and, since L_1 is also nonsingular, zero is not among them. Since all arguments go in both directions, we have proved that $\mathcal{L}$ is invertible if and only if (i) holds.

(ii) Conditions (i) and (ii) are equivalent, since interchanging A and B we have the same operator $\mathcal{L}$.

(iii) Suppose that (i) (and hence (ii)) is fulfilled. Then $\alpha_i(A, B) \neq 0$ and

$$\begin{aligned} 0 &\neq \lambda_i(A^{-1}B) + \lambda_k(A^{-1}B) = \frac{\beta_i(A,B)}{\alpha_i(A,B)} + \frac{\beta_k(A,B)}{\alpha_k(A,B)} \\ &= \frac{\alpha_k(A,B)\beta_i(A,B) + \alpha_i(A,B)\beta_k(A,B)}{\alpha_i(A,B)\alpha_k(A,B)} \end{aligned}$$

for all $i, k = 1, \ldots, n$, which proves (iii).

Let finally (iii) be valid. Then $\alpha_i(A, B) \neq 0$ and A is nonsingular. Dividing the inequality in (iii) by $\alpha_i(A, B)\alpha_k(A, B)$ we obtain (i). □

To use the conditions (i) or (ii) of Proposition 11.12 it is not necessary to explicitly invert A or B but rather to solve n vector equations $Ax_i = b_i$ for the columns x_i of $A^{-1}B$, where $B = [b_1, \ldots, b_n]$. Note also that the invertibility of both A and B is a necessary condition for the invertibility of $\mathcal{L}$.

Theorem 11.13 *The norm-wise perturbation bound for equation (11.9) is*

$$\delta_X \leq \frac{\operatorname{est}(\delta_C, \delta_A, \delta_B, \Lambda, N_A, N_B) + 2\|X\|_2\|\Lambda\|_2^*\delta_A\delta_B}{1 - l_A\delta_A - l_B\delta_B - 2\|\Lambda\|_2^*\delta_A\delta_B}, \tag{11.11}$$

where

$$\begin{aligned} \Lambda &= \left(A^\top \otimes B^\top + B^\top \otimes A^\top\right)^{-1}, \\ N_A &= -\Lambda(I_{n^2} + P_{n^2})\left(I_n \otimes \left(B^\top X\right)\right), \\ N_B &= -\Lambda(I_{n^2} + P_{n^2})\left(I_n \otimes \left(A^\top X\right)\right), \\ l_A &= \left\|\Lambda(I_{n^2} + P_{n^2})\left(I_n \otimes B^\top\right)\right\|_2, \\ l_B &= \left\|\Lambda(I_{n^2} + P_{n^2})\left(I_n \otimes A^\top\right)\right\|_2 \end{aligned}$$

and the quantities δ_A, δ_B satisfy the inequality

$$l_A\delta_A + l_B\delta_B + 2\|\Lambda\|_2^*\delta_A\delta_B < 1.$$

Proof. The expressions for N_A, N_B and l_A, l_B are obtained as follows. First we note that $P_{n^2}^2 = I_{n^2}$ and

$$P_{n^2}(A \otimes B)P_{n^2} = B \otimes A,$$

see e.g. [107]. Then the matrix N_A is defined by considering the vector

$$v_1 = v_1(\delta A) := -\Lambda\operatorname{vec}(\delta A^\top XB + B^\top X\delta A)$$

in the vectorized expression

$$\mathrm{vec}(\delta X) = \mathrm{vec}(\Phi(\delta X, \delta D)), \ \delta D := (\delta C, \delta A, \delta B), \tag{11.12}$$

of the operator equation

$$\delta X = \Phi(\delta X, \delta D)$$

for δX. We have

$$\begin{aligned} v_1 &= -\Lambda\left(\left((B^\top X)\otimes I_n\right)\mathrm{vec}(\delta A^\top) + \left(I_n \otimes (B^\top X)\right)\mathrm{vec}(\delta A)\right) \\ &= -\Lambda\left(\left((B^\top X)\otimes I_n\right)P_{n^2} + I_n \otimes (B^\top X)\right)\mathrm{vec}(\delta A) \\ &= -\Lambda\left(P_{n^2}\left(P_{n^2}\left((B^\top X)\otimes I_n\right)P_{n^2}\right) + I_n \otimes (B^\top X)\right)\mathrm{vec}(\delta A) \\ &= -\Lambda\left(P_{n^2} + I_{n^2}\right)\left(I_n \otimes (B^\top X)\right)\mathrm{vec}(\delta A) \\ &= N_A \mathrm{vec}(\delta A). \end{aligned}$$

In turn, the quantity l_A is obtained from

$$l_A = \|\widetilde{N}_A\|_2,$$

where $\widetilde{N}_A$ is the matrix N_A with X replaced by I_n,

$$\widetilde{N}_A = -\Lambda(I_{n^2} + P_{n^2})\left(I_n \otimes B^\top\right).$$

The expression for l_A is obtained by estimating the norm of the vector

$$v_2 = v_2(\delta X, \delta A) := -\Lambda \mathrm{vec}(\delta A^\top \delta X B + B^\top \delta X \delta A)$$

in the right-hand side of (11.12). Similarly to the expression for v_1 we have

$$\begin{aligned} v_2 &= -\Lambda((\delta X \delta A)^\top B + B^\top(\delta X \delta A)) \\ &= -\Lambda\left(I_{n^2} + P_{n^2}\right)\left(I_n \otimes B^\top\right)\mathrm{vec}(\delta X \delta A) \\ &= \widetilde{N}_A \mathrm{vec}(\delta X \delta A) \end{aligned}$$

and hence,

$$\|v_2\|_2 \le l_A \delta_A \delta_X.$$

The quadratic terms in the numerator and denominator of (11.11) are obtained by bounding from above the norms of the matrices

$$-\Lambda(\delta A^\top X \delta B + \delta B^\top X A)$$

and

$$-\Lambda(\delta A^\top \delta X \delta B + \delta B^\top \delta X A)$$

respectively, and using the concept of the symmetrized norm of Λ. $\square$

The component-wise perturbation bound for equation (11.9) is

$$|\delta X| \preceq (I_s - \Psi_1(\Delta) - \Theta_2(\Delta))^{-1}(\Theta_1(\Delta) + \Theta_2(\Delta))\mathrm{vec}(|X|), \tag{11.13}$$

where

$$\begin{aligned}
\Theta_1(\Delta) &= |\Lambda|\Delta_C + |N_A|\Delta_A + |N_B|\Delta_B, \\
\Theta_2(\Delta) &= |\Lambda| \left(W_B^\top \otimes W_A^\top + W_A^\top \otimes W_B^\top\right), \\
\Psi_1(\Delta) &= \left|\Lambda \left(B^\top \otimes I_n\right)\right| \left(I_n \otimes W_A^\top\right) + \left|\Lambda \left(I_n \otimes B^\top\right)\right| \left(W_A^\top \otimes I_n\right) \\
&+ \left|\Lambda \left(A^\top \otimes I_n\right)\right| \left(I_n \otimes W_B^\top\right) + \left|\Lambda \left(I_n \otimes A^\top\right)\right| \left(W_B^\top \otimes I_n\right).
\end{aligned}$$

For the complex continuous-time descriptor Lyapunov equation

$$\mathcal{L}_c(X) := A^{\mathrm{H}}XB + B^{\mathrm{H}}XA = C \tag{11.14}$$

we obtain the following result.

Proposition 11.14 *The operator $\mathcal{L}_c$, defined by (11.14), is invertible if and only if at least one of the following three equivalent conditions holds:*

(i) *The matrix A is nonsingular and*

$$\lambda_i(A^{-1}B) + \overline{\lambda}_k(A^{-1}B) \neq 0;\ i,k = 1,\ldots,n.$$

(ii) *The matrix B is nonsingular and*

$$\lambda_i(B^{-1}A) + \overline{\lambda}_k(B^{-1}A) \neq 0;\ i,k = 1,\ldots,n.$$

(iii)

$$\alpha_i(A,B)\overline{\beta}_k(A,B) + \overline{\alpha}_k(A,B)\beta_i(A,B) \neq 0,\ i,k = 1,\ldots,n.$$

Theorem 11.15 *The norm-wise perturbation bound for equation (11.14) is*

$$\delta_X \leq \frac{\mathrm{est}(\delta_C, \delta_A, \delta_B, \Lambda^{\mathbb{R}}, N_A^{\mathbb{R}}, N_B^{\mathbb{R}}) + 2\|X\|_2\|\Lambda_c\|_2^*\delta_A\delta_B}{1 - l_A^{\mathbb{R}}\delta_A - l_B^{\mathbb{R}}\delta_B - 2\|\Lambda_c\|_2^*\delta_A\delta_B}, \tag{11.15}$$

provided that

$$l_A^{\mathbb{R}}\delta_A + l_B^{\mathbb{R}}\delta_B + 2\|\Lambda_c\|_2^*\delta_A\delta_B < 1.$$

Here

$$\begin{aligned}
\Lambda^{\mathbb{R}} &= \begin{bmatrix} \Lambda_0 & -\Lambda_1 \\ \Lambda_1 & \Lambda_0 \end{bmatrix}, \\
N_A^{\mathbb{R}} &= -\mathrm{Mat}\left(\Lambda_c(I_n \otimes (B^{\mathrm{H}}X)), \Lambda_c\left((XB)^\top \otimes I_n\right)P_{n^2}\right), \\
N_B^{\mathbb{R}} &= -\mathrm{Mat}\left(\Lambda_c(I_n \otimes (A^{\mathrm{H}}X)), \Lambda_c\left((XA)^\top \otimes I_n\right)P_{n^2}\right), \\
l_A^{\mathbb{R}} &= \left\|\mathrm{Mat}\left(\Lambda_c(I_n \otimes B^{\mathrm{H}}), \Lambda_c\left(B^\top \otimes I_n\right)P_{n^2}\right)\right\|_2, \\
l_B^{\mathbb{R}} &= \left\|\mathrm{Mat}\left(\Lambda_c(I_n \otimes A^{\mathrm{H}}), \Lambda_c\left(A^\top \otimes I_n\right)P_{n^2}\right)\right\|_2
\end{aligned}$$

and

$$\Lambda_c := \left(B^\top \otimes A^{\mathrm{H}} + A^\top \otimes B^{\mathrm{H}}\right)^{-1} = \Lambda_0 + \imath\Lambda_1.$$

Theorem 11.16 *The component-wise bound for the complex equation (11.14) has the form (11.13) with* Θ_i *and* Ψ_1 *defined by*

$$\begin{aligned}
\Theta_1(\Delta) &= |\Lambda_c|\Delta_C + \left(\left|\Lambda_c\left(I_n \otimes (B^{\mathrm{H}}X)\right)\right| + \left|\Lambda_c\left((XB)^\top \otimes I_n\right)P_{n^2}\right|\right)\Delta_A \\
&\quad + \left(\left|\Lambda_c\left(I_n \otimes (A^{\mathrm{H}}X)\right)\right| + \left|\Lambda_c\left((XA)^\top \otimes I_n\right)P_{n^2}\right|\right)\Delta_B, \\
\Theta_2(\Delta) &= |\Lambda_c|\left(W_B^\top \otimes W_A^\top + W_A^\top \otimes W_B^\top\right), \\
\Psi_1(\Delta) &= \left|\Lambda_c\left(B^\top \otimes I_n\right)\right|\left(I_n \otimes W_A^\top\right) + \left|\Lambda_c\left(I_n \otimes B^{\mathrm{H}}\right)\right|\left(W_A^\top \otimes I_n\right) \\
&+ \left|\Lambda_c\left(A^\top \otimes I_n\right)\right|\left(I_n \otimes W_B^\top\right) + \left|\Lambda_c\left(I_n \otimes A^{\mathrm{H}}\right)\right|\left(W_B^\top \otimes I_n\right).
\end{aligned}$$

11.5 Discrete-time equations

The spectrum of the Lyapunov operator $\mathcal{L}$ in the real discrete-time Lyapunov equation

$$\mathcal{L}(X) := A^\top XA - \alpha X = C \qquad (11.16)$$

is

$$\operatorname{spect}(\mathcal{L}) = \{\lambda_i(A)\lambda_k(A) - \alpha : i, k = 1, \ldots, n\}.$$

Theorem 11.17 *The norm-wise perturbation bound for equation (11.16) is*

$$\delta_X \le \frac{\operatorname{est}(\delta_C, \delta_A, \Lambda, N_A) + \|X\|_2\|\Lambda\|_2^*\delta_A^2}{1 - 2l_A\delta_A - \|\Lambda\|_2^*\delta_A^2}, \qquad (11.17)$$

where

$$\delta_A < \frac{2}{l_A + \sqrt{l_A^2 + 4\|\Lambda\|_2^*}}$$

and

$$\begin{aligned}
\Lambda &= \left(A^\top \otimes A^\top - \alpha I_{n^4}\right)^{-1}, \\
N_A &= -\Lambda(I_{n^2} + P_{n^2})\left(I_n \otimes \left(A^\top X\right)\right), \\
l_A &= \left\|\Lambda(I_{n^2} + P_{n^2})\left(I_n \otimes A^\top\right)\right\|_2.
\end{aligned}$$

Theorem 11.18 *The component-wise bound for equation (11.16) is*

$$|\delta X| \preceq (I_{n^2} - \Psi_1(\Delta) - \Theta_2(\Delta))\,(\Theta_1(\Delta) + \Theta_2(\Delta))\operatorname{vec}(|X|)), \qquad (11.18)$$

where

$$\begin{aligned}
\Theta_1(\Delta) &= |\Lambda|\Delta_C + |N_A|\Delta_A, \\
\Theta_2(\Delta) &= |\Lambda|\left(W_A^\top \otimes W_A^\top\right), \\
\Psi_1(\Delta) &= \left|\Lambda\left(A^\top \otimes I_n\right)\right|\left(I_n \otimes W_A^\top\right) + \left|\Lambda\left(I_n \otimes A^\top\right)\right|\left(W_A^\top \otimes I_n\right).
\end{aligned}$$

The spectrum of the linear operator $\mathcal{L}_c$ in the complex discrete-time Lyapunov equation

$$\mathcal{L}_c(X) := A^{\mathrm{H}}XA - \alpha X = C \tag{11.19}$$

is

$$\operatorname{spect}(\mathcal{L}_c) = \{\lambda_i(A)\overline{\lambda}_k(A) - \alpha : i, k = 1, \ldots, n\}.$$

Theorem 11.19 *The norm-wise perturbation bound for equation (11.19) is*

$$\delta_X \leq \frac{\operatorname{est}(\delta_C, \delta_A, \Lambda^{\mathbb{R}}, N_A^{\mathbb{R}}) + \|X\|_2\|\Lambda_c\|_2^*\delta_A^2}{1 - 2l_A^{\mathbb{R}}\delta_A - \|\Lambda_c\|_2^*\delta_A^2}, \tag{11.20}$$

where

$$\begin{aligned}
\Lambda^{\mathbb{R}} &= \begin{bmatrix} \Lambda_0 & -\Lambda_1 \\ \Lambda_1 & \Lambda_0 \end{bmatrix}, \\
N_A^{\mathbb{R}} &= -\operatorname{Mat}\left(\Lambda_c(I_n \otimes (A^{\mathrm{H}}X)), \Lambda_c\left((XA)^\top \otimes I_n\right)P_{n^2}\right), \\
l_A^{\mathbb{R}} &= \left\|\operatorname{Mat}\left(\Lambda_c\left(I_n \otimes A^{\mathrm{H}}\right), \Lambda_c\left(A^\top \otimes I_n\right)P_{n^2}\right)\right\|_2
\end{aligned}$$

and

$$\Lambda_c = L_c^{-1} = \left(A^\top \otimes A^{\mathrm{H}} - \alpha I_{n^4}\right)^{-1} = \Lambda_0 + \imath\Lambda_1.$$

Theorem 11.20 *The component-wise bound for the complex equation (11.19) has the form (11.18) with Θ_i and Ψ_1 defined by*

$$\begin{aligned}
\Theta_1(\Delta) &= |\Lambda_c|\Delta_C + \left(\left|\Lambda_c\left(I_n \otimes (A^{\mathrm{H}}X)\right)\right| + \left|\Lambda_c\left((XA)^\top \otimes I_n\right)P_{n^2}\right|\right)\Delta_A, \\
\Theta_2(\Delta) &= |\Lambda_c|\left(W_A^\top \otimes W_A^\top\right), \\
\Psi_1(\Delta) &= \left|\Lambda_c\left(A^\top \otimes I_n\right)\right|\left(I_n \otimes W_A^\top\right) + \left|\Lambda_c\left(I_n \otimes A^{\mathrm{H}}\right)\right|\left(W_A^\top \otimes I_n\right).
\end{aligned}$$

11.6 Discrete-time equations in descriptor form

Again, no simple expression for the spectrum $\operatorname{spect}(\mathcal{L})$ of the Lyapunov operator $\mathcal{L}$ in the descriptor discrete-time Lyapunov equation

$$\mathcal{L}(X) := A^\top XA - B^\top XB = C \tag{11.21}$$

is available in the nontrivial case when neither A nor B are multiples of I_n. Hence, $\operatorname{spect}(\mathcal{L})$ may be computed (if necessary) as $\operatorname{spect}(L)$, where

$$L := A^\top \otimes A^\top - B^\top \otimes B^\top \tag{11.22}$$

is the matrix of $\mathcal{L}$. In this case we again have a test for invertibility of $\mathcal{L}$ in terms of $n \times n$ matrices instead of the $n^2 \times n^2$ matrix L.

Proposition 11.21 *The operator $\mathcal{L}$ is invertible if and only if at least one of the following two conditions is fulfilled:*

(i) *At least one of the matrices A or B is nonsingular and either*

$$\lambda_i(A^{-1}B)\lambda_k(A^{-1}B) \neq 1;\ i,k = 1,\dots,n$$

(if A is nonsingular), or

$$\lambda_i(B^{-1}A)\lambda_k(B^{-1}A) \neq 1;\ i,k = 1,\dots,n$$

(if B is nonsingular).

(ii)

$$\alpha_i(A,B)\beta_k(A,B) \neq \alpha_k(A,B)\beta_i(A,B);\ i,k = 1,\dots,n.$$

Proof. (i) We show first that if $\mathcal{L}$ is invertible then at least one of the matrices A or B is nonsingular. If both A and B are nonsingular there is nothing to prove, so suppose that A is singular. Then there exists $V \in \mathcal{O}(n)$ such that

$$AV = [A_1, 0_{n\times 1}].$$

Partition the matrix $BV = [B_1, b]$ accordingly, where $b \in \mathbb{R}^n$.

In view of the invertibility of the operator $\mathcal{L}$ its matrix L, given in (11.22), is nonsingular. Hence, the matrix

$$L_1 := (V^\top \otimes V^\top)L$$

is also nonsingular. We have

$$\begin{aligned} L_1 &= (AV)^\top \otimes (AV)^\top - (BV)^\top \otimes (BV)^\top \\ &= \begin{bmatrix} A_1^\top \\ 0 \end{bmatrix} \otimes (AV)^\top - \begin{bmatrix} B_1^\top \\ b^\top \end{bmatrix} \otimes (BV)^\top \\ &= \begin{bmatrix} A_1^\top \otimes (AV)^\top - B_1^\top \otimes (BV)^\top \\ -b^\top \otimes (BV)^\top \end{bmatrix}. \end{aligned}$$

Therefore, the matrix

$$b^\top \otimes (BV)^\top \in \mathbb{R}^{n\times n^2}$$

must be of full row rank, equal to n. Since

$$\operatorname{rank}(b^\top \otimes (BV)^\top) = \operatorname{rank}(b)\operatorname{rank}(B),$$

we see that $b \neq 0$ and $\operatorname{rank}(B)$ must be n, i.e., B is nonsingular.

Next we show that in this case

$$\lambda_i(B^{-1}A)\lambda_k(B^{-1}A) \neq 1;\ i,k = 1,\dots,n. \tag{11.23}$$

Indeed, the matrix

$$L_2 := L(B\otimes B)^{-\top} = (B^{-1}A)^\top \otimes (B^{-1}A)^\top - I_{n^4}$$

is nonsingular together with the matrix L from (11.22), since its eigenvalues

$$\lambda_j(L_2) = \lambda_i(B^{-1}A)\lambda_k(B^{-1}A) - 1$$

are nonzero.

That (i) implies the invertibility of $\mathcal{L}$ is checked by repeating the above arguments in reverse order.

(ii) Suppose that (i) holds and that e.g. B is nonsingular. Then the identities

$$\lambda_i(B^{-1}A) = \frac{\alpha_i(A,B)}{\beta_i(A,B)}, \quad i = 1, \ldots, n,$$

together with (11.23) yield (ii). In turn, if B is nonsingular then (ii) yields (11.23). □

Theorem 11.22 *The norm-wise perturbation estimate for equation (11.21) is*

$$\delta_X \le \frac{\mathrm{est}(\delta_C, \delta_A, \delta_B, \Lambda, N_A, N_B) + \|X\|_2 \|\Lambda\|_2^* \left(\delta_A^2 + \delta_B^2\right)}{1 - l_A\delta_A - l_B\delta_B - \|\Lambda\|_2^* \left(\delta_A^2 + \delta_B^2\right)}, \tag{11.24}$$

where

$$\begin{aligned}
\Lambda &= \left(A^\top \otimes A^\top - B^\top \otimes B^\top\right)^{-1}, \\
N_A &= -\Lambda(I_{n^2} + P_{n^2})\left(I_n \otimes \left(A^\top X\right)\right), \\
N_B &= \Lambda(I_{n^2} + P_{n^2})\left(I_n \otimes \left(B^\top X\right)\right), \\
l_A &= \left\|\Lambda(I_{n^2} + P_{n^2})\left(I_n \otimes A^\top\right)\right\|_2, \\
l_B &= \left\|\Lambda(I_{n^2} + P_{n^2})\left(I_n \otimes B^\top\right)\right\|_2.
\end{aligned}$$

The domain for δ_A, δ_B in (11.24) is determined by

$$l_A\delta_A + l_B\delta_B + \|\Lambda\|_2^* \left(\delta_A^2 + \delta_B^2\right) < 1. \tag{11.25}$$

If X is nonnegative or nonpositive definite, then the expression $\delta_A^2 + \delta_B^2$ may be replaced by $\max\left\{\delta_A^2, \delta_B^2\right\}$ in both the numerator and the denominator of the norm-wise bound (11.24) as well as in the left-hand side of (11.25).

Theorem 11.23 *The component-wise bound for equation (11.21) is*

$$|\delta X| \preceq \left(I_{n^2} - \Psi_1(\Delta) - \Theta_2(\Delta)\right)\left(\Theta_1(\Delta) + \Theta_2(\Delta)\right)\mathrm{vec}(|X|)), \tag{11.26}$$

where

$$\begin{aligned}
\Theta_1(\Delta) &= |\Lambda|\delta_C + |N_A|\Delta_A + |N_B|\Delta_B, \\
\Theta_2(\Delta) &= |\Lambda|\left(W_A^\top \otimes W_A^\top + W_B^\top \otimes W_B^\top\right), \\
\Psi_1(\Delta) &= \left|\Lambda\left(A^\top \otimes I_n\right)\right|\left(I_n \otimes W_A^\top\right) + \left|\Lambda\left(I_n \otimes A^\top\right)\right|\left(W_A^\top \otimes I_n\right) \\
&+ \left|\Lambda\left(B^\top \otimes I_n\right)\right|\left(I_n \otimes W_B^\top\right) + \left|\Lambda\left(I_n \otimes B^\top\right)\right|\left(W_B^\top \otimes I_n\right).
\end{aligned}$$

Consider finally the complex discrete-time descriptor Lyapunov equation

$$\mathcal{L}_d(X) := A^{\mathrm{H}}XA - B^{\mathrm{H}}XB = C. \tag{11.27}$$

A result, similar to Proposition 11.14 is the following.

Proposition 11.24 *The operator $\mathcal{L}_d$ is invertible if at least one of the following conditions is fulfilled:*

(i) *At least one of the matrices A or B is nonsingular and either*

$$\lambda_i(A^{-1}B)\overline{\lambda}_k(A^{-1}B) \neq 1;\ i,k = 1,\ldots,n$$

(if A is nonsingular), or

$$\lambda_i(B^{-1}A)\overline{\lambda}_k(B^{-1}A) \neq 1;\ i,k = 1,\ldots,n$$

(if B is nonsingular).

(ii)

$$\alpha_i(A,B)\overline{\beta}_k(A,B) \neq \overline{\alpha}_k(A,B)\beta_i(A,B);\ i,k = 1,\ldots,n.$$

Theorem 11.25 *The norm-wise perturbation bound for equation (11.27) is*

$$\delta_X \leq \frac{\mathrm{est}(\delta_C, \delta_A, \delta_B, \Lambda^{\mathbb{R}}, N_A^{\mathbb{R}}, N_B^{\mathbb{R}}) + \|X\|_2 \|\Lambda_c\|_2^* \left(\delta_A^2 + \delta_B^2\right)}{1 - l_A^{\mathbb{R}}\delta_A - l_B^{\mathbb{R}}\delta_B - \|\Lambda_c\|_2^* \left(\delta_A^2 + \delta_B^2\right)}, \tag{11.28}$$

where

$$\begin{aligned}
\Lambda^{\mathbb{R}} &= \begin{bmatrix} \Lambda_0 & -\Lambda_1 \\ \Lambda_1 & \Lambda_0 \end{bmatrix}, \\
N_A^{\mathbb{R}} &= -\mathrm{Mat}\left(\Lambda_c(I_n \otimes (A^{\mathrm{H}}X)), \Lambda_c\left((XA)^{\top} \otimes I_n\right) P_{n^2}\right), \\
N_B^{\mathcal{R}} &= \mathrm{Mat}\left(\Lambda_c(I_n \otimes (B^{\mathrm{H}}X)), \Lambda_c\left((XB)^{\top} \otimes I_n\right) P_{n^2}\right), \\
l_A^{\mathbb{R}} &= \left\|\mathrm{Mat}\left(\Lambda_c(I_n \otimes A^{\mathrm{H}}), \Lambda_c\left(A^{\top} \otimes I_n\right) P_{n^2}\right)\right\|_2, \\
l_B^{\mathbb{R}} &= \left\|\mathrm{Mat}\left(\Lambda_c(I_n \otimes B^{\mathrm{H}}), \Lambda_c\left(B^{\top} \otimes I_n\right) P_{n^2}\right)\right\|_2
\end{aligned}$$

and

$$\Lambda_c = \left(A^{\top} \otimes A^{\mathrm{H}} - B^{\top} \otimes B^{\mathrm{H}}\right)^{-1} = \Lambda_0 + \imath\Lambda_1.$$

The domain for δ_A, δ_B in (11.28) is given by

$$l_A^{\mathbb{R}}\delta_A + l_B^{\mathbb{R}}\delta_B + \|\Lambda_c\|_2^* \left(\delta_A^2 + \delta_B^2\right) < 1. \tag{11.29}$$

If X is positive or negative semidefinite we may, as in the real case, replace $\delta_A^2 + \delta_B^2$ by $\max\left\{\delta_A^2, \delta_B^2\right\}$ in both the numerator and the denominator of the norm-wise bound (11.28) as well as in the left-hand side of (11.29).

Theorem 11.26 *The component-wise bound for the complex equation (11.27) has the form (11.26) with* Θ_i *and* Ψ_1 *defined by*

$$\begin{aligned}
\Theta_1(\Delta) &= |\Lambda_c|\Delta_C + \left(\left|\Lambda_c\left(I_n \otimes (A^{\mathrm{H}}X)\right)\right| + \left|\Lambda_c\left((XA)^\top \otimes I_n\right) P_{n^2}\right|\right)\Delta_A \\
&\quad + \left(\left|\Lambda_c\left(I_n \otimes (B^{\mathrm{H}}X)\right)\right| + \left|\Lambda_c\left((XB)^\top \otimes I_n\right) P_{n^2}\right|\right)\Delta_B, \\
\Theta_2(\Delta) &= |\Lambda_c|\left(W_A^\top \otimes W_A^\top + W_B^\top \otimes W_B^\top\right), \\
\Psi_1(\Delta) &= \left|\Lambda_c\left(A^\top \otimes I_n\right)\right|\left(I_n \otimes W_A^\top\right) + \left|\Lambda_c\left(I_n \otimes A^{\mathrm{H}}\right)\right|\left(W_A^\top \otimes I_n\right) \\
&+ \left|\Lambda_c\left(B^\top \otimes I_n\right)\right|\left(I_n \otimes W_B^\top\right) + \left|\Lambda_c\left(I_n \otimes B^{\mathrm{H}}\right)\right|\left(W_B^\top \otimes I_n\right).
\end{aligned}$$

11.7 Notes and references

The presented results are partially published in the literature for particular classes of Lyapunov equations, see e.g. [95, 3, 134, 132, 136, 125]. The presentation above follows the paper [132]. Residual bounds for the standard discrete-time Lyapunov equation are given in [78].

Descriptor Lyapunov equations are studied in [207], while condition and error estimates for the solution of Lyapunov equations are given in [179, 146].

Chapter 12

General quadratic equations

12.1 Introductory remarks

In this chapter we present a complete perturbation analysis for general quadratic matrix equations. We also briefly consider symmetric quadratic matrix equations, particular cases of which are the continuous-time Riccati equations, arising in the optimal control and filtering (including $\mathcal{H}_\infty$ control and filtering) of continuous time-invariant control systems.

12.2 Problem statement

Consider the general quadratic matrix equation

$$F(X, P) := A(X, P_1) + Q(X, P_2) = 0, \tag{12.1}$$

where $X \in \mathbb{F}^{m\times n}$ is the unknown matrix. The function

$$F(\cdot, P) : \mathbb{F}^{m\times n} \to \mathbb{F}^{p\times q}$$

is a quadratic matrix operator, depending on the matrix collection $P = (P_1, P_2)$. In (12.1)

$$A(\cdot, P_1) : \mathbb{F}^{m\times n} \to \mathbb{F}^{p\times q}$$

is an affine operator,

$$A(X, P_1) := A_0 + \sum_{i=1}^{r_1} A_i X B_i, \tag{12.2}$$

depending on the matrix collection

$$P_1 := (A_0, A_1, B_1, \ldots, A_{r_1}, B_{r_1}),$$

where

$$A_0 \in \mathbb{F}^{p\times q},\ A_i \in \mathbb{F}^{p\times m},\ B_i \in \mathbb{F}^{n\times q}$$

are given matrix coefficients. Furthermore,

$$Q(\cdot, P_2) : \mathbb{F}^{m\times n} \to \mathbb{F}^{p\times q}$$

is a homogeneous quadratic operator,

$$Q(X, P_2) := \sum_{k=1}^{r_2} C_k X D_k X E_k, \tag{12.3}$$

depending on the matrix collection

$$P_2 := (C_1, D_1, E_1, \ldots, C_{r_2}, D_{r_2}, E_{r_2}),$$

where

$$C_k \in \mathbb{F}^{p\times m},\ D_k \in \mathbb{F}^{n\times m},\ E_k \in \mathbb{F}^{n\times q}$$

are given matrices.

It is assumed that $mn = pq := l$. The matrix $(2r_1+1)$-tuple P_1 depends on $pq + r_1(mp + nq)$ parameters – the elements of the matrices A_0, A_i and B_i, while the $3r_2$-tuple P_2 depends on $r_2(pm + nm + nq)$ parameters.

Denote by

$$F_Z(X, P) : \mathbb{F}^{r\times t} \to \mathbb{F}^{p\times q}$$

the partial Fréchet derivative of F in the corresponding $r \times t$ matrix argument

$$Z \in \mathcal{P} := \{A_0, A_1, B_1, \ldots, C_{r_2}, D_{r_2}, E_{r_2}\}, \tag{12.4}$$

computed at the point (X, P).

We assume that equation (12.1) has a solution X, such that the linear operator

$$F_X := F_X(X, P) : \mathbb{F}^{m\times n} \to \mathbb{F}^{p\times q}$$

is invertible (we recall that $mn = pq$ and hence the matrix spaces $\mathbb{F}^{m\times n}$ and $\mathbb{F}^{p\times q}$ are isomorphic). Then according to the implicit function theorem (see [117, 173] and Appendix A) the solution X is isolated, i.e., there exists $\varepsilon > 0$ such that equation (12.1) has no other solution $\widetilde{X}$ with $\|\widetilde{X} - X\| < \varepsilon$. The problems of existence and uniqueness of the solution of quadratic matrix equations are of independent interest but they are not the subject of this monograph. We only mention that the general solution (i.e., the set of all solutions) of equation (12.1) is the intersection of l quadrics and is thus a closed algebraic variety in the Zariski topology of $\mathbb{F}^{p\times q} \simeq \mathbb{F}^l$. For the geometry of such sets see [198, 199].

The perturbation problem for equation (12.1) is stated as follows. Let the matrices from $\mathcal{P}$ be perturbed as

$$\begin{aligned} A_0 &\mapsto A_0 + \delta A_0,\ A_i \mapsto A_i + \delta A_i,\ B_i \mapsto B_i + \delta B_i, \\ C_k &\mapsto C_k + \delta C_k,\ D_k \mapsto D_k + \delta D_k,\ E_k \mapsto E_k + \delta E_k. \end{aligned}$$

Denote by $P + \delta P$ the perturbed collection P, in which every matrix $Z \in \mathcal{P}$ is replaced by $Z + \delta Z$. Then for a given solution X, the perturbed equation is

$$F(X + \delta X, P + \delta P) = 0. \tag{12.5}$$

Typically, some of the matrices from $\mathcal{P}$ are not perturbed. Denote by

$$\widetilde{\mathcal{P}} := \{Z_1, Z_2, \dots, Z_r\} \subset \mathcal{P}$$

the set of matrices from $\mathcal{P}$, which are perturbed.

Since the operator F_X is invertible, equation (12.5) has a unique isolated solution $X + \delta X$ in the neighborhood of X if the perturbation δP is sufficiently small. Moreover, in this case the elements of δX are analytic functions of the elements of δP.

Denote by

$$\begin{aligned} \delta^0 &:= \left[\delta_1^0, \delta_2^0, \delta_3^0, \dots, \delta_{\nu-2}^0, \delta_{\nu-1}^0, \delta_\nu^0\right]^\top \\ &:= \left[\delta_{A_0}, \delta_{A_1}, \delta_{B_1}, \dots, \delta_{C_{r_2}}, \delta_{D_{r_2}}, \delta_{E_{r_2}}\right]^\top \in \mathbb{R}_+^\nu \end{aligned} \tag{12.6}$$

the full vector of absolute norm perturbations $\delta_Z := \|\delta Z\|_{\mathrm{F}}$ in the data matrices (12.4), where

$$\nu := 1 + 2r_1 + 3r_2.$$

Let also

$$\delta := [\delta_1, \delta_2, \dots, \delta_r]^\top := [\delta_{Z_1}, \delta_{Z_2}, \dots, \delta_{Z_r}]^\top \in \mathbb{R}_+^r \tag{12.7}$$

be the vector of nonzero norm perturbations of the matrices δZ for $Z \in \widetilde{\mathcal{P}}$.

The perturbation problem is to find a bound

$$\delta_X \le f(\delta), \ \delta \in \Omega \subset \mathbb{R}_+^r, \tag{12.8}$$

for the perturbation

$$\delta_X := \|\delta X\|_{\mathrm{F}},$$

where Ω is a given set and f is a continuous function, nondecreasing in each of its arguments and satisfying

$$f(0) = 0.$$

We first derive a local bound

$$\delta_X \le f_1(\delta) + O(\|\delta\|^2), \ \delta \to 0,$$

which is then incorporated in the nonlocal bound (12.8). The inclusion $\delta \in \Omega$ guarantees that the perturbed equation (12.5) has a unique solution $X + \delta X$ in the neighborhood of the solution X of the original equation (12.1).

Estimates in terms of relative perturbations

$$\rho_j := \frac{\|\delta Z_j\|_{\mathrm{F}}}{\|Z_j\|_{\mathrm{F}}}, \ Z_j \in \widetilde{\mathcal{P}},$$

for

$$\rho_X := \frac{\|\delta X\|_{\rm F}}{\|X\|_{\rm F}}$$

are straightforward when $X \neq 0$. Indeed, we have

$$\rho_X \leq \frac{f\left(\|Z_1\|_{\rm F}\rho_1, \ldots, \|Z_r\|_{\rm F}\rho_r\right)}{\|X\|_{\rm F}}.$$

Suppose now that $p = q = m = n$ and that we have a *symmetric* quadratic matrix equation of type (12.1).

Definition 12.1 A *symmetric* operator

$$G(\cdot, R) : \mathbb{F}^{n\times n} \to \mathbb{F}^{n\times n},$$

depending on a collection of matrices R, satisfies

$$G^\top(X, R) = G(X^\top, R)$$

in the real case and

$$G^{\rm H}(X, R) = G(X^{\rm H}, R)$$

in the complex case.

Hence, in symmetric quadratic matrix equations we must assume that A and Q are symmetric operators. Thus, in the real case the symmetric operator A is of the form

$$A(X, R_1) = A_0 + \sum_{i=1}^{l_1} \left(A_i X B_i + B_i^\top X A_i^\top\right) + \sum_{i=1}^{k_1} \varepsilon_i M_i X M_i^\top,$$

where

$$A_0 = A_0^\top,\ \varepsilon_i = \pm 1,\ 2l_1 + k_1 = r_1$$

and

$$R_1 := \{A_0, A_1, B_1, \ldots, A_{l_1}, B_{l_1}, M_1, \ldots, M_{k_1}\}.$$

Similarly, the symmetric operator Q is determined by

$$Q(X, R_2) = \sum_{k=1}^{l_2} \left(C_k X D_k X E_k + E_k^\top X D_k^\top X C_k^\top\right) + \sum_{k=1}^{k_2} \varepsilon_k T_k X S_k X T_k^\top,$$

where

$$S_k = S_k^\top,\ \varepsilon_k = \pm 1,\ 2l_2 + k_2 = r_2$$

and

$$R_2 := \{C_1, D_1, E_1, \ldots, C_{l_2}, D_{l_2}, E_{l_2}, T_1, S_1, \ldots, S_{k_2}, T_{k_2}\}.$$

In the complex case we have

$$A(X,R_1) = A_0 + \sum_{i=1}^{l_1} \left(A_i X B_i + B_i^{\mathrm{H}} X A_i^{\mathrm{H}}\right) + \sum_{i=1}^{k_1} \varepsilon_i M_i X M_i^{\mathrm{H}}$$

and

$$Q(X,R_2) = \sum_{k=1}^{l_2} \left(C_k X D_k X E_k + E_k^{\mathrm{H}} X D_k^{\mathrm{H}} X C_k^{\mathrm{H}}\right) + \sum_{k=1}^{k_2} \varepsilon_k T_k X S_k X T_k^{\mathrm{H}},$$

where $A_0 = A_0^{\mathrm{H}}$ and $S_k = S_k^{\mathrm{H}}$.

The symmetric quadratic matrix equations, arising in the optimal control and filtering of continuous time-invariant linear systems, are called *continuous-time algebraic Riccati equations.* An example of a real continuous-time algebraic Riccati equation is given in the next section.

12.3 Motivating example

Nonsymmetric quadratic matrix equations arise in the analysis of continuous time-invariant systems as shown in the following example.

Example 12.2 Consider the continuous time-invariant dynamic system

$$\dot{x}(t) = Ax(t), \ t > 0,$$

with initial condition $x(0) = x_0$, where $x(t) \in \mathbb{F}^n$. Let $n = q + p$, where $p, q \geq 1$ are integers. We may write the system in a partitioned form

$$\begin{aligned} \dot{x}_1(t) &= A_{11}x_1(t) + A_{12}x_2(t), \\ \dot{x}_2(t) &= A_{21}x_1(t) + A_{22}x_2(t), \ t > 0, \end{aligned}$$

with initial conditions $x_i(0) = x_{i0}$, where

$$x(t) = \left[x_1^{\top}(t), x_2^{\top}(t)\right]^{\top}, \ x_1(t) \in \mathbb{F}^q, \ x_2(t) \in \mathbb{F}^p.$$

The system admits a *mutual observation property* [128] if there exists a matrix $X \in \mathbb{F}^{p\times q}$, such that for every $x_0 \in \mathbb{F}^n$ one has

$$x_2(t) = Xx_1(t) + v_2(t), \ t \geq 0,$$

where $v_2(t) \in \mathbb{F}^p$ is the state of a time-invariant system,

$$\dot{v}_2(t) = B_2 v_2(t), \ t > 0, \ v_2(0) = x_{20} - Xx_{10},$$

and

$$\lim_{t\to\infty} \sup \left\{ \frac{\|v_2(t)\|}{\|x(t)\|} : x_0 \in \mathbb{F}^n \right\} < \infty.$$

Using the change of variables

$$x(t) = U(X)v(t), \; U(X) := \left[\begin{array}{cc} I_q & 0 \\ X & I_p \end{array} \right],$$

where

$$v(t) = \left[v_1^\top(t), v_2^\top(t) \right]^\top,$$

we see that

$$\dot{v}(t) = B(X)v(t),$$

where

$$B(X) := U^{-1}(X)AU(X) = \left[\begin{array}{cc} A_{11} + A_{12}X & A_{12} \\ A_{21} + A_{22}X - XA_{11} - XA_{12}X & A_{22} - XA_{12} \end{array} \right].$$

Hence, the mutual observation property will be valid with

$$B_2 = A_{22} - XA_{12}$$

if the matrix X satisfies the nonsymmetric Riccati equation

$$A_{21} + A_{22}X - XA_{11} - XA_{12}X = 0$$

under the additional condition $\mathrm{Re}(\lambda_2) < \mathrm{Re}(\lambda_1)$ for every $\lambda_1 \in \mathrm{spect}(A_{11} + A_{12}X)$ and $\lambda_2 \in \mathrm{spect}(A_{22} - XA_{12})$.

Consider also two types of descriptor systems,

$$E_1\dot{x}(t) = Ax(t), \; E_1 := \mathrm{diag}\left(F_1, I_p\right),$$

and

$$E_2\dot{x}(t) = Ax(t), \; E_2 := \mathrm{diag}\left(I_q, F_2\right),$$

where the matrices F_i are nonsingular and the matrices $E_i^{-1}A$ are stable. In the first case, to ensure the mutual observation property, we have to solve the nonsymmetric Riccati equation

$$A_{21} + A_{22}YF_1 - YA_{11} - YA_{12}YF_1 = 0$$

for $Y := TF_1^{-1}$. In the second case the nonsymmetric Riccati equation is

$$A_{21} + A_{22}X - F_2XA_{11} - F_2XA_{12}X = 0.$$

◇

Similar nonsymmetric quadratic matrix equations arise in the problem of determining invariant subspaces of the form $\mathrm{Rg}\left[\begin{array}{c} U_1 \\ U_2 \end{array} \right]$ for the matrix $A \in \mathbb{F}^{n\times n}$, where the matrix $U_1 \in \mathbb{F}^{q\times q}$ is nonsingular.

12.4 Local perturbation analysis

In this section we derive local perturbation bounds for the general quadratic matrix equation (12.1). A detailed study of its symmetric real and complex versions is given in Chapter 13.

12.4.1 Condition numbers

Consider the conditioning of equation (12.1). Recall that $\mathbf{Lin}(p, m, n, q, \mathbb{F})$ is the space of linear operators $\mathcal{F}^{m\times n} \to \mathcal{F}^{p\times q}$. Having in mind that $F(X, P) = 0$, the perturbed equation (12.5) may be written as

$$F(X + \delta X, P + \delta P) := F_X(\delta X) + \sum_{Z\in\widetilde{\mathcal{P}}} F_Z(\delta Z) + G(\delta X, \delta P) = 0, \qquad (12.9)$$

where

$$\begin{aligned}
F_X(\cdot) &:= F_X(X,P)(\cdot) \in \mathbf{Lin}(p,m,n,q),\\
F_{A_0}(\cdot) &:= F_{A_0}(X,P)(\cdot) \in \mathbf{Lin}(p,p,q,q,\mathbb{F}),\\
F_{A_i}(\cdot) &:= F_{A_i}(X,P)(\cdot) \in \mathbf{Lin}(p,p,m,q,\mathbb{F}),\\
F_{B_i}(\cdot) &:= F_{B_i}(X,P)(\cdot) \in \mathbf{Lin}(p,n,q,q,\mathbb{F}),\\
F_{C_k}(\cdot) &:= F_{C_k}(X,P)(\cdot) \in \mathbf{Lin}(p,p,m,q,\mathbb{F}),\\
F_{D_k}(\cdot) &:= F_{D_k}(X,P)(\cdot) \in \mathbf{Lin}(p,n,m,q,\mathbb{F}),\\
F_{E_k}(\cdot) &:= F_{E_k}(X,P)(\cdot) \in \mathbf{Lin}(p,n,q,q,\mathbb{F})
\end{aligned}$$

are the Fréchet derivatives of $F(X, P)$ in the corresponding matrix arguments, evaluated at the solution X and the matrix $G(\delta X, \delta P)$ contains second and higher order terms in δX, δP. A straightforward calculation leads to

$$\begin{aligned}
F_X(Z) &= \sum_{i=1}^{r_1} A_i Z B_i + \sum_{k=1}^{r_2} (C_k X D_k Z E_k + C_k Z D_k X E_k),\\
F_{A_0}(Z) &= Z,\\
F_{A_i}(Z) &= Z X B_i,\\
F_{B_i}(Z) &= A_i X Z, \qquad (12.10)\\
F_{C_k}(Z) &= Z X D_k X E_k,\\
F_{D_k}(Z) &= C_k X Z X E_k,\\
F_{E_k}(Z) &= C_k X D_k X Z.
\end{aligned}$$

Since the operator $F_X(.)$ is invertible we get

$$\delta X = -\sum_{Z\in\widetilde{\mathcal{P}}} F_X^{-1} \circ F_Z(\delta Z) - F_X^{-1}(G(\delta X, \delta P)). \qquad (12.11)$$

Relation (12.11) gives

$$\delta_X \leq \sum_{Z\in\widetilde{\mathcal{P}}} K_Z\delta_Z + O(\|\delta\|^2),\ \delta\to 0, \tag{12.12}$$

where the quantities

$$K_Z := \left\|F_X^{-1}\circ F_Z\right\|,\ Z\in\widetilde{\mathcal{P}}, \tag{12.13}$$

are the *absolute individual condition numbers* [188] of the quadratic matrix equation (12.1). Here $\|.\|$ is the norm, induced by the Frobenius norm in the corresponding space of linear operators, i.e.,

$$\|\mathcal{F}\| := \max\{\|\mathcal{F}(Y)\|_{\mathrm{F}} : \|Y\|_{\mathrm{F}} = 1\}.$$

If $X \neq 0$, then an estimate in terms of relative perturbations is

$$\rho_X := \frac{\|\delta X\|_{\mathrm{F}}}{\|X\|_{\mathrm{F}}} \leq \sum_{Z\in\widetilde{\mathcal{P}}} k_Z\rho_Z + O(\|\delta\|^2),\ \delta\to 0,$$

where the scalars

$$k_Z := K_Z\frac{\|Z\|_{\mathrm{F}}}{\|X\|_{\mathrm{F}}},\ Z\in\widetilde{\mathcal{P}},$$

are the *relative individual condition numbers* with respect to perturbations in the matrix coefficients $Z\in\widetilde{\mathcal{P}}$.

A calculation of the condition numbers K_Z is straightforward. Denote by $L_Z\in\mathbb{F}^{pq\times rt}$ the matrix representation of the operator $F_Z(\cdot)\in\mathbf{Lin}(p,r,t,q)$. We have

$$\begin{aligned}
L_X &= \sum_{i=1}^{r_1} B_i^\top\otimes A_i + \sum_{k=1}^{r_2}\left(E_k^\top\otimes(C_kXD_k) + (D_kXE_k)^\top\otimes C_k\right),\\
L_{A_0} &= I_l,\\
L_{A_i} &= (XB_i)^\top\otimes I_p,\\
L_{B_i} &= I_q\otimes(A_iX),\\
L_{C_k} &= (XD_kXE_k)^\top\otimes I_p,\\
L_{D_k} &= (XE_k)^\top\otimes(C_kX),\\
L_{E_k} &= I_q\otimes(C_kXD_kX).
\end{aligned} \tag{12.14}$$

Thus, the absolute condition numbers are

$$K_Z = \left\|L_X^{-1}L_Z\right\|_2,\ Z\in\widetilde{\mathcal{P}}. \tag{12.15}$$

A drawback of this approach is the large size of the involved matrices. Condition and accuracy estimates, avoiding the formation and analysis of large matrices, are proposed in [179].

An overall relative condition number may be defined as follows. Denote by

$$\Theta = (Z_1, \ldots, Z_r)$$

the r-tuple of matrix coefficients from $\widetilde{\mathcal{P}}$. The matrix collection Θ may be considered as an element of a linear space (the Cartesian product of the matrix spaces, to whom the matrices Z_i belong). Hence, we may define the product

$$\alpha\Theta = (\alpha Z_1, \alpha Z_2, \ldots, \alpha Z_r)$$

of Θ and $\alpha \in \mathbb{F}$ as well as the sum

$$\Theta + \widetilde{\Theta} = \left(Z_1 + \widetilde{Z}_1, Z_2 + \widetilde{Z}_2, \ldots, Z_r + \widetilde{Z}_r\right)$$

of two r-tuples Θ and $\widetilde{\Theta}$. We also introduce the generalized norm

$$\|\Theta\|_g := [\|Z_1\|_{\mathrm{F}}, \|Z_2\|_{\mathrm{F}}, \ldots, \|Z_r\|_{\mathrm{F}}]^\top \in \mathbb{R}_+^r$$

of the r-tuple Θ.

Let

$$\delta X = \delta X(\delta\Theta)$$

be the perturbation in the solution, where

$$\delta\Theta := [\delta Z_1, \delta Z_2, \ldots, \delta Z_r],$$

and $\gamma \in \mathbb{R}^r$ is a vector with positive elements.

Definition 12.3 The *absolute overall condition number with respect to* γ is

$$\kappa(\gamma) := \lim_{\varepsilon \to 0} \max \left\{ \|\delta X(\delta\Theta)\|_{\mathrm{F}} : \|\delta\Theta\|_g \preceq \varepsilon\gamma \right\}.$$

We have (see Chapter 8)

$$\kappa(\gamma) = \max \left\{ \left\| \sum_{Z \in \widetilde{\mathcal{P}}} F_X^{-1} \circ F_Z(\delta Z) \right\|_2 : \|\delta\Theta\|_g \preceq \gamma \right\}. \tag{12.16}$$

When $\gamma_j = \|Z_j\|_{\mathrm{F}}$ and $\gamma_i = 0$ for $i \neq j$, then the quantity $\kappa(\gamma)$ is the individual absolute condition number with respect to the matrix $Z_j \in \widetilde{\mathcal{P}}$, determined above. When $\gamma = \|\Theta\|_g$ then $\kappa(\gamma)$ is the overall norm-wise relative condition number of equation (12.1).

In general there does not exist a closed form expression for $\kappa(\gamma)$. However, we will derive bounds for $\kappa(\gamma)$ in the next section.

12.4.2 First order homogeneous bounds

In this section we derive local first order homogeneous estimates.

The perturbed equation may be written in vector form as

$$\operatorname{vec}(\delta X) = \sum_{Z \in \widetilde{\mathcal{P}}} N_Z \operatorname{vec}(\delta Z) - L_X^{-1} \operatorname{vec}(G(\delta X, \delta P)), \tag{12.17}$$

where

$$N_Z := -L_X^{-1} L_Z, \ Z \in \widetilde{\mathcal{P}}.$$

The absolute condition number based estimate is a corollary of (12.17). When using the Frobenius norm, the estimate is obtained as follows. Set

$$\delta_X = \|\delta X\|_{\mathrm{F}} = \|\operatorname{vec}(\delta X)\|_2.$$

Since $\delta_Z \geq \|\operatorname{vec}(\delta Z)\|_2$, we have

$$\delta_X \leq \operatorname{est}_1(\delta) + O(\|\delta\|^2), \ \delta \to 0,$$

where

$$\operatorname{est}_1(\delta) := \sum_{Z \in \widetilde{\mathcal{P}}} \|N_Z\|_2 \delta_Z = \sum_{Z \in \widetilde{\mathcal{P}}} K_Z \delta_Z.$$

Relation (12.17) also gives

$$\delta_X \leq \operatorname{est}_2(\delta) + O(\|\delta\|^2), \delta \to 0, \tag{12.18}$$

where

$$\operatorname{est}_2(\delta) := \|N\|_2 \|\delta\|_2 \tag{12.19}$$

and

$$N := [N_1, N_2, \ldots, N_r] := [N_{Z_1}, N_{Z_2}, \ldots, N_{Z_r}]. \tag{12.20}$$

The bounds $\operatorname{est}_1(\delta)$ and $\operatorname{est}_2(\delta)$ are alternative, i.e., which one is of less value depends on the particular value of δ.

Again, a third bound, which is always less than or equal to $\operatorname{est}_1(\delta)$, is given by

$$\delta_X \leq \operatorname{est}_3(\delta) + O(\|\delta\|^2), \ \delta \to 0, \tag{12.21}$$

where

$$\operatorname{est}_3(\delta) := \sqrt{\delta^\top M \delta} \tag{12.22}$$

and $M = [m_{ij}]$ is an $r \times r$ matrix with elements $[m_{ij}] := \|N_i^{\mathrm{H}} N_j\|_2$.

Hence, we have the overall estimate

$$\delta_X \leq \operatorname{est}(\delta) + O(\|\delta\|^2), \ \delta \to 0, \tag{12.23}$$

where

$$\operatorname{est}(\delta) := \min\{\operatorname{est}_2(\delta), \operatorname{est}_3(\delta)\}. \tag{12.24}$$

The local bound est in (12.23), (12.24) is a nonlinear, first order homogeneous and piece-wise real analytic function in δ.

12.4.3 Component-wise bounds

A local component-wise perturbation bound follows from (12.17):

$$|\mathrm{vec}(\delta X)| \preceq \sum_{Z\in\widetilde{\mathcal{P}}} \left|L_X^{-1}L_Z\right| |\mathrm{vec}(\delta Z)| + O(\|\delta\|^2),\ \delta \to 0.$$

Its implementation requires information about the perturbations in the components of the data such as $|\mathrm{vec}(Z)| \preceq W_Z$, $Z \in \widetilde{\mathcal{P}}$, where $W_Z \succeq 0$ are given vectors.

12.5 Nonlocal perturbation analysis

In this section use nonlinear perturbation analysis to determine a domain $\Omega \subset \mathbb{R}_+^r$ and a nonlinear function $f : \Omega \to \mathbb{R}_+$ such that

$$\delta_X \leq f(\delta)$$

for all $\delta \in \Omega$.

The inclusion $\delta \in \Omega$ guarantees that the perturbed equation has a solution. Also, the estimate $\delta_X \leq f(\delta)$ is rigorous, i.e., it is true for perturbations with $\delta \in \Omega$, unlike the local bounds.

Let the collections P_i be perturbed to $P_i + \delta P_i$ and hence $P \mapsto P + \delta P$. Set $Y = X + \delta X$ for the solution of the perturbed equation (12.5). In what follows we shall mark the dependence of certain quantities only on the perturbations δX, δP_i and δP, recalling that they are evaluated at the nominal collection P, and that X is a fixed solution of (12.1).

The perturbed equation (12.5) may be rewritten as an operator equation

$$\delta X = \Phi(\delta X, \delta P) := \Phi_0(\delta P) + \Phi_1(\delta X, \delta P) + \Phi_2(\delta X, \delta P), \tag{12.25}$$

where

$$\begin{aligned}
\Phi_0(\delta P) &:= -F_X^{-1}(X,P)(G_0(\delta P)),\\
\Phi_i(\delta X, \delta P) &:= -F_X^{-1}(X,P)(G_i(\delta X, \delta P)),\ i = 1, 2,
\end{aligned}$$

are homogeneous functions of order i in δX. Here the quantity $G_0(\delta P)$ depends only on δP, while $G_i(\delta X, \delta P)$, $i = 1, 2$, depend on both δX and δP, as shown below. We have

$$\begin{aligned}
G_0(\delta P) &= G_{01}(\delta P) + G_{02}(\delta P) + G_{03}(\delta P_2),\\
G_1(\delta X, \delta P) &= G_{11}(\delta X, \delta P) + G_{12}(\delta X, \delta P) + G_{13}(\delta X, \delta P_2),\\
G_2(\delta X, \delta P) &= G_{20}(\delta X, \delta P) + G_{21}(\delta X, \delta P) + G_{22}(\delta X, \delta P) + G_{23}(\delta X, \delta P_2),
\end{aligned}$$

where the matrices $G_{ij}(\delta X, \delta P)$, $i = 0, 1, 2$, are sums of perturbed terms of order j with regard to δP and are defined as follows.

In the case $i = 0$ we have

$$
\begin{aligned}
G_{01}(\delta P) &:= \delta A_0 + \sum_{i=1}^{r_1} (\delta A_i X B_i + A_i X \delta B_i) \\
&\quad + \sum_{k=1}^{r_2} (\delta C_k X D_k X E_k + C_k X \delta D_k X E_k + C_k X D_k X \delta E_k), \\
G_{02}(\delta P) &:= \sum_{i=1}^{r_1} \delta A_i X \delta B_i \\
&\quad + \sum_{k=1}^{r_2} (\delta C_k X \delta D_k X E_k + C_k X \delta D_k X \delta E_k + \delta C_k X D_k X \delta E_k), \\
G_{03}(\delta P_2) &:= \sum_{k=1}^{r_2} \delta C_k X \delta D_k X \delta E_k.
\end{aligned}
$$

In the case $i = 1$ it holds that

$$
\begin{aligned}
G_{11}(Z, \delta P) &:= \sum_{i=1}^{r_1} (\delta A_i Z B_i + A_i Z \delta B_i) + \sum_{k=1}^{r_2} \delta C_k (X D_k Z + Z D_k X) E_k \\
&\quad + \sum_{k=1}^{r_2} C_k (X \delta D_k Z + Z \delta D_k X) E_k, \\
&\quad + \sum_{k=1}^{r_2} C_k (X D_k Z + Z D_k X) \delta E_k \\
G_{12}(Z, \delta P) &:= \sum_{i=1}^{r_1} \delta A_i Z \delta B_i \\
&\quad + \sum_{k=1}^{r_2} \delta C_k (X \delta D_k Z + Z \delta D_k X) E_k \\
&\quad + \sum_{k=1}^{r_2} C_k (X \delta D_k Z + Z \delta D_k X) \delta E_k \\
&\quad + \sum_{k=1}^{r_2} \delta C_k (X D_k Z + Z D_k X) \delta E_k, \\
G_{13}(Z, \delta P_2) &:= \sum_{k=1}^{r_2} (\delta C_k X \delta D_k Z \delta E_k + \delta C_k Z \delta D_k X \delta E_k).
\end{aligned}
$$

Finally, in the case $i = 2$ the corresponding expressions are

$$
\begin{aligned}
G_{20}(Z) &:= \sum_{k=1}^{r_2} C_k Z D_k Z E_k, \\
G_{21}(Z, \delta P_2) &:= \sum_{k=1}^{r_2} (\delta C_k X D_k Z E_k + C_k Z \delta D_k Z E_k + C_k Z D_k Z \delta E_k),
\end{aligned}
$$

$$
\begin{aligned}
G_{22}(Z,\delta P_2) &:= \sum_{k=1}^{r_2}(\delta C_k X\delta D_k Z E_k + C_k Z\delta D_k Z\delta E_k + \delta C_k Z D_k Z\delta E_k),\\
G_{23}(Z,\delta P_2) &:= \sum_{k=1}^{r_2}\delta C_k Z\delta D_k Z\delta E_k.
\end{aligned}
$$

Suppose that $\|Z\|_{\mathrm{F}} \le \rho$. Then we have

$$
\begin{aligned}
\|\Phi_0(\delta P)\|_{\mathrm{F}} &\le a_0(\delta),\\
\|\Phi_1(Z,\delta P)\|_{\mathrm{F}} &\le a_1(\delta)\rho,\\
\|\Phi_2(Z,\delta P)\|_{\mathrm{F}} &\le a_2(\delta)\rho^2,
\end{aligned}
$$

where

$$
\begin{aligned}
a_0(\delta) &:= a_{01}(\delta) + a_{02}(\delta) + a_{03}(\delta),\\
a_1(\delta) &:= a_{11}(\delta) + a_{12}(\delta) + a_{13}(\delta),\\
a_2(\delta) &:= a_{20} + a_{21}(\delta) + a_{22}(\delta) + a_{23}(\delta).
\end{aligned}
\tag{12.26}
$$

The quantities $a_{ij}(\delta)$ are of order $O(\|\delta\|^j)$ for $\delta \to 0$ and are given by the following formulae.

In the case $i = 0$ we have

$$
\begin{aligned}
a_{01}(\delta) &:= \mathrm{est}(\delta),\\
a_{02}(\delta) &:= \left\|L_X^{-1}\right\|_2 \|X\|_2 \sum_{i=1}^{r_1} \delta_{A_i}\delta_{B_i}\\
&\quad + \|X\|_2 \sum_{k=1}^{r_2} \left\|L_X^{-1}\left((XE_k)^\top \otimes I_p\right)\right\|_2 \delta_{C_k}\delta_{D_k}\\
&\quad + \|X\|_2 \sum_{k=1}^{r_2} \left\|L_X^{-1}\left(I_q \otimes (C_k X)\right)\right\|_2 \delta_{D_k}\delta_{E_k}\\
&\quad + \left\|L_X^{-1}\right\|_2 \|X\|_2 \sum_{k=1}^{r_2} \delta_{C_k}\delta_{E_k},\\
a_{03}(\delta) &:= \left\|L_X^{-1}\right\|_2 \|X\|_2^2 \sum_{k=1}^{r_2} \delta_{C_k}\delta_{D_k}\delta_{E_k}.
\end{aligned}
\tag{12.27}
$$

The case $i = 1$ is characterized by

$$
\begin{aligned}
a_{11}(\delta) &:= \sum_{i=1}^{r_1} \left\|\left(L_X^{-1}\left(B_i^\top \otimes I_p\right)\right\|_2 \delta_{A_i} + \left(L_X^{-1}\left(I_q \otimes A_i\right)\right\|_2 \delta_{B_i}\right)\\
&\quad + \sum_{k=1}^{r_2} \left\|\left(L_X^{-1}\left(E_k^\top \otimes I_p\right)\right\|_2 \left\|I_n \otimes (XD_k) + (D_k X)^\top \otimes I_m\right\|_2 \delta_{C_k}
\end{aligned}
\tag{12.28}
$$

$$
\begin{aligned}
&+ 2\|X\|_2 \sum_{k=1}^{r_2} \left\|(L_X^{-1}\left(E_k^\top \otimes C_k\right)\right\|_2 \delta_{D_k} \\
&+ \sum_{k=1}^{r_2} \left\|(L_X^{-1}\left(I_q \otimes C_k\right)\right\|_2 \left\|I_n \otimes (XD_k) + (D_k X)^\top \otimes I_m\right\|_2 \delta_{E_k},
\end{aligned}
$$

$$
\begin{aligned}
a_{12}(\delta) \quad &:= \quad \left\|L_X^{-1}\right\|_2 \sum_{i=1}^{r_1} \delta_{A_i}\delta_{B_i} \\
&+ \sum_{k=1}^{r_2} \left\|(L_X^{-1}\left(E_k^\top \otimes I_p\right)\right\|_2 \left\|I_n \otimes X + X^\top \otimes I_m\right\|_2 \delta_{C_k}\delta_{D_k} \\
&+ \sum_{k=1}^{r_2} \left\|(L_X^{-1}\left(I_q \otimes C_k\right)\right\|_2 \left\|I_n \otimes X + X^\top \otimes I_m\right\|_2 \delta_{D_k}\delta_{E_k} \\
&+ \left\|L_X^{-1}\right\|_2 \sum_{k=1}^{r_2} \left\|I_n \otimes (XD_k) + (D_k X)^\top \otimes I_m\right\|_2 \delta_{C_k}\delta_{E_k}, \\
a_{13}(\delta) \quad &:= \quad 2\left\|L_X^{-1}\right\|_2 \|X\|_2 \sum_{k=1}^{r_2} \delta_{C_k}\delta_{D_k}\delta_{E_k}.
\end{aligned}
$$

Finally, in the case $i = 2$ the expressions are

$$
\begin{aligned}
a_{20} \quad &:= \quad \sum_{k=1}^{r_2} \left\|L_X^{-1}\left(E_k^\top \otimes C_k\right)\right\|_2 \|D_k\|_2, \qquad (12.29) \\
a_{21}(\delta) \quad &:= \quad \sum_{k=1}^{r_2} \left\|L_X^{-1}\left(E_k^\top \otimes I_p\right)\right\|_2 \|D_k\|_2 \delta_{C_k} \\
&+ \sum_{k=1}^{r_2} \left\|L_X^{-1}\left(E_k^\top \otimes C_k\right)\right\|_2 \delta_{D_k} \\
&+ \sum_{k=1}^{r_2} \left\|L_X^{-1}\left(I_q \otimes C_k\right)\right\|_2 \|D_k\|_2 \delta_{E_k}, \\
a_{22}(\delta) \quad &:= \quad \sum_{k=1}^{r_2} \left\|L_X^{-1}\left(E_k^\top \otimes I_p\right)\right\|_2 \delta_{C_k}\delta_{D_k} \\
&+ \sum_{k=1}^{r_2} \left\|L_X^{-1}\left(I_q^\top \otimes C_k\right)\right\|_2 \delta_{D_k}\delta_{E_k} \\
&+ \left\|L_X^{-1}\right\|_2 \sum_{k=1}^{r_2} \|D_k\|_2 \delta_{C_k}\delta_{E_k}, \\
a_{23}(\delta) \quad &:= \quad \left\|L_X^{-1}\right\|_2 \sum_{k=1}^{r_2} \|D_k\|_2 \delta_{C_k}\delta_{D_k}\delta_{E_k}.
\end{aligned}
$$

In the following we apply again the technique of Lyapunov majorants and Banach fixed point principle in order to show that the operator equation (12.25) has a (unique) solution and to estimate its norm.

Let

$$\|Z\|_F, \|\widetilde{Z}\|_F \leq \rho.$$

The Lyapunov majorant (see Chapter 5) for equation (12.25) is a function $(\delta, \rho) \mapsto h(\delta, \rho)$, defined on a subset of $\mathbb{R}_+^r \times \mathbb{R}_+$, and satisfying the conditions

$$\|\Phi(Z, \delta P)\|_F \leq h(\delta, \rho)$$

and

$$\|\Phi(Z, \delta P) - \Phi(\widetilde{Z}, \delta P)\|_F \leq h'_\rho(\delta, \rho)\|Z - \widetilde{Z}\|_F.$$

The Lyapunov majorant here is

$$h(\delta, \rho) = a_0(\delta) + a_1(\delta)\rho + a_2(\delta)\rho^2$$

and the majorant equation

$$h(\delta, \rho) = \rho$$

for determining the nonlocal bound $\rho = \rho(\delta)$ for δ_X is quadratic,

$$a_2(\delta)\rho^2 - (1 - a_1(\delta))\rho + a_0(\delta) = 0. \tag{12.30}$$

Suppose that $\delta \in \Omega$, where

$$\Omega := \left\{\delta \succeq 0 : a_1(\delta) + 2\sqrt{a_0(\delta)a_2(\delta)} \leq 1\right\} \subset \mathbb{R}_+^r. \tag{12.31}$$

Then equation (12.30) has nonnegative roots $\rho_1(\delta) \leq \rho_2(\delta)$ with

$$\rho_1(\delta) := f(\delta) := \frac{2a_0(\delta)}{1 - a_1(\delta) + \sqrt{(1 - a_1(\delta))^2 - 4a_0(\delta)a_2(\delta)}}. \tag{12.32}$$

The operator $\Phi(\cdot, \delta P)$ maps the closed convex ball

$$\mathcal{B}(\delta) := \left\{Z \in \mathbb{F}^{m\times n} : \|Z\|_F \leq f(\delta)\right\} \subset \mathbb{F}^{m\times n}$$

into itself. According to the Schauder fixed point principle there exists a solution

$$\delta X \in \mathcal{B}(\delta)$$

of equation (12.25), for which

$$\delta_X = \|\delta X\|_F \leq f(\delta), \ \ \delta \in \Omega. \tag{12.33}$$

In addition, the elements of δX are continuous functions of the elements of δP and hence of those of $\delta\Theta$.

If $\delta \in \Omega_1$, where

$$\Omega_1 := \left\{\delta \succeq 0 : a_1(\delta) + 2\sqrt{a_0(\delta)a_2(\delta)} < 1\right\} \subset \Omega,$$

then

$$\rho_1(\delta) < \rho_2(\delta), \ h'_\rho(\rho_1, \delta) < 1$$

and the operator $\Phi(\cdot, \delta)$ is a contraction on $\mathcal{B}(\delta)$. Hence, according to the Banach fixed point principle the solution δX, for which the estimate (12.33) holds, is unique. This means that the perturbed equation has an isolated solution $X + \delta X$. Moreover, in this case the elements of δX are analytic functions of the elements of δP.

As a result of the nonlocal perturbation analysis we have the perturbation bound (12.31)–(12.33), where the involved quantities are determined via the relations (12.26) – (12.29).

12.6 Notes and references

Results similar to those that we have presented were obtained in [149, 120, 150, 134, 211]. The perturbation bounds, given in this chapter, are derived in [131] and are an improvement over the results from [150] .

Chapter 13

Continuous-time Riccati equations

13.1 Introductory remarks

In this chapter we present perturbation bounds for continuous-time matrix Riccati equations as they arise in control and filtering of linear multivariable systems. Both standard and descriptor, real and complex equations are considered. As before, we derive condition numbers, first order local bounds and nonlinear nonlocal bounds.

13.2 Motivating example

Consider the stabilizable and detectable continuous-time control system

$$\begin{aligned} x'(t) &= Ax(t) + Bu(t),\ t > 0,\ x(0) = x_0, \\ y(t) &= Cx(t), \end{aligned} \tag{13.1}$$

where $x(t) \in \mathbb{F}^n$, $u(t) \in \mathbb{F}^m$ and $y(t) = \mathbb{F}^r$ are the state, control and output vectors, respectively, and $A \in \mathbb{F}^{n\times n}$, $B \in \mathbb{F}^{n\times m}$, $C \in \mathbb{F}^{r\times n}$ are constant matrices. The system is real or complex if the underlying field $\mathbb{F}$ is $\mathbb{R}$ or $\mathbb{C}$, respectively.

We recall that the system (13.1), or the pair $[A, B)$, is *stabilizable* if there exists a gain matrix $H \in \mathbb{F}^{m\times n}$ such that the closed-loop system matrix $A+BH$ is *stable*, i.e., has its spectrum in the left open complex half-plane. The system (13.1), or the pair $(C, A]$, is *detectable* if the pair $[A^{\mathrm{H}}, C^{\mathrm{H}})$ is stabilizable. Systems of type (13.1), or triples (C, A, B), that are both stabilizable and detectable are called *regular*.

Let the quadratic performance index

$$J(u, x_0) := \int_0^\infty (y^{\mathrm{H}}(t)y(t) + u^{\mathrm{H}}(t)Ru(t))\mathrm{d}t \to \min \tag{13.2}$$

be given, where $R = R^{\mathrm{H}} > 0$ is a positive definite weighting matrix. The control that minimizes $J(u, x_0)$ for every initial state $x_0 \in \mathbb{F}^n$ can be realized as a state feedback $u(t) = -R^{-1}B^{\mathrm{H}}X_0x(t)$, where $X_0 = X_0^{\mathrm{H}}$ is the nonnegative definite solution of the *standard continuous-time matrix Riccati equation*

$$Q + A^{\mathrm{H}}X + XA - XSX = 0, \; Q := C^{\mathrm{H}}C, \; S := BR^{-1}B^{\mathrm{H}}. \tag{13.3}$$

In this case $J(u, x_0) = x_0^{\mathrm{H}}X_0x_0$. It follows from the regularity of (C, A, B) that equation (13.3) has a unique symmetric (in the sense $X_0 = X_0^{\mathrm{H}}$) nonnegative definite stabilizing solution X_0. At the same time the Riccati equation may have other solutions (which necessarily are not nonnegative definite and not stabilizing), including nonsymmetric ones.

Consider also the *descriptor system*

$$\begin{aligned} Ex'(t) &= Ax(t) + Bu(t), \; t > 0, \; x(0) = x_0, \\ y(t) &= Cx(t), \end{aligned} \tag{13.4}$$

with the same performance index (13.2). Here the matrix $E \in \mathbb{F}^{n\times n}$ is nonsingular but may be ill-conditioned with respect to inversion. Formally we have

$$x'(t) = E^{-1}Ax(t) + E^{-1}Bu(t).$$

If the triple $(C, E^{-1}A, E^{-1}B)$ is regular then the optimal control in (13.4), (13.2) may again be realized by a feedback $u(t) = -R^{-1}B^{\mathrm{H}}E^{-\mathrm{H}}\widehat{X}_0x(t)$, where $\widehat{X}_0$ is the nonnegative stabilizing solution of the Riccati equation

$$Q + (E^{-1}A)^{\mathrm{H}}\widehat{X} + \widehat{X}E^{-1}A - \widehat{X}E^{-1}SE^{-\mathrm{H}}\widehat{X} = 0. \tag{13.5}$$

There are two ways to deal with equation (13.5). First, setting $\widehat{Q} = E^{\mathrm{H}}QE$, $\widehat{A} := E^{-1}AE$ and $\widehat{S} := E^{-1}SE^{-\mathrm{H}}$ we have the Riccati equation

$$\widehat{Q} + \widehat{A}^{\mathrm{H}}\widehat{X}E + E^{\mathrm{H}}\widehat{X}\widehat{A} - E^{\mathrm{H}}\widehat{X}\widehat{S}\widehat{X}E = 0. \tag{13.6}$$

To avoid even the formal inversion of E one may also set $\widehat{X}_0 = E^{\mathrm{H}}X_0E$. Then it follows from (13.5) that the matrix X_0 is the nonnegative solution to the *descriptor continuous-time matrix Riccati equation*

$$Q + A^{\mathrm{H}}XE + E^{\mathrm{H}}XA - E^{\mathrm{H}}XSXE = 0 \tag{13.7}$$

In the following we will work with equation (13.7) instead of (13.6).

Matrix Riccati equations of the types considered arise also in many other areas of linear control systems theory. For example, in the so called *filtering problem* the standard Riccati equation is in the *dual form* $S + AX + XA^{\mathrm{H}} - XQX = 0$.

In $\mathcal{H}_\infty$ control problems Riccati equations of type (13.3) arise without the assumptions that Q is nonnegative definite and/or R is positive definite, e.g., the matrix R may be symmetric nonsingular and indefinite. With regard to the perturbation analysis the real and complex cases are treated similarly. $A \mapsto A^{\mathrm{H}}$ is not linear (it is additive but not homogeneous).

13.3 Standard equation

13.3.1 Statement of the problem

Consider first the real standard equation

$$F(P,X) := Q + A^\top X + XA - XSX = 0, \; P := (Q,A,S), \tag{13.8}$$

under the assumption that it has a symmetric solution X_0 such that the linear matrix operator $\mathcal{K} : \mathbb{R}^{n\times n} \to \mathbb{R}^{n\times n}$, defined by

$$\mathcal{K}(Z) = (A - SX_0)^\top Z + Z(A - SX_0),$$

is invertible. The eigenvalues of $\mathcal{K}$ are the eigenvalues of its matrix

$$K := I_n \otimes (A - SX_0)^\top + (A - SX_0)^\top \otimes I_n \in \mathbb{R}^{n^2\times n^2}$$

and are equal (with multiplicity counted) to $\lambda_i(A - SX_0) + \lambda_j(A - SX_0)$, $i, j = 1, 2, \ldots, n$. We recall that the matrix K of a linear matrix operator $\mathcal{K}$ is defined by $\mathrm{vec}(\mathcal{K}(Z)) = K\mathrm{vec}(Z)$ for all Z.

Note that if Q, S are nonnegative definite and the triple (Q, A, S) is regular then there is a (unique) nonnegative definite solution X_0 such that the matrix $A - SX_0$ is stable and hence, the operator $\mathcal{K}$ is invertible. This latter case is interesting from the point of view of applications but the perturbation analysis given below holds also under the weaker assumption that a solution $X_0 = X_0^\top$ exists with $\mathcal{K}$ invertible. There are also other sets of sufficient conditions for invertibility of $\mathcal{K}$ which are not considered here.

The matrix parameter P may be regarded as a matrix triple (Q, A, S) from $\mathbb{R}^{n\times n} \times \mathbb{R}^{n\times n} \times \mathbb{R}^{n\times n}$ or as a matrix $[Q, A, S]$ from $\mathbb{R}^{n\times 3n}$. Hence, we may introduce a norm and a generalized norm of P by

$$\|P\| := \|[Q,A,S]\| \in \mathbb{R}_+, \; |\!|\!|P|\!|\!| := [\|Q\|, \|A\|, \|S\|]^\top \in \mathbb{R}^3_+,$$

where $\|\cdot\|$ is any matrix norm.

Let the matrix coefficients in (13.8) be subject to perturbations $Q \mapsto Q + \delta Q$, $A \mapsto A + \delta A$, $S \mapsto S + \delta S$. If $Q = C^\top C$ and $S = BR^{-1}B^\top$ and C, B, R are perturbed as $C \mapsto C + \delta C$, $B \mapsto B + \delta B$, $R \mapsto R + \delta R$ with $\delta R = \delta R^\top$ and $R + \delta R$ invertible, then the perturbations $\delta Q = C^\top \delta C + \delta C^\top C + \delta C^\top \delta C$ and $\delta S = BR^{-1}\delta B^\top + \delta BR^{-1}B^\top + \delta BR^{-1}\delta B^\top - (B+\delta B)R^{-1}\delta R(R+\delta R)^{-1}(B+\delta B)^\top$ in Q and S respectively, are also symmetric.

The analysis given below applies to both symmetric and nonsymmetric perturbations in the matrices Q and S. The aim of perturbation analysis here is to find computable bounds for the norm

$$\delta_X := \|\delta X\|_{\mathrm{F}}$$

of the perturbation in the solution X_0 as a function of the *perturbation vector*

$$\delta := [\delta_1, \delta_2, \delta_3]^\top := [\delta_Q, \delta_A, \delta_S]^\top \in \mathbb{R}_+^3$$

whose components

$$\delta_Q := \|\delta Q\|_F, \ \delta_A := \|\delta A\|_F, \ \delta_S := \|\delta S\|_F$$

are the Frobenius norms of the perturbations in the data matrices Q, A, S. Thus, $\delta = |||\delta P|||$.

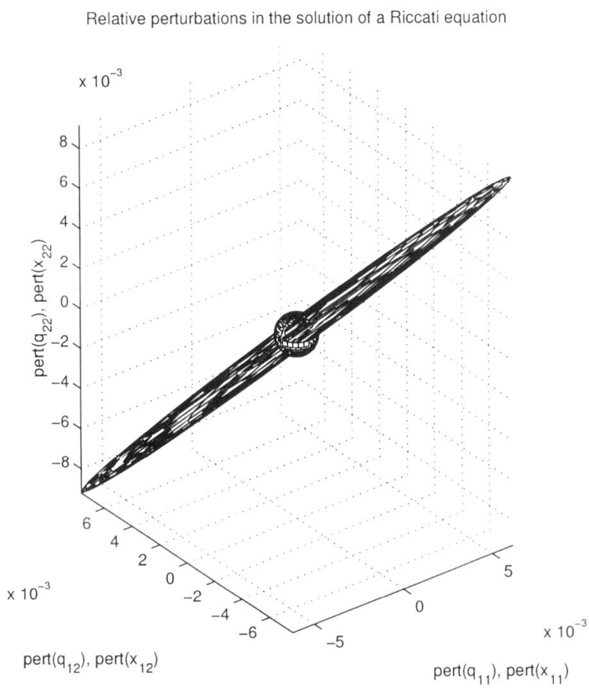

Figure 13.1: Perturbed solutions of Riccati equation

In Figure 13.1 we show the elements of the perturbed solutions $\delta X/\|X\|$ of a 2nd order Riccati equation, generated by perturbations in the elements q_{11}, q_{12} and q_{22} of Q. The perturbations in the data are represented by a sphere and the changes in the solution are represented by an ellipsoid.

13.3.2 Perturbed equation

The *perturbed equation* is obtained from (13.8) by replacing a nominal value $P = (Q, A, S)$ of the collection of data matrices with $P+\delta P = (Q+\delta Q, A+\delta A, S+\delta S)$:

$$F(P + \delta P, X + \delta X) = 0. \tag{13.9}$$

A priori it is not clear whether, given a perturbation δP, the perturbed equation (13.9) has a solution at all. So, formally, we have to assume that a solution to (13.9) exists for the given δP. However, from the nonlinear perturbation analysis presented below, we will obtain conditions for the solvability of equation (13.9).

We may rewrite (13.9) as an equivalent equation $F(P, X) = 0$, where $X \in \mathbb{R}^{n\times n}$ is the unknown matrix and $P = (P_1, \ldots, P_k)$ a k–tuple of matrix parameters $P_1, \ldots, P_k$, and use the general scheme, that has been described in Chapter 12. We have, for any $P, \delta P = (\delta P_1, \ldots, \delta P_k)$ and $X, Z \in \mathbb{R}^{n\times n}$, that

$$F(P + \delta P, X + Z) = F(P, X) + F_X(P, X)(Z) + F_P(P, X)(\delta P) + \mathcal{F}(P, X)(\delta P, Z),$$

where $F_X(P, X) : \mathbb{R}^{n\times n} \to \mathbb{R}^{n\times n}$ is the partial Fréchet derivative of F in X calculated at the point (P, X). Similarly,

$$F_P(P, X)(\delta P) = \sum_{i=1}^{k} F_{P_i}(P, X)(\delta P_i).$$

In the complex case the operators $F_{P_i}(P, X)$ are not Fréchet derivatives but some related additive operators constructed as follows. Suppose that $F(P, X)$ is written in the form $F(P_1, \overline{P}_1, \ldots, P_k, \overline{P}_k, X)$ and, for X fixed, consider the function

$$(Y_1, Z_1, \ldots, Y_k, Z_k) \mapsto F(Y_1, Z_1, \ldots, Y_k, Z_k, X).$$

Assume that the partial Fréchet derivatives $F_{Y_i}(P, X)$, $F_{Z_i}(P, X)$ of this function exist. Then we set

$$F_P(P, X)(\delta P) = \sum_{i=1}^{k} \left(F_{Y_i}(P, X)(\delta P_i) + F_{Z_i}(P, X)(\delta \overline{P}_i) \right).$$

The operator $F_P(P, X) : \mathbb{C}^{n\times n} \to \mathbb{C}^{n\times n}$ is additive in the sense that $F_P(U + V, X) = F_P(U, X) + F_P(V, X)$ but it is not homogeneous.

The term $\mathcal{F}(P, X)(\delta P, Z)$ contains second and higher order terms in δP, Z,

$$\|(\mathcal{F}(P, X)(\delta P, Z)\| = O(\|\delta P\|^2 + \|Z\|^2), \ \|\delta P\| + \|Z\| \to 0.$$

Suppose that the linear operator $F_X(P, X_0)$ is invertible, where $F(P, X_0) = 0$. Then we may rewrite the perturbed equation $F(P + \delta P, X_0 + \delta X) = 0$ as

$$\delta X = -F_X^{-1}(P, X_0) \circ F_P(P, X_0)(\delta P) - F_X^{-1}(P, X_0) \circ \mathcal{F}(P, X_0)(\delta P, \delta X_0). \quad (13.10)$$

Note that $F_P(P, X_0)(0) = 0$ and $\mathcal{F}(P, X_0)(0, 0) = 0$. This guarantees that for small δP the perturbed equation (13.10) has a "small" solution δX in the sense that

$$\delta X = -F_X^{-1}(P, X_0) \circ F_P(P, X_0)(\delta P) + O(\|\delta P\|^2) = O(\|\delta P\|), \ \delta P \to 0.$$

In the following we abbreviate $F_X(P, X_0)$ as F_X, etc., thus, omitting the dependence on the fixed quantities P, X_0 whenever appropriate.

For equation (13.9) we then obtain

$$\mathcal{K} = F_X, \ F_P(\delta P) = \delta Q + \delta A^\top X_0 + X_0 \delta A - X_0 \delta S X_0.$$

Therefore

$$\mathcal{K}(\delta X) = U_1(\delta P) + U_2(\delta P, \delta X),$$

where

$$\begin{aligned} U_1(\delta P) &:= X_0 \delta S X_0 - \delta Q - \delta A^\top X_0 - X_0 \delta A, \\ U_2(\delta P, Z) &:= Z(S + \delta S) Z + Z \delta S X_0 + X_0 \delta S Z - \delta A^\top Z - Z \delta A. \end{aligned}$$

Note that $\|U_1(\delta P)\| = O(\|\delta P\|)$, $\delta P \to 0$.

Since $\mathcal{K}$ is invertible we have

$$\delta X = \Pi(\delta P, \delta X) := \Pi_1(\delta P) + \Pi_2(\delta P, \delta X), \tag{13.11}$$

where

$$\Pi_1(\delta P) := \mathcal{K}^{-1}(U_1(\delta P)), \ \Pi_2(\delta P, \delta X) := \mathcal{K}^{-1}(U_2(\delta P, \delta X)).$$

The *equivalent operator equation* (13.11) and its vectorized counterpart are the basis of the local and nonlocal perturbation analysis presented next.

It is convenient to rewrite (13.11) in vectorized form using the formulae

$$\mathrm{vec}(ABC) = (C^\top \otimes A)\mathrm{vec}(B), \ (B \otimes A) P_{n^2} = P_{n^2}(A \otimes B),$$

where $P_{n^2} \in \mathbb{R}^{n^2 \times n^2}$ is the *vec-permutation matrix* such that $\mathrm{vec}(A^\top) = P_{n^2}\mathrm{vec}(A)$. Introducing

$$\begin{aligned} \xi &:= \mathrm{vec}(\delta X), \\ \mu_1 &:= \mathrm{vec}(\delta Q), \ \mu_2 := \mathrm{vec}(\delta A), \ \mu_3 := \mathrm{vec}(\delta S) \in \mathbb{R}^{n^2}, \\ \mu &:= \mathrm{vec}(\delta P) = [\mu_1^\top, \mu_2^\top, \mu_3^\top]^\top \in \mathbb{R}^{3n^2}, \end{aligned} \tag{13.12}$$

we have $\delta X = \mathrm{vec}^{-1}(\xi)$, $\delta P = \mathrm{vec}^{-1}(\mu)$ and

$$\xi = \pi(\mu, \xi) := \pi_1(\mu) + \pi_2(\mu, \xi), \tag{13.13}$$

where

$$\begin{aligned} \pi_1(\mu) &:= K^{-1}\mathrm{vec}(U_1(\delta P)) = M_1\mu_1 + M_2\mu_2 + M_3\mu_3, \\ M_1 &:= -K^{-1}, \\ M_2 &:= -K^{-1}(I_{n^2} + P_{n^2})(I_n \otimes X_0), \\ M_3 &:= K^{-1}(X_0 \otimes X_0) \end{aligned}$$

and

$$\begin{aligned}\pi_2(\mu,\xi) \ := \ & K^{-1}(U_2(\delta P,\delta X) = K^{-1}\text{vcc}(\delta X(S+\delta S)\delta X) \\ & + K^{-1}(X_0\otimes I_n)\text{vec}(\delta X\delta S) + K^{-1}(I_n\otimes X_0)\text{vec}(\delta S\delta X) \\ & - K^{-1}(\text{vec}(\delta A^\top\delta X)+\text{vec}(\delta X\delta A)).\end{aligned} \tag{13.14}$$

For the complex equation

$$Q + A^{\mathrm{H}}X + XA - XSX = 0. \tag{13.15}$$

we obtain

$$\widetilde{\mathcal{K}}(\delta X) = \widetilde{U}_1(\delta P) + \widetilde{U}_2(\delta P,\delta X),$$

where the linear matrix operator $\widetilde{\mathcal{K}} : \mathbb{C}^{n\times n} \to \mathbb{C}^{n\times n}$ is defined by

$$\widetilde{\mathcal{K}}(Z) = (A - SX_0)^{\mathrm{H}}Z + Z(A - SX_0)$$

and

$$\begin{aligned}\widetilde{U}_1(\delta P) \ &:= \ X_0\delta SX_0 - \delta Q - \delta A^{\mathrm{H}}X_0 - X_0\delta A, \\ \widetilde{U}_2(\delta P, Z) \ &:= \ Z(S+\delta S)Z + Z\delta SX_0 + X_0\delta SZ - \delta A^{\mathrm{H}}Z - Z\delta A.\end{aligned}$$

Since $\widetilde{\mathcal{K}}$ is invertible we have

$$\delta X = \widetilde{\Pi}(\delta P,\delta X) := \widetilde{\Pi}_1(\delta P) + \widetilde{\Pi}_2(\delta P,\delta X), \tag{13.16}$$

where

$$\widetilde{\Pi}_1(\delta P) := \widetilde{\mathcal{K}}^{-1}(\widetilde{U}_1(\delta P)),\ \widetilde{\Pi}_2(\delta P,\delta X) := \widetilde{\mathcal{K}}^{-1}(\widetilde{U}_2(\delta P,\delta X)).$$

As in the real case we rewrite the equivalent operator equation (13.16) in vectorized form

$$\xi = \widetilde{\pi}(\mu,\xi) := \widetilde{\pi}_1(\mu) + \widetilde{\pi}_2(\mu,\xi).$$

Here we have used the substitutions (13.12) (having in mind that now $\xi, \mu_i \in \mathbb{C}^{n^2}$ and $\mu \in \mathbb{C}^{3n^2}$) as well as

$$\begin{aligned}\widetilde{\pi}_1(\mu) \ &:= \ \widetilde{M}_1\mu_1 + \widetilde{M}_{21}\mu_2 + \widetilde{M}_{22}\overline{\mu}_2 + \widetilde{M}_3\mu_3, \\ \widetilde{M}_1 \ &:= \ -\widetilde{K}^{-1},\ \widetilde{M}_{21} := -\widetilde{K}^{-1}(I_n\otimes X_0), \\ \widetilde{M}_{22} \ &:= \ -\widetilde{K}^{-1}(\overline{X}_0\otimes I_n)P_{n^2},\ \widetilde{M}_3 := \ \widetilde{K}^{-1}(\overline{X}_0\otimes X_0)\end{aligned} \tag{13.17}$$

and

$$\begin{aligned}\widetilde{\pi}_2(\mu,\xi) \ := \ & \widetilde{K}^{-1}\text{vec}(\delta X(S+\delta S)\delta X) + \widetilde{K}^{-1}(\overline{X}_0\otimes I_n)\text{vec}(\delta X\delta S) \\ & + \widetilde{K}^{-1}(I_n\otimes X_0)\text{vec}(\delta S\delta X) - \widetilde{K}^{-1}(\text{vec}(\delta A^{\mathrm{H}}\delta X)+\text{vec}(\delta X\delta A)).\end{aligned} \tag{13.18}$$

In (13.17), (13.18) we have utilized the fact that X_0 is Hermitian.

13.3.3 Condition numbers and local bounds

In this section we use the results from Section 13.3.2 to determine the condition numbers and to derive local, first order perturbation bounds for the perturbation $\delta_X = \|\delta X\|_{\mathrm{F}}$ in solution X_0 of the standard Riccati equations (13.8) and (13.15).

If we suppose for the moment that the solution ξ of (13.13) exists, when this is the case will be proved in the next section, then based on (13.13) we have

$$\xi = \pi_1(\mu) + O(\|\mu\|^2 + \|\xi\|^2),\ \|\mu\| + \|\xi\| \to 0.$$

Since $\|\xi\| = O(\|\mu\|)$, $\mu \to 0$, this is equivalent to

$$\xi = M_1\mu_1 + M_2\mu_2 + M_3\mu_3 + O(\|\mu\|^2),\ \mu \to 0.$$

Hence, using the fact that $\delta_X = \|\xi\|_2$, we have the following theorem.

Theorem 13.1 *In Frobenius norm the absolute condition numbers K_Z for the solution X_0 of the real equation (13.8) relative to the matrix coefficients $Z = Q, A, S$ are*

$$\begin{aligned} K_Q &= \|M_1\|_2 = \|K^{-1}\|_2, \\ K_A &= \|M_2\|_2 = \left\|K^{-1}(I_{n^2} + P_{n^2})(I_n \otimes X_0)\right\|_2, \\ K_S &= \|M_3\|_2 = \left\|K^{-1}(X_0 \otimes X_0)\right\|_2. \end{aligned}$$

In particular, if only one matrix Z from the set $\mathcal{P} := \{Q, A, S\}$ is perturbed, we have

$$\delta_X \le K_Z\|\delta Z\|_{\mathrm{F}} + O(\delta_Z^2),\ \delta Z \to 0.$$

Note, however, that if more than one matrix coefficient is perturbed, then the condition number based linear bound

$$\delta_X \le \mathrm{est}_1(\delta) + O(\|\delta\|^2) := \sum_{Z\in\mathcal{P}} K_Z\delta_Z + O(\|\delta\|^2),\ \delta \to 0,$$

may not give good results.

In addition to the condition number based estimates we also have

$$\delta_X \le \mathrm{est}_2(\delta) + O(\|\delta\|^2),\ \delta \to 0,$$

where

$$\mathrm{est}_2(\delta) := \|[M_1, M_2, M_3]\|_2\|\delta\|_2.$$

Another perturbation bound is

$$\delta_X \le \mathrm{est}_3(\delta) + O(\|\delta\|^2),\ \delta \to 0,$$

where

$$\mathrm{est}_3(\delta) := \sqrt{\delta^\top M\delta}$$

and $M = [m_{ij}] \in \mathbb{R}_+^{3\times 3}$ is the matrix with elements

$$m_{ij} = \left\| M_i^\top M_j \right\|_2, \ i,j = 1,2,3.$$

The bounds est_2 and est_3 are again *alternative*, in the sense that in general both inequalities $\mathrm{est}_2(\delta) \leq \mathrm{est}_3(\delta)$ and $\mathrm{est}_2(\delta) > \mathrm{est}_3(\delta)$ are possible. Thus, we obtain the following theorem.

Theorem 13.2 *The perturbation δ_X in the solution X_0 of the real equation (13.8) satisfies the local perturbation estimate*

$$\delta_X \leq \mathrm{est}(\delta) + O(\|\delta\|^2), \ \delta \to 0,$$

where

$$\mathrm{est}(\delta) := \min\{\mathrm{est}_2(\delta), \mathrm{est}_3(\delta)\}.$$

Fr the complex Riccati equation (13.15) we obtain similar results.

Recall some results about nonlinear additive operators. Let $\Gamma = \Gamma_0 + \imath\Gamma_1$, $\Delta = \Delta_0 + \imath\Delta_1$ be complex $m\times n$ matrices (with $\Gamma_0, \Gamma_1, \Delta_0, \Delta_1$ real), and $z = z_0 + \imath z_1$ be a complex n–vector (with z_0, z_1 real). Then according to Chapter 10 we have

$$\max\{\|\Gamma z + \Delta\overline{z}\|_2 : \|z\|_2 \leq a\} = a\|\Theta(\Gamma,\Delta)\|_2,$$

where

$$\Theta(\Gamma,\Delta) := \begin{bmatrix} \Gamma_0 + \Delta_0 & \Delta_1 - \Gamma_1 \\ \Gamma_1 + \Delta_1 & \Gamma_0 - \Delta_0 \end{bmatrix}. \tag{13.19}$$

Theorem 13.3 *In Frobenius norm the absolute condition numbers $\widetilde{K}_Z$ for the solution X_0 of the complex equation (13.15) relative to the matrix coefficients $Z = Q, A, S$ are*

$$\widetilde{K}_Q = \|\widetilde{K}^{-1}\|_2, \ \widetilde{K}_A = \|\Theta(\widetilde{M}_{21}, \widetilde{M}_{22})\|_2, \ \widetilde{K}_S = \left\|\widetilde{K}^{-1}(\overline{X}_0 \otimes X_0)\right\|_2,$$

where the matrices $\widetilde{M}_{21}$, $\widetilde{M}_{22}$ are displayed in (13.17).

To derive local first order bounds in the complex case observe that

$$\xi = \widetilde{M}_1\mu_1 + \widetilde{M}_{21}\mu_2 + \widetilde{M}_{22}\overline{\mu}_2 + \widetilde{M}_3\mu_3 + O(\|\mu\|^2), \ \mu \to 0. \tag{13.20}$$

For the product Γz of a complex matrix $\Gamma = \Gamma_0 + \imath\Gamma_1 \in \mathbb{C}^{m\times n}$ and a complex vector $z = z_0 + \imath z_1 \in \mathbb{C}^n$ with Γ_0, Γ_1 and z_0, z_1 real, we have the real versions

$$(\Gamma z)^{\mathbb{R}} := \Gamma^{\mathbb{R}} z^{\mathbb{R}} \in \mathbb{R}^{2m},$$

where

$$z^{\mathbb{R}} := \begin{bmatrix} z_0 \\ z_1 \end{bmatrix} \in \mathbb{R}^{2n}, \ \Gamma^{\mathbb{R}} := \begin{bmatrix} \Gamma_0 & -\Gamma_1 \\ \Gamma_1 & \Gamma_0 \end{bmatrix} \in \mathbb{R}^{2m\times 2n}.$$

Note that $(\Gamma z + \Delta \overline{z})^{\mathbb{R}} = \Theta(\Gamma, \Delta) z^{\mathbb{R}}$ and $\Theta(\Gamma, 0) = \Gamma^{\mathbb{R}}$.

Now it follows from (13.20) that

$$\xi^{\mathbb{R}} = \widetilde{M}_1^{\mathbb{R}} \mu_1^{\mathbb{R}} + \Theta(\widetilde{M}_{21}, \widetilde{M}_{22}) \mu_2^{\mathbb{R}} + \widetilde{M}_3^{\mathbb{R}} \mu_3^{\mathbb{R}} + O(\|\mu^{\mathbb{R}}\|^2), \ \mu^{\mathbb{R}} \to 0.$$

Since $\|\xi\|_2 = \|\xi^{\mathbb{R}}\|_2$, we have

$$\|\xi\|_2 \le \widetilde{\mathrm{est}}_2(\delta) + O(\|\delta\|^2), \ \delta \to 0,$$

where

$$\widetilde{\mathrm{est}}_2(\delta) := \left\| \left[M_1^0, M_2^0, M_3^0 \right] \right\|_2 \|\delta\|_2$$

and

$$M_1^0 := \widetilde{M}_1^{\mathbb{R}}, \ M_2^0 := \Theta(\widetilde{M}_{21}, \widetilde{M}_{22}), \ M_3^0 := \widetilde{M}_3^{\mathbb{R}}.$$

Similarly, it is also fulfilled that

$$\|\xi\|_2 \le \widetilde{\mathrm{est}}_3(\delta) + O(\|\delta\|^2), \ \delta \to 0,$$

where

$$\widetilde{\mathrm{est}}_3(\delta) := \sqrt{\delta^\top M^0 \delta}$$

and $M^0 = \left[m_{ij}^0 \right] \in \mathbb{R}_+^{3 \times 3}$ is the matrix with elements

$$m_{ij}^0 = \left\| (M_i^0)^\top M_j^0 \right\|_2, \ i, j = 1, 2, 3.$$

Thus, we get the following theorem.

Theorem 13.4 *The perturbation δ_X in the solution X_0 of the complex equation (13.15) satisfies the local perturbation estimate*

$$\delta_X \le \widetilde{\mathrm{est}}(\delta) + O(\|\delta\|^2), \ \delta \to 0,$$

where

$$\widetilde{\mathrm{est}}(\delta) := \min \left\{ \widetilde{\mathrm{est}}_2(\delta), \widetilde{\mathrm{est}}_3(\delta) \right\}.$$

Note that the bounds given in Theorems 13.2 and 13.4 may be tight. Indeed, let $w \in \mathbb{F}^n$ be the right singular vector of the matrix $W \in \mathbb{F}^{m \times n}$, corresponding to its maximum singular value $\|W\|_2$. Then $\|Ww\|_2 = \|W\|_2$. Also, for $v = \alpha w$, where $\alpha \in \mathbb{F}$, we have $\|Wv\|_2 = |\alpha| \, \|W\|_2$. Suppose now that the vector μ is proportional to the singular vector of the matrix $[M_1, M_2, M_3]$, corresponding to its 2-norm. Then $\|\pi_1(\mu)\|_2$ is equal to $\mathrm{est}_2(\delta)$ and hence, to $\mathrm{est}(\delta)$. Similarly, the quantity $\|\widetilde{\pi}_1(\mu)\|_2$ may be equal to its bound $\widetilde{\mathrm{est}}(\delta)$.

13.3.4 Nonlocal bounds

For the nonlocal perturbation analysis we show that, for δ from a certain small set Ω, the equivalent operator $\pi(\mu,\cdot)$ in (13.13) maps a closed convex set $\mathcal{B} \subset \mathbb{R}^{n^2}$ into itself. The set $\mathcal{B}$ is also small, of diameter $f(\delta) = O(\|\delta\|)$. Then according to the Schauder fixed point principle, see Appendix D, there exists a solution $\xi \in \mathcal{B}$ of (13.13) and hence, $\delta_X = \|\xi\|_2 \leq f(\delta)$. It even turns out that for $\delta \in \Omega\backslash\Omega_1$, where Ω_1 is a part of the boundary $\partial\Omega$, the operator $\pi(\mu,\cdot)$ is a contraction and according to the Banach fixed point principle (see Chapter D) the solution to the perturbed equation is unique.

Consider first the real equation (13.8) which is equivalent to the operator equation (13.13). Suppose that $\|\xi\|_2 = \delta_X \leq \rho$ for some $\rho > 0$. Estimating the right-hand side of (13.13) we get

$$\|\pi(\mu,\xi)\|_2 \leq \|\pi_1(\mu)\|_2 + \|\pi_2(\mu,\xi)\|_2.$$

Since $\|\pi_1(\mu)\|_2 \leq \operatorname{est}(\delta)$ and, in view of (13.14),

$$\begin{aligned}\|\pi_2(\mu,\xi)\|_2 &\leq \|K^{-1}\|_2(\|S\|_2 + \delta_S)\delta_X^2 \\ &+ \left(\left\|K^{-1}(X_0\otimes I_n)\right\|_2 + \left\|K^{-1}(I_n\otimes X_0)\right\|_2\right)\delta_S\delta_X + 2\|K^{-1}\|_2\delta_A\delta_X,\end{aligned}$$

we obtain

$$\|\pi(\mu,\xi)\|_2 \leq h(\delta,\rho) := a_0(\delta) + a_1(\delta)\rho + a_2(\delta)\rho^2, \tag{13.21}$$

where

$$\begin{aligned} a_0(\delta) &:= \operatorname{est}(\delta), \\ a_1(\delta) &:= \left(\left\|K^{-1}(X_0\otimes I_n)\right\|_2 + \left\|K^{-1}(I_n\otimes X_0)\right\|_2\right)\delta_S + 2\|K^{-1}\|_2\delta_A, \\ a_2(\delta) &:= \|K^{-1}\|_2(\|S\|_2 + \delta_S). \end{aligned}$$

The functions $a_i : \mathbb{R}_+^3 \to \mathbb{R}_+$, $i = 1,2,3$, are nondecreasing in the sense that $0 \preceq \delta \preceq \widehat{\delta}$ implies $a_i(\delta) \leq a_i(\widehat{\delta})$ (here $\preceq$ is the component-wise partial order relation in $\mathbb{R}^3$).

The function h is a *Lyapunov majorant* (Chapter 5) for the operator equation (13.13). It is a quadratic polynomial in ρ and we may apply directly the results from Chapter 5.

Consider the domain

$$\Omega := \left\{\delta \in \mathbb{R}_+^3 : a_1(\delta) + 2\sqrt{a_0(\delta)a_2(\delta)} \leq 1\right\}. \tag{13.22}$$

Since $a_0(0) = a_1(0) = 0$, a continuity argument shows that for some δ with positive elements it is fulfilled $a_1(\delta) + 2\sqrt{a_0(\delta)a_2(\delta)} < 1$. Hence, the set $\Omega \subset \mathbb{R}_+^3$ is well defined and has a nonempty interior. This set is bounded by the coordinate planes and by part of the surface $\mathcal{S} \subset \mathbb{R}^3$ given by $(1 - a_1(\delta))^2 = 4a_0(\delta)a_2(\delta)$. Due to the nonlinearity of a_0 the set Ω may have a complex geometry. In particular it may not be convex. However, it has the property that $\delta \in \Omega$ and $0 \preceq \widehat{\delta} \preceq \delta$ implies $\widehat{\delta} \in \Omega$. If one chooses a linear $a_0(\cdot)$, say $a_0(\delta) = \mathrm{est}_1(\delta)$, then $\mathcal{S}$ is a quadric.

If $\delta \in \Omega$ then the *majorant equation* $\rho = h(\delta, \rho)$, equivalent to

$$a_2(\delta)\rho^2 - (1 - a_1(\delta))\rho + a_0(\delta) = 0,$$

has a root

$$\rho(\delta) = f(\delta) := \frac{2a_0(\delta)}{1 - a_1(\delta) + \sqrt{(1 - a_1(\delta))^2 - 4a_0(\delta)a_2(\delta)}}. \tag{13.23}$$

Hence, for $\delta \in \Omega$ the operator $\pi(\mu, \cdot)$ maps the set $\mathcal{B}_{f(\delta)}$ into itself, where

$$\mathcal{B}_r := \left\{\xi \in \mathbb{R}^{n^2} : \|\xi\|_2 \leq r\right\}$$

is the closed central ball of radius $r \geq 0$. Then according to the Schauder fixed point principle (Appendix D) there exists a solution $\xi \in \mathcal{B}_{f(\delta)}$ of equation (13.13) and we have the following result.

Theorem 13.5 *Let $\delta \in \Omega$, where Ω is given in (13.22). Then the nonlocal perturbation bound*

$$\delta_X \leq f(\delta)$$

is valid for the real equation (13.8), where $f(\delta)$ is determined by (13.23).

Note that if the perturbation vector δ is in the subdomain of Ω defined by the strict inequality in (13.22) then $\pi(\mu, \cdot)$ is a contraction and the operator equation (13.13) (and hence, the perturbed equation (13.9)) has a unique solution.

In the complex case we have a similar nonlocal result. The quantities $a_i(\delta)$ in the expressions determining the domain Ω and the bound $f(\delta)$ need to be replaced by $\widetilde{a}_i(\delta)$, where

$$\begin{aligned}
\widetilde{a}_0(\delta) &:= \widetilde{\mathrm{est}}(\delta), \\
\widetilde{a}_1(\delta) &:= \left(\left\|\widetilde{K}^{-1}(\overline{X} \otimes I_n)\right\|_2 + \left\|\widetilde{K}^{-1}(I_n \otimes X)\right\|_2\right)\delta_S + 2\|\widetilde{K}^{-1}\|_2\delta_A, \\
\widetilde{a}_2(\delta) &:= \|\widetilde{K}^{-1}\|_2(\|S\|_2 + \delta_S).
\end{aligned}$$

As a result we can formulate the following nonlocal bound.

Theorem 13.6 *For the complex equation (13.15) the nonlocal bound*

$$\delta_X \leq \frac{2\widetilde{a}_0(\delta)}{1 - \widetilde{a}_1(\delta) + \sqrt{(1 - \widetilde{a}_1(\delta))^2 - 4\widetilde{a}_0(\delta)\widetilde{a}_2(\delta)}}$$

is valid provided that $\delta \in \mathbb{R}^3_+$ is small enough to ensure

$$\widetilde{a}_1(\delta) + 2\sqrt{\widetilde{a}_0(\delta)\widetilde{a}_2(\delta)} \leq 1.$$

To illustrate the perturbation bounds we present some examples.

Example 13.7 Consider the scalar version of (13.8) $Q + 2AX - SX^2 = 0$, where $S > 0$ and $Q \geq 0$. Let the nominal values of the parameters be $Q = S = 1$ and $A = 0$, which gives the positive solution $X_0 = 1$. We have $K = -2$, $M_1 = 0.5$, $M_2 = 1$, $M_3 = -0.5$ and

$$M = \frac{1}{4}\begin{bmatrix} 1 & 2 & 1 \\ 2 & 4 & 2 \\ 1 & 2 & 1 \end{bmatrix}.$$

The bounds $\mathrm{est}_2(\delta)$ and $\mathrm{est}_3(\delta)$ are

$$\begin{aligned} \mathrm{est}_2(\delta) &= \|[0.5, 1, -0.5]\|_2\|\delta\|_2 = \sqrt{1.5}\sqrt{\delta_1^2 + \delta_2^2 + \delta_3^2}, \\ \mathrm{est}_3(\delta) &= \sqrt{\delta^\top M\delta} = 0.5(\delta_1 + \delta_3) + \delta_2. \end{aligned}$$

Here the bound $\mathrm{est}_3(\delta)$ is always better than $\mathrm{est}_2(\delta)$ since

$$\mathrm{est}_2^2(\delta) - \mathrm{est}_3^2(\delta) = \frac{(\delta_1 - \delta_3)^2 + (2\delta_1 - \delta_2)^2 + (\delta_2 - 2\delta_3)^2}{4} \geq 0.$$

The two bounds are equal only when $\delta_1 = \delta_3 = \delta_2/2$.

A right singular value of the matrix $[M_1, M_2, M_3]$ corresponding to its norm $\sqrt{1.5}$ is $[1, 2, -1]^\top/\sqrt{6}$ and this suggests that the corresponding perturbations may be taken as $\delta Q = \sigma \geq 0$, $\delta S = -\sigma \leq 0$ and $\delta A = 2\sigma \geq 0$, i.e., $\delta P = (\sigma, 2\sigma, -\sigma)$ and $\delta = [\sigma, 2\sigma, \sigma]^\top$. For $\sigma < 1$ the positive solution to the perturbed equation $1 + \sigma + 4\sigma(1 + \delta X) - (1 - \sigma)(1 + \delta X)^2 = 0$ is

$$\delta X = \frac{2\sigma + \sqrt{1 + 3\sigma^2}}{1 - \sigma} - 1.$$

At the same time $a_0(\delta) = 3\sigma$, $a_1(\delta) = 3\sigma$ and $a_2(\delta) = (1 + \sigma)/2$. Thus, the local and nonlocal bounds are $\mathrm{est}(\delta) = 3\sigma$ and

$$f(\delta) = \frac{6\sigma}{1 - 3\sigma + \sqrt{1 - 12\sigma + 3\sigma^2}}$$

respectively, where the nonlocal bound is valid for $1 - 12\sigma + 3\sigma^2 \geq 0$ or $\sigma \leq 2 - \sqrt{11/3} \simeq 0.0851$. Since for $\sigma > 0$ it is fulfilled

$$\delta X > \frac{2\sigma + 1}{1 - \sigma} - 1 = \frac{3\sigma}{1 - \sigma} > 3\sigma,$$

we see that the local bound always *underestimates* the true perturbation for this particular structure of the perturbation δP.

In Table 13.1 we give the exact perturbation $\delta_X = \delta X$, the local bound est and the nonlocal bound f as a function of $\sigma \geq 0$. The cases when the nonlocal bound does not exist are marked by asterisk.

Table 13.1: Exact perturbation, local and nonlocal perturbation bounds

σ	δ_X	local	nonlocal
0.01	0.03046	0.03000	0.03144
0.02	0.06184	0.06000	0.06621
0.03	0.09417	0.09000	0.10516
0.04	0.12750	0.12000	0.14959
0.05	0.16183	0.15000	0.20156
0.06	0.19722	0.18000	0.26485
0.07	0.23368	0.21000	0.34769
0.08	0.27125	0.24000	0.47842
0.09	0.30997	0.27000	*
0.10	0.34988	0.30000	*

We see the main drawback of the nonlocal bounds – their relatively small domain of applicability. On the other hand in this case the local bound is *not* an upper bound for the perturbation in the solution but only gives information for its order of magnitude. ◇

Example 13.8 Consider the same equation as in Example 13.7 but with no perturbation in A. Here the exact perturbation in the solution is

$$\delta X = \sqrt{\frac{1+\sigma}{1-\sigma}} - 1 = \sigma + \frac{\sigma^2}{2} + O(\sigma^3).$$

The local perturbation bound is $\mathrm{est}(\delta) = \sigma < \delta X$ for $\sigma > 0$, while the nonlocal one is

$$f(\delta) = \frac{2\sigma}{1-\sigma+\sqrt{1-4\sigma-\sigma^2}}, \quad \sigma \leq \sqrt{5}-2 \simeq 0.2361.$$

The corresponding results are presented in Table 13.2.

Here again the local bound always underestimates the true perturbation. ◇

The case $n = 2$ already reveals some nontrivial sensitivity properties of Riccati equations for multivariable systems. The following two examples are devoted to this case.

Example 13.9 Consider the standard equation (13.8) for $n = 2$ and

$$Q = \begin{bmatrix} 0 & 0 \\ 0 & q \end{bmatrix}, \quad A = \begin{bmatrix} 0 & 0 \\ a & 0 \end{bmatrix}, \quad S = \begin{bmatrix} s & 0 \\ 0 & 0 \end{bmatrix},$$

where $q, a, s > 0$. The positive definite solution here is $X_0 = \begin{bmatrix} x_1 & x \\ x & x_2 \end{bmatrix}$, where

$$x = q^{1/2}s^{-1/2}, \; x_1 = \sqrt{2}\,q^{1/4}a^{1/2}s^{-3/4}, \; x_2 = \sqrt{2}\,q^{3/4}a^{-1/2}s^{-1/4}.$$

Table 13.2: Exact perturbation, local and nonlocal perturbation bounds

σ	δ_X	local	nonlocal
0.03	0.03046	0.03000	0.03145
0.06	0.06191	0.06000	0.06631
0.09	0.09444	0.09000	0.10558
0.12	0.12815	0.12000	0.15084
0.15	0.16316	0.15000	0.20486
0.18	0.19959	0.18000	0.27323
0.21	0.23760	0.21000	0.37154
0.24	0.27733	0.24000	*
0.27	0.31899	0.27000	*
0.30	0.36277	0.30000	*

We take the nominal parameters to be $q = s = 1$, $a = 2$, which gives $x_1 = 2$, $x = x_2 = 1$. Hence, $A - SX_0 = \begin{bmatrix} -2 & -1 \\ 2 & 0 \end{bmatrix}$ and

$$M_1 = \frac{1}{8}\begin{bmatrix} 2 & 0 & 0 & 4 \\ 0 & 2 & -2 & 4 \\ 0 & -2 & 2 & 4 \\ 1 & -2 & -2 & 6 \end{bmatrix}, \quad M_2 = \frac{1}{4}\begin{bmatrix} 4 & 2 & 4 & 4 \\ 0 & 0 & 4 & 4 \\ 0 & -0 & 4 & 4 \\ 0 & -1 & 2 & 6 \end{bmatrix},$$

$$M_3 = -\frac{1}{8}\begin{bmatrix} 12 & 8 & 8 & 6 \\ 4 & 6 & 2 & 4 \\ 4 & 2 & 6 & 4 \\ 2 & 2 & 2 & 3 \end{bmatrix}.$$

The condition numbers are $K_Q = 1.18596$, $K_A = 2.73749$, $K_S = 2.64920$. Furthermore, we have

$$M = \begin{bmatrix} 1.40651 & 3.04150 & 2.57588 \\ 3.04150 & 7.49386 & 6.89220 \\ 2.57588 & 6.89220 & 7.01826 \end{bmatrix}$$

and $\|[M_1, M_2, M_3]\|_2 = 3.90524$. Taking the perturbations as

$$\delta Q = \varepsilon \begin{bmatrix} 6 & -2 \\ -2 & 27 \end{bmatrix}, \quad \delta A = \varepsilon \begin{bmatrix} 18 & 45 \\ 7 & 49 \end{bmatrix}, \quad \delta S = \varepsilon \begin{bmatrix} -67 & 0 \\ 0 & 0 \end{bmatrix},$$

where $\varepsilon > 0$ is a small parameter, then $\delta_Q = 27.8029\varepsilon$, $\delta_A = 69.2748\varepsilon$, $\delta_S = 67\varepsilon$ and

$$\mathrm{est}_1(\delta) = 400.109\varepsilon, \ \mathrm{est}_2(\delta) = 391.713\varepsilon, \mathrm{est}_3(\delta) = 392.233\varepsilon.$$

Table 13.3: Exact perturbation, local and nonlocal perturbation bounds

k	δ_X	local	nonlocal
10	3.2388×10^{-8}	3.9171×10^{-8}	3.9171×10^{-8}
9	3.2388×10^{-7}	3.9171×10^{-7}	3.9171×10^{-7}
8	3.2388×10^{-6}	3.9171×10^{-6}	3.9172×10^{-6}
7	3.2388×10^{-5}	3.9171×10^{-5}	3.9175×10^{-5}
6	3.2390×10^{-4}	3.9171×10^{-4}	3.9204×10^{-4}
5	3.2411×10^{-3}	3.9171×10^{-3}	3.9503×10^{-3}
4	3.2626×10^{-2}	3.9171×10^{-2}	4.2966×10^{-2}
3	3.4924×10^{-1}	3.9171×10^{-1}	*
2	10.2532	3.9171	*

We see that here $\mathrm{est}_2(\delta)$ gives best results and hence, the local perturbation bound according to Theorem 13.2 is $a_0(\delta) = \mathrm{est}(\delta) = \mathrm{est}_2(\delta)$. Furthermore, $a_1(\delta) = 370.229\varepsilon$ and $a_2(\delta) = 1.18596 + 79.4593\varepsilon$. Therefore, the nonlocal perturbation bound described in Theorem 13.5 is

$$f(\delta) = \frac{783.426\varepsilon}{1 - 370.229\varepsilon + \sqrt{1 - 2598.682\varepsilon + 12568.549\varepsilon^2}}.$$

The results are illustrated in Table 13.3 for $\varepsilon = 10^{-k}$ and $k = 10, 9, \ldots, 2$. $\diamondsuit$

We see that for $k = 2$ the local bound underestimates the true perturbation more than twice. However, in this case the relative perturbation in S is 67 percent and is not small at all.

Example 13.10 Consider again the standard equation for $n = 2$ from Example 13.9. Let now $q = 1$, $a = 2$ and $s = \sigma^{-4}$, where $\sigma > 0$ is a parameter. In this example we study the conditioning of the standard equation as a function of the parameter σ. For this case the stabilizing solution $X(\sigma)$ and the corresponding closed-loop system matrix are

$$X(\sigma) = \begin{bmatrix} 2\sigma^3 & \sigma^2 \\ \sigma^2 & \sigma \end{bmatrix}, \; A - S(\sigma)X(\sigma) = \begin{bmatrix} -2\sigma & -1/\sigma^2 \\ 2 & 0 \end{bmatrix}.$$

In Table 13.4 we give the individual condition numbers $K_Q(\sigma)$, $K_A(\sigma)$, $K_S(\sigma)$ as well as the quantity $K_*(\sigma) := \|[M_1(\sigma), M_2(\sigma), M_3(\sigma)]\|_2$ for $\sigma = 10^k$ and $k = -5, -4, \ldots, 1, 2$.

$\diamondsuit$

Table 13.4: Individual condition numbers

k	K_Q	K_A	K_S	K_*
-4	1.2500×10^3	2.5000×10^{-5}	3.7500×10^{-13}	1.2500×10^3
-3	1.2500×10^2	2.5000×10^{-4}	3.7500×10^{-10}	1.2500×10^2
-2	1.2505×10^1	2.5020×10^{-3}	3.7508×10^{-7}	1.2505×10^1
-1	1.3014×10^0	2.6940×10^{-2}	3.8344×10^{-4}	1.3017×10^0
0	1.1860×10^0	2.7375×10^0	2.6492×10^0	3.9052×10^0
1	5.0504×10^2	1.0198×10^5	1.5084×10^7	1.5084×10^7
2	5.0005×10^5	1.0001×10^{10}	1.5001×10^{14}	1.5001×10^{14}

13.4 Descriptor equation

13.4.1 Statement of the problem

In this section we consider the descriptor Riccati equation (13.7). The perturbation analysis for this equation is similar to that for the standard Riccati equation except that the calculations (and corresponding expressions) are more involved. So we follow the scheme from Section 13.3 but omit some of the details.

Consider first the real descriptor equation

$$G(T, X) := Q + A^\top XE + E^\top XA - E^\top XSXE = 0, \ T := (Q, E, A, S), \quad (13.24)$$

where the matrix E is nonsingular and $Q = Q^\top$, $S = S^\top$. We assume that equation (13.24) has a symmetric solution X_0 such that the linear matrix operator $\mathcal{L} : \mathbb{R}^{n\times n} \to \mathbb{R}^{n\times n}$, defined by

$$\mathcal{L}(Z) := (A - SX_0E)^\top ZE + E^\top Z(A - SX_0E),$$

is invertible. The eigenvalues of $\mathcal{L}$ are the eigenvalues of its matrix

$$L := E^\top \otimes (A - SX_0E)^\top + (A - SX_0E)^\top \otimes E^\top \in \mathbb{R}^{n^2\times n^2}.$$

The operator $\mathcal{L}$ is invertible if and only if the matrix $(A - SX_0E)E^{-1} = AE^{-1} - SX_0$ has no eigenvalues of opposite signs, i.e.,

$$\lambda_i(AE^{-1} - SX_0) + \lambda_j(AE^{-1} - SX_0) \neq 0, \ i, j = 1, 2, \ldots, n.$$

Note that if $Q, S \geq 0$ and the triple $(Q, E^{-1}A, E^{-1}SE^{-\top})$ is regular, then there is a (unique) stabilizing solution $X_0 \geq 0$ such that the matrix $AE^{-1} - SX_0$ is stable and hence, the operator $\mathcal{L}$ is invertible.

13.4.2 Perturbed equation

Let the matrix coefficients in (13.24) be subject to perturbations $Q \mapsto Q + \delta Q$, $E \mapsto E + \delta E$, $A \mapsto A + \delta A$, $S \mapsto S + \delta S$.

The *perturbed equation* is obtained by replacing T with $T + \delta T = (Q + \delta Q, E + \delta E, A + \delta A, S + \delta S)$ in (13.24):

$$G(T + \delta T, X + \delta X) = 0. \tag{13.25}$$

This equation is quite technical, since its left-hand side contains 50 terms (a product of k perturbed matrices produces 2^k terms), which after some manipulations reduce fortunately to only 26.

We can rewrite (13.25) as an equation for the perturbation δX,

$$\mathcal{L}(\delta X) = V_1(\delta T) + V_2(\delta T, \delta X),$$

where

$$\begin{aligned}
V_1(\delta T) &:= V_{11}(\delta T) + V_{12}(\delta T), \\
V_{11}(\delta T) &:= -\delta Q - \delta E^\top X_0(A - SX_0E) - (A - SX_0E)^\top X\delta E \\
&\quad - \delta A^\top X_0 E - E^\top X_0 \delta A + E^\top X_0 \delta S X_0 E, \\
V_{12}(\delta T) &:= -\delta A^\top X_0 \delta E - \delta E^\top X_0 \delta A + \delta E^\top X_0 S X_0 \delta E \\
&\quad + \delta E^\top X_0 \delta S X_0 E + E^\top X_0 \delta S X_0 \delta E + \delta E^\top X_0 \delta S X_0 \delta E
\end{aligned}$$

and

$$\begin{aligned}
V_2(\delta T, Z) &:= V_{21}(\delta T, Z) + V_{22}(\delta T, Z), \\
V_{21}(\delta T, Z) &:= -\delta E^\top Z A - A^\top Z \delta E - \delta A^\top Z E - E^\top Z \delta A \\
&\quad + E^\top(Z(S + \delta S)X_0 + X_0(S + \delta S)Z)\delta E \\
&\quad + \delta E^\top(Z(S + \delta S)X_0 + X_0(S + \delta S)Z)E \\
&\quad + E^\top(Z\delta S X_0 + X_0 \delta S Z)E + \delta E^\top(Z(S + \delta S)X_0 \\
&\quad + X_0(S + \delta S)Z)\delta E, \\
V_{22}(\delta T, Z) &:= \delta E^\top Z(S + \delta S)ZE + E^\top Z(S + \delta S)Z\delta E \\
&\quad + E^\top Z(S + \delta S)ZE + \delta E^\top Z(S + \delta S)Z\delta E.
\end{aligned}$$

Note that $\|V_{1i}(\delta T)\| = O(\|\delta T\|^i)$, $\delta T \to 0$ and $\|V_{2i}(\delta T, Z)\| = O(\|Z\|^i)$, $Z \to 0$, $i = 1, 2$.

The aim of perturbation analysis here is to find computable bounds for the norm $\delta_X := \|\delta X\|_F$ of the perturbation in the solution X_0 as a function of the perturbation vector

$$\eta := [\eta_1, \eta_2, \eta_3, \eta_4]^\top := [\delta_Q, \delta_E, \delta_A, \delta_S]^\top \in \mathbb{R}_+^4$$

with elements $\delta_Q := \|\delta Q\|_{\mathrm{F}}$, $\delta_E := \|\delta E\|_{\mathrm{F}}$, $\delta_A := \|\delta A\|_{\mathrm{F}}$, $\delta_S := \|\delta S\|_{\mathrm{F}}$, which are the norms of the perturbations in the data matrices Q, E, A, S.

Since the operator $\mathcal{L}$ is invertible we have

$$\delta X = \Phi(\delta T, \delta X) := \Phi_1(\delta T) + \Phi_2(\delta T, \delta X), \tag{13.26}$$

where

$$\Phi_1(\delta T) := \mathcal{L}^{-1}(V_1(\delta T)),\ \Phi_2(\delta T, \delta X) = \mathcal{L}^{-1}(V_2(\delta T, \delta X)).$$

The *equivalent operator equation* (13.26) is

$$\begin{aligned} \xi &:= \mathrm{vec}(\delta X),\ \nu_1 := \mathrm{vec}(\delta Q),\ \nu_2 := \mathrm{vec}(\delta E),\ \nu_3 := \mathrm{vec}(\delta A), \\ \nu_4 &:= \mathrm{vec}(\delta S) \in \mathbb{R}^{n^2}, \nu := \mathrm{vec}(\delta T) = [\nu_1^\top, \nu_2^\top, \nu_3^\top, \nu_4^\top]^\top \in \mathbb{R}^{4n^2}. \end{aligned} \tag{13.27}$$

We have $\delta X = \mathrm{vec}^{-1}(\xi)$, $\delta T = \mathrm{vec}^{-1}(\nu)$ and

$$\xi = \varphi(\nu, \xi) := \varphi_1(\nu) + \varphi_2(\nu, \xi), \tag{13.28}$$

where

$$\varphi_1(\nu) := L^{-1}\mathrm{vec}(V_1(\delta T)),\ \varphi_2(\nu, \xi) := L^{-1}\mathrm{vec}(V_2(\delta T, \delta X)).$$

We now may represent φ_1 and φ_2 as

$$\varphi_1(\nu) := \varphi_{11}(\nu) + \varphi_{12}(\nu),\ \varphi_2(\nu, \xi) := \varphi_{21}(\nu, \xi) + \varphi_{22}(\nu, \xi),$$

where

$$\varphi_{1i}(\nu) := L^{-1}\mathrm{vec}(V_{1i}(\delta T)),\ \varphi_{2i}(\nu, \xi) := L^{-1}\mathrm{vec}(V_{2i}(\delta T, \delta X)),\ i = 1, 2.$$

After some computations we obtain

$$\varphi_{11}(\nu) = N_1\nu_1 + N_2\nu_2 + N_3\nu_3 + N_4\nu_4,$$

where

$$\begin{aligned} N_1 &:= -L^{-1},\ N_2 := -L^{-1}(I_{n^2} + P_{n^2})\left(I_n \otimes (A - SX_0E)^\top X_0\right), \\ N_3 &:= -L^{-1}(I_{n^2} + P_{n^2})\left(I_n \otimes E^\top X_0\right),\ N_4 := \ L^{-1}(E^\top X_0 \otimes E^\top X_0). \end{aligned} \tag{13.29}$$

We also obtain the bound

$$\|\varphi_{12}(\nu)\|_2 \leq \delta_E(\beta_1\delta_E + \beta_2\delta_A + \beta_3\delta_S + \beta_4\delta_E\delta_S), \tag{13.30}$$

where

$$\begin{aligned} \beta_1 &:= \|L^{-1}\|_2\|X_0\|_2^2\|S\|_2,\ \beta_2 := 2\|L^{-1}\|_2\|X_0\|_2, \\ \beta_3 &:= \|X_0\|_2\left(\left\|L^{-1}\left(E^\top X_0 \otimes I_n\right)\right\|_2 + \left\|L^{-1}\left(I_n \otimes E^\top X_0\right)\right\|_2\right), \\ \beta_4 &:= \|L^{-1}\|_2\|X_0\|_2^2. \end{aligned}$$

For the complex descriptor equation

$$Q + A^{\mathrm{H}}XE + E^{\mathrm{H}}XA - E^{\mathrm{H}}XSXE = 0. \tag{13.31}$$

we obtain

$$\widetilde{\mathcal{L}}(\delta X) = \widetilde{V}_1(\delta T) + \widetilde{V}_2(\delta T, \delta X),$$

where the linear matrix operator $\widetilde{\mathcal{L}} : \mathbb{C}^{n\times n} \to \mathbb{C}^{n\times n}$ is defined by

$$\widetilde{\mathcal{L}}(Z) = (A - SX_0E)^{\mathrm{H}}ZE + E^{\mathrm{H}}Z(A - SX_0E),$$

with

$$\begin{aligned}
\widetilde{V}_1(\delta T) &:= \widetilde{V}_{11}(\delta T) + \widetilde{V}_{12}(\delta T),\\
\widetilde{V}_{11}(\delta T) &:= -\,\delta Q - \delta E^{\mathrm{H}}X_0(A - SX_0E) - (A - SX_0E)^{\mathrm{H}}X_0\delta E\\
&\quad -\,\delta A^{\mathrm{H}}X_0E - E^{\mathrm{H}}X_0\delta A + E^{\mathrm{H}}X_0\delta SX_0E,\\
\widetilde{V}_{12}(\delta T) &:= -\,\delta A^{\mathrm{H}}X_0\delta E - \delta E^{\mathrm{H}}X_0\delta A + \delta E^{\mathrm{H}}X_0SX_0\delta E\\
&\quad +\,\delta E^{\mathrm{H}}X_0\delta SX_0E + E^{\mathrm{H}}X_0\delta SX_0\delta E + \delta E^{\mathrm{H}}X_0\delta SX_0\delta E
\end{aligned}$$

and

$$\begin{aligned}
\widetilde{V}_2(\delta T, Z) &:= \widetilde{V}_{21}(\delta T, Z) + \widetilde{V}_{22}(\delta T, Z),\\
\widetilde{V}_{21}(\delta T, Z) &:= -\,\delta E^{\mathrm{H}}ZA - A^{\mathrm{H}}Z\delta E - \delta A^{\mathrm{H}}ZE - E^{\mathrm{H}}Z\delta A\\
&\quad +\,E^{\mathrm{H}}(Z(S + \delta S)X_0 + X_0(S + \delta S)Z)\delta E\\
&\quad +\,\delta E^{\mathrm{H}}(Z(S + \delta S)X_0 + X_0(S + \delta S)Z)E\\
&\quad +\,E^{\mathrm{H}}(Z\delta SX_0 + X_0\delta SZ)E + \delta E^{\mathrm{H}}(Z(S + \delta S)X_0\\
&\quad +\,X_0(S + \delta S)Z)\delta E,\\
V_{22}(\delta T, Z) &:= \delta E^{\mathrm{H}}Z(S + \delta S)ZE + E^{\mathrm{H}}Z(S + \delta S)Z\delta E\\
&\quad +\,E^{\mathrm{H}}Z(S + \delta S)ZE + \delta E^{\mathrm{H}}Z(S + \delta S)Z\delta E.
\end{aligned}$$

Since $\widetilde{\mathcal{L}}$ is invertible, we have

$$\delta X = \widetilde{\Phi}(\delta T, \delta X) := \widetilde{\Phi}_1(\delta T) + \widetilde{\Phi}_2(\delta T, \delta X), \tag{13.32}$$

where

$$\widetilde{\Phi}_1(\delta T) := \widetilde{\mathcal{L}}^{-1}(\widetilde{V}_1(\delta T)),\ \widetilde{\Phi}_2(\delta T, \delta X) := \widetilde{\mathcal{L}}^{-1}(\widetilde{V}_2(\delta T, \delta X)).$$

As in the real case we rewrite the equivalent operator equation (13.32) in vectorized form

$$\xi = \widetilde{\varphi}(\nu, \xi) := \widetilde{\varphi}_1(\nu) + \widetilde{\varphi}_2(\nu, \xi).$$

Here we have used the substitutions (13.27) noting that now $\xi, \nu_i \in \mathbb{C}^{n^2}$ and $\nu \in \mathbb{C}^{4n^2}$. Then,

$$\widetilde{\varphi}_1(\nu) = \widetilde{\varphi}_{11}(\nu) + \widetilde{\varphi}_{12}(\nu), \widetilde{\varphi}_2(\nu, \xi) = \widetilde{\varphi}_{21}(\nu, \xi) + \widetilde{\varphi}_{22}(\nu, \xi),$$

where $\widetilde{\varphi}_{ij}(\cdot) := L^{-1}\text{vec}(\widetilde{\Phi}_{ij}(\cdot))$, $i,j = 1,2$. In particular

$$\begin{aligned}
\widetilde{\varphi}_{11}(\nu) &= \widetilde{N}_1\mu_1 + \widetilde{N}_{21}\mu_2 + \widetilde{N}_{22}\overline{\mu}_2 + \widetilde{N}_{31}\mu_3 + \widetilde{N}_{32}\overline{\nu}_3 + \widetilde{N}_4\nu_4, \qquad (13.33)\\
\widetilde{N}_1 &:= -\widetilde{L}^{-1},\ \widetilde{N}_{21} := -\widetilde{L}^{-1}\left(I_n \otimes (A - SX_0E)^{\mathrm{H}}X_0\right),\\
\widetilde{N}_{22} &:= -\widetilde{L}^{-1}\left((A - SX_0E)^{\top}\overline{X}_0 \otimes I_n\right)P_{n^2},\ \widetilde{N}_{31} := -\widetilde{L}^{-1}\left(I_n \otimes E^{\mathrm{H}}X_0\right),\\
\widetilde{N}_{32} &:= -\widetilde{L}^{-1}\left(E^{\top}\overline{X}_0 \otimes I_n\right)P_{n^2},\ \widetilde{N}_4 := \ \widetilde{L}^{-1}\left(E^{\top}\overline{X}_0 \otimes E^{\mathrm{H}}X_0\right),
\end{aligned}$$

where in (13.33) we have used the fact that $X_0^{\top} = \overline{X}_0$.

Finally, we obtain the bound

$$\|\widetilde{\varphi}_{12}(\nu)\|_2 \le \delta_E(\widetilde{\beta}_1\delta_E + \widetilde{\beta}_2\delta_A + \widetilde{\beta}_3\delta_S + \widetilde{\beta}_4\delta_E\delta_S), \qquad (13.34)$$

where

$$\begin{aligned}
\widetilde{\beta}_1 &:= \|\widetilde{L}^{-1}\|_2\|X_0\|_2^2\|S\|_2,\ \widetilde{\beta}_2 := 2\|\widetilde{L}^{-1}\|_2\|X_0\|_2,\\
\widetilde{\beta}_3 &:= \|X\|_2\left(\left\|\widetilde{L}^{-1}\left(E^{\top}\overline{X}_0 \otimes I_n\right)\right\|_2 + \left\|\widetilde{L}^{-1}\left(I_n \otimes E^{\mathrm{H}}X_0\right)\right\|_2\right),\\
\widetilde{\beta}_4 &:= \|\widetilde{L}^{-1}\|_2\|X_0\|_2^2.
\end{aligned}$$

13.4.3 Condition numbers and local bounds

In this section we use the results from Section 13.4.2 in order to derive the condition numbers and to derive local, first order bounds for the perturbation $\delta_X = \|\delta X\|_{\mathrm{F}}$ in the solution X of the descriptor Riccati equation (13.7).

Based on (13.28) we have

$$\xi = \varphi_1(\nu) + O(\|\nu\|^2 + \|\xi\|^2),\ \|\nu\| + \|\xi\| \to 0.$$

Since $\|\xi\| = O(\|\nu\|)$, $\nu \to 0$, this is equivalent to

$$\xi = N_1\nu_1 + N_2\nu_2 + N_3\nu_3 + N_4\nu_4 + O(\|\nu\|^2),\ \nu \to 0.$$

Hence, using the fact that $\delta_X = \|\xi\|_2$, and having in mind (13.29), we see that the following result holds.

Theorem 13.11 *In Frobenius norm the absolute condition numbers K_Z for the solution X of the real equation (13.24) relative to the matrix coefficients $Z = Q, E, A, S$ are*

$$\begin{aligned}
K_Q &= \|L^{-1}\|_2,\ K_E = \left\|L^{-1}\left(I_{n^2} + P_{n^2}\right)\left(I_n \otimes (A - SX_0E)^{\top}X_0\right)\right\|_2,\\
K_A &= \left\|L^{-1}\left(I_{n^2} + P_{n^2}\right)\left(I_n \otimes E^{\top}X_0\right)\right\|_2,\ K_S = \left\|L^{-1}\left(E^{\top}X_0 \otimes E^{\top}X_0\right)\right\|_2.
\end{aligned}$$

In particular, if only one matrix Z from the set $\{Q, E, A, S\}$ is perturbed, we have

$$\delta_X \le K_Z\|\delta Z\|_{\mathrm{F}} + O(\delta_Z^2),\ \delta Z \to 0.$$

We again have two more bounds. First observe that

$$\delta_X \leq \mathrm{est}_2(\eta) + O(\|\eta\|^2), \ \eta \to 0,$$

where

$$\mathrm{est}_2(\eta) := \|[N_1, N_2, N_3, N_4]\|_2 \|\eta\|_2$$

and the matrices N_i are displayed in (13.29).

The other perturbation bound is

$$\delta_X \leq \mathrm{est}_3(\eta) + O(\|\eta\|^2), \ \eta \to 0,$$

where

$$\mathrm{est}_3(\eta) := \sqrt{\eta^\top N \eta}$$

and $N = [n_{ij}] \in \mathbb{R}_+^{4\times 4}$ is the matrix with elements

$$n_{ij} = \left\| N_i^\top N_j \right\|, \ i, j = 1, 2, 3, 4.$$

The bounds est_2 and est_3 are again *alternative*, since both inequalities $\mathrm{est}_2(\eta) \leq \mathrm{est}_3(\eta)$ and $\mathrm{est}_2(\eta) > \mathrm{est}_3(\eta)$ are possible. Thus, we have the following theorem.

Theorem 13.12 *The perturbation δ_X in the solution X of the real equation (13.24) satisfies the local perturbation estimate*

$$\delta_X \leq \mathrm{est}(\eta) + O(\|\eta\|^2), \ \eta \to 0,$$

where

$$\mathrm{est}(\eta) := \min\{\mathrm{est}_2(\eta), \mathrm{est}_3(\eta)\}.$$

For the complex descriptor Riccati equation (13.31), we give only the final results, since the technique for their derivation had already been described in detail.

Theorem 13.13 *In Frobenius norm the absolute condition numbers $\widetilde{K}_Z$ for the solution X_0 of the complex equation (13.31) relative to the matrix coefficients $Z = Q, E, A, S$ are*

$$\begin{aligned} \widetilde{K}_Q &= \|\widetilde{L}^{-1}\|_2, \ \widetilde{K}_E = \left\| \Theta(\widetilde{N}_{21}, \widetilde{N}_{22}) \right\|_2, \\ \widetilde{K}_A &= \left\| \Theta(\widetilde{N}_{31}, \widetilde{N}_{32}) \right\|_2, \ \widetilde{K}_S = \left\| \widetilde{L}^{-1} \left(E^\top \overline{X}_0 \otimes E^{\mathrm{H}} X_0 \right) \right\|_2. \end{aligned}$$

Define the real $2n \times 2n^2$ matrices

$$N_1^0 := \widetilde{N}_1^{\mathbb{R}}, \ N_2^0 := \Theta(\widetilde{N}_{21}, \widetilde{N}_{22}), \ N_3^0 := \Theta(\widetilde{N}_{31}, \widetilde{N}_{32}), \ N_4^0 := \widetilde{N}_4^{\mathbb{R}}$$

and let

$$\widetilde{N} := [\widetilde{n}_{ij}] \in \mathbb{R}_+^{4\times 4}, \ \widetilde{n}_{ij} := \left\| (N_i^0)^{\mathrm{H}} N_j^0 \right\|_2, \ i, j = 1, 2, 3, 4.$$

As in the real case, set

$$\widetilde{\mathrm{est}}_2(\eta) := \left\| \left[N_1^0, N_2^0, N_3^0, N_4^0 \right] \right\|_2 \|\eta\|_2, \ \widetilde{\mathrm{est}}_3(\eta) := \sqrt{\eta^\top \widetilde{N} \eta}.$$

Thus, we have the following result.

Theorem 13.14 *The perturbation δ_X in the solution X_0 of the complex equation (13.31) satisfies the local perturbation estimate*

$$\delta_X \leq \widetilde{\mathrm{est}}(\eta) + O(\|\eta\|^2), \ \eta \to 0,$$

where

$$\widetilde{\mathrm{est}}(\eta) := \min \left\{ \widetilde{\mathrm{est}}_2(\eta), \widetilde{\mathrm{est}}_3(\eta) \right\}.$$

13.4.4 Nonlocal bounds

The nonlocal perturbation analysis is similar to that of Section 13.3.4. Consider first the real case. Suppose that $\|\xi\|_2 \leq \rho$ for some $\rho > 0$. Estimating the right-hand side of (13.28) we get

$$\|\varphi(\nu, \xi)\|_2 \leq \|\varphi_1(\nu)\|_2 + \|\varphi_2(\nu, \delta)\|_2.$$

Furthermore,

$$\|\varphi_1(\nu)\|_2 \leq b_0(\eta) := \mathrm{est}(\eta) + \delta_E(\beta_1 \delta_E + \beta_2 \delta_A + \beta_3 \delta_S + \beta_4 \delta_E \delta_S)$$

and

$$\|\varphi_2(\nu, \xi)\|_2 \leq b_1(\eta)\rho + b_2(\eta)\rho^2,$$

where

$$\begin{aligned} b_1(\eta) &:= \alpha_1 \delta_E + \alpha_2 \delta_A + \alpha_3 \delta_S + \alpha_4 \delta_E^2, \\ b_2(\eta) &:= (\|S\|_2 + \delta_S)\left(\gamma_0 + \gamma_1 \delta_E + \gamma_2 \delta_E^2\right). \end{aligned}$$

Here the constants α_i and γ_j are given by

$$\begin{aligned} \alpha_1 &:= \left\| L^{-1}\left(A^\top \otimes I_n\right) \right\|_2 + \left\| L^{-1}\left(I_n \otimes A^\top\right) \right\|_2 \\ &\quad + (\|S\|_2 + \delta_S)\left(\left\| L^{-1}\left(E^\top X_0 \otimes I_n\right) \right\|_2 + \left\| L^{-1}\left(I_n \otimes E^\top X_0\right) \right\|_2 \right) \\ &\quad + \|X_0\|_2(\|S\|_2 + \delta_S)\left(\left\| L^{-1}\left(E^\top \otimes I_n\right) \right\|_2 + \left\| L^{-1}\left(I_n \otimes E^\top\right) \right\|_2 \right), \\ \alpha_2 &:= \left\| L^{-1}\left(E^\top \otimes I_n\right) \right\|_2 + \left\| L^{-1}\left(I_n \otimes E^\top\right) \right\|_2, \\ \alpha_3 &:= \left\| L^{-1}\left(E^\top X_0 \otimes E^\top\right) \right\|_2 + \left\| L^{-1}\left(E^\top \otimes E^\top X_0\right) \right\|_2, \\ \alpha_4 &:= 2\|L^{-1}\|_2 \|X_0\|_2 (\|S\|_2 + \delta_S). \end{aligned}$$

and

$$\begin{aligned} \gamma_0 &:= \left\| L^{-1}(E^\top \otimes E^\top) \right\|_2, \\ \gamma_1 &:= \left\| L^{-1}\left(E^\top \otimes I_n\right) \right\|_2 + \left\| L^{-1}\left(I_n \otimes E^\top\right) \right\|_2, \\ \gamma_2 &:= \|L^{-1}\|_2. \end{aligned}$$

Hence, we get

$$\|\varphi(\nu,\xi)\|_2 \le l(\eta,\rho) := b_0(\eta) + b_1(\eta)\rho + b_2(\eta)\rho^2. \tag{13.35}$$

The Lyapunov majorant function l is a quadratic polynomial function in ρ and we may apply directly the results from Chapter 5. Consider the domain

$$\Psi := \left\{\eta \in \mathbb{R}^4_+ : b_1(\eta) + 2\sqrt{b_0(\eta)b_2(\eta)} \le 1\right\}. \tag{13.36}$$

If $\eta \in \Psi$, then the majorant equation $\rho = l(\eta,\rho)$, equivalent to

$$b_2(\eta)\rho^2 - (1 - b_1(\eta))\rho + b_0(\eta) = 0,$$

has a root

$$\rho(\eta) = g(\eta) := \frac{2b_0(\eta)}{1 - b_1(\eta) + \sqrt{(1-b_1(\eta))^2 - 4b_0(\eta)b_2(\eta)}}. \tag{13.37}$$

Hence, for $\eta \in \Psi$ the operator $\varphi(\nu,\cdot)$ maps the set $\mathcal{B}_{g(\delta)}$ into itself. Applying the Schauder fixed point principle we obtain the following result.

Theorem 13.15 *Let $\eta \in \Psi$, where Ψ is given in (13.36). Then, for the real equation (13.24), the nonlocal perturbation bound*

$$\delta_X \le g(\eta)$$

holds, where $g(\eta)$ is determined by (13.37).

In the complex case we have a similar nonlocal result. Let

$$\begin{aligned}
\widetilde{b}_0(\eta) &:= \widetilde{\mathrm{est}}(\eta) + \delta_E(\widetilde{\beta}_1\delta_E + \widetilde{\beta}_2\delta_A + \widetilde{\beta}_3\delta_S + \widetilde{\beta}_4\delta_E\delta_S),\\
\widetilde{b}_1(\eta) &:= \widetilde{\alpha}_1\delta_E + \widetilde{\alpha}_2\delta_A + \widetilde{\alpha}_3\delta_S + \widetilde{\alpha}_4\delta_E^2,\\
\widetilde{b}_2(\eta) &:= (\|S\|_2 + \delta_S)\left(\widetilde{\gamma}_0 + \widetilde{\gamma}_1\delta_E + \widetilde{\gamma}_2\delta_E^2\right),
\end{aligned}$$

where

$$\begin{aligned}
\widetilde{\alpha}_1 &:= \left\|\widetilde{L}^{-1}\left(A^\top \otimes I_n\right)\right\|_2 + \left\|\widetilde{L}^{-1}\left(I_n \otimes A^{\mathrm{H}}\right)\right\|_2\\
&\quad + (\|S\|_2 + \delta_S)\left(\left\|\widetilde{L}^{-1}\left(E^\top \overline{X}_0 \otimes I_n\right)\right\|_2 + \left\|\widetilde{L}^{-1}\left(I_n \otimes E^{\mathrm{H}}X_0\right)\right\|_2\right)\\
&\quad + \|X_0\|_2(\|S\|_2 + \delta_S)\left(\left\|\widetilde{L}^{-1}\left(E^\top \otimes I_n\right)\right\|_2 + \left\|\widetilde{L}^{-1}\left(I_n \otimes E^{\mathrm{H}}\right)\right\|_2\right),\\
\widetilde{\alpha}_2 &:= \left\|\widetilde{L}^{-1}\left(E^\top \otimes I_n\right)\right\|_2 + \left\|\widetilde{L}^{-1}\left(I_n \otimes E^{\mathrm{H}}\right)\right\|_2,\\
\widetilde{\alpha}_3 &:= \left\|\widetilde{L}^{-1}\left(E^\top \overline{X}_0 \otimes E^{\mathrm{H}}\right)\right\|_2 + \left\|\widetilde{L}^{-1}\left(E^\top \otimes E^{\mathrm{H}}X_0\right)\right\|_2,\\
\widetilde{\alpha}_4 &:= 2\|\widetilde{L}^{-1}\|_2\|X_0\|_2(\|S\|_2 + \delta_S),
\end{aligned}$$

and

$$\begin{aligned}\widetilde{\gamma}_0 &:= \left\|\widetilde{L}^{-1}(E^\top \otimes E^{\mathrm{H}})\right\|_2,\\ \widetilde{\gamma}_1 &:= \left\|\widetilde{L}^{-1}\left(E^\top \otimes I_n\right)\right\|_2 + \left\|\widetilde{L}^{-1}\left(I_n \otimes E^{\mathrm{H}}\right)\right\|_2,\\ \widetilde{\gamma}_2 &:= \|\widetilde{L}^{-1}\|_2,\end{aligned}$$

then we obtain the following theorem.

Theorem 13.16 *For the complex equation (13.31) the nonlocal bound*

$$\delta_X \le \frac{2\widetilde{b}_0(\eta)}{1-\widetilde{b}_1(\eta)+\sqrt{(1-\widetilde{b}_1(\eta))^2-4\widetilde{b}_0(\eta)\widetilde{b}_2(\eta)}}$$

holds, provided that $\eta \in \mathbb{R}_+^4$ *is such that*

$$\widetilde{b}_1(\eta)+2\sqrt{\widetilde{b}_0(\eta)\widetilde{b}_2(\eta)} \le 1.$$

To illustrate the perturbation bounds consider the following example of a 2×2 descriptor equation under special perturbations.

Example 13.17 Consider the descriptor 2×2 Riccati equation with matrices

$$Q = \begin{bmatrix} 0 & 0 \\ 0 & q \end{bmatrix},\ E = \begin{bmatrix} e_1 & 0 \\ 0 & e_2 \end{bmatrix},\ A = \begin{bmatrix} 0 & 0 \\ a & 0 \end{bmatrix},\ S = \begin{bmatrix} s & 0 \\ 0 & 0 \end{bmatrix},$$

where $q, e_1, e_2, a, s > 0$. Setting $X_0 = \begin{bmatrix} x_1 & x \\ x & x_2 \end{bmatrix}$, the element-wise version of the equation becomes

$$G(T,X) = \begin{bmatrix} e_1(2ax - e_1 s x_1^2) & e_2(ax_2 - e_1 s x_1) \\ e_2(ax_2 - e_1 s x x_1) & q - e_2^2 s x^2 \end{bmatrix} = 0_{2\times 2}.$$

The positive definite solution is given by

$$\begin{aligned} x &= q^{1/2} e_2^{-1} s^{-1/2},\ x_1 = \sqrt{2}\, q^{1/4} e_1^{-1/2} e_2^{-1/2} a^{1/2} s^{-3/4},\\ x_2 &= \sqrt{2}\, q^{3/4} e_1^{1/2} e_2^{-3/2} a^{-1/2} s^{-1/4}.\end{aligned}$$

Note that $x_1 x_2 - x^2 = q e_2^{-1} s^{-1}$ and the matrix $(AE^{-1} - SX_0) = \begin{bmatrix} -sx_1 & -sx \\ a/e_1 & 0 \end{bmatrix}$ has eigenvalues $\sqrt{2}\, q^{1/4} e_1^{-1/2} e_2^{-1/2} a^{1/2} s^{1/4}(-1 \pm \imath)$.

We choose nominal values of the data as $q = e_1 = e_2 = s = 1$, $a = 2$, which gives $X_0 = \begin{bmatrix} 2 & 1 \\ 1 & 1 \end{bmatrix}$, $A - SX_0E = \begin{bmatrix} -2 & -1 \\ 2 & 0 \end{bmatrix}$ and

$$N_1 = \frac{1}{8}\begin{bmatrix} 2 & 0 & 0 & 4 \\ 0 & 2 & -2 & 4 \\ 0 & -2 & 2 & 4 \\ 1 & -2 & -2 & 6 \end{bmatrix},\ N_2 = \frac{1}{2}\begin{bmatrix} -2 & 0 & -4 & -2 \\ 0 & 0 & -4 & -2 \\ 0 & 0 & -4 & -2 \\ 1 & 1 & -4 & -3 \end{bmatrix},$$

$$N_3 = \frac{1}{4}\begin{bmatrix} 4 & 2 & 4 & 4 \\ 0 & 0 & 4 & 4 \\ 0 & -0 & 4 & 4 \\ 0 & -1 & 2 & 6 \end{bmatrix}, \quad N_4 = -\frac{1}{8}\begin{bmatrix} 12 & 8 & 8 & 6 \\ 4 & 6 & 2 & 4 \\ 4 & 2 & 6 & 4 \\ 2 & 2 & 2 & 3 \end{bmatrix}.$$

The condition numbers are $K_Q = 1.18596$, $K_E = 4.60750$, $K_A = 2.73749$, $K_S = 2.64920$. Furthermore, we have

$$N = \begin{bmatrix} 1.40651 & 5.37067 & 3.04150 & 2.57588 \\ 5.37067 & 21.22907 & 12.40750 & 10.83322 \\ 3.04150 & 12.40750 & 7.49386 & 6.89220 \\ 2.57588 & 10.83322 & 6.89220 & 7.01826 \end{bmatrix}$$

and $\|[N_1, N_2, N_3, N_4]\|_2 = 5.9781$.

Let the perturbations in the data be $\delta q = \sigma \geq 0$, $\delta e_1 = \delta e_2 = \delta s = -\sigma$, $\delta a = 2\sigma$, which gives $\delta = \sigma[1, \sqrt{2}, 2, 1]^\top$. Then the perturbation in the solution is $\delta X = \begin{bmatrix} \delta x_1 & \delta x \\ \delta x & \delta x_2 \end{bmatrix}$, where

$$\begin{aligned} \delta x &= (1+\sigma)^{1/2}(1-\sigma)^{-3/2} - 1, \ \delta x_1 = 2(1+\sigma)^{3/4}(1-\sigma)^{-7/4} - 2, \\ \delta x_2 &= (1+\sigma)^{1/4}(1-\sigma)^{-5/4} - 1. \end{aligned}$$

We also have

$$\text{est}_1(\delta) = 15.8261\sigma, \ \text{est}_2(\delta) = 16.9087\sigma, \ \text{est}_3(\delta) = 15.5487\sigma.$$

Thus, $\text{est}(\delta) = \text{est}_3(\delta) = 15.5487\sigma$. The quantities $b_i(\delta)$ in the nonlocal bound from Theorem 13.15 are

$$\begin{aligned} b_0(\delta) &= 15.5487\sigma, \ b_1(\delta) = \sigma(23.4487 + 25.5479\sigma), \\ b_2(\delta) &= 1.1860 + 4.5404\sigma + 5.7263\sigma^2 + 2.3719\sigma^3. \end{aligned}$$

The results for this example are given in Table 13.5.
◊

Example 13.18 Consider the descriptor equation from Example 13.17 with the same nominal data but now with perturbations

$$\delta Q = \sigma \begin{bmatrix} 3 & -2 \\ -2 & 19 \end{bmatrix}, \ \delta E = -\sigma \begin{bmatrix} 6 & 67 \\ -4 & 37 \end{bmatrix}, \ \delta A = \sigma \begin{bmatrix} 10 & 29 \\ 3 & 33 \end{bmatrix},$$

$$\delta S = -\sigma \begin{bmatrix} 42 & 0 \\ 0 & 0 \end{bmatrix},$$

where $\sigma > 0$ is a small parameter. Hence, $\delta = \sigma[19.4422, 76.8765, 45.1553, 42.0]^\top$ and $\|\delta\| = 100.454\sigma$. Furthermore, we have $\text{est}_1(\delta) = 612.1449\sigma$, $\text{est}_2(\delta) =$

Table 13.5: Exact perturbation, local and nonlocal perturbation bounds

σ	δ_X	local	nonlocal
0.001	0.005945	0.01555	0.01624
0.002	0.01191	0.03110	0.03409
0.003	0.01789	0.04665	0.05394
0.004	0.02388	0.06219	0.07643
0.005	0.02989	0.07774	0.1025
0.006	0.03592	0.09329	0.1341
0.007	0.04197	0.1088	0.1752
0.008	0.04803	0.1244	0.2410
0.009	0.05411	0.1399	*
0.010	0.06020	0.1555	*

600.5259σ, $\mathrm{est}_3(\delta) = 601.2477\sigma$. Thus, $\mathrm{est}(\delta) = \mathrm{est}_2(\delta) = 600.5259\sigma$. The quantities $a_i(\delta)$ in the nonlocal bound are

$$\begin{aligned} a_0(\delta) &= 10^3\sigma(0.60053 + 95.577\sigma + 1997.2\sigma^2), \\ a_1(\delta) &= 10^3\sigma(1.0859 + 36.7\sigma + 1541.4\sigma^2), \\ a_2(\delta) &= (1 + 42\sigma)(1.186 + 182.343\sigma + 7009.3\sigma^2). \end{aligned}$$

The results for these perturbations are presented in Table 13.6. $\diamondsuit$

13.5 Notes and references

There are several studies in the literature on perturbation analysis of continuous-time Riccati equations arising in linear control theory [35, 149, 87, 66, 150, 120, 4, 237, 131, 211], see also [184, 186, 185]. Until recently, however, the results for the complex case had not been clarified. Here the treatment in [211] had to be complemented with the analysis from [145]. The analysis for the descriptor case is new [127].

Condition and error estimates for the solution of Riccati equations are given in [182, 183, 179, 146].

Residual bounds for algebraic Riccati equations are given in [210].

Perturbation analysis of pairs of Riccati equations arising in the H_∞ control is done in [144].

Backward errors for the Riccati equation are derived in [77].

Computational methods for Riccati equations are considered in [37, 181, 167, 159, 200, 32, 21, 22, 23, 168]. General theory of algebraic Riccati equations is presented in [81, 61, 154, 156, 108].

Table 13.6: Exact perturbation, local and nonlocal perturbation bounds

σ	δ_X	local	nonlocal
0.00002	0.0112	0.0120	0.0125
0.00004	0.0225	0.0240	0.0261
0.00006	0.0338	0.0360	0.0411
0.00008	0.0452	0.0480	0.0577
0.00010	0.0566	0.0601	0.0764
0.00012	0.0681	0.0721	0.0979
0.00014	0.0797	0.0841	0.1233
0.00016	0.0913	0.0961	0.1550
0.00018	0.1030	0.1081	0.1989
0.00020	0.1147	0.1201	*

In this book we do not consider differential and difference matrix equations. Perturbation bounds for matrix differential and difference Riccati equations are given in [120, 143, 139, 138, 123].

Chapter 14

Coupled Riccati equations

In this chapter we present the perturbation theory for coupled systems of continuous-time Riccati equations

$$
\begin{aligned}
F_1(X_1, X_2, P_1) &:= (A_1 + B_1X_2)^\top X_1 + X_1(A_1 + B_1X_2) \\
&\quad + C_1 - X_1D_1X_1 = 0, \\
F_2(X_1, X_2, P_2) &:= (A_2 + X_1B_2)X_2 + X_2(A_2 + X_1B_2)^\top \\
&\quad + C_2 - X_2D_2X_2 = 0,
\end{aligned}
\tag{14.1}
$$

where $X_i \in \mathcal{R}$ are the unknown matrices, $A_i, B_i \in \mathcal{R}$, $C_i, D_i \in \mathcal{S}$, $i = 1, 2$, are given matrix coefficients and $P_i := (A_i, B_i, C_i, D_i) \in \mathcal{R}^4$. Here we use the abbreviations $\mathcal{R} = \mathbb{R}^{n\times n}$ and $\mathcal{S} = \{A \in \mathcal{R} : A = A^\top\} \subset \mathcal{R}$.

Equations of this type arise in the $\mathcal{H}_2/\mathcal{H}_\infty$ analysis and design of linear multivariate control systems [27, 227, 118] and in differential games [5].

14.1 Problem statement

For the perturbation analysis we set

$$
\begin{aligned}
P &:= (P_1, P_2) = (A_1, B_1, C_1, D_1, A_2, B_2, C_2, D_2) \\
&=: (E_1, E_2, E_3, E_4, E_5, E_6, E_7, E_8) \in \mathcal{R}^8,
\end{aligned}
$$

i.e., the individual matrix members of P are denoted as $E_1, \ldots, E_8$. The *generalized norm* of the matrix 8-tuple P is the vector

$$
\|P\| := [\|E_1\|_\mathrm{F}, \ldots, \|E_8\|_\mathrm{F}]^\top \in \mathbb{R}^8_+. \tag{14.2}
$$

Although the matrices C_i, D_i are symmetric, system (14.1) may have solutions (X_1, X_2) in which one (or both) of the matrices X_i is not symmetric. In this work we are interested only in symmetric solutions of system (14.1), i.e., $(X_1, X_2) \in \mathcal{S}^2$. The nonsymmetric case may be treated similarly.

Note that the system (14.1) may be written as one matrix equation. This may be done in several ways. Set, for example,

$$X := [X_1, X_2] \in \mathbb{R}^{n \times 2n}, \ C := [C_1, C_2] \in \mathbb{R}^{n \times 2n}.$$

Then we have the single equation

$$\begin{aligned} & C + X \begin{bmatrix} A_1 & 0 \\ 0 & A_2^\top \end{bmatrix} + \left(A_1^\top + X \begin{bmatrix} -D_1 \\ B_1^\top \end{bmatrix} \right) X \begin{bmatrix} I_n & 0 \\ 0 & 0 \end{bmatrix} \\ & + \left(A_2 + X \begin{bmatrix} B_2 \\ -D_2^\top \end{bmatrix} \right) X \begin{bmatrix} 0 & 0 \\ 0 & I_n \end{bmatrix} \\ & + X \left(\begin{bmatrix} B_1 \\ 0 \end{bmatrix} X \begin{bmatrix} 0 & 0 \\ I_n & 0 \end{bmatrix} + \begin{bmatrix} 0 \\ B_2^\top \end{bmatrix} X \begin{bmatrix} 0 & I_n \\ 0 & 0 \end{bmatrix} \right) = 0 \end{aligned}$$

and the condition $(X_1, X_2) \in \mathcal{S}^2$ may be expressed as

$$X \begin{bmatrix} I_n \\ 0 \end{bmatrix}, \ X \begin{bmatrix} 0 \\ I_n \end{bmatrix} \in \mathcal{S}.$$

Thus, we may apply the general perturbation theory for quadratic equations of Chapter 12. However, in this case it is difficult to take into account the special structure of the coefficient matrices and the resulting perturbation bounds will not be tight.

Another desired property of the solutions of (14.1) is whether they stabilize the corresponding closed-loop system matrices.

Definition 14.1 *The solution pair $(X_1, X_2) \in \mathcal{S}^2$ is called* stabilizing *if the matrices*

$$\begin{aligned} G_1 &:= A_1 + B_1 X_2 - D_1 X_1, \\ G_2 &:= A_2 + X_1 B_2 - X_2 D_2 \end{aligned}$$

are stable.

Note that F_i as defined by (14.1) are functions from $\mathcal{R} \times \mathcal{R} \times \mathcal{R}^4 = \mathcal{R}^6$ to $\mathcal{R}$. It will be convenient to write the (14.1) as one operator equation. For this purpose we set

$$X := (X_1, X_2), \ F := (F_1, F_2)$$

and obtain

$$F(X, P) = 0. \tag{14.3}$$

Here F is considered as a mapping $\mathcal{R}^{10} \to \mathcal{R}^2$, or equivalently, as a mapping $\mathbb{R}^{n \times 2n} \times \mathcal{R}^8 \to \mathbb{R}^{n \times 2n}$.

Finding conditions for existence of solutions $(X_1, X_2) \in \mathcal{S}^2$ with X_i nonnegative of system (14.1), as well as of stabilizing solutions, is difficult. Even if both triples (C_i, A_i, B_i) are controllable and observable the system may have no solution in $\mathcal{S}^2$ with nonnegative definite X_1, X_2, nor a stabilizing solution, see Example 14.2.

Example 14.2 Consider the simplest case $n = 1$,

$$\begin{aligned} 2(a_1 + b_1 x_2)x_1 + c_1 - d_1 x_1^2 &= 0, \\ 2(a_2 + b_2 x_1)x_2 + c_2 - d_2 x_2^2 &= 0, \end{aligned} \tag{14.4}$$

where a_i, b_i, c_i, d_i and the unknowns x_i are scalars. Suppose that $b_1 b_2 c_1 c_2 \neq 0$, thus excluding trivial solutions $x_i = 0$ as well as cases of decoupling. If in addition $d_1 d_2 \neq 0$ then the system (14.4) is equivalent to a quartic equation. Geometrically, the solution is given by the intersection points of two hyperbolas with two branches each. We have

$$x_j = \frac{d_i x_i^2 - 2a_i x_i - c_i}{2b_i x_i}, \ i \neq j,$$

and

$$\sum_{k=0}^{4} \alpha_{ik} x_i^k = 0,$$

where

$$\begin{aligned} \alpha_{i0} &:= -c_i^2 d_j, \\ \alpha_{i1} &:= -4(b_i c_i a_j + a_i c_i d_j), \\ \alpha_{i2} &:= 2\left(c_i d_i d_j + 2b_i^2 c_j - 2a_i^2 d_j - 2b_i c_i b_j - 4a_i b_i a_j\right), \\ \alpha_{i3} &:= 4\left(b_i d_i a_j + a_i d_i d_j - 2a_i b_i b_j\right), \\ \alpha_{i4} &:= d_i(4b_1 b_2 - d_1 d_2). \end{aligned}$$

Let $a_i = b_i = c_i = d_i = 1$, $i = 1, 2$. Then we have a double root $(x_1, x_2) = (-1, -1)$ and two more roots (α, β), (β, α), where $\alpha := 1 - 2/\sqrt{3} \approx -0.1547$ and $\beta := 1 + 2/\sqrt{3} \approx 2.1547$. Hence, the system has no solution with $x_1, x_2 \geq 0$. Also, the closed loop system matrices $a_i + b_i x_j - d_i x_i = 1 + x_j - x_i$ are not simultaneously negative on any of the solution pairs (x_1, x_2). Hence, there do not exist stabilizing solutions as well. $\diamondsuit$

In what follows we assume that (14.1) has a solution $X = (X_1, X_2) \in \mathcal{S}^2$ such that the partial Fréchet derivative $F_X(X, P)(\cdot)$ of F in X at the point (X, P) is invertible.

The partial Fréchet derivative of F in X at (X, P) is a linear operator $\mathcal{R}^2 \to \mathcal{R}^2$, calculated as follows. Let $Y = (Y_1, Y_2) \in \mathcal{R}^2$ be arbitrary. We have

$$F_X(X, P)(Y) = (F_{1,X}(X, P_1)(Y), F_{2,X}(X, P_2)(Y))$$

and

$$F_{i,X}(X,P_i)(Y) = F_{i,X_1}(X,P_i)(Y_1) + F_{i,X_2}(X,P_i)(Y_2).$$

A direct calculation gives

$$\begin{array}{rcl} F_{1,X_1}(X,P_1)(Z) &=& G_1^\top Z + ZG_1, \\ F_{1,X_2}(X,P_1)(Z) &=& X_1B_1Z + Z^\top B_1^\top X_1, \\ F_{2,X_1}(X,P_2)(Z) &=& X_2B_2^\top Z^\top + ZB_2X_2, \\ F_{2,X_2}(X,P_2)(Z) &=& G_2Z + ZG_2^\top. \end{array} \tag{14.5}$$

We use the following abbreviations for the partial Fréchet derivatives of F and F_i

$$\begin{array}{rcl} \mathbf{L}(\cdot) &:=& F_X(X,P)(\cdot) \in \mathbf{Lin}(\mathcal{R}^2,\mathcal{R}^2), \\ \mathbf{L}_i(\cdot) &:=& F_{i,X}(X,P_i)(\cdot) \in \mathbf{Lin}(\mathcal{R}^2,\mathcal{R}), \\ \mathbf{L}_{ij}(\cdot) &:=& F_{i,X_j}(X,P_i)(\cdot) \in \mathbf{Lin}(\mathcal{R},\mathcal{R}). \end{array}$$

Thus, we have

$$F_X(X,P)(Y) = (\mathbf{L}_1(Y),\mathbf{L}_2(Y)) = (\mathbf{L}_{11}(Y_1)+\mathbf{L}_{12}(Y_2),\mathbf{L}_{21}(Y_1)+\mathbf{L}_{22}(Y_2)).$$

Note that $\mathbf{L}_{ii}(\cdot)$ are Lyapunov operators. At the same time $\mathbf{L}_{ij}(\cdot)$, $i \neq j$, are associated Lyapunov operators when $X_i \in \mathcal{S}$, see [125].

Applying the vec operation to the pair $F_X(X,P)(Y)$ and using the identity $(A\otimes B)P_{n^2} = P_{n^2}(B\otimes A)$ we obtain that the matrix representation of the linear operator $\mathbf{L}(\cdot)$ is

$$L := \mathrm{Mat}(\mathbf{L}(\cdot)) = \left[\begin{array}{cc} L_{11} & L_{12} \\ L_{21} & L_{22} \end{array}\right] \in \mathbb{R}^{2n^2\times 2n^2},$$

where

$$\begin{array}{rcl} L_{11} &:=& I_n\otimes G_1^\top + G_1^\top \otimes I_n, \\ L_{12} &:=& (I_{n^2}+P_{n^2})(I_n\otimes (X_1B_1)), \\ L_{21} &:=& (I_{n^2}+P_{n^2})\left((B_2X_2)^\top \otimes I_n\right), \\ L_{22} &:=& I_n\otimes G_2 + G_2\otimes I_n. \end{array}$$

Here $L_{ij}\in\mathbb{R}^{n^2\times n^2}$ is the matrix representation of the operator $\mathbf{L}_{ij}(\cdot)$, $i,j=1,2$.

Example 14.3 *For the system from Example 14.2 we have*

$$L = 2\left[\begin{array}{cc} a_1+b_1x_2-d_1x_1 & b_1x_1 \\ b_2x_2 & a_2+b_2x_1-d_2x_2 \end{array}\right] \in \mathbb{R}^{2\times 2}$$

and

$$\det(L) = 4((a_1-d_1x_1)(a_2-d_2x_2)+b_1x_2(a_2-d_2x_2)+b_2x_1(a_1-d_1x_1)).$$

$\diamondsuit$

By assumption and using the implicit function theorem [173] it follows that the solution X is isolated, i.e., there exists $\varepsilon > 0$ such that equation (14.3) has no other solution $\widetilde{X}$ with $\|\widetilde{X} - X\| < \varepsilon$.

In the following, with a certain abuse of notation, we consider P_i both as an ordered pair (and hence, as an element of the linear space $\mathcal{R}^4$) and as a collection.

The perturbation problem for (14.1) is formulated as follows. Let the matrices from P_i be perturbed as

$$A_i \mapsto A_i + \delta A_i, \; B_i \mapsto B_i + \delta B_i, \; C_i \mapsto C_i + \delta C_i, \; D_i \mapsto D_i + \delta D_i$$

(if some of the above matrices are not perturbed then the corresponding perturbations are assumed to be zero).

We assume that the perturbations δC_i and δD_i are symmetric. This assumption is necessary to ensure that the perturbed equation, considered below, also has a solution in $\mathcal{S}^2$.

Denote by $P_i + \delta P_i$ the perturbed collection P_i, in which every matrix $Z \in P_i$ is replaced by $Z + \delta Z$ and let $\delta P = (\delta P_1, \delta P_2)$. Then the perturbed version of equation (14.3) is

$$F(X + \delta X, P + \delta P) = 0. \tag{14.6}$$

The invertibility of the operator F_X and the symmetry of the matrices $C_i + \delta C_i$, $D_i + \delta D_i$ implies that equation (14.6) has a unique isolated solution $X + \delta X \in \mathcal{S}^2$ in the neighborhood of X if the perturbation δP is sufficiently small. Moreover, in this case the elements of δX are analytic functions of the elements of δP. Setting

$$\delta := \begin{bmatrix} \delta_1 \\ \delta_2 \end{bmatrix} \in \mathbb{R}^8_+,$$

where

$$\delta_i := [\delta_{A_i}, \delta_{B_i}, \delta_{C_i}, \delta_{D_i}]^\top \in \mathbb{R}^4_+,$$

the vector of absolute Frobenius norm perturbations $\delta_Z := \|\delta Z\|_F$ in the data matrices $Z \in P$, then the perturbation problem for (14.1) is to find bounds

$$\delta_{X_i} \leq f_i(\delta), \; \delta \in \Omega \subset \mathbb{R}^8_+, \; i = 1, 2, \tag{14.7}$$

for the perturbations $\delta_{X_i} := \|\delta X_i\|_{\mathrm{F}}$. Here Ω is a certain set and f_i are continuous functions, nondecreasing in each of their arguments and satisfying $f_i(0) = 0$. The inclusion $\delta \in \Omega$ guarantees that the perturbed equation (14.6) has a unique solution $X + \delta X$ in a neighborhood of the unperturbed solution X such that the elements of δX_1, δX_2 are analytic functions of the elements of the matrices δZ, $Z \in P$, provided δ is in the interior of Ω.

In the next section, first order local bounds

$$\delta_{X_i} \leq \mathrm{est}_i(\delta) + O(\|\delta\|^2), \; \delta \to 0, \; i = 1, 2, \tag{14.8}$$

are derived with $\mathrm{est}_i(\delta) = O(\|\delta\|)$, $\delta \to 0$, which are then incorporated (see Section 14.3) in the nonlocal bounds (14.7). Here the functions $\mathrm{est}_i : \mathbb{R}_+^8 \to \mathbb{R}_+$ are nonlinear first order homogeneous, i.e., $\mathrm{est}_i(\lambda\delta) = \lambda \mathrm{est}_i(\delta)$ for every $\lambda \geq 0$.

Estimates in terms of relative perturbations

$$\varepsilon_Z := \frac{\|\delta Z\|_F}{\|Z\|_F}, \ 0 \neq Z \in P,$$

for

$$\varepsilon_{X_i} := \frac{\|\delta X_i\|_F}{\|X_i\|_F}, \ i = 1, 2,$$

are straightforward when $X_1 \neq 0$, $X_2 \neq 0$, and are therefore not given in detail.

14.2 Local perturbation analysis

In this section we present a local perturbation analysis for the system (14.1) by determining the functions est_i in (14.8).

14.2.1 Condition numbers

Consider first the conditioning of (14.1). Let

$$\mathbf{Lin} := \mathbf{Lin}(\mathcal{R}, \mathcal{R})$$

ge the space of linear operators $\mathcal{R} \to \mathcal{R}$. Since we want that $F_i(X, P_i) = 0$, the perturbed equations may be written as

$$F_i(X + \delta X_i, P_i + \delta P_i) = \sum_{j=1}^{2} \mathbf{L}_{ij}(\delta X_j) + \sum_{Z \in P_i} F_{i,Z}(\delta Z) + H_i(\delta X, \delta P_i) = 0, \ i = 1, 2,$$

where

$$F_{i,Z}(\cdot) := F_{i,Z}(X, P_i)(.) \in \mathbf{Lin}, \ Z \in P_i,$$

are the Fréchet derivatives of $F_i(X, P_i)$ in the matrix argument Z, evaluated at the point (X, P_i). The matrix expressions

$$H_i(\delta X, \delta P_i) = O\left(\|[\delta X, \delta P_i]\|^2\right), \ \delta X \to 0, \ \delta P_i \to 0,$$

contain second and higher order terms in δX, δP_i. In fact, for $Y = (Y_1, Y_2) \in \mathcal{S}^2$, we have

$$\begin{aligned} H_1(Y, \delta P_1) \quad = \quad & (\delta B_1 Y_2 - \delta D_1 Y_1)^\top X_1 + X_1(\delta B_1 Y_2 - \delta D_1 Y_1) \qquad (14.9) \\ & + Y_1 \delta B_1 X_2 + X_2 \delta B_1^\top Y_1 \\ & - Y_1(D_1 + \delta D_1) Y_1 + Y_1 \delta A_1 + \delta A_1^\top Y_1 \\ & + Y_1(B_1 + \delta B_1) Y_2 + Y_2(B_1 + \delta B_1)^\top Y_1 \end{aligned}$$

and

$$\begin{aligned} H_2(Y,\delta P_2) \quad = \quad & X_2\left(Y_1\delta B_2 - Y_2\delta D_2\right)^\top + \left(Y_1\delta B_2 - Y_2\delta D_2\right)X_2 \\ & + X_1\delta B_2 Y_2 + Y_2\delta B_2^\top X_1 \\ & - Y_2(D_2+\delta D_2)Y_2 + \delta A_2 Y_2 + Y_2\delta A_2^\top \\ & + Y_2(B_2+\delta B_2)^\top Y_1 + Y_1(B_2+\delta B_2)Y_2. \end{aligned} \tag{14.10}$$

We stress that the first four terms in the right-hand sides of (14.9) and (14.10) have extra structure that will be exploited later in the derivation of tighter nonlocal bounds. Indeed, suppose that we want to bound from above the 2-norms of the vector $A\mathrm{vec}(BZC)$, where A, B and C are given matrices and the only information about the matrix Z is that

$$\|Z\|_{\mathrm{F}} = \|\mathrm{vec}(Z)\|_2 \le \delta_Z.$$

Then we have the rough bound

$$\begin{aligned} \|A\mathrm{vec}(BZC)\|_2 \quad \le \quad & \|A\|_2\|\mathrm{vec}(BZC)\|_2 = \|A\|_2\|BZC\|_{\mathrm{F}} \\ \le \quad & \|A\|_2\|B\|_2\|C\|_2\|Z\|_{\mathrm{F}} \le \|A\|_2\|B\|_2\|C\|_2\delta_Z. \end{aligned} \tag{14.11}$$

But we have also the bound

$$\|A\mathrm{vec}(BZC)\|_2 = \|A(C^\top \otimes B)\mathrm{vec}(Z)\|_2 \le \|A(C^\top \otimes B)\|_2\delta_Z, \tag{14.12}$$

and since

$$\|A(C^\top \otimes B)\|_2 \le \|A\|_2\|B\|_2\|C\|_2$$

and since the strict inequality is possible, we see that the bound (14.12) is tighter than (14.11).

Note that we have already calculated the operator $F_X(X,P)(\cdot) = \mathbf{L}(\cdot)$ via the operators $F_{i,X_j}(X,P_i)(\cdot) = \mathbf{L}_i(\cdot)$ and $F_{i,X_j}(X,P_i)(\cdot) = \mathbf{L}_{ij}(\cdot)$, $i,j = 1,2$, namely

$$F_X(X,P)(Y) = (\mathbf{L}_1(Y), \mathbf{L}_2(Y)),$$

where

$$\mathbf{L}_i(Y) = \mathbf{L}_{ii}(Y_i) + \mathbf{L}_{ij}(Y_j).$$

Recalling that the matrix representation of $\mathbf{L}_{ij}(\cdot)$ is denoted by L_{ij}, we have for $(X_1, X_2) \in \mathcal{S}^2$ that

$$\begin{aligned} F_{1,A_1}(Z) &= X_1 Z + Z^\top X_1, \\ F_{1,B_1}(Z) &= X_1 Z X_2 + X_2 Z^\top X_1, \\ F_{1,C_1}(Z) &= Z, \\ F_{1,D_1}(Z) &= -X_1 Z X_1, \\ F_{2,A_2}(Z) &= Z X_2 + X_2 Z^\top, \\ F_{2,B_2}(Z) &= X_1 Z X_2 + X_2 Z^\top X_1, \\ F_{2,C_2}(Z) &= Z, \\ F_{2,D_2}(Z) &= -X_2 Z X_2. \end{aligned}$$

The inverse

$$\mathbf{M}(\cdot) := \mathbf{L}(\cdot)^{-1} \in \mathbf{Lin}(\mathcal{R}^2 \times \mathcal{R}^2)$$

of the operator $\mathbf{L} = F_X(X, P)(\cdot)$ may be represented as

$$\mathbf{L}^{-1}(\cdot) = (\mathbf{M}_1(\cdot), \mathbf{M}_2(\cdot)),$$

where, for $Z := (Z_1, Z_2) \in \mathcal{R}^2$,

$$\mathbf{M}_i(Z) = \mathbf{M}_{i1}(Z_1) + \mathbf{M}_{i2}(Z_2), \;\; M_{ij}(\cdot) \in \mathbf{Lin}, \;\; i = 1, 2.$$

Hence, we have

$$\delta X = -\mathbf{M}(W_1(\delta X, \delta P_1), W_2(\delta X, \delta P_2)), \tag{14.13}$$

where

$$W_i(Y, \delta P_i) := \sum_{Z \in P_i} F_{i,Z}(\delta Z) + H_i(Y, \delta P_i).$$

In this way we obtain

$$\delta X_i = -\sum_{j=1}^{2} \mathbf{M}_{ij}(W_j(\delta X, \delta P_j)), \;\; i = 1, 2,$$

which gives

$$\begin{aligned} \delta X_i \quad &= \quad -\sum_{j=1}^{2} \sum_{Z \in P_j} \mathbf{M}_{ij} \circ F_{j,Z}(\delta Z) \\ & \qquad -\sum_{j=1}^{2} \mathbf{M}_{ij}(H_j(\delta X, \delta P_j)), \;\; i = 1, 2. \end{aligned} \tag{14.14}$$

Therefore, we have the bounds

$$\delta_{X_i} \leq \sum_{j=1}^{2} \sum_{Z \in P_j} K_{ij,Z} \delta_Z + O(\|\delta\|^2), \;\; \delta \to 0,$$

where the quantity

$$K_{ij,Z} := \|\mathbf{M}_{ij} \circ F_{j,Z}\|_{\mathbf{Lin}}, \;\; i, j = 1, 2,$$

is the *absolute condition number* of the solution component X_i with respect to the matrix coefficient $Z \in P_j$. Here $\|.\|_{\mathbf{Lin}}$ is the induced norm in the space $\mathbf{Lin}$ of linear operators $\mathcal{R} \to \mathcal{R}$.

If $X_i \neq 0$, then estimates in terms of relative perturbations are given by

$$\varepsilon_{X_i} \leq \sum_{j=1}^{2} \sum_{Z \in P_i} k_{ij,Z} \varepsilon_Z + O(\|\delta\|^2), \;\; \delta \to 0,$$

where the quantity

$$k_{ij,Z} := K_{ij,Z} \frac{\|Z\|_{\mathrm{F}}}{\|X_i\|_{\mathrm{F}}}, \ \ i,j = 1,2,$$

is the *relative condition number* of the solution component X_i with respect to the matrix coefficient $0 \neq Z \in P_j$.

The calculation of the condition numbers $K_{ij,Z}$ is straightforward when the Frobenius norm is used in $\mathcal{R}$. Indeed, for $U \in \mathbf{Lin}$ we have

$$\begin{aligned} \|U\|_{\mathbf{Lin}} &:= \max\|\|U(Z)\|_{\mathrm{F}} : \|Z\|_{\mathrm{F}} = 1\} \\ &= \max\{\|\mathrm{vec}(U(Z))\|_2 : \|\mathrm{vec}(Z)\|_2 = 1\} \\ &= \max\{\|\mathrm{Mat}(U)\mathrm{vec}(Z)\|_2 : \|\mathrm{vec}(Z)\|_2 = 1\} \\ &= \|\mathrm{Mat}(U)\|_2 = \sigma_{\max}(\mathrm{Mat}(U)), \end{aligned}$$

where $\sigma_{\max}(A)$ is the maximum singular value of the matrix A.

Let $L_{i,Z} \in \mathbb{R}^{n^2 \times n^2}$ be the matrix of the operator $F_{i,Z} \in \mathbf{Lin}$. Then a direct calculation yields

$$\begin{aligned} L_{1,A_1} &= (P_{n^2} + I_{n^2})(I_n \otimes X_1), \\ L_{2,A_2} &= (P_{n^2} + I_{n^2})(X_2 \otimes I_n), \\ L_{1,B_1} &= (P_{n^2} + I_{n^2})(X_2 \otimes X_1), \\ L_{2,B_2} &= (P_{n^2} + I_{n^2})(X_2 \otimes X_1), \\ L_{1,C_1} &= I_{n^2}, \\ L_{2,C_2} &= I_{n^2}, \\ L_{1,D_1} &= X_1 \otimes X_1, \\ L_{2,D_2} &= X_2 \otimes X_2. \end{aligned}$$

Denoting the matrix representation of the operator

$$\mathbf{M}(\cdot) = F_X^{-1}(X,P)(\cdot) \in \mathbf{Lin}(\mathcal{R}^2, \mathcal{R}^2)$$

by

$$M := \mathrm{Mat}(\mathbf{M}) = L^{-1} := \begin{bmatrix} M_{11} & M_{12} \\ M_{21} & M_{22} \end{bmatrix}, \ M_{ij} \in \mathbb{R}^{n^2 \times n^2},$$

the absolute condition numbers are calculated from

$$K_{ij,Z} = \|M_{ij} L_{j,Z}\|_2, \ Z \in P_j, \ i,j = 1,2.$$

A possible disadvantage of this approach may again be the large size $n^2 \times n^2$ of the involved matrices.

14.2.2 First order homogeneous estimates

If we rewrite equations (14.14) in vectorized form as

$$\begin{aligned} \mathrm{vec}(\delta X_i) &= \sum_{j=1}^{2} \sum_{Z \in P_j} N_{i,Z} \mathrm{vec}(\delta Z) \\ &\quad - \sum_{j=1}^{2} M_{ij} \mathrm{vec}(H_j(\delta X, \delta P_j)), \ i = 1, 2, \end{aligned} \tag{14.15}$$

where

$$N_{i,Z} := -M_{ij} L_{j,Z} \in \mathbb{R}^{n^2 \times n^2}, \ Z \in P_j, \ i, j = 1, 2,$$

then the condition number based perturbation bounds may be derived as an immediate consequence of (14.15), as

$$\delta_{X_i} = \|\delta X_i\|_{\mathrm{F}} = \|\mathrm{vec}(\delta X_i)\|_2 \le \mathrm{est}_i^{(1)}(\delta) + O(\|\delta\|^2), \ \delta \to 0,$$

where

$$\mathrm{est}_i^{(1)}(\delta) := \sum_{j=1}^{2} \sum_{Z \in P_j} \|N_{i,Z}\|_2 \delta_Z.$$

Note that the bounds $\mathrm{est}_i^{(1)}(\cdot)$ are linear functions in the perturbation vector $\delta \in \mathbb{R}^8$.

Relations (14.15) also give the second perturbation bound

$$\delta_{X_i} \le \mathrm{est}_i^{(2)}(\delta) + O(\|\delta\|^2), \ \delta \to 0,$$

where

$$\mathrm{est}_i^{(2)}(\delta) := \|N_i\|_2 \|\delta\|_2$$

and

$$\begin{aligned} N_i &:= [N_{i,1}, N_{i,2}] \in \mathbb{R}^{n^2 \times 8n^2}, \\ N_{i,j} &:= [N_{i,A_j}, N_{i,B_j}, N_{i,C_j}, N_{i,D_j}] \in \mathbb{R}^{n^2 \times 4n^2}, \ i = 1, 2. \end{aligned}$$

The bounds $\mathrm{est}_i^{(1)}(\delta)$ and $\mathrm{est}_i^{(2)}(\delta)$ are again alternative and we also have the third bound, which is always less or equal to $\mathrm{est}_1^{(1)}(\delta)$. We have

$$\delta_{X_i}^2 = \mathrm{vec}^\top(\delta X_i)\mathrm{vec}(\delta X_i) = \eta^\top N_i^\top N_i \eta + O(\|\delta\|^2), \ \delta \to 0,$$

where

$$\eta := \left[\mathrm{vec}^\top(\delta A_1), \mathrm{vec}^\top(\delta B_1), \ldots, \mathrm{vec}^\top(\delta D_2)\right]^\top \in \mathbb{R}^{8n^2}.$$

We will rcpresent the matrix

$$N_i^\top N_i \in \mathbb{R}_+^{8n^2 \times 8n^2}$$

as a 8×8 block matrix with $n^2 \times n^2$ blocks as follows. Let the $n^2 \times n^2$ blocks of N_i be denoted as $\widehat{N}_{i,k}$, $k = 1, \ldots, 8$, i.e.,

$$N_i = \left[\widehat{N}_{i,1}, \widehat{N}_{i,2}, \ldots, \widehat{N}_{i,8}\right], \; \widehat{N}_{i,k} \in \mathbb{R}^{n^2 \times n^2},$$

where

$$\widehat{N}_{i,1} := N_{i,A_1}, \; \widehat{N}_{i,2} := N_{i,B_1}, \ldots, \widehat{N}_{i,8} := N_{i,D_2}.$$

Then we have that

$$\eta^\top N_i^\top N_i \eta \le \delta^\top \widehat{N}_i \delta,$$

where $\widehat{N}_i = [n_{i,pq}] \in \mathbb{R}_+^{8\times 8}$, $i = 1, 2$, is a matrix with elements

$$n_{i,pq} := \left\| \widehat{N}_{i,p}^\top \widehat{N}_{i,q} \right\|_2, \; p, q = 1, \ldots, 8.$$

In this way we obtain

$$\delta_{X_i} \le \mathrm{est}_i^{(3)}(\delta) + O(\|\delta\|^2), \; \delta \to 0,$$

where

$$\mathrm{est}_i^{(3)}(\delta) := \sqrt{\delta^\top \widehat{N}_i \delta},$$

and since

$$\left\| \widehat{N}_{i,p}^\top \widehat{N}_{i,q} \right\|_2 \le \left\| \widehat{N}_{i,p} \right\|_2 \left\| \widehat{N}_{i,q} \right\|_2,$$

then $\mathrm{est}_i^{(3)}(\delta) \le \mathrm{est}_i^{(1)}(\delta)$ and we have the overall estimates

$$\delta_{X_i} = \mathrm{est}_i(\delta) + O(\|\delta\|^2), \; \delta \to 0, \; i = 1, 2, \tag{14.16}$$

where

$$\mathrm{est}_i(\delta) := \min\left\{ \mathrm{est}_i^{(2)}(\delta), \mathrm{est}_i^{(3)}(\delta) \right\}, \; i = 1, 2. \tag{14.17}$$

The local bounds considered in this section are continuous, first order homogeneous, nonlinear functions in δ. Also, for $\delta \ne 0$ these functions are real analytic.

All three bounds $\mathrm{est}_i^{(k)}$ are in fact majorants for the solution of a complicated optimization problem, defining the conditioning of the problem as follows. Set

$$\xi_i := \mathrm{vec}(\delta X_i), \; i = 1, 2,$$

and

$$\delta := [\delta_1, \ldots, \delta_8]^\top := [\delta_{A_1}, \ldots, \delta_{D_2}]^\top \in \mathbb{R}_+^8.$$

Then we have

$$\xi_i = \sum_{k=1}^{8} \widehat{N}_{i,k} \eta_k + O(\|\delta\|^2), \; \delta \to 0$$

and

$$\delta_{X_i} = \|\xi_i\|_2 \le K_i(\delta) + O(\|\delta\|^2), \; \delta \to 0,$$

where

$$K_i(\delta) := \max\left\{ \left\| \sum_{k=1}^{8} \widehat{N}_{i,k}\eta_k \right\|_2 : \|\eta_k\| \leq \delta_k,\ k = 1, \ldots, 8 \right\}$$

is the exact upper bound for the first order term in the perturbation bound for the solution component X_i (note that $K_i(\delta)$ is well defined, since the minimization in η is carried out over a compact set).

The calculation of $K_i(\delta)$ is a difficult task. Instead, one can use again a bound such as $\mathrm{est}_i(\delta) \geq K_i(\delta)$.

For a given vector $\gamma \in \mathbb{R}_+^8$ we may define the relative conditioning of the problem as follows.

Definition 14.4 *Let $X_i \neq 0$. The quantity*

$$\kappa_i(\gamma) := \frac{K_i(\gamma)}{\|X_i\|_{\mathrm{F}}}$$

is the relative condition number of X_i with respect to γ.

If $|||P|||$ is the generalized norm (14.2) of P, then $\kappa_i(|||P|||)$ is the relative normwise condition number *of X_i.*

Note that if all elements γ_k of γ are zero except one, equal to $\|E_l\|_{\mathrm{F}}$ in the l-th position, then the quantity $\kappa_i(\gamma)$ is the individual relative condition number of X_i with respect to perturbations in the matrix coefficient E_l.

14.3 Nonlocal perturbation analysis

14.3.1 Implicit bounds

As in the previous section, we obtain nonlinear bounds by using the techniques of nonlinear perturbation analysis. As a result, we get a domain $\Omega \subset \mathcal{R}_+^8$ and two nonlinear continuous functions $f_1, f_2 : \Omega \to \mathcal{R}_+$, satisfying

$$f_1(0) = f_2(0) = 0,$$

and such that

$$\delta_{X_i} \leq f_i(\delta),\ \delta \in \Omega,\ i = 1, 2. \tag{14.18}$$

The inclusion $\delta \in \Omega$ guarantees that the perturbed equation has a unique solution in a neighborhood of the unperturbed solution. Furthermore, the estimate (14.18) is rigorous, i.e., the inequality holds for all perturbations with $\delta \in \Omega$. To get the nonlinear nonlocal bounds the perturbed equation

$$F(X + \delta X, P + \delta P) = 0$$

is again rewritten as an operator equation for the perturbation δX

$$\delta X = \Pi(\delta X, \delta P), \ \Pi = (\Pi_1, \Pi_2), \tag{14.19}$$

where

$$\Pi(Y, \delta P) := -\mathbf{M}(F_P(X, P)(\delta P) + H(Y, \delta P)).$$

Here

$$H(Y, \delta P) := (H_1(Y, \delta P_1), H_2(Y, \delta P_2))$$

contains second and third order terms in Y and δP, see (14.9), (14.10).

Equation (14.19) comprises two equations, namely

$$\delta X_i = \Pi(\delta X, \delta P_i), \ i = 1, 2, \tag{14.20}$$

where the right-hand side of (14.20) is defined by relations (14.14). Setting

$$\begin{aligned} \xi_i &:= \operatorname{vec}(\delta X_i) \in \mathbb{R}^{n^2}, \ i = 1, 2, \\ \xi &:= \begin{bmatrix} \xi_1 \\ \xi_2 \end{bmatrix} \in \mathbb{R}^{2n^2}, \end{aligned}$$

we obtain the vector operator equation

$$\xi = \pi(\xi, \eta) \tag{14.21}$$

in $\mathbb{R}^{2n^2}$ which is reduced to two coupled vector equations

$$\xi_i = \pi_i(\xi, \eta), \ i = 1, 2,$$

in $\mathbb{R}^{n^2}$, respectively.

To obtain Lyapunov majorants we define generalized norms in $\mathbb{R}^{2n^2}$ and $\mathbb{R}^{8n^2}$ by

$$|||\xi||| := \begin{bmatrix} \|\xi_1\|_2 \\ \|\xi_2\|_2 \end{bmatrix} \in \mathbb{R}_+^2$$

and

$$|||\eta||| := \begin{bmatrix} \|\eta_1\|_2 \\ \vdots \\ \|\eta_8\|_2 \end{bmatrix} \in \mathbb{R}_+^8.$$

For $\rho \in \mathbb{R}_+^2$ let

$$\mathcal{B}_\rho := \left\{ \xi \in \mathbb{R}^{2n^2} : |||\xi||| \preceq \rho \right\}$$

be the ball centered at the origin and of generalized radius ρ. We determine

$$h = \begin{bmatrix} h_1 \\ h_2 \end{bmatrix} : \mathbb{R}_+^2 \times \mathbb{R}_+^8 \to \mathbb{R}_+^2,$$

such that the functions $h_i : \mathbb{R}_+^2 \times \mathbb{R}_+^8 \to \mathbb{R}_+$ are nondecreasing in all of their scalar arguments; for all $\delta \in \mathbb{R}_+^8$ the function $h(\cdot, \delta) : \mathbb{R}_+^2 \to \mathbb{R}_+^2$ is differentiable; and the relations

$$h(0,0) = 0, \ \mathrm{rad}(h_\rho(0,0)) < 1$$

hold. Here $h_\rho(\rho, \delta)$ is the Jacobi matrix of the function $\rho \mapsto h(\rho, \delta)$ for a fixed value of δ. In our case the matrix $h_\rho(\rho, \delta)$ is nonnegative and according to the Perron-Frobenius theorem [26] its spectral radius is equal to its maximum (nonnegative) eigenvalue.

Suppose that for all $\rho \in \mathbb{R}_+^2$, all $\xi, \widetilde{\xi} \in \mathcal{B}_\rho$ and all $\eta \in \mathbb{R}^{8n^2}$ with $\|\eta\| \preceq \delta$, the inequalities

$$\|\pi(\xi, \eta)\| \preceq h(\rho, \delta)$$

and

$$\|\pi(\xi, \eta) - \pi(\widetilde{\xi}, \eta)\| \preceq h_\rho(\rho, \delta)\|\xi - \widetilde{\xi}\|$$

hold, then the function h is a Lyapunov majorant and there exists a domain $\Omega \subset \mathbb{R}_+^8$ such that for $\delta \in \Omega$ the vector majorant equation

$$\rho = h(\rho, \delta)$$

has a solution

$$\rho = f(\delta) = \begin{bmatrix} f_1(\delta) \\ f_2(\delta) \end{bmatrix} \in \mathbb{R}_+^2.$$

Here $f : \Omega \to \mathbb{R}_+^2$ is a continuous function, the components f_i of f are nondecreasing in each of their scalar arguments (i.e., $\delta \preceq \widetilde{\delta}$ implies $f(\delta) \preceq f(\widetilde{\delta})$), and $f(0) = 0$.

For $\delta \in \Omega$ the operator $\pi(\cdot, \delta) : \mathbb{R}^{2n^2} \to \mathbb{R}^{2n^2}$ maps the closed convex set $\mathcal{B}_{f(\delta)}$ into itself. Hence, according to the Schauder fixed point principle (Appendix D), there exists a solution $\xi \in \mathcal{B}_{f(\delta)}$ of the operator equation (14.21) and the desired nonlocal perturbation bounds for the solution are

$$\delta_{X_i} = \|\xi_i\|_2 \leq f_i(\delta), \ \delta \in \Omega.$$

We have

$$\pi_i(\xi, \eta) = N_i \eta_i + \psi_i(\xi, \eta),$$

where

$$\psi_i(\xi, \eta) := -\mathrm{vec}\left(\sum_{j=1}^{2} M_{ij} \mathrm{vec}\left(H_j\left(\mathrm{vec}^{-1}(\xi), \mathrm{vec}^{-1}(\eta_j)\right)\right)\right).$$

The next step is to show that the operator $\pi(\cdot, \delta) : \mathcal{R}^{2n^2} \to \mathcal{R}^{2n^2}$ is a contraction on a certain small set of diameter that vanishes together with δ. An estimate of this set in terms of the perturbation vector δ will give us the desired nonlocal perturbation bound.

The vectorizations of the matrices $H_i(Y, \delta P_i)$ are given by

$$\begin{aligned} \mathrm{vec}(H_1(Y, \delta P_1)) &= (I_n \otimes X_1)(I_{n^2} + P_{n^2})\,\mathrm{vec}(\delta B_1 Y_2 - \delta D_1 Y_1) \\ &+ (X_2 \otimes I_n)(I_{n^2} + P_{n^2})\,\mathrm{vec}(Y_1 \delta B_1) \\ &- \mathrm{vec}(Y_1(D_1 + \delta D_1)Y_1) + \mathrm{vec}\left(Y_1 \delta A_1 + \delta A_1^\top Y_1\right) \\ &+ \mathrm{vec}\left(Y_1(B_1 + \delta B_1)Y_2 + Y_2(B_1 + \delta B_1)^\top Y_1\right) \end{aligned} \tag{14.22}$$

and

$$\begin{aligned} \mathrm{vec}(H_2(Y, \delta P_2)) &= (X_2 \otimes I_n)(I_{n^2} + P_{n^2})\,\mathrm{vec}\left(Y_1 \delta B_2 - Y_2 \delta D_2\right) \\ &+ (I_n \otimes X_1)(I_{n^2} + P_{n^2})\,\mathrm{vec}(\delta B_2 Y_2) \\ &- \mathrm{vec}(Y_2(D_2 + \delta D_2)Y_2) + \mathrm{vec}\left(\delta A_2 Y_2 + Y_2 \delta A_2^\top\right) \\ &+ \mathrm{vec}\left(Y_2(B_2 + \delta B_2)^\top Y_1 + Y_1(B_2 + \delta B_2)Y_2\right). \end{aligned} \tag{14.23}$$

Let

$$\|Y_i\|_{\mathrm{F}} \le \rho_i, \; i = 1, 2,$$

where ρ_i are nonnegative constants. Then it follows from (14.22), (14.23) that

$$\begin{aligned} \|\psi_i(\xi, \eta)\|_2 &\le \mathrm{est}_i(\delta) + \left\| \sum_{j=1}^{2} M_{ij} \mathrm{vec}(H_j(Y, \delta P_j)) \right\|_2 \\ &\le \mathrm{est}_i(\delta) + \sum_{j=1}^{2} \left\| M_{ij} \mathrm{vec}(H_j(Y, \delta P_j)) \right\|_2 \\ &\le h_i(\rho, \delta), \end{aligned}$$

where

$$\rho = \begin{bmatrix} \rho_1 \\ \rho_2 \end{bmatrix} \in \mathbb{R}_+^2$$

and

$$\begin{aligned} h_i(\rho_1, \rho_2, \delta) &:= \mathrm{est}_i(\delta) + a_{i1}(\delta)\rho_1 + a_{i2}(\delta)\rho_2 \\ &+ 2b_i(\delta)\rho_1\rho_2 + c_{i1}(\delta)\rho_1^2 + c_{i2}(\delta)\rho_2^2, \; i = 1, 2. \end{aligned}$$

Here

$$\begin{aligned} a_{i1}(\delta) &:= 2\|M_{i1}\|_2 \delta_{A_1} + \nu_{i1}(\delta_{B_1} + \delta_{D_1}) + \nu_{i2}\delta_{B_2}, \\ a_{i2}(\delta) &:= 2\|M_{i2}\|_2 \delta_{A_2} + \nu_{i2}(\delta_{B_2} + \delta_{D_2}) + \nu_{i1}\delta_{B_2}, \\ b_i(\delta) &:= \|M_{i1}\|_2(\|B_1\|_2 + \delta_{B_1}) + \|M_{i2}\|_2(\|B_2\|_2 + \delta_{B_2}), \\ c_{i1}(\delta) &:= \|M_{i1}\|_2(\|D_1\|_2 + \delta_{D_1}), \\ c_{i2}(\delta) &:= \|M_{i2}\|_2(\|D_2\|_2 + \delta_{D_2}), \; i = 1, 2, \end{aligned}$$

and

$$
\begin{aligned}
\nu_{i1} &:= \|M_{i1}(I_n \otimes X_1)\,(I_{n^2} + P_{n^2})\|_2\,, \\
\nu_{i2} &:= \|M_{i2}(X_2 \otimes I_n)\,(I_{n^2} + P_{n^2})\|_2\,.
\end{aligned}
$$

The function $h : \mathbb{R}^2_+ \times \mathbb{R}^8_+ \to \mathbb{R}^2_+$ that we have constructed is a vector Lyapunov majorant for the operator equation (14.21) and the majorant system of two scalar quadratic equations

$$
\rho_i = h_i(\rho_1, \rho_2, \delta),\ i = 1, 2, \tag{14.24}
$$

may also be written in vector form as

$$
\rho = h(\rho, \delta),
$$

where

$$
h(\rho, \delta) := \begin{bmatrix} h_1(\rho, \delta) \\ h_2(\rho, \delta) \end{bmatrix}.
$$

We have

$$
h(0, \delta) = \begin{bmatrix} \mathrm{est}_1(\delta) \\ \mathrm{est}_2(\delta) \end{bmatrix},
$$

$$
h_\rho(\rho, \delta) = \begin{bmatrix} a_{11}(\delta) + 2b_1(\delta)\rho_2 + 2c_{11}(\delta)\rho_1 & a_{12}(\delta) + 2b_1(\delta)\rho_1 + 2c_{12}(\delta)\rho_2 \\ a_{21}(\delta) + 2b_2(\delta)\rho_2 + 2c_{21}(\delta)\rho_1 & a_{22}(\delta) + 2b_2(\delta)\rho_1 + 2c_{22}(\delta)\rho_2 \end{bmatrix},
$$

and

$$
h(0, 0) = 0,\ h_\rho(0, 0) = 0.
$$

Therefore, for δ sufficiently small, the system (14.24) has a solution

$$
\rho = f(\delta) = \begin{bmatrix} f_1(\delta) \\ f_2(\delta) \end{bmatrix},
$$

which is continuous, real analytic in $\delta \neq 0$ and satisfies $f(0) = 0$. The function $f(\cdot)$ is defined in a domain $\Omega \subset \mathbb{R}^8_+$ whose boundary $\partial\Omega$ may be obtained by excluding ρ from the system of equations

$$
\rho = h(\rho, \delta),\ \det(I_2 - h_\rho(\rho, \delta)) = 0. \tag{14.25}
$$

The second equation in (14.25) implies that the Jacobi matrix $h_\rho(\rho, \delta)$ of h in ρ has an eigenvalue 1. In fact, in this case the spectral radius of $h_\rho(\rho, \delta)$ is equal to 1. Relations (14.25) form a system of 3 scalar functionally independent equations of 4-th degree in 10 unknowns (the elements of ρ and δ). This defines a 7-dimensional algebraic variety $\widehat{\Omega} \subset \mathbb{R}^{10}_+$. In a neighborhood of the origin the variety $\widehat{\Omega}$ may be parametrized as

$$
\rho = \widehat{\rho}(t),\ \delta = \widehat{\delta}(t),\ t \in \mathbb{R}^7,
$$

where $\widehat{\rho}(\cdot) : \mathbb{R}^7 \rightarrow \mathbb{R}^2_+$ and $\widehat{\rho}(\cdot) : \mathbb{R}^7_+ \rightarrow \mathbb{R}^8_+$ are algebraic functions. In turn, the surface (an algebraic variety of co-dimension 1) in $\mathbb{R}^8_+$, parametrized by

$$\delta = \widehat{\delta}(t), \ t \in \mathbb{R}^7,$$

forms part of the boundary of the set $\Omega \subset \mathbb{R}^8_+$.

The second equation in (14.25) is equivalent to

$$\begin{aligned}\omega(\rho, \delta) \ &:= \ 1 - \varepsilon(\delta) + \alpha_1(\delta)\rho_1 + \alpha_2(\delta)\rho_2 + 2\beta(\delta)\rho_1\rho_2 \\ &\quad + \gamma_1(\delta)\rho_1^2 + \gamma_2(\delta)\rho_2^2 = 0,\end{aligned}$$

where

$$\begin{aligned}\varepsilon(\delta) \ &:= \ a_{11}(\delta) + a_{22}(\delta) - a_{11}(\delta)a_{22}(\delta) + a_{12}(\delta)a_{21}(\delta), \\ \alpha_1(\delta) \ &:= \ -2(c_{11}(\delta)(1 - a_{22}(\delta)) + b_2(\delta)(1 - a_{11}(\delta)) \\ &\quad + a_{12}(\delta)c_{21}(\delta) + b_1(\delta)a_{21}(\delta)), \\ \alpha_2(\delta) \ &:= \ -2(c_{22}(\delta)(1 - a_{11}(\delta)) + b_1(\delta)(1 - a_{22}(\delta)) \\ &\quad + a_{21}(\delta)c_{12}(\delta) + b_2(\delta)a_{12}(\delta)), \\ \beta(\delta) \ &:= \ 4(c_{11}(\delta)c_{22}(\delta) - c_{12}(\delta)c_{21}(\delta)), \\ \gamma_1(\delta) \ &:= \ 4(b_2(\delta)c_{11}(\delta) - b_1(\delta)c_{21}(\delta)), \\ \gamma_2(\delta) \ &:= \ 4(b_1(\delta)c_{22}(\delta) - b_2(\delta)c_{12}(\delta)).\end{aligned}$$

Thus, for the determination of (part of) the boundary $\partial\Omega$ of the set Ω we have a system of 3 scalar full 2-nd degree equations in ρ_1, ρ_2, whose coefficients are 2-nd degree polynomials in δ. For $\delta \in \Omega$ denote by $\rho = f(\delta)$ the smallest nonnegative solution of the majorant system (14.24). As a result, we have the nonlocal nonlinear perturbation bounds

$$\delta_{X_i} \leq f_i(\delta), \ \delta \in \Omega, \ i = 1, 2. \tag{14.26}$$

Note that if δ is not on the boundary of Ω, in the sense that $\omega(\rho, \delta) > 0$, then

$$\mathrm{rad}(h_\rho(\rho, \delta)) < 1.$$

In this case $\pi(\cdot, \delta)$ is a generalized contraction on $\mathcal{B}_\rho$ and, according to the Banach fixed point principle, the solution for δX is locally unique. Moreover, its elements are real analytic functions in the elements of the perturbations in the coefficient matrices.

14.3.2 Explicit bounds

In practice, it is not necessary to explicitly determine the domain Ω and the functions f_i. It suffices, for a given δ, to solve numerically the majorant system

(14.24) and then to check the condition $\omega(\widetilde{\rho}, \delta) \geq 0$, where $\widetilde{\rho}$ is the computed solution. Then, if such solutions exist (which is guaranteed for δ sufficiently small), one has to choose the smallest nonnegative solution of the system (14.24). We can again avoid the numerical solution by finding a new Lyapunov majorant g, such that

$$h(\rho, \delta) \preceq g(\rho, \delta)$$

and for which the equation

$$\rho = g(\rho, \delta) \tag{14.27}$$

has an explicit form solution. This can be done in many different ways. The sharpest result is obtained by considering Consider the function $l = [l_1, l_2]^\top$ with components

$$l_i(\delta, \rho) := e_i + a_{i1}\rho_1 + a_{i2}\rho_2 + 2b\rho_1\rho_2 + c_1\rho_1^2 + c_2\rho_2^2$$

together with the majorant equations

$$\rho_i = l_i(\rho, \delta), \; i = 1, 2.$$

Subtracting both sides of these equations we get

$$\rho_1 - \rho_2 = a_{11}\rho_1 + a_{12}\rho_2 - a_{21}\rho_1 - a_{22}\rho_2 + e_1 - e_2.$$

If we assume that $a_{ii} < 1 + a_{ji}$, then we have

$$\rho_1 = \lambda\rho_2 + \mu,$$

where

$$\lambda := \frac{1 + a_{12} - a_{22}}{1 + a_{21} - a_{11}}, \quad \mu := \frac{1}{1 + a_{21} - a_{11}}.$$

Substituting this expression in any of the equations $\rho_i = l_i(\rho, \delta)$ we get the quadratic equation

$$\beta_2\rho_2^2 - (1 - \beta_1)\rho_2 + \beta_0 = 0$$

for ρ_2, where the coefficients $\beta_k = \beta_k(\delta)$ are given by

$$\begin{aligned}
\beta_0 &:= \mathrm{est}_1 + c_1\mu^2(e_1 - e_2)^2, \\
\beta_1 &:= \lambda a_{21} + a_{22} + 2\mu(b + c_1\lambda)(e_1 - e_2), \\
\beta_2 &:= c_2 + c_1\lambda^2 + 2b\lambda.
\end{aligned}$$

If

$$\delta \in \Omega_2 := \left\{\delta \in \mathbb{R}_+^8 : \beta_1(\delta) + 2\sqrt{\beta_0(\delta)\beta_2(\delta)} \leq 1\right\}$$

then we obtain the perturbation bound

$$\rho_2 \leq \varphi_2(\delta) := \frac{2\beta_0(\delta)}{1 - \beta_1(\delta) + \sqrt{(1 - \beta_1(\delta))^2 - 4\beta_0(\delta)\beta_2(\delta)}}.$$

Hence, we also have the bound

$$\rho_1 \leq \mu(\delta)(e_1(\delta) - e_2(\delta)) + \lambda(\delta)\varphi_2(\delta).$$

14.4 Notes and references

Coupled linear and quadratic matrix equations arise in many areas of control theory, see [27, 227, 57, 118]. Their sensitivity analysis, however, is less developed. Perturbation analysis of coupled Lyapunov equations is done in [47]. Complete local and nonlocal perturbation analysis of coupled Riccati equations, as presented above, is published in [124], see also [2]. The test examples presented in [124] show that the perturbation bounds presented above can be very tight.

Chapter 15

General fractional-affine equations

15.1 Introductory remarks

In this chapter we present the perturbation analysis for general fractional affine matrix equations of the form

$$F_1 + F_2 F_3^{-1} F_4 = 0,$$

where the F_i are affine matrix expressions in an unknown matrix X. We also briefly discuss symmetric fractional affine matrix equations, particular cases of which are the discrete-time Riccati equations. A detailed treatment of this equation is given in Chapter 16.

Each fractional affine term in a fractional affine matrix equation includes the inversion of a matrix, depending on the solution. Thus, in general the equation is not defined over the whole space of matrix arguments. This significantly complicates the proof of existence theorems for the solution, and still little is known in this area for general fractional affine matrix equations.

15.2 Problem statement

Consider the general fractional affine matrix equation

$$F(X,P) := F_1(X,P_1) + F_2(X,P_2)F_3^{-1}(X,P_3)F_4(X,P_4) = 0, \tag{15.1}$$

where $X \in \mathbb{F}^{m\times n}$ is the unknown matrix. The function

$$F(\cdot,P) : \mathbb{F}^{m\times n} \to \mathbb{F}^{p\times q}$$

is a fractional affine matrix operator, depending on the matrix collection

$$P = (P_1, P_2, P_3, P_4)$$

and

$$F_i(\cdot, P_i) : \mathbb{F}^{m\times n} \to \mathbb{F}^{p_i\times q_i}$$

are affine operators,

$$F_i(X, P_i) := C_i + \sum_{k=1}^{r_i} A_{ik} X B_{ik}, \tag{15.2}$$

depending on the matrix collections

$$P_i := (C_i, A_{i1}, B_{i1}, \ldots, A_{i,r_i}, B_{i,r_i}), \ i = 1, 2, 3, 4.$$

Here

$$C_i \in \mathbb{F}^{p_i\times q_i}, \ A_{ik} \in \mathbb{F}^{p_i\times m}, \ B_{ik} \in \mathbb{F}^{n\times q_i}$$

are given matrix coefficients. It is assumed that $mn = pq := l$ and

$$p_1 = p_2 = p, \ p_3 = p_4 = s, \ q_1 = q_4 = q, \ q_2 = q_3 = s.$$

The matrix $(2r_i + 1)$-tuple P_i depends on $p_i q_i + r_i(mp_i + nq_i)$ parameters – the elements of the matrices C_i, A_{ik} and B_{ik}.

The most general fractional affine matrix equation

$$F_1(X, P_1) + \sum_{j=1}^{r} F_{2j}(X, P_{2j}) F_{3j}^{-1}(X, P_{3j}) F_{4j}(X, P_{4j})$$

includes $r \geq 1$ fractional affine terms $F_{2j} F_{3j}^{-1} F_{4j}$.

Denote by

$$F_Z(X, P) : \mathbb{F}^{r\times t} \to \mathbb{F}^{p\times q}$$

the partial Fréchet derivative of F in the corresponding $r \times t$ matrix argument

$$Z \in \mathcal{P} := \{C_1, A_{11}, B_{11}, \ldots, C_4, \ldots, A_{4,r_4}, B_{4,r_4}\}, \tag{15.3}$$

computed at the point (X, P).

We assume that equation (15.1) has a solution X, such that the linear operator

$$F_X := F_X(X, P) : \mathbb{F}^{m\times n} \to \mathbb{F}^{p\times q}$$

is invertible. We recall that we assume in general that $mn = pq$ and hence, the matrix spaces $\mathbb{F}^{m\times n}$ and $\mathbb{F}^{p\times q}$ are isomorphic.

The problem of existence and uniqueness of the solution of general fractional affine matrix equations is of independent interest but it is not the subject of this monograph.

According to the implicit function theorem, (see Appendix A) the solution X is isolated, i.e., there exists $\varepsilon > 0$ such that equation (15.1) has no other solution $\widetilde{X}$ with $\|\widetilde{X} - X\| < \varepsilon$. The matrix $F_3(\widehat{X}, \widehat{P}_3)$ is invertible in an open neighborhood of the pair (X, P_3) and hence, the functions $F_3(\cdot,\cdot)$ and $F(\cdot,\cdot)$ are properly defined and even analytic in some neighborhoods of (X, P_3) and (X, P), respectively.

The perturbation problem for equation (15.1) is stated as follows. Let the matrices from $\mathcal{P}$ be perturbed as

$$C_i \mapsto C_i + \delta C_i, \;\; A_{ik} \mapsto A_{ik} + \delta A_{ik}, \;\; B_{ik} \mapsto B_{ik} + \delta B_{ik}$$

(if some of the above matrices are not perturbed then the corresponding perturbations are assumed to be zero). Denote by $P + \delta P$ the perturbed collection P, in which each matrix $Z \in \mathcal{P}$ is replaced by $Z + \delta Z$. Then the perturbed equation is

$$F(Y, P + \delta P) = 0. \tag{15.4}$$

In general some of the coefficient matrices from $\mathcal{P}$ may not be perturbed. For instance, some of the matrices C_i may be zero, or some A_{ik} or B_{ik} may be unit matrices as in the symmetric fractional affine matrix equations discussed below. To treat such cases we shall need some more notation. Denote by

$$\widetilde{\mathcal{P}} := \{Z_1, Z_2, \ldots, Z_r\} \subset \mathcal{P}$$

the set of all matrices from $\mathcal{P}$, which are perturbed, and let $\chi^* : \mathcal{P} \to \{0, 1\}$ be the characteristic function of the subset $\widetilde{\mathcal{P}}$, i.e.,

$$\chi^*(Z) = \begin{cases} 1 & \text{if} \;\; Z \in \widetilde{\mathcal{P}}, \\ 0 & \text{if} \;\; Z \in \mathcal{P} \backslash \widetilde{\mathcal{P}}. \end{cases}$$

Consider for example the following equation in $\mathbb{F}^{n \times n}$

$$C_1 + A_1 X + X B_2 (I_n + X)^{-1} A_4 X = 0.$$

Then

$$\mathcal{P} = \{C_1, A_1, I_n; 0, I_n, B_2; I_n, I_n, I_n; 0, A_4, I_n\}$$

and

$$\widetilde{\mathcal{P}} = \{C_1, A_1, B_2, A_4\}$$

if perturbations in C_1, A_1, B_2 and A_4 are considered.

Since the operator F_X is invertible, equation (15.4) has a unique isolated solution $X + \delta X$ in the neighborhood of X if the perturbation δP is sufficiently small. Moreover, in this case the elements of δX are analytic functions of the elements of δP.

Denote by

$$\begin{aligned}\delta^0 &:= \left[\delta_1^0, \delta_2^0, \delta_3^0, \ldots, \delta_{4+2(r_1+r_2+r_3)}^0, \ldots, \delta_{\nu-1}^0, \delta_\nu^0\right]^\top \\ &:= \left[\delta_{C_1}, \delta_{A_{11}}, \delta_{B_{11}}, \ldots, \delta_{C_4}, \ldots, \delta_{A_{4,r_4}}, \delta_{B_{4,r_4}}\right]^\top \in \mathbb{R}_+^\nu\end{aligned} \tag{15.5}$$

the full vector of absolute norm perturbations $\delta_Z := \|\delta Z\|_{\mathrm{F}}$ in the data matrices (15.3), where

$$\nu := 4 + 2(r_1 + r_2 + r_3 + r_4).$$

Similarly, let

$$\delta := [\delta_1, \delta_2, \ldots, \delta_r]^\top := [\delta_{Z_1}, \delta_{Z_2}, \ldots, \delta_{Z_r}]^\top \in \mathbb{R}_+^r \tag{15.6}$$

be the vector of non-zero absolute norm perturbations in the data matrices $Z \in \widetilde{\mathcal{P}}$. Thus, some of the quantities $\delta_i^0 \geq 0$ may be zero, while all δ_j are positive.

The perturbation problem for equation (15.1) is to find a bound

$$\delta_X \leq f(\delta), \ \delta \in \Omega \subset \mathbb{R}_+^r, \tag{15.7}$$

for the perturbation $\delta_X := \|\delta X\|_{\mathrm{F}}$. Here f is a continuous function, non-decreasing in each of its arguments δ_j and satisfying

$$f(0) = 0.$$

The inclusion $\delta \in \Omega$ guarantees that the perturbed equation (15.4) has a unique solution $X + \delta X$ in a neighborhood of the unperturbed solution X, such that the elements of δX are analytic functions of the elements of the matrices δZ, $Z \in \widetilde{\mathcal{P}}$, provided δ is in the interior of Ω. We derive a first order local bound

$$\delta_X \leq f_1(\delta) + O(\|\delta\|^2), \ \delta \to 0,$$

which is then incorporated in the non-local bound (15.7), where

$$f_1(\delta) = O(\|\delta\|), \ \delta \to 0.$$

Estimates in terms of relative perturbations

$$\rho_Z := \frac{\|\delta Z\|_{\mathrm{F}}}{\|Z\|_{\mathrm{F}}}, \ Z \in \widetilde{\mathcal{P}},$$

for $\rho_X := \|\delta X\|_{\mathrm{F}}/\|X\|_{\mathrm{F}}$ are straightforward when $X \neq 0$, and are not given in detail.

An important special case of fractional affine equations are symmetric fractional affine matrix equations of type (15.1). Symmetry means that the operator F satisfies

$$F^\top(X, P) = F(X^\top, P)$$

in the real case and

$$F^{\mathrm{H}}(X,P) - F(X^{\mathrm{H}},P)$$

in the complex case. Hence, we must assume that F_1 and $F_2 F_3^{-1} F_4$ are symmetric operators. This will be the case when the operators F_1 and F_3 are symmetric and, since they are affine, that their linear parts

$$Z \mapsto \sum_{k=1}^{r_i} A_{ik} Z B_{ik}, \ i = 1,3,$$

are Lyapunov operators, see Appendix F and that the operator F_4 is the transpose or the complex conjugate transpose of the operator F_2. Hence, in the real symmetric case

$$F_i(X,P_i) = C_i + \sum_{k=1}^{r_i} \left(A_{ik} X B_{ik}^{\top} + B_{ik} X A_{ik}^{\top} + \varepsilon_{ik} D_{ik} X D_{ik}^{\top} \right), \ i = 1,3,$$

with $C_1 = C_1^{\top}$, $C_3 = C_3^{\top}$, $\varepsilon_{ik} = \pm 1$, and

$$F_2(X,P_2) = C_2 + \sum_{k=1}^{r_2} A_{2k} X B_{2k}^{\top}, \ F_4(X,P_2) = C_2^{\top} + \sum_{k=1}^{r_2} B_{2k} X A_{2k}^{\top}.$$

In the complex case we have

$$F_i(X,P_i) = C_i + \sum_{k=1}^{r_i} \left(A_{ik} X B_{ik}^{\mathrm{H}} + B_{ik} X A_{ik}^{\mathrm{H}} + \varepsilon_{ik} D_{ik} X D_{ik}^{\mathrm{H}} \right), \ i = 1,3,$$

with $C_1 = C_1^{\mathrm{H}}$, $C_3 = C_3^{\mathrm{H}}$, and

$$F_2(X,P_2) = C_2 + \sum_{k=1}^{r_2} A_{2k} X B_{2k}^{\mathrm{H}}, \ F_4(X,P_2) = C_2^{\mathrm{H}} + \sum_{k=1}^{r_2} B_{2k} X A_{2k}^{\mathrm{H}}.$$

Note that the above conditions on F_2, F_3 and F_4 imply symmetry of the fractional affine term $F_2 F_3^{-1} F_4$ but they are not necessary for symmetry to occur as shown next.

Symmetric fractional affine matrix equations, as they arise in optimal control and filtering of discrete-time linear systems, are often called *discrete-time algebraic Riccati equations*, see Chapter 16.

15.3 Local perturbation analysis

In this section we present the local perturbation analysis of equation (15.1).

15.3.1 Condition numbers

Consider the conditioning of equation (15.1). The perturbed equation (15.4) may be written as

$$\begin{aligned} F(X+\delta X, P+\delta P) &:= F_X(\delta X) + \sum_{Z\in\mathcal{P}} F_Z(\delta Z) + G(\delta X, \delta P) \qquad (15.8) \\ &= F_X(\delta X) + \sum_{Z\in\mathcal{P}} \chi^*(Z) F_Z(\delta Z) + G(\delta X, \delta P) \\ &= F_X(\delta X) + \sum_{Z\in\widetilde{\mathcal{P}}} F_Z(\delta Z) + G(\delta X, \delta P) = 0, \end{aligned}$$

where

$$F_X(\cdot) := F_X(X,P)(\cdot) \in \mathbf{Lin}(p,m,n,q,\mathbb{F})$$

and

$$\begin{aligned} F_{C_i}(\cdot) &:= F_{C_i}(X,P)(\cdot) \in \mathbf{Lin}(p,p_i,q_i,q,\mathbb{F}), \\ F_{A_{ik}}(\cdot) &:= F_{A_{ik}}(X,P)(\cdot) \in \mathbf{Lin}(p,p_i,m,q,\mathbb{F}), \\ F_{B_{ik}}(\cdot) &:= F_{B_{ik}}(X,P)(\cdot) \in \mathbf{Lin}(p,n,q_i,q,\mathbb{F}) \end{aligned}$$

are the Fréchet derivatives of $F(X,P)$ in the corresponding matrix arguments, evaluated at the solution X (see Appendix A), and the matrix $G(\delta X, \delta P)$ contains second and higher order terms in δX, δP.

A straightforward calculation leads to

$$\begin{aligned} F_X(Z) &= \sum_{k=1}^{r_1} A_{1k} Z B_{1k} + \left(\sum_{k=1}^{r_2} A_{2k} Z B_{2k}\right) N \qquad (15.9) \\ &\quad - M\left(\sum_{k=1}^{r_3} A_{3k} Z B_{3k}\right) N + M\left(\sum_{k=1}^{r_4} A_{4k} Z B_{4k}\right), \end{aligned}$$

and

$$\begin{aligned} F_{C_1}(Z) &= Z,\ F_{A_{1k}}(Z) = ZXB_{1k},\ F_{B_{1k}}(Z) = A_{1k}XZ, \\ F_{C_2}(Z) &= ZN,\ F_{A_{2k}}(Z) = ZXB_{2k}N,\ F_{B_{2k}}(Z) = A_{2k}XZN, \\ F_{C_3}(Z) &= -MZN,\ F_{A_{3k}}(Z) = -MZXB_{3k}N,\ F_{B_{3k}}(Z) = -MA_{3k}XZN, \\ F_{C_4}(Z) &= MZ,\ F_{A_{4k}}(Z) = MZXB_{4k},\ F_{B_{4k}}(Z) = MA_{4k}XZ, \end{aligned}$$

where

$$M := F_2(X,P)F_3^{-1}(X,P),\ N := F_3^{-1}(X,P)F_4(X,P). \qquad (15.10)$$

Since the operator $F_X(.)$ is invertible, we get

$$\delta X = -\sum_{Z\in\widetilde{\mathcal{P}}} F_X^{-1} \circ F_Z(\delta Z) - F_X^{-1}(G(\delta X, \delta P)). \qquad (15.11)$$

Relation (15.11) gives

$$\delta_X \leq \sum_{Z\in\widetilde{\mathcal{P}}} K_Z \delta_Z + O(\|\delta\|^2),\ \delta \to 0, \tag{15.12}$$

where the quantities

$$K_Z := \left\|F_X^{-1} \circ F_Z\right\|,\ Z \in \widetilde{\mathcal{P}}, \tag{15.13}$$

are the *absolute individual condition numbers* [188] of equation (15.1). Here $\|.\|$ is the induced norm in the corresponding space of linear operators.

If $X \neq 0$, then an estimate in terms of relative perturbations is given by

$$\rho_X := \frac{\|\delta X\|_{\mathrm{F}}}{\|X\|_{\mathrm{F}}} \leq \sum_{Z\in\widetilde{\mathcal{P}}} k_Z \rho_Z + O(\|\delta\|^2),\ \delta \to 0,$$

where the scalars

$$k_Z := K_Z \frac{\|Z\|_{\mathrm{F}}}{\|X\|_{\mathrm{F}}},\ Z \in \widetilde{\mathcal{P}},$$

are the *relative condition numbers* with respect to perturbations in the matrix coefficients $Z \in \widetilde{\mathcal{P}}$.

The calculation of the condition numbers K_Z is straightforward. Denote by $L_Z \in \mathbb{F}^{pq\times rt}$ the matrix representation of the operator $F_Z(\cdot) \in \mathbf{Lin}(p,r,t,q)$. We have

$$\begin{aligned} L_X &= \sum_{k=1}^{r_1} B_{1k}^\top \otimes A_{1k} + \sum_{k=1}^{r_2} (B_{2k}N)^\top \otimes A_{2k} \\ &\quad - \sum_{k=1}^{r_3} (B_{3k}N)^\top \otimes (MA_{3k}) + \sum_{k=1}^{r_4} B_{4k}^\top \otimes (MA_{4k}) \end{aligned} \tag{15.14}$$

and

$$\begin{aligned} L_{C_1} &= I_l,\ L_{A_{1k}} = (XB_{1k})^\top \otimes I_p,\ L_{B_{1k}} = I_q \otimes (A_{1k}X), \\ L_{C_2} &= N^\top \otimes I_p,\ L_{A_{2k}} = (XB_{2k}N)^\top \otimes I_p,\ L_{B_{2k}} = N^\top \otimes (A_{2k}X), \\ L_{C_3} &= -N^\top \otimes M,\ L_{A_{3k}} = -(XB_{3k}N)^\top \otimes M,\ L_{B_{3k}} = -N^\top \otimes (MA_{3k}X), \\ L_{C_4} &= I_q \otimes M,\ L_{A_{4k}} = (XB_{4k})^\top \otimes M,\ L_{B_{4k}} = I_q \otimes (MA_{4k}X). \end{aligned}$$

With these expressions, the absolute condition numbers are calculated from

$$K_Z = \|L_X^{-1} L_Z\|_2,\ Z \in \widetilde{\mathcal{P}}. \tag{15.15}$$

A possible disadvantage of this approach may again be the large size of the involved matrices L_X and L_Z. Condition and accuracy estimates, avoiding the formation and analysis of large matrices, are proposed in [179].

15.3.2 First order homogeneous bounds

As in Section 12, we derive local first order homogeneous estimates. For this we rewrite the perturbed equation in vector form as

$$\mathrm{vec}(\delta X) = \sum_{Z \in \widetilde{\mathcal{P}}} N_Z \mathrm{vec}(\delta Z) - L_X^{-1} \mathrm{vec}(G(\delta X, \delta P)), \tag{15.16}$$

where

$$N_Z := -L_X^{-1} L_Z, \ Z \in \widetilde{\mathcal{P}}.$$

The condition number based estimate is an immediate consequence of (15.16), taking into account that $\delta_Z \geq \|\mathrm{vec}(\delta Z)\|_2$,

$$\delta_X = \|\delta X\|_{\mathrm{F}} = \|\mathrm{vec}(\delta X)\|_2 \leq \mathrm{est}_1(\delta) + O(\|\delta\|^2), \ \delta \to 0,$$

where

$$\mathrm{est}_1(\delta) := \sum_{Z \in \widetilde{\mathcal{P}}} \|N_Z\|_2 \delta_Z.$$

Relation (15.16) also gives

$$\delta_X \leq \mathrm{est}_2(\delta) + O(\|\delta\|^2), \ \delta \to 0,$$

where

$$\mathrm{est}_2(\delta) := \|N\|_2 \|\delta\|_2$$

and

$$N := [N_1, N_2, \ldots, N_r] := [N_{Z_1}, N_{Z_2}, \ldots, N_{Z_r}].$$

The bounds $\mathrm{est}_1(\delta)$ and $\mathrm{est}_2(\delta)$ are again alternative, i.e., which one is better depends on the particular value of Δ. Again, there is a third bound, given by

$$\delta_X \leq \mathrm{est}_3(\delta) + O(\|\delta\|^2), \ \delta \to 0,$$

where

$$\mathrm{est}_3(\delta) := \sqrt{\delta^\top M \delta}$$

and M is a $r \times r$ matrix with elements

$$m_{ij} := \left\|N_i^{\mathrm{H}} N_j\right\|_2.$$

Since

$$m_{ij} \leq \|N_i\|_2 \|N_j\|_2$$

then

$$\mathrm{est}_3(\delta) \leq \mathrm{est}_1(\delta).$$

Therefore we have the overall estimate

$$\delta_X \leq \mathrm{est}(\delta) + O(\|\delta\|^2), \ \delta \to 0, \tag{15.17}$$

where

$$\operatorname{est}(\delta) := \min\{\operatorname{est}_2(\delta), \operatorname{est}_3(\delta)\}. \qquad (15.18)$$

The local bound est in (15.17), (15.18) is a non-linear, first order homogeneous and piece-wise real analytic function in δ.

15.3.3 Component-wise bounds

A local component-wise bound follows directly from relation (15.16) as

$$|\operatorname{vec}(\delta X)| \preceq \sum_{Z\in\widetilde{\mathcal{P}}} \left|L_X^{-1}L_Z\right| |\operatorname{vec}(\delta Z)| + O(\|\delta\|^2),\ \delta \to 0.$$

To compute the componentwise estimate one must have information about the perturbations in the components of the data of the form $|\operatorname{vec}(Z)| \preceq \Delta_Z$, $Z \in \widetilde{\mathcal{P}}$, where $\Delta_Z \succeq 0$ are given vectors.

15.4 Non-local perturbation analysis

To derive non-local bounds we use the matrix Taylor expansion of $(A+E)^{-1}$ in E, where A is invertible and $\operatorname{rad}(A^{-1}E) = \operatorname{rad}(EA^{-1}) < 1$. It is a generalization of the scalar Taylor expansion

$$\frac{1}{a+e} = \frac{1}{a} - \frac{e}{a^2} + \frac{e^2}{a^3} - \cdots = t_m(a,e) + r_m(a,e),$$

where $m \in \mathbb{N}$ and

$$\begin{aligned} t_m(a,e) &:= \frac{1}{a} - \frac{e}{a^2} + \cdots + \frac{(-1)^m e^m}{a^{m+1}} = \frac{1}{a}\sum_{k=0}^{m}(-1)^k\left(\frac{e}{a}\right)^k, \\ r_m(a,e) &:= (-1)^{m+1}\left(\frac{e}{a}\right)^{m+1}\frac{1}{a+e}, \end{aligned}$$

which is valid for $a \neq 0$ and $|e| < |a|$. The generalization to the matrix case is straightforward

$$\begin{aligned} (A+E)^{-1} &= A^{-1} - A^{-1}EA^{-1} + A^{-1}EA^{-1}EA^{-1} - \cdots \\ &= T_m(A,E) + R_m(A,E), \end{aligned}$$

where

$$\begin{aligned} T_m(A,E) &:= A^{-1}\sum_{k=0}^{m}(-1)^k(EA^{-1})^k, \\ R_m(A,E) &:= (-1)^{m+1}(A^{-1}E)^l(A+E)^{-1}(EA^{-1})^{m+1-k}. \end{aligned}$$

Here k may take any of the values $l = 0, 1, \ldots, m+1$, since the value of R_m in fact does not depend on k. Thus, we have $m+2$ different forms for the remainder R_m. In particular for $m = 0$ and $m = 1$ we have

$$(A+E)^{-1} = A^{-1} - (A+E)^{-1}EA^{-1} = A^{-1} - A^{-1}E(A+E)^{-1} \tag{15.19}$$

and

$$\begin{aligned} (A+E)^{-1} &= A^{-1} - A^{-1}EA^{-1} + (A+E)^{-1}(EA^{-1})^2 \\ &= A^{-1} - A^{-1}EA^{-1} + A^{-1}E(A+E)^{-1}EA^{-1} \\ &= A^{-1} - A^{-1}EA^{-1} + (A^{-1}E)^2(A+E)^{-1}, \end{aligned} \tag{15.20}$$

respectively.

Let now the collections P_i be perturbed to $P_i + \delta P_i$. Then in equation (15.4) we may represent the perturbed quantities $F_i(X+\delta X, P_i+\delta P_i)$ as described below. In what follows we mark only the dependence on the perturbations δX and δP_i, recalling that X is a fixed solution of (15.1). We have

$$\widetilde{F}_i := F_i(X+\delta X, P_i+\delta P_i) = F_i + E_i(\delta X, \delta P_i),$$

where $F_i := F_i(X, P_i)$ and

$$E_i(\delta X, \delta P_i) := L_i(\delta X) + K_i(\delta P_i) + Q_i(\delta X, \delta P_i).$$

Here $L_i(\cdot) : \mathbb{F}^{m\times n} \to \mathbb{F}^{p_i \times q_i}$ is a linear operator, defined by

$$L_i(Z) := \sum_{k=1}^{r_i} A_{ik} Z B_{ik}.$$

The term

$$K_i(\delta P_i) := \delta C_i + \sum_{k=1}^{r_i} (\delta A_{ik} X B_{ik} + A_{ik} X \delta B_{ik})$$

contains the first order perturbations in P_i, and $Q_i(\cdot, \delta P_i)$ is the affine operator

$$Q_i(Z, \delta P_i) := \sum_{k=1}^{r_i} (\delta A_{ik} Z B_{ik} + A_{ik} Z \delta B_{ik} + \delta A_{ik}(X+Z)\delta B_{ik}).$$

Thus, the expression $Q_i(\delta X, \delta P_i)$ contains the second and third order terms in δX and δP_i. The perturbed equation

$$\widetilde{F}_1 + \widetilde{F}_2 \widetilde{F}_3^{-1} \widetilde{F}_4 = 0$$

may be written as

$$\widetilde{F}(\mathcal{E}) := F_1 + \mathcal{E}_1 + (F_2 + \mathcal{E}_2)(F_3 + \mathcal{E}_3)^{-1}(F_4 + \mathcal{E}_4) = 0, \tag{15.21}$$

where $\mathcal{E} := (\mathcal{E}_1, \mathcal{E}_2, \mathcal{E}_3, \mathcal{E}_4)$ and $\mathcal{E}_i := E_i(\delta X, \delta P_i)$. We may represent $\widetilde{F}(\mathcal{E})$ as the sum $L(\mathcal{E})$ of its partial Fréchet derivatives $\widetilde{F}_{\mathcal{E}_i}(0)$ in $\mathcal{E}_i$ at $\mathcal{E} = 0$ plus the second and higher order terms $Q(\mathcal{E})$ in $\mathcal{E}$,

$$\widetilde{F}(\mathcal{E}) = L(\mathcal{E}) + Q(\mathcal{E}).$$

Then we have

$$\begin{aligned} \widetilde{F}_{\mathcal{E}_1}(0)(Z) &= Z, \ \widetilde{F}_{\mathcal{E}_2}(0)(Z) = ZN, \\ \widetilde{F}_{\mathcal{E}_3}(0)(Z) &= -MZN, \ \widetilde{F}_{\mathcal{E}_4}(0)(Z) = MZ, \end{aligned}$$

where the matrices M and N are defined in (15.10). Hence,

$$L(\mathcal{E}) = \mathcal{E}_1 + \mathcal{E}_2 N - M\mathcal{E}_3 N + M\mathcal{E}_4. \tag{15.22}$$

The expression for $Q(\mathcal{E})$ is more tricky. It may be written in six different forms which lead to 12 possible norm estimates. We present only one of them, which is based on the two representations in (15.19) and the three representations in (15.20). We use the 'most symmetric' form, using both representations in (15.19) and the second one in (15.20). This gives

$$\begin{aligned} Q(\mathcal{E}) &= \mathcal{E}_2(F_3 + \mathcal{E}_3)^{-1}\mathcal{E}_4 - \mathcal{E}_2(F_3 + \mathcal{E}_3)^{-1}\mathcal{E}_3 N \\ &\quad + M\mathcal{E}_3(F_3 + \mathcal{E}_3)^{-1}\mathcal{E}_3 N - M\mathcal{E}_3(F_3 + \mathcal{E}_3)^{-1}\mathcal{E}_4 \\ &= (\mathcal{E}_2 - M\mathcal{E}_3)(F_3 + \mathcal{E}_3)^{-1}(\mathcal{E}_4 - \mathcal{E}_3 N). \end{aligned} \tag{15.23}$$

In the following we give an estimate for

$$\varphi(\mathcal{E}) := \left\| F_X^{-1}(Q(\mathcal{E})) \right\|_{\mathrm{F}}.$$

When estimating the Frobenius norm of the expression $\mathcal{E}_i(F_3 + \mathcal{E}_3)^{-1}\mathcal{E}_j$, we get two different bounds based on the representations

$$\mathcal{E}_i(F_3 + \mathcal{E}_3)^{-1}\mathcal{E}_j = \mathcal{E}_i\left(I_s + F_3^{-1}\mathcal{E}_3\right)^{-1} F_3^{-1}\mathcal{E}_j = \mathcal{E}_i F_3^{-1}\left(I_s + \mathcal{E}_3 F_3^{-1}\right)^{-1}\mathcal{E}_j,$$

namely

$$\left\| \mathcal{E}_i(F_3 + \mathcal{E}_3)^{-1}\mathcal{E}_j \right\|_{\mathrm{F}} \le \frac{\|\mathcal{E}_i\|_2 \left\| F_3^{-1}\mathcal{E}_j \right\|_2}{1 - \left\| F_3^{-1}\mathcal{E}_3 \right\|_{\mathrm{F}}} \tag{15.24}$$

and

$$\left\| \mathcal{E}_i(F_3 + \mathcal{E}_3)^{-1}\mathcal{E}_j \right\|_{\mathrm{F}} \le \frac{\left\| \mathcal{E}_i F_3^{-1} \right\|_2 \|\mathcal{E}_j\|_2}{1 - \left\| \mathcal{E}_3 F_3^{-1} \right\|_{\mathrm{F}}}. \tag{15.25}$$

In order to have equal denominators we must choose the first (15.24) or the second (15.25) option in all four terms in the first equality in (15.23). Let us for example choose the first option (15.24). Then we have

$$\begin{aligned}\varphi(\mathcal{E}) &\leq \left\|L_X^{-1}\right\|_2 \frac{\|\mathcal{E}_2\|_2 \left\|F_3^{-1}\mathcal{E}_4\right\|_2}{1-\left\|F_3^{-1}\mathcal{E}_3\right\|_{\mathrm{F}}} + \left\|L_X^{-1}\left(N^\top \otimes I_p\right)\right\|_2 \frac{\|\mathcal{E}_2\|_2 \left\|F_3^{-1}\mathcal{E}_3\right\|_2}{1-\left\|F_3^{-1}\mathcal{E}_3\right\|_{\mathrm{F}}} \\ &+ \left\|L_X^{-1}(I_q \otimes M)\right\|_2 \frac{\|\mathcal{E}_3\|_2 \left\|F_3^{-1}\mathcal{E}_4\right\|_2}{1-\left\|F_3^{-1}\mathcal{E}_3\right\|_{\mathrm{F}}} \\ &+ \left\|L_X^{-1}(N^\top \otimes M)\right\|_2 \frac{\|\mathcal{E}_3\|_2 \left\|F_3^{-1}\mathcal{E}_3\right\|_2}{1-\left\|F_3^{-1}\mathcal{E}_3\right\|_{\mathrm{F}}}.\end{aligned}$$

Using the second equality in (15.23) we also get

$$\varphi(\mathcal{E}) \leq \left\|L_X^{-1}\right\|_2 \frac{\|\mathcal{E}_2 - M\mathcal{E}_3\|_2 \left\|F_3^{-1}(\mathcal{E}_4 - \mathcal{E}_3 N)\right\|_2}{1-\left\|F_3^{-1}\mathcal{E}_3\right\|_{\mathrm{F}}}. \tag{15.26}$$

The implementation of (15.26) and (15.26) may produce different overall perturbation bounds. This, however, will result in small (second and higher order) changes, so we shall consider only bounds based on (15.26).

The perturbed equation may be rewritten as an equivalent operator equation for δX,

$$\delta X = \Phi(\delta X, \delta P) := \Phi_1(\delta P) + \Phi_2(\delta X, \delta P) + \Psi(\delta X, \delta P), \tag{15.27}$$

where

$$\begin{aligned}\Phi_1(\delta P) &:= -F_X^{-1}(K_1(\delta P_1) + K_2(\delta P_2)N \\ &\quad - MK_3(\delta P_3)N + MK_4(\delta P_4)), \\ \Phi_2(Z, \delta P) &:= -F_X^{-1}(Q_1(Z, \delta P_1) + Q_2(Z, \delta P_2)N \\ &\quad - MQ_3(Z, \delta P_3)N + MQ_4(Z, \delta P_4)), \\ \Psi(Z, \delta P) &:= -F_X^{-1}((E_2(Z, \delta P_2) - ME_3(Z, \delta P_3)) \\ &\quad \times (F_3 + E_3(Z, \delta P_3))^{-1}(E_4(Z, \delta P_4) - E_3(Z, \delta P_3)N)).\end{aligned} \tag{15.28}$$

We again apply the technique of Lyapunov majorants and fixed point principles (Chapter 5) in order to derive non-local perturbation bounds.

Let $\|Z\|_{\mathrm{F}} \leq \rho$. After some straightforward calculations we get

$$\begin{aligned}\|E_2(Z, \delta P_2) - ME_3(Z, \delta P_3)\|_2 &\leq \alpha_2(\delta) + \beta_2(\delta)\rho, \\ \left\|F_3^{-1}E_3(Z, \delta P_3)\right\|_{\mathrm{F}} &\leq \alpha_3(\delta) + \beta_3(\delta)\rho, \\ \left\|F_3^{-1}(E_4(Z, \delta P_4) - E_3(Z, \delta P_3)N)\right\|_2 &\leq \alpha_4(\delta) + \beta_4(\delta)\rho,\end{aligned} \tag{15.29}$$

where, for $i = 2, 3, 4$,

$$\begin{aligned}\alpha_i(\delta) &:= \alpha_{i1}(\delta) + \alpha_{i2}(\delta), \\ \beta_i(\delta) &:= \beta_{i0}(\delta) + \beta_{i1}(\delta) + \beta_{i2}(\delta).\end{aligned} \tag{15.30}$$

The quantities

$$\alpha_{ij}(\delta) = O(\|\delta\|^j);\ \beta_{ik}(\delta) = O(\|\delta\|^k),\ \delta \to 0,$$

are determined as follows.

Case $i = 2$:

$$\begin{aligned} \alpha_{21}(\delta) &:= \delta_{C_2} + \sum_{k=1}^{r_2}(\|XB_{2k}\|_2\delta_{A_{2k}} + \|A_{2k}X\|_2\delta_{B_{2k}}) \\ &\quad + \|M\|_2\delta_{C_3} + \sum_{k=1}^{r_3}(\|M\|_2\|XB_{3k}\|_2\delta_{A_{3k}} + \|MA_{3k}X\|_2\delta_{B_{3k}}), \\ \alpha_{22}(\delta) &:= \|X\|_2\beta_{22}(\Delta), \\ \beta_{20}(\delta) &:= \left\|\sum_{k=1}^{r_2} B_{2k}^\top \otimes A_{2k} - \sum_{k=1}^{r_3} B_{3k}^\top \otimes (MA_{3k})\right\|_2, \\ \beta_{21}(\delta) &:= \sum_{k=1}^{r_2}(\|B_{2k}\|_2\delta_{A_{2k}} + \|A_{2k}\|_2\delta_{B_{2k}}) \\ &\quad + \sum_{k=1}^{r_3}(\|M\|_2\|B_{3k}\|_2\delta_{A_{3k}} + \|MA_{3k}\|_2\delta_{B_{3k}}), \\ \beta_{22}(\delta) &:= \sum_{k=1}^{r_2}\delta_{A_{2k}}\delta_{B_{2k}} + \|M\|_2\sum_{k=1}^{r_3}\delta_{A_{3k}}\delta_{B_{3k}}. \end{aligned} \tag{15.31}$$

Case $i = 3$:

$$\begin{aligned} \alpha_{31}(\delta) &:= \left\|F_3^{-1}\right\|_2\delta_{C_3} \\ &\quad + \sum_{k=1}^{r_3}\left(\left\|F_3^{-1}\right\|_2\|XB_{3k}\|_2\delta_{A_{3k}} + \left\|F_3^{-1}A_{3k}X\right\|_2\delta_{B_{3k}}\right), \\ \alpha_{32}(\delta) &:= \|X\|_2\beta_{32}(\Delta), \\ \beta_{30}(\delta) &:= \left\|\sum_{k=1}^{r_3} B_{3k}^\top \otimes \left(F_3^{-1}A_{3k}\right)\right\|_2, \\ \beta_{31}(\delta) &:= \sum_{k=1}^{r_3}\left(\left\|F_3^{-1}\right\|_2\|B_{3k}\|_2\delta_{A_{3k}} + \left\|F_3^{-1}A_{3k}\right\|_2\delta_{B_{3k}}\right), \\ \beta_{32}(\delta) &:= \left\|F_3^{-1}\right\|_2\sum_{k=1}^{r_3}\delta_{A_{3k}}\delta_{B_{3k}}. \end{aligned} \tag{15.32}$$

Case $i = 4$:

$$\begin{aligned} \alpha_{41}(\delta) &:= \left\|F_3^{-1}\right\|_2\delta_{C_4} \\ &\quad + \sum_{k=1}^{r_4}\left(\left\|F_3^{-1}\right\|_2\|XB_{4k}\|_2\delta_{A_{4k}} + \|F_3^{-1}A_{4k}X\|_2\delta_{B_{4k}}\right) \end{aligned} \tag{15.33}$$

$$
\begin{aligned}
& + \left\|F_3^{-1}\right\|_2 \|N\|_2 \delta_{C_3} \\
& + \sum_{k=1}^{r_3} \left(\left\|F_3^{-1}\right\|_2 \|XB_{3k}N\|_2 \delta_{A_{3k}} + \|F_3^{-1}A_{3k}X\|_2 \|N\|_2 \delta_{B_{3k}} \right), \\
\alpha_{42}(\delta) &:= \|X\|_2 \beta_{42}(\Delta), \\
\beta_{40}(\delta) &:= \left\| \sum_{k=1}^{r_4} B_{4k}^\top \otimes \left(F_3^{-1}A_{4k}\right) - \sum_{k=1}^{r_3} (B_{3k}N)^\top \otimes \left(F_3^{-1}A_{3k}\right) \right\|_2, \\
\beta_{41}(\delta) &:= \sum_{k=1}^{r_4} \left(\left\|F_3^{-1}\right\|_2 \|B_{4k}\|_2 \delta_{A_{4k}} + \left\|F_3^{-1}A_{4k}\right\|_2 \delta_{B_{4k}} \right) \\
& + \sum_{k=1}^{r_3} \left(\left\|F_3^{-1}\right\|_2 \|B_{3k}N\|_2 \delta_{A_{3k}} + \left\|F_3^{-1}A_{3k}\right\|_2 \|N\|_2 \delta_{B_{3k}} \right), \\
\beta_{42}(\delta) &:= \left\|F_3^{-1}\right\|_2 \left(\sum_{k=1}^{r_4} \delta_{A_{4k}} \delta_{B_{4k}} + \|N\|_2 \sum_{k=1}^{r_3} \delta_{A_{3k}} \delta_{B_{3k}} \right).
\end{aligned}
$$

It follows from (15.29)–(15.33) that

$$
\begin{aligned}
\|\Phi_1(\delta P) + \Phi_2(Z, \delta P)\|_{\mathrm{F}} &\leq a_0(\delta) + a_1(\delta)\rho, \\
\|\Psi(Z, \delta P)\|_{\mathrm{F}} &\leq \frac{b_0(\delta) + b_1(\delta)\rho + b_2(\delta)\rho^2}{1 - \alpha_3(\delta) - \beta_3(\delta)\rho},
\end{aligned}
\tag{15.34}
$$

provided that

$$
\rho < \frac{1 - \alpha_3(\delta)}{\beta_3(\delta)}. \tag{15.35}
$$

Here

$$
\begin{aligned}
a_0(\delta) &:= a_{01}(\delta) + a_{02}(\delta) := \mathrm{est}(\delta) + \|X\|_2 a_{12}(\delta), \\
a_1(\delta) &:= a_{11}(\delta) + a_{12}(\delta), \\
a_{11}(\delta) &:= \sum_{k=1}^{r_1} \left\|L_X^{-1}\left(B_{1k}^\top \otimes I_p\right)\right\|_2 \delta_{A_{1k}} + \sum_{k=1}^{r_1} \left\|L_X^{-1}\left(I_q \otimes A_{1k}\right)\right\|_2 \delta_{B_{1k}} \\
& + \sum_{k=1}^{r_2} \left\|L_X^{-1}\left((B_{2k}N)^\top \otimes I_p\right)\right\|_2 \delta_{A_{2k}} + \sum_{k=1}^{r_2} \left\|L_X^{-1}\left(N^\top \otimes A_{2k}\right)\right\|_2 \delta_{B_{2k}} \\
& + \sum_{k=1}^{r_3} \left\|L_X^{-1}\left((B_{3k}N)^\top \otimes M\right)\right\|_2 \delta_{A_{3k}} \\
& + \sum_{k=1}^{r_3} \left\|L_X^{-1}\left(N^\top \otimes (MA_{3k})\right)\right\|_2 \delta_{B_{3k}} \\
& + \sum_{k=1}^{r_4} \left\|L_X^{-1}\left(B_{4k}^\top \otimes M\right)\right\|_2 \delta_{A_{4k}} + \sum_{k=1}^{r_4} \left\|L_X^{-1}\left(I_q \otimes (MA_{4k})\right)\right\|_2 \delta_{B_{4k}}, \\
a_{12}(\delta) &:= \sum_{k=1}^{r_1} \left\|L_X^{-1}\right\|_2 \delta_{A_{1k}} \delta_{B_{1k}} + \sum_{k=1}^{r_2} \left\|L_X^{-1}\left(N^\top \otimes I_p\right)\right\|_2 \delta_{A_{2k}} \delta_{B_{2k}}
\end{aligned}
\tag{15.36}
$$

$$+\sum_{k=1}^{r_3} \left\|L_X^{-1}(N^\top \otimes M)\right\|_2 \delta_{A_{3k}}\delta_{B_{3k}} + \sum_{k-1}^{r_4} \left\|L_X^{-1}(I_q \otimes M)\right\|_2 \delta_{A_{4k}}\delta_{B_{4k}}$$

and

$$\begin{aligned} b_0(\delta) &:= \left\|L_X^{-1}\right\|_2 \alpha_2(\delta)\alpha_4(\delta), \\ b_1(\delta) &:= \left\|L_X^{-1}\right\|_2 (\alpha_2(\delta)\beta_4(\delta) + \alpha_4(\Delta)\beta_2(\Delta)), \\ b_2(\delta) &:= \left\|L_X^{-1}\right\|_2 \beta_2(\delta)\beta_4(\delta). \end{aligned} \tag{15.37}$$

Using (15.34) we see that the Lyapunov majorant $h(\delta, \rho)$ for equation (15.27), such that

$$\|\Phi(Z, \delta P)\|_{\mathrm{F}} \leq h(\delta, \rho),$$

is

$$h(\delta, \rho) = a_0(\delta) + a_1(\delta)\rho + \frac{b_0(\delta) + b_1(\delta)\rho + b_2(\delta)\rho^2}{1 - \alpha_3(\delta) - \beta_3(\delta)\rho}.$$

Thus, the fundamental majorant equation $h(\delta, \rho) = \rho$ for determining the non-local bound $\rho = \rho(\delta)$ for δ_X is quadratic:

$$d_2(\delta)\rho^2 - d_1(\delta)\rho + d_0(\delta) = 0, \tag{15.38}$$

where

$$\begin{aligned} d_0(\delta) &:= b_0(\delta) + a_0(\delta)(1 - \alpha_3(\delta)), \\ d_1(\delta) &:= a_0(\delta)\beta_3(\delta) + (1 - \alpha_3(\delta))(1 - a_1(\delta)) - b_1(\beta), \\ d_2(\Delta) &:= b_2(\delta) + \beta_3(\delta)(1 - a_1(\delta)). \end{aligned} \tag{15.39}$$

Suppose that $\delta \in \Omega$, where

$$\Omega := \left\{\delta \succeq 0 : 2\sqrt{d_0(\delta)d_2(\delta)} \leq d_1(\delta)\right\} \subset \mathbb{R}_+^r. \tag{15.40}$$

Note that the inclusion $\delta \in \Omega$ guarantees that inequality (15.35) is fulfilled. Then equation (15.38) has non-negative roots $\rho_1(\rho) \leq \rho_2(\delta)$,

$$\rho_1(\delta) = f(\delta) := \frac{2d_0(\delta)}{d_1(\delta) + \sqrt{d_1^2(\delta) - 4d_0(\delta)d_2(\delta)}}. \tag{15.41}$$

Hence, the operator $\Phi(\cdot, \delta P)$ maps the closed convex ball

$$\mathcal{B}(\delta) := \{Z \in \mathbb{F}^{m\times n} : \|Z\|_{\mathrm{F}} \leq f(\delta)\} \subset \mathbb{F}^{m\times n}$$

into itself. According to the Schauder fixed point principle there exists a solution $\delta X \in \mathcal{B}(\delta)$ of equation (15.27), for which

$$\delta_X = \|\delta X\|_{\mathrm{F}} \leq f(\delta),\ \delta \in \Omega. \tag{15.42}$$

If $\delta \in \Omega_1$, where

$$\Omega_1 := \left\{ \delta \succeq 0 : 2\sqrt{d_0(\delta) d_2(\delta)} < d_1(\delta) \right\} \subset \Omega,$$

then $\rho_1(\delta) < \rho_2(\delta)$ and the operator $\Phi(\cdot, \delta P)$ is a contraction on $\mathcal{B}(\delta)$. Hence, the solution δX, for which the estimate (15.42) holds true, is unique. This means that the perturbed equation has an isolated solution $X + \delta X$, where the elements of δX are analytical functions of the elements of δP.

As a result of the non-local perturbation analysis, presented above, we have the perturbation bound (15.40)–(15.42), where the involved quantities are determined via the relations (15.30)–(15.31), (15.37) and (15.39).

15.5 Notes and references

Local and nonlocal perturbation bounds for general fractional-affine matrix equations have been derived in [130]. The problem of existence of solutions of certain classes of nonsymmetric matrix quadratic equations is addessed in [217].

Chapter 16

Symmetric fractional-affine equations

16.1 Introductory remarks

Symmetric fractional-affine equation in general form may be described as

$$Q + \mathcal{L}_0(X) + \sum_{i=1}^{k} \mathcal{S}_i(X)\mathcal{L}_i^{-1}(X)\mathcal{S}_i^*(X^*) = 0,$$

where Q is symmetric ($Q^* = Q$), $\mathcal{L}_0, \mathcal{L}_1, \ldots \mathcal{L}_k$ are Lyapunov operators and $\mathcal{S}_i$ are Sylvester operators (see Appendices F and E).

The corresponding general perturbation results are cumbersome and we shall not give them here. Instead, we shall consider two important classes of symmetric fractional-affine equations: the descriptor discrete-time Riccati equation arising in the theory of optimal control and filtering, and a special equation arising in some applications.

16.2 Discrete-time Riccati equations

16.2.1 Statement of the problem

In this section we present perturbation bounds for the decriptor discrete-time matrix Riccati equations arising in the control and filtering of linear multivariable systems. Both real and complex equations are considered. We derive condition numbers, first order local bounds and nonlinear nonlocal bounds.

We note that a complete perturbation analysis for the descriptor discrete-time Riccati equation has not been published up to the moment.

We use the notations $\mathcal{F}_4 = \mathbb{F}^{n\times n} \times \mathbb{F}^{n\times n} \times \mathbb{F}^{n\times n} \times \mathbb{F}^{n\times n}$ and accordingly $\mathcal{R}_4 = \mathbb{R}^{n\times n} \times \mathbb{R}^{n\times n} \times \mathbb{R}^{n\times n} \times \mathbb{R}^{n\times n}$, $\mathcal{C}_4 = \mathbb{C}^{n\times n} \times \mathbb{C}^{n\times n} \times \mathbb{C}^{n\times n} \times \mathbb{C}^{n\times n}$.

16.2.2 Motivating example

Consider the stabilizable and detectable discrete-time control system

$$\begin{aligned} Ex(t+1) &= Ax(t) + Bu(t),\ t = 0, 1, \ldots,\ x(0) = x_0, \\ y(t) &= Cx(t), \end{aligned} \tag{16.1}$$

where $x(t) \in \mathbb{F}^n$, $u(t) \in \mathbb{F}^m$ and $y(t) = \mathbb{F}^r$ are the state, control and output vectors, respectively, and $E, A \in \mathbb{F}^{n\times n}$, $B \in \mathbb{F}^{n\times m}$, $C \in \mathbb{F}^{r\times n}$ are constant matrices. It is supposed that the matrix E is nonsingular but may be ill-conditioned with respect to inversion. The system is real or complex if the underlying field $\mathbb{F}$ is $\mathbb{R}$ or $\mathbb{C}$. Accordingly, we use A^* to denote $A^\top$ in the real case and A^{H} in the complex case.

We recall that the system (16.1), or the pair $[E^{-1}A, E^{-1}B)$, is *stabilizable* if there exists a gain matrix $H \in \mathbb{F}^{m\times n}$ such that the closed-loop system matrix $E^{-1}(A + BH)$ is *convergent*, i.e., has its spectrum in the central open unit disc in the complex plane. The system (16.1), or the pair $(C, E^{-1}A]$, is *detectable* if the pair $[A^*E^{-*}, C^*)$ is stabilizable. Systems of type (16.1), or triples $(C, E^{-1}A, E^{-1}B)$, that are both stabilizable and detectable are called *regular*.

Let the quadratic performance index

$$J(u, x_0) := \sum_{t=0}^{\infty} (y^*(t)y(t) + u^*(t)u(t)) \to \min$$

be given. The control sequence $u = \{u(t)\}$, $t = 0, 1, \ldots$, that minimizes the quantity $J(u, x_0)$ for each initial state $x_0 \in \mathbb{F}^n$ can be realized in the form of a state feedback

$$u(t) = -(I_m + B^*X_0B)^{-1}B^*X_0Ax(t),$$

where $X_0 = X_0^* \geq 0$ is the solution of the descriptor discrete-time Riccati equation

$$E^*XE = C^*C + A^*XA - A^*XB(I_m + B^*XB)^{-1}B^*XA. \tag{16.2}$$

In this case $J(u, x_0) = x_0^* X_0 x_0$.

The closed-loop system is described by the equation $x(t+1) = E^{-1}A_0x(t)$, where

$$A_0 := A - B(I_m + B^*X_0B)^{-1}B^*X_0A$$

and the matrix $E^{-1}A_0$ is convergent.

Matrix Riccati equations of this type arise also in other areas of control and filtering theory for discrete-time linear systems.

16.2.3 Statement of the problem

Recall that for arbitrary matrices U, V over $\mathbb{F}$ such that the products UV and VU are defined, the nonzero eigenvalues of UV and VU coincide (the eigenvalues are counted according to their algebraic multiplicity). Hence for $B \in \mathbb{F}^{n\times m}$ and $S := BB^*$ the spectra of $I_n + SX$ and $I_n + XS$ coincide since $\mathrm{spect}(I_n + M) = 1 + \mathrm{spect}(M)$ for an arbitrary matrix $M \in \mathbb{F}^{n\times n}$. Moreover, for $m \leq n$ the eigenvalues of SX are those of B^*XB plus $n - m$ zero eigenvalues. In particular the matrices $I_n + SX$ and $I_m + B^*XB$ are simultaneously singular (if B^*XB has an eigenvalue -1) or nonsingular. Suppose further on that $-1 \notin \mathrm{spect}(B^*XB)$.

Using the identities

$$\begin{aligned}(I_n + SX)^{-1} &= I_n - B(I_m + B^*XB)^{-1}B^*X,\\ (I_n + XS)^{-1} &= I_n - XB(I_m + B^*XB)^{-1}B^*\end{aligned}$$

we can rewrite equation (16.2) in the equivalent form

$$R(P, X) := E^*XE - Q - A^*X(I_n + SX)^{-1}A = 0, \tag{16.3}$$

where $P := (Q, E, A, S) \in \mathcal{F}_4$, $X \in \mathbb{F}^{n\times n}$. We also have the form

$$E^*XE - Q - A^*(I_n + XS)^{-1}XA = 0.$$

It follows from the regularity of the system that equation (16.3) has a unique symmetric (in the sense $X_0 = X_0^*$) nonnegative stabilizing solution X_0. At the same time the Riccati equation may have other solutions (which necessarily are not nonnegative and not stabilizing), including nonsymmetric ones. Note also that

$$A_0 := A - B(I_m + B^*X_0B)^{-1}B^*X_0A = (I_n + SX_0)^{-1}A.$$

In many applications the systems under considerations are real ($\mathbb{F} = \mathbb{R}$) and the corresponding equation is also real,

$$\begin{aligned}&E^\top XE - Q - A^\top X(I_n + SX)^{-1}A = 0, \\ &P := (Q, E, A, S) \in \mathcal{R}_4,\ X \in \mathbb{R}^{n\times n}.\end{aligned} \tag{16.4}$$

In the complex case $\mathbb{F} = \mathbb{C}$ the descriptor equation is

$$\begin{aligned}&E^\mathrm{H}XE - Q - A^\mathrm{H}X(I_n + SX)^{-1}A = 0, \\ &P := (Q, E, A, S) \in \mathcal{C}_4,\ X \in \mathbb{C}^{n\times n}.\end{aligned} \tag{16.5}$$

The real and complex cases are treated similarly as a whole with one exception. In calculating condition numbers and constructing first order estimates the technique from [147, 139] must be used which is based on the theory of additive complex operators. The reason is that the function $A \mapsto A^\mathrm{H}$ is not linear (it is additive but not homogeneous).

Consider equation (16.3) under the assumption that is has a solution $X_0 = X_0^*$ such that the linear matrix operator $\mathcal{L} : \mathbb{F}^{n\times n} \to \mathbb{F}^{n\times n}$, defined by

$$\mathcal{L}(Z) = E^* Z E - A_0^* Z A_0, \ Z \in \mathbb{F}^{n\times n},$$

is invertible. The eigenvalues of $\mathcal{L}$ are the eigenvalues of its matrix

$$L := E^\top \otimes E^* - A_0^\top \otimes A_0^* \in \mathbb{F}^{n^2\times n^2}.$$

We recall that the matrix L of a linear matrix operator $\mathcal{L}$ is defined by the relation $\mathrm{vec}(\mathcal{L}(Z)) = L\mathrm{vec}(Z)$ for all Z.

It is easy to show that $\mathcal{L}$ is invertible if and only if $\lambda_{ij} := \lambda_i(E^{-1}A_0)\overline{\lambda}_j(E^{-1}A_0) - 1 \neq 0$, $i, j = 1, \ldots, n$. Indeed, the matrix L is invertible if and only if so is the matrix

$$L(E^\top \otimes E^*)^{-1} = I_{n^2} - \left(E^{-1}A_0\right)^\top \otimes \left(E^{-1}A_0\right)^*. \tag{16.6}$$

But the eigenvalues of the matrix (16.6) are exactly the numbers λ_{ij}, $i, j = 1, \ldots, n$.

Note that if $Q, S \geq 0$ and the triple $(Q, E^{-1}A, E^{-1}SE^{-*})$ is regular then there is a (unique) stabilizing solution $X_0 \geq 0$ such that the matrix $E^{-1}A_0$ is convergent and hence the operator $\mathcal{L}$ is invertible. This latter case is interesting from point of view of applications but the perturbation analysis given below holds also under the weaker assumption that only a solution $X_0 = X_0^*$ with $\mathcal{L}$ invertible exists.

Let the matrix coefficients in (16.3) be subject to perturbations $Q \mapsto Q + \delta Q$, $E \mapsto E + \delta E$, $A \mapsto A + \delta A$, $S \mapsto S + \delta S$. If $Q = C^*C$ and $S = BB^*$ and C, B are perturbed as $C \mapsto C + \delta C$, $B \mapsto B + \delta B$, then the perturbations $\delta Q = C^*\delta C + \delta C^* C + \delta C^* \delta C$, $\delta S = B\delta B^* + \delta B B^* + \delta B \delta B^*$ are also symmetric (here symmetry means $Q = Q^*$, etc.).

The analysis given below is different for symmetric and nonsymmetric perturbations in the matrices Q and S. We shall consider symmetric perturbations only. The nonsymmetric case can be treated by the scheme proposed in [153].

The aim of the norm-wise perturbation analysis is to find computable bounds for the norm

$$\delta_X := \|\delta X\|_{\mathrm{F}}$$

of the perturbation in the solution X_0 as a function of the *perturbation vector*

$$\delta := [\delta_1, \delta_2, \delta_3, \delta_4]^\top := [\delta_Q, \delta_E, \delta_A, \delta_S]^\top \in \mathbb{R}_+^4$$

whose elements $\delta_Q := \|\delta Q\|_{\mathrm{F}}$, $\delta_E = \|\delta E\|_{\mathrm{F}}$, $\delta_A := \|\delta A\|_{\mathrm{F}}$, $\delta_S := \|\delta S\|_{\mathrm{F}}$ are the Frobenius norms of the perturbations in the data matrices Q, E, A, S.

Having a perturbation estimate

$$\delta_X \leq f(\delta)$$

in absolute perturbations $\delta_Z = \|\delta Z\|_{\mathrm{F}}$, a perturbation bound in relative perturbations

$$\varepsilon_Z := \frac{\delta_Z}{\|Z\|_{\mathrm{F}}}, \ Z \neq 0,$$

is straightforward, namely

$$\varepsilon_X \leq \frac{f(D\varepsilon)}{\|X\|_{\mathrm{F}}},$$

where $\varepsilon := [\varepsilon_Q, \varepsilon_A, \varepsilon_E, \varepsilon_S] \in \mathbb{R}^4_+$, $D := \mathrm{diag}(\|Q\|_{\mathrm{F}}, \|A\|_{\mathrm{F}}, \|E\|_{\mathrm{F}}, \|S\|_{\mathrm{F}})$.

16.2.4 Perturbed equation

General case

The *perturbed equation* is obtained from (16.3) replacing the nominal value $P = (Q, E, A, S)$ of the collection of data matrices with

$$P + \delta P = (Q + \delta Q, E + \delta E, A + \delta A, S + \delta S),$$

namely

$$R(P + \delta P, Y) = 0. \tag{16.7}$$

A priori it is not clear whether the perturbed equation (16.7) has a solution with the required properties. So we shall assume that a solution to (16.7) exists for the given δP. However, from the nonlinear perturbation analysis presented below we shall find conditions for solvability of equation (16.7), see also Chapter 13.

Setting $Y = X_0 + \delta X$ we may rewrite (16.7) as an equivalent equation for the perturbation δX in X_0.

The construction of the equivalent perturbed equation is based on the following scheme, described in Chapter 13.

Suppose that the linear operator $R_X(P, X_0)$ is invertible, where $R(P, X_0) = 0$. Then we may rewrite the perturbed equation

$$R(P + \delta P, X_0 + \delta X) = 0$$

as

$$\delta X = -R_X^{-1}(P, X_0) \circ R_P(P, X_0)(\delta P) - R_X^{-1}(P, X_0) \circ \mathcal{R}(P, X_0)(\delta P, \delta X). \tag{16.8}$$

Note that $R_P(P, X_0)(0) = 0$ and $\mathcal{R}(P, X_0)(0, 0) = 0$. This guarantees that for small δP equation (16.8) has a small solution δX in the sense that

$$\delta X = -R_X^{-1}(P, X_0) \circ R_P(P, X_0)(\delta P) + O(\|\delta P\|^2), \ \delta P \to 0.$$

Further on we shall abbreviate $R_X(P, X_0)$ as R_X, etc., omitting the dependence on the fixed quantities P, X_0 whenever appropriate. We shall also write the unperturbed solution as X.

Real case

In the real case $\mathcal{L} := R_X$ is a real Lyapunov operator $\mathbb{R}^{n\times n} \to \mathbb{R}^{n\times n}$, defined by

$$\mathcal{L}(Z) = E^\top Z E - A_0^\top Z A_0, \; Z \in \mathbb{R}^{n\times n},$$

and has a matrix

$$L = E^\top \otimes E^\top - A_0^\top \otimes A_0^\top \in \mathbb{R}^{n^2\times n^2}.$$

We also have

$$\begin{aligned} R_Q(Z) &= -Z, \; R_E(Z) = Z^\top X E + E^\top X Z, \\ R_A(Z) &= -Z^\top X A_0 - A_0^\top X Z, \; R_S(Z) = A_0^\top X Z X A_0 \end{aligned}$$

and hence

$$R_P(\delta P) = -\delta Q + \delta E^\top X E + E^\top X \delta E - \delta A^\top X A_0 - A_0^\top X \delta A + A_0^\top X \delta S X A_0.$$

Thus we can write equation (16.8) as

$$\delta X = \Pi(\delta P, \delta X) = \Pi_1(\delta P) + \Pi_2(\delta P, \delta X), \tag{16.9}$$

where $\Pi_1(\delta P) := -\mathcal{L}^{-1} \circ R_P(U_1(\delta P))$, $\Pi_2(\delta P, \delta X) := -\mathcal{L}^{-1}(\mathcal{R}(\delta P, \delta X))$.

Set

$$M := I_n + SX, \; \widetilde{M} := M + H, \; H := SZ + \delta S(X + Z) \tag{16.10}$$

and

$$N := M^{-1} - \widetilde{M}^{-1} = M^{-1} H \widetilde{M}^{-1} = \widetilde{M}^{-1} H M^{-1}. \tag{16.11}$$

Note also that

$$A^\top X N = A_0^\top X H \widetilde{M}^{-1}, \; NA = \widetilde{M}^{-1} H A_0. \tag{16.12}$$

Using the inequalities (16.10)–(16.12), the term

$$\mathcal{R}(\delta P, Z) = R(P + \delta P, X + Z) - R(P, X) - R_X(Z) - R_P(\delta P)$$

can be written in the form

$$\begin{aligned} \mathcal{R}(\delta P, Z) \; := \; & A_0^\top X \delta S Z A_0 - A_0^\top X H \widetilde{M}^{-1} H A_0 \\ & + \delta E Z E + E^\top Z \delta E + \delta E^\top X \delta E + \delta E^\top Z \delta E \\ & - A^\top Z M^{-1} \delta A - \delta A Z A_0 - \delta A^\top X M^{-1} \delta A - \delta A^\top Z M^{-1} \delta A \\ & + \delta A^\top (X + Z) N \delta A + A^\top (X + Z) N \delta A \\ & + A^\top Z N A + \delta A^\top (X + Z) A. \end{aligned}$$

We shall rewrite (16.9) in a vector form. Denote

$$\begin{aligned} \xi &:= \operatorname{vec}(\delta X), \; \Delta_1 := \operatorname{vec}(\delta Q), \; \Delta_2 := \operatorname{vec}(\delta E), \; \Delta_3 := \operatorname{vec}(\delta A), \quad (16.13) \\ \Delta_4 &:= \operatorname{vec}(\delta S) \in \mathbb{R}^{n^2}, \; \Delta := \operatorname{vec}(\delta P) = [\Delta_1^\top, \Delta_2^\top, \Delta_3^\top, \Delta_4^\top]^\top \in \mathbb{R}^{4n^2}. \end{aligned}$$

We have

$$\xi = \pi(\Delta, \xi) := \pi_1(\Delta) + \pi_2(\Delta, \xi), \tag{16.14}$$

where $\pi_1(\Delta) := L^{-1}\mathrm{vec}(U_1(\delta P))$, $\pi_2(\Delta, \xi) := L^{-1}\mathrm{vec}(U_2(\delta P, \delta X))$.

After some computations we obtain

$$\pi_1(\Delta) = M_1\Delta_1 + M_2\Delta_2 + M_3\Delta_3 + M_4\Delta_4,$$

where

$$\begin{aligned} M_1 &:= L^{-1},\ M_2 := L^{-1}(I_{n^2} + P_{n^2})\left(I_n \otimes E^\top X\right), \\ M_3 &:= -\ L^{-1}(I_{n^2} + P_{n^2})\left(I_n \otimes A_0^\top X\right), \\ M_4 &:= -\ L^{-1}(A_0^\top X \otimes A_0^\top X). \end{aligned} \tag{16.15}$$

For $\delta_X \leq \rho$ the Frobenius norm of $\Pi_2(\delta P, \delta X)$ can be estimated as

$$\|\Pi_2(\delta P, \delta X\|_{\mathrm{F}} \leq \alpha_0 + a_1\rho + \frac{b_0 + b_1\rho + b_2\rho^2}{c_0 - c_1\rho},$$

where $c_0 > 0$ is a constant and the coefficients α_0, a_1, b_i and c_1 are nondecreasing nonnegative functions of δ with $\alpha_0(0) = a_1(0) = b_0(0) = b_1(0) = 0$.

We do not present here the coefficients in explicit form since this can be done immediately using the expression for $\vec{(\mathcal{L}(\mathcal{R}))} = L\vec{(\mathcal{R})}$.

Complex case

In the complex case of equation (16.5) there are certain modifications. For properties of nonlinear complex additive operators see Chapter 13. We recall that for complex $m \times n$ matrices $G = G_0 + \imath G_1$, $H = H_0 + \imath H_1$ we set

$$\Theta(G, H) := \begin{bmatrix} G_0 + H_0 & H_1 - G_1 \\ G_1 + H_1 & G_0 - H_0 \end{bmatrix}.$$

We may define the *real version* $z^{\mathbb{R}} \in \mathbb{R}^{2n}$ of the vector $z \in \mathbb{C}^n$ as

$$z^{\mathbb{R}} := \begin{bmatrix} z_0 \\ z_1 \end{bmatrix} \in \mathbb{R}^{2n}.$$

This gives $(Gz)^{\mathbb{R}} := G^{\mathbb{R}}z^{\mathbb{R}} \in \mathbb{R}^{2m}$, where

$$G^{\mathbb{R}} := \begin{bmatrix} G_0 & -G_1 \\ G_1 & G_0 \end{bmatrix} \in \mathbb{R}^{2m\times 2n}$$

is the real version of G. Note that

$$(Gz + H\overline{z})^{\mathbb{R}} = \Theta(G, H)z^{\mathbb{R}}$$

and $\Theta(G, 0) = G^{\mathbb{R}}$.

The complex Lyapunov operator

$$\mathcal{K} := R_X : \mathbb{C}^{n\times n} \to \mathbb{C}^{n\times n}$$

acts as

$$\mathcal{K}(Z) = E^{\mathrm{H}} Z E - A_0^{\mathrm{H}} Z A_0,\ \ Z \in \mathbb{C}^{n\times n},$$

and has a matrix

$$K = E^\top \otimes E^{\mathrm{H}} - A_0^\top \otimes A_0^{\mathrm{H}}.$$

Due to the invertibility of $\mathcal{K}$ we have

$$\delta X = \Psi(\delta P, \delta X) := \Psi_1(\delta P) + \Psi_2(\delta P, \delta X), \tag{16.16}$$

where

$$\Psi_1(\delta P) := \mathcal{K}^{-1}(V_1(\delta P)),\ \ \Psi_2(\delta P, \delta X) := \mathcal{K}^{-1}(V_2(\delta P, \delta X)).$$

The expressions V_i are similar to U_i in the real case, with transposition replaced by Hermitian conjugation.

As in the real case we rewrite the equivalent operator equation (16.16) in a vector form

$$\xi = \psi(\Delta, \xi) := \psi_1(\Delta) + \psi_2(\Delta, \xi).$$

Here we have used the substitutions (16.13) having in mind that $\xi, \Delta_i \in \mathbb{C}^{n^2}$ and $\Delta \in \mathbb{C}^{4n^2}$. We have $\psi_i := K^{-1}\mathrm{vec}(\Psi_i)$. In particular

$$\begin{aligned}
\psi_1(\Delta) &= N_1\mu_1 + N_{21}\mu_2 + N_{22}\overline{\mu}_2 + N_{31}\mu_3 + N_{32}\overline{\Delta}_3 + N_4\Delta_4, \qquad (16.17)\\
N_1 &:= K^{-1},\ N_{21} := K^{-1}\left(I_n \otimes A_0^{\mathrm{H}} X\right),\\
N_{22} &:= K^{-1}\left(A_0^\top \overline{X} \otimes I_n\right) P_{n^2},\ N_{31} := -K^{-1}\left(I_n \otimes E^{\mathrm{H}} X\right),\\
N_{32} &:= -K^{-1}\left(E^\top \overline{X} \otimes I_n\right) P_{n^2},\ N_4 := -K^{-1}\left(A_0^\top \overline{X} \otimes A_0^{\mathrm{H}} X\right).
\end{aligned}$$

16.2.5 Condition numbers and local bounds

In this section we give condition numbers and derive local first order bounds for the perturbation $\delta_X = \|\delta X\|_{\mathrm{F}}$ in the solution X of the descriptor Riccati equation (16.3).

Real equation

Based on (16.14) we get

$$\xi = \pi_1(\Delta) + O(\|\Delta\|^2 + \|\xi\|^2),\ \ \|\Delta\| + \|\xi\| \to 0.$$

Since $\|\xi\| = O(\|\Delta\|)$, $\Delta \to 0$, this is equivalent to

$$\xi = M_1\Delta_1 + M_2\Delta_2 + M_3\Delta_3 + M_4\Delta_4 + O(\|\Delta\|^2),\ \ \Delta \to 0.$$

Hence, using the fact that $\delta_X = \|\xi\|_2$, and having in mind (16.15), we see that the following result is valid.

Theorem 16.1 *In Frobenius norm the absolute condition numbers C_Z for the solution X of the real equation (16.4) relative to the matrix coefficients $Z = Q$, E, A, S are*

$$C_Q = \|M_1\|_2, \ C_E = \|M_2\|_2, \ C_A = \|M_3\|_2, \ C_S = \|M_4\|_2.$$

The determination of the relative condition numbers $c_Z := C_Z\|Z\|_{\mathrm{F}}/\|X\|_{\mathrm{F}}$ is straightforward provided $Z, X \neq 0$.

Except condition number based estimates we also have

$$\delta_X \leq \mathrm{est}_2(\delta) + O(\|\delta\|^2) := \|[M_1, M_2, M_3, M_4]\|_2\|\delta\|_2 + O(\|\delta\|^2), \ \delta \to 0,$$

where the matrices M_i are displayed in (16.15).

Another perturbation bound is [133]

$$\delta_X \leq \mathrm{est}_3(\delta) + O(\|\delta\|^2) := \sqrt{\delta^\top M_0 \delta} + O(\|\delta\|^2), \ \delta \to 0,$$

where

$$M_0 := [m_{ij}] \in \mathbb{R}_+^{4\times 4}, \ m_{ij} := \left\|M_i^\top M_j\right\|_2, \ i, j = 1, 2, 3, 4.$$

We stress that the matrix M_0 may not be nonnegative definite, i.e., it may have negative eigenvalues. At the same time $\delta^\top M_0 \delta \geq 0$ for $\delta \in \mathbb{R}_+^4$.

The bounds est_2 and est_3 are *alternative* since both inequalities $\mathrm{est}_2(\delta) \leq \mathrm{est}_3(\delta)$ and $\mathrm{est}_2(\delta) > \mathrm{est}_3(\delta)$ are possible.

Thus we see that the following theorem is valid.

Theorem 16.2 *The perturbation δ_X in the solution X of the real equation (16.4) satisfies the local perturbation estimate*

$$\delta_X \leq \mathrm{est}(\delta) + O(\|\delta\|^2), \ \delta \to 0,$$

where

$$\mathrm{est}(\delta) := \min\{\mathrm{est}_2(\delta), \mathrm{est}_3(\delta)\}.$$

Complex equation

Consider the complex descriptor Riccati equation (16.5). We give only the final results.

Theorem 16.3 *In Frobenius norm the absolute condition numbers $K_Z S$ for the solution X of the complex equation (16.5) relative to the matrix coefficients Q, E, A, S are*

$$\begin{aligned} K_Q &= \|N_1\|_2, \ K_E = \|\Theta(N_{21}, N_{22})\|_2, \\ K_A &= \|\Theta(N_{31}, N_{32})\|_2, \ K_S = \|N_4\|_2. \end{aligned}$$

If $Z, X \neq 0$ the relative condition number for X relative to Z is $k_Z := K_Z\|Z\|_{\rm F}/\|X\|_{\rm F}$.

Define the matrices

$$T_1 := N_1^{\mathbb{R}},\ T_2 := \Theta(N_{21}, N_{22}),\ T_3 := \Theta(N_{31}, N_{32}),\ T_4 := N_4^{\mathbb{R}}$$

from $\mathbb{R}^{2n^2\times 2n^2}$ and let

$$T := [t_{ij}] \in \mathbb{R}_+^{4\times 4},\ t_{ij} := \left\|T_i^{\rm H} T_j\right\|_2,\ i,j = 1,2,3,4.$$

As in the real case, set

$$\mathrm{Est}_2(\delta) := \|[T_1, T_2, T_3, T_4]\|_2\, \|\delta\|_2,\ \mathrm{Est}_3(\delta) := \sqrt{\delta^\top T \delta}.$$

Then we have the following result.

Theorem 16.4 *The perturbation δ_X in the solution X of the complex equation (16.5) satisfies the local perturbation estimate*

$$\delta_X \leq \mathrm{Est}(\delta) + O(\|\delta\|^2),\ \delta \to 0,$$

where

$$\mathrm{Est}(\delta) := \min\left\{\mathrm{Est}_2(\delta), \mathrm{Est}_3(\delta)\right\}.$$

The bounds given in Theorems 16.2 and 16.4 have the properties of the similar bounds in the real case. In particular they may be very accurate.

Suppose that the perturbations in the coefficient matrices $Z \in \{Q, E, A, S\}$ satisfy $\|\delta Z\|_{\rm F} = \varepsilon\|Z\|_{\rm F}$ for some $\varepsilon > 0$. Let $d_i(P, X)$ be the quantity $\mathrm{est}_i(|\!|\!|P|\!|\!|)/\|X\|_{\rm F}$ in the real case or $\mathrm{Est}_i(|\!|\!|P|\!|\!|)/\|X\|_{\rm F}$ in the complex case, $i = 2, 3$. The quantities $\varepsilon d_i(P, X)\varepsilon$ are first order bounds for the relative perturbation $\delta_X/\|X{\rm F}$ in the solution X. Thus we may define the overall relative conditioning of X as

$$d(P, X) := \min\{d_2(P, X), d_3(P, X)\}.$$

16.2.6 Nonlocal bounds

The local estimates from Theorems 16.2 and 16.4 are valid for asymptotically small perturbations.

To avoid the disadvantages of the local bounds one can apply the methods of nonlocal perturbation analysis. As a result one gets nonlocal (and in general nonlinear) perturbation bounds of the form $\delta_X \leq f(\delta)$ for $\delta \in \Omega$, where Ω is a certain domain in the space of the norms of the perturbations in the coefficient matrices, see Chapter 13. Here the inclusion $\delta \in \Omega$ guarantees that the perturbed equation (16.7) indeed has a solution $Y = X + \delta X$ for which the bound $\delta_X \leq f(\delta)$ holds true.

The nonlocal perturbation analysis is based on the techniques of Lyapunov majorants and fixed point principles.

Real equation

Set $a_0(\delta) := \alpha_0(\delta) + \mathrm{est}(\delta)$. Then we have the Lyapunov majorant

$$h(\delta, \rho) = a_0(\delta) + a_1(\delta) + \frac{b_0(\delta) + b_1(\delta)\rho + b_2(\delta)\rho^2}{c_0 - c_1(\delta)\rho}.$$

For $a_1(\delta) < 1$ the majorant equation $\rho = h(\delta, \rho)$ reduces to the quadratic equation

$$m_2(\delta)\rho^2 - m_1(\delta)\rho + m_0(\delta) = 0,$$

where

$$\begin{aligned} m_2(\delta) &:= b_2(\delta) + c_1(\delta)(1 - a_1(\delta)), \\ m_1(\delta) &:= a_0(\delta)c_1(\delta) + c_0(1 - a_1(\delta)), \\ m_0(\delta) &:= b_0(\delta) + c_0 a_0(\delta). \end{aligned}$$

Thus we come to the following statement.

Theorem 16.5 *Let δ is small enough in order to satisfy the inequality*

$$m_1(\delta) \geq 2\sqrt{m_0(\delta)m_2(\delta)}.$$

Then the nonlocal bound

$$\delta_X \leq f(\delta) := \frac{2m_0(\delta)}{m_1(\delta) + \sqrt{m_1^2(\delta) - 4m_0(\delta)m_2(\delta)}}$$

ia valid for the perturbation δX in the solution X.

16.2.7 Complex equation

In the complex case we have similar nonlocal result with some differences, e.g. $\mathcal{L}$ is replaced by $\mathcal{K}$, $\mathrm{est}(\delta)$ – by $\mathrm{Est}(\delta)$, and the transposition – by Hermitian conjugation.

16.2.8 An alternative approach

An alternative approach to the construction of Lyapunov majorant for the operator equation is given in [153]. It is based on the following considerations. Suppose that the matrices $S + \delta S$ and $Q + \delta Q$ are symmetric and nonnegative definite and that the perturbed system is regular. Then the perturbed equation has a nonnegative definite solution $Y = X + \delta X$. It is shown in [153] that for every nonnegaive definite $S + \delta S$ and Y one has

$$\begin{aligned} &\|Y(I_n + SY)^{-1}\|_{\mathrm{F}} = \|(I_n + YS)^{-1}Y\| \leq \|Y\|, \qquad (16.18) \\ &\|Y(I_n + (S + \delta S)Y)^{-1}\|_{\mathrm{F}} \leq \|Y\|, \\ &\|(I_n + SY)^{-1}S\| \leq \|S\| \end{aligned}$$

for both the spectral and Frobenius norms. Indeed, the first inequality is obvious if Y is nonsingular since $Y(I_n+SY)^{-1} = (Y^{-1}+S)^{-1}$ and $Y^{-1}+S \geq Y^{-1}$. This gives $(Y^{-1}+S)^{-1} \leq Y$. If Y is singular we may consider the matrix $Y(\mu) := Y + \mu I_n$, $\mu > 0$, and pass to the limit $\mu \to 0$.

Consider for simplicity only the real case. The complex case is treated similarly using the expressions for the induced norms of additive operators.

We can rewrite the perturbed equation as

$$R(P,Y) + R(P+\delta P, Y) - R(P,Y) = 0. \tag{16.19}$$

Furthermore we have

$$R(P,Y) = R(P,X) + R_X(\delta X) + \mathcal{B}(\delta P), \tag{16.20}$$

where

$$\mathcal{B}(Z) := A_0^\top Z(I_n + S(X+Z))^{-1} SZA_0. \tag{16.21}$$

In turn, we have

$$\begin{aligned} R(P+\delta P, Y) - R(P,Y) &= A^\top Y(I_n + SY)^{-1}\delta SY(I_n + (S+\delta S)Y)^{-1}A \\ &- A^\top Y(I_n + (S+\delta S)Y)^{-1}\delta A - \delta A^\top Y(I_n + (S+\delta S)Y)^{-1}A \\ &- \delta A^\top Y(I_n + (S+\delta S)Y)^{-1}\delta A - \delta Q \\ &+ \delta E^\top YE + E^\top Y\delta E + \delta E^\top Y\delta E. \end{aligned} \tag{16.22}$$

It follows from (16.21), (16.22) and(16.18) that

$$\begin{aligned} \|\mathcal{L}^{-1}\mathcal{B}(Z)\|_{\mathrm{F}} &\leq \left\|L^{-1}\left(A_0^\top \otimes A_0^\top\right)\right\|_2 \|Z(I_n + S(X+Z))^{-1}SZ\|_{\mathrm{F}} \\ &\leq \left\|L^{-1}\left(A_0^\top \otimes A_0^\top\right)\right\|_2 \|S\|_2\|Z\|_{\mathrm{F}}^2 \end{aligned} \tag{16.23}$$

and

$$\begin{aligned} &\|\mathcal{L}^{-1}(R(P+\delta P,Y) - R(P,Y))\|_{\mathrm{F}} \leq \|L^{-1}\|_2(\delta_Q + (\|X\|_2 + \delta_X)\delta_E^2) \\ &+ \left\|L^{-1}(I_{n^2} + P_{n^2})\left(I_n \otimes E^\top\right)\right\|_2 \delta_E(\|X\|_2 + \delta_X) \\ &+ \|L^{-1}(A^\top \otimes A^\top)\|_2\delta_S(\|X\|_2 + \delta_X)^2 \\ &+ \left\|L^{-1}(I_{n^2} + P_{n^2})\left(I_n \otimes A^\top\right)\right\|_2 \delta_A(\|X\|_2 + \delta_X)^2. \end{aligned} \tag{16.24}$$

Relations (16.23) and (16.24) yield the Lyapunov majorant

$$h(\delta,\rho) = \alpha_0(\delta) + \alpha_1(\delta)\rho + \alpha_2(\delta)\rho^2,$$

where

$$\begin{aligned} \alpha_0(\delta) \ &:= \ \|L \in \|_2\delta_Q + \left\|L^{-1}(I_{n^2} + P_{n^2})\left(I_n \otimes E^\top\right)\right\|_2 \|X\|_2\delta_E \\ &+ \ \left\|L^{-1}(I_{n^2} + P_{n^2})\left(I_n \otimes A^\top\right)\right\|_2 \|X\|_2\delta_A \end{aligned}$$

$$\begin{aligned}
&\quad + \|L^{-1}(A^\top \otimes A^\top)\|_2 \|X\|_2^2 \delta_S \\
&\quad + \|L^{-1}\|_2 \|X\|_2 \left(\delta_E^2 + \delta_A^2\right) \\
\alpha_1(\delta) \quad &:= \quad \left\|L^{-1}(I_{n^2} + P_{n^2})\left(I_n \otimes E^\top\right)\right\|_2 \delta_E \\
&\quad + \left\|L^{-1}(I_{n^2} + P_{n^2})\left(I_n \otimes A^\top\right)\right\|_2 \delta_A + 2\|L^{-1}(A^\top \otimes A^\top)\|_2 \delta_S \\
&\quad + \|L^{-1}\|_2 \left(\delta_E^2 + \delta_A^2\right), \\
a_2(\delta) \quad &:= \quad \left\|L^{-1}\left(A_0^\top \otimes A_0^\top\right)\right\|_2 \|S\|_2 + \|L^{-1}(A^\top \otimes A^\top)\|_2 \delta_S.
\end{aligned}$$

Hence we have the following result.

Theorem 16.6 *Let*

$$\delta \in \Gamma := \left\{\alpha_1(\delta) + 2\sqrt{\alpha_0(\delta)\alpha_2(\delta)} \le 1\right\}.$$

Then the nonlocal nonlinear perturbation estimate

$$\delta_X \le g(\delta) := \frac{2\alpha_0(\delta)}{1 - \alpha_1(\delta) + \sqrt{\gamma(\delta)}}$$

is valid, where

$$\gamma(\delta) := \alpha_1^2(\delta) - 4\alpha_0(\delta)\alpha_2(\delta).$$

16.2.9 Numerical example

We shall illustrate the implementation of Theorem 16.6.

Consider a third order standard discrete-time matrix algebraic Riccati equation

$$X - A^\top X(I_3 + SX)^{-1}A - Q = 0$$

with matrices $Q = VQ_0V$, $A = VA_0V$, $S = VS_0V$, where V is an elementary reflection, $V = I_3 - 2vv^\top/3$, $v = [1, 1, 1]^\top$, and

$$Q_0 = \mathrm{diag}(10^k, 1, 10^{-k}), \ A_0 = \mathrm{diag}(0, 10^{-k}, 1), \ S_0 = \mathrm{diag}(10^{-k}, 10^{-k}, 10^{-k})$$

for some positive integer k. The sensitivity of this equation increases with the increasing of k.

Due to the diagonal form of the matrices Q_0, A_0 and S_0, the solution is given by $X = VX_0V$, $X_0 := \mathrm{diag}(x_1, x_2, x_3)$, where

$$x_i = \frac{a_i^2 + q_i s_i - 1 + ((a_i^2 + q_i s_i - 1)^2 + 4q_i s_i)^{1/2}}{2s_i}$$

and q_i, a_i and s_i are the corresponding diagonal elements of Q_0, A_0 and S_0.

The perturbations in the data are taken as $\Delta Q = V\Delta Q_0 V$, $\Delta A = V\Delta A_0 V$, $\Delta S = V\Delta S_0 V$, where

$$\Delta Q_0 = \begin{bmatrix} 10^k & -5 & 7 \\ -5 & 1 & 3 \\ 7 & 3 & 10^k \end{bmatrix} \times 10^{-j},$$

Table 16.1: Exact perturbation, local and nonlocal perturbation bounds for $k = 0$

j	$\|\delta X\|_F/\|X\|_F$	est2	est3	$g(\delta)$
10	7.51×10^{-10}	1.72×10^{-9}	6.21×10^{-9}	1.02×10^{-8}
9	7.51×10^{-9}	1.72×10^{-8}	6.21×10^{-8}	1.02×10^{-7}
8	7.51×10^{-8}	1.72×10^{-7}	6.21×10^{-7}	1.02×10^{-6}
7	7.51×10^{-7}	1.72×10^{-6}	6.21×10^{-6}	1.02×10^{-5}
6	7.51×10^{-6}	1.72×10^{-5}	6.21×10^{-5}	1.02×10^{-4}
5	7.51×10^{-5}	1.72×10^{-4}	6.22×10^{-4}	1.03×10^{-3}
4	7.51×10^{-4}	1.72×10^{-3}	6.27×10^{-3}	1.06×10^{-2}
3	7.51×10^{-3}	1.72×10^{-2}	6.83×10^{-2}	1.70×10^{-1}
2	7.51×10^{-2}	1.72×10^{-1}	1.72×10^{-1}	*

$$\Delta A_0 = \begin{bmatrix} 3 & -4 & 8 \\ -6 & 2 & -9 \\ 2 & 7 & 5 \end{bmatrix} \times 10^{-j}$$

$$\Delta S_0 = \begin{bmatrix} 10^{-k} & -10^{-k} & 2\times 10^{-k} \\ -10^{-k} & 5\times 10^{-k} & -10^{-k} \\ 2\times 10^{-k} & -10^{-k} & 3\times 10^{-k} \end{bmatrix} \times 10^{-j}$$

for $j = 10, 9, ..., 2$.

The perturbation $\|\delta X\|_F$ in the solution is estimated by the local bounds $\mathrm{est}_2(\delta)$, $\mathrm{est}_3(\delta)$ and the nonlocal bound $g(\delta)$.

The cases when the nonlocal estimate is not valid since the existence condition $\delta \in \Gamma$ is violated, are denoted by asterisk.

The results obtained for different values of k and j are shown at Tables 16.1–16.2.

16.3 Symmetric fractional-linear equation

16.3.1 Statement of the problem

In this section we present perturbation bounds for the complex matrix equation

$$F(X, A) := X - A_1 - \sigma A_2^{\mathrm{H}} X^{-1} A_2 = 0, \tag{16.25}$$

where $A_1 \in \mathbb{C}^{n\times n}$ and the solution $X \in \mathbb{C}^{n\times n}$ are symmetric matrices, and $A := (A_1, A_2)$. Real equations of type (16.25) are formally obtained replacing $\mathbb{C}$ by $\mathbb{R}$, and the complex conjugation A_2^{H} – by the transposition $A_2^{\top}$.

Table 16.2: Exact perturbation, local and nonlocal perturbation bounds for $k = 1$

j	$\|\delta X\|_F/\|X\|_F$	est2	est3	$g(\delta)$
10	1.07×10^{-10}	3.26×10^{-9}	2.84×10^{-9}	2.36×10^{-8}
9	1.07×10^{-9}	3.26×10^{-8}	2.84×10^{-8}	2.36×10^{-7}
8	1.07×10^{-8}	3.26×10^{-7}	2.84×10^{-7}	2.36×10^{-6}
7	1.07×10^{-7}	3.26×10^{-6}	2.84×10^{-6}	2.36×10^{-5}
6	1.07×10^{-6}	3.26×10^{-5}	2.84×10^{-5}	2.36×10^{-4}
5	1.07×10^{-5}	3.26×10^{-4}	2.84×10^{-4}	2.39×10^{-3}
4	1.08×10^{-4}	3.26×10^{-3}	2.84×10^{-3}	2.78×10^{-2}
3	1.09×10^{-3}	3.26×10^{-2}	2.84×10^{-2}	*
2	1.17×10^{-2}	3.26×10^{-1}	2.84×10^{-1}	*

First order local bounds and nonlinear nonlocal bounds are derived for equation (16.25) following the general scheme described in this book. The technique used is based on Lyapunov majorants and fixed point principles [137]. The perturbations in the data A and the solution X are estimated in terms of the Frobenius matrix norm $\|\cdot\|_F$. The use of this norm allows to obtain explicit expressions for the individual condition numbers of X relative to perturbations in A_k. The perturbation bounds allow to derive condition and accuracy estimates for the computed solution when a numerically stable algorithm is applied to solve (16.25). To avoid trivial results we assume that $A_2 \neq 0$.

16.3.2 Existence and uniqueness of the solution

We do not consider in detail the problems of existence and (local) uniqueness of the solution of equation (16.25) which may be quite complicated. In particular this equation may have no solutions or may have both symmetric (in the sense $X^H = H$) and nonsymmetric solutions. In turn the solutions may be isolated (or locally unique) or belong to certain algebraic manifolds. An idea of these problems is illustrated in the following low order examples for real equations.

Example 16.7 For $n = 1$ the equation $X = A_1 + A_2^2/X$ with $A_2 \neq 0$ is equivalent to the quadratic equation

$$X^2 - A_1X - A_2^2 = 0$$

and has real solutions

$$\frac{A_1 \pm \sqrt{A_1^2 + 4A_2^2}}{2}.$$

$\diamondsuit$

Example 16.8 Let $n = 2$, $A_1 = 0$ and $A_2 = \operatorname{diag}(1, \omega)$, where $\omega \in \mathbb{R}$ is a parameter. If X is a solution of the equation then

$$\det X = (\det A_2)^2 / \det X = \omega^2 / \det X$$

and hence for $\omega = 0$ the equation has no solution. For $\omega \neq 0$ we have $\det X = \pm\omega$ and the equation has two isolated solutions $X_{1,2} = \pm A_2$, two 1-parametric families of solutions

$$X(t) = \begin{bmatrix} 1 & t \\ 0 & -\omega \end{bmatrix}, -X(t),\ t \in \mathbb{R},$$

and one 2-parametric family of solutions

$$X(t_1, t_2) = \begin{bmatrix} t_1 & \omega(1 - t_1^2)/t_2 \\ t_2 & -\omega t_1 \end{bmatrix},\ t_1 \in \mathbb{R},\ t_2 \in \mathbb{R}\backslash\{0\}.$$

The isolated solutions are symmetric, each 1-parametric family of solutions contains one symmetric matrix (take $t = 0$) and the 2-parametric family of solutions contains a 1-parametric family of symmetric matrices (take $t_1^2 + t_2^2/\omega = 1$). $\diamondsuit$

We assume that equation (16.25) has an (local) unique solution. Conditions for existence and uniqueness of extremal solutions to (E.3) are given in [58] and [63]. Applications of this equation to a number of problems in control theory, networks, dynamic programming, filtering and statistics are considered in [111, 242], while a computational algorithms for its solution is proposed in [171].

In the following the subindexes k, l take values 1,2.

Consider the matrix equation (16.25) under the assumption that for a particular value A^0 of A it has a symmetric solution X^0 such that the partial Fréchet derivative of F in X at the point (X^0, A^0) is invertible. Further on we omit the superindex "0" and denote the matrix parameter and the particular solution as A and X.

The perturbed equation is obtained from (E.3) by replacing a nominal value $A = (A_1, A_2)$ of the collection of data matrices with $A + \delta A = (A_1 + \delta A_1, A_2 + \delta A_2)$:

$$F(X + \delta X, A + \delta A) = 0. \tag{16.26}$$

Let $\delta_k \geq 0$ and suppose that

$$\alpha_k := \|\delta A_k\|_{\mathrm{F}} \leq \delta_k.$$

Set

$$\delta A := (\delta A_1, \delta A_2).$$

For δ_k sufficiently small the perturbed equation (16.26) has a solution

$$\delta X = \Upsilon(\delta A),$$

depending on δA and such that $\Upsilon(0) = 0$. The solution with this property is unique and, moreover, the elements of the matrix valued function Υ are analytic functions of the elements of δA_k.

Denote

$$\xi := \|\delta X\|_{\mathrm{F}}$$

and

$$\delta := [\delta_1, \delta_2]^\top = [\|\delta A_1\|_{\mathrm{F}}, \|\delta A_2\|_{\mathrm{F}}]^\top \in \mathbb{R}_+^2.$$

Then we have

$$\xi \le \Gamma(\delta),$$

where

$$\Gamma(\delta) := \sup\{\|\Upsilon(\delta A)\|_{\mathrm{F}} : \alpha_k \le \delta_k\}.$$

Thus the aim of perturbation analysis is to estimate the quantity $\Gamma(\delta)$ from above since its exact determination is a hopeless task. In particular the local perturbation analysis produces the *individual condition numbers* c_k which are defined by

$$\Gamma(\delta) = c_1\delta_1 + c_2\delta_2 + O(\|\delta\|^2),\ \delta \to 0.$$

In what follows we shall also find a first order homogeneous function $g : \mathbb{R}_+^2 \to \mathbb{R}_+$ such that

$$\Gamma(\delta) \le g(\delta) + O(\|\delta\|^2),\ \delta \to 0,$$

and

$$g(\delta) \le c_1\delta_1 + c_2\delta_2.$$

However, the use of the "chopped" bounds

$$\xi \le c_1\delta_1 + c_2\delta_2$$

or

$$\xi \le g(\delta)$$

may be misleading since for some values of δ the opposite inequalities may in fact hold.

To obtain rigorous perturbation bounds one can use the techniques of nonlocal perturbation analysis. As a result one gets a domain $\Omega \subset \mathbb{R}_+^2$ and a function $f : \Omega \to \mathbb{R}_+$ such that

$$\xi \le f(\delta),\ \delta \in \Omega,$$

where f is nondecreasing in each of its arguments and $f(0) = 0$.

16.3.3 Local perturbation analysis

We may rewrite (16.26) as an equivalent equation for the perturbation δX in X. We have

$$\begin{aligned} F(X+\delta X, A+\delta A) &= F(X,A) + F_X(X,A)(\delta X) + F_{A_1}(X,A)(\delta A_1) \\ &\quad F_{A_2}(X,A)(\delta A_2) + F_{\overline{A}_2}(X,A)(\delta \overline{A}_2) + G(X,A)(\delta X, \delta A), \end{aligned}$$

where $F_X(X,A) : \mathbb{C}^{n\times n} \to \mathbb{C}^{n\times n}$ is the partial Fréchet derivative of F in X calculated at the point (X,A), and $F_{A_1}(X,A) : \mathbb{C}^{n\times n} \to \mathbb{C}^{n\times n}$ is the partial Fréchet derivative of F in A_1 calculated at the point (X,A). Similarly,

$$F_{A_2}(X,A)(\delta A_2) + F_{\overline{A}_2}(X,A)(\delta \overline{A}_2),$$

is the partial Fréchet derivative of $F(X+\delta X, A+\delta A)$ in A_2. The operator $F_{A_2}(X,A)$ is additive but not homogeneous. This specific difficulty arises due to the fact that complex conjugation (and hence the map $A \to A^{\mathrm{H}}$) is not a linear operation.

Set

$$\mathcal{L} := F_X,\ \mathcal{L}_1 := F_{A_1},\ \mathcal{L}_2 = \mathcal{L}_{21} + \mathcal{L}_{22} := F_{A_2} + F_{\overline{A}_2},$$

than

$$F(X+\delta X, A+\delta A) = F(X,A) + \mathcal{L}(\delta X) + \mathcal{L}_1(\delta A_1) + \mathcal{L}_{21}(\delta A_2) + \mathcal{L}_{22}(\delta \overline{A}_2) + G(\delta X, \delta A),$$

where G contains second and higher order terms in $\delta X, \delta A$,

$$G(\delta X, \delta A) = O(\xi^2 + \alpha_1^2 + \alpha_2^2),\ \xi + \alpha_1 + \alpha_2 \to 0.$$

Having in mind that $F(X,A) = 0$ and supposing that the operator $\mathcal{L}$ is invertible we obtain

$$\delta X = -\mathcal{L}^{-1} \circ \mathcal{L}_1(\delta A) - \mathcal{L}^{-1} \circ \mathcal{L}_{21}(\delta A_2) - \mathcal{L}^{-1} \circ \mathcal{L}_{22}(\delta \overline{A}_2) + O(\|a\|^2),\ a \to 0$$

and

$$x = M_1 a_1 + M_{21} a_2 + M_{22}\ a_2 + O(\|a\|^2),\ a \to 0. \tag{16.27}$$

Here

$$x := \mathrm{vec}(\delta X),\ a_k := \mathrm{vec}(\delta A_k)$$

are n^2-vectors,

$$a := \mathrm{vec}(\delta A) = \begin{bmatrix} a_1 \\ a_2 \end{bmatrix} \in \mathbb{C}^{2n^2},$$

$M_1 := -L^{-1}L_1 \in \mathbb{C}^{n^2\times n^2}$, is the matrix of the operator $-\mathcal{L}^{-1} \circ \mathcal{L}_1$, $M_{2k} = -L^{-1}L_{2k} \in \mathbb{C}^{n^2\times n^2}$ is the matrix of the operator $-\mathcal{L}^{-1} \circ \mathcal{L}_{2k}$, and $L, L_1, L_{2k} \in \mathbb{R}^{n^2\times n^2}$ are the matrices of the operators $\mathcal{L}$, $\mathcal{L}_1$, $\mathcal{L}_{2k}$ respectively.

Recall the fact that for nonlinear additive operators the following is fulfilled, see [139].

For the complex $m \times n$ matrices $H = H_0 + \imath H_1$, $\Delta = \Delta_0 + \imath \Delta_1$ (with $H_0, H_1, \Delta_0, \Delta_1$ real), and the complex $n-$vector $z = z_0 + \imath z_1$ (with z_0, z_1 real) we have

$$\max\{\|Hz + \Delta\bar{z}\|_2 : \|z\|_2 \leq a\} = a\|\Theta(H, \Delta)\|_2,$$

where

$$\Theta(H, \Delta) := \begin{bmatrix} H_0 + \Delta_0 & \Delta_1 - H_1 \\ H_1 + \Delta_1 & H_0 - \Delta_0 \end{bmatrix}. \tag{16.28}$$

Hence for the product Hz of a complex matrix $H = H_0 + \imath H_1 \in \mathbb{C}^{n\times n}$ and a complex vector $z = z_0 + \imath z_1 \in \mathbb{C}^n$ with H_0, H_1 and z_0, z_1 real, we have the real versions

$$z^{\mathbb{R}} := \begin{bmatrix} z_0 \\ z_1 \end{bmatrix} \in \mathcal{R}^{2n}, \ H^{\mathbb{R}} := \begin{bmatrix} H_0 & -H_1 \\ H_1 & H_0 \end{bmatrix} \in \mathbb{R}^{2m\times 2n}$$

$$(Hz + \Delta\bar{z})^{\mathbb{R}} = \Theta(H, \Delta)z^{\mathbb{R}}$$

and $\Theta(H, 0) = H^{\mathbb{R}}$.

Now it follows from (16.27) that

$$x^{\mathbb{R}} = M_1^{\mathbb{R}} a_1^{\mathbb{R}} + \Theta(M_{21}, M_{22})a_2^{\mathbb{R}} + O(\|a^{\mathbb{R}}\|^2), \ a^{\mathbb{R}} \to 0.$$

Since

$$\mathcal{L}(Y) = Y + \sigma A_2^{\mathrm{H}} X^{-1} Y X^{-1} A_2$$

it follows that

$$L = I_{n^2} + \sigma(X^{-1}A_2)^{\top} \otimes (A_2^{\mathrm{H}} X^{-1}). \tag{16.29}$$

By definition, the eigenvalues of the operator $\mathcal{L}$ are the eigenvalues (counted according to their algebraic multiplicites) of its matrix L which in turn are

$$1 + \sigma\lambda_i(X^{-1}A_2)\lambda_j(A_2^{\mathrm{H}} X^{-1}),$$

where $\lambda_i(Z)$ are the eigenvalues of the matrix Z. Hence the operator $\mathcal{L}$ and its matrix L are invertible if and only if

$$\sigma\lambda_i(X^{-1}A_2)\lambda_j(A_2^{\mathrm{H}} X^{-1}) \neq -1. \tag{16.30}$$

In what follows we assume that the inequalities (16.30) hold true.

Furthermore we have

$$\mathcal{L}_1(Y) = -Y, \ \mathcal{L}_{21}(Y) = -\sigma A_2^{\mathrm{H}} X^{-1} Y, \ \mathcal{L}_{22} = -\sigma Y^{\mathrm{H}} X^{-1} A_2$$

and hence

$$\begin{aligned} L_1 &= -I_{n^2}, \\ L_{21} &= -\sigma I_n \otimes (A_2^{\mathrm{H}} X^{-1}), \\ L_{22} &= -\sigma((X^{-1}A_2)^{\top} \otimes I_n)P_{n^2}, \end{aligned} \tag{16.31}$$

where $P_{n^2} \in \mathbb{C}^{n^2 \times n^2}$ is the so called vec-permutation matrix such that

$$\mathrm{vec}(Y^\top) = P_{n^2}\mathrm{vec}(Y)$$

for each $Y \in \mathbb{C}^{n\times n}$. For the matrices M_1, M_{2k} we obtain

$$\begin{aligned} M_1 &= -L^{-1}, \\ M_{21} &= -\sigma L^{-1}(I_n \otimes (A_2^{\mathrm{H}} X^{-1})), \\ M_{22} &= -\sigma L^{-1}(((X^{-1}A_2)^\top \otimes I_n))P_{n^2} \end{aligned} \tag{16.32}$$

Recalling that $\xi = \|\delta X\|_{\mathrm{F}} = \|x\|_2$ and since $\|x\|_2 = \|x^{\mathcal{R}}\|_2$ we have

$$\xi \le c\delta + O(\|\delta\|^2) = c_1\delta_1 + c_2\delta_2 + O(\|\delta\|^2), \ \delta \to 0,$$

where

$$c := [c_1, c_2] \in \mathcal{R}^{1\times 2}$$

and

$$\begin{aligned} c_1 &= \|M_1^0\|_2, \ M_1^0 := M_1^{\mathbb{R}}, \\ c_2 &= \|M_2^0\|_2, \ M_2^0 := \Theta(M_{21}, M_{22}). \end{aligned}$$

Hence the absolute individual condition numbers are calculated from

$$c_1 = \left\|M_1^0\right\|_2, \ c_2 = \left\|M_2^0\right\|_2, \tag{16.33}$$

where the matrices M_1^0, M_2^0 are the real version of the matrices M_1, $\Theta(M_{21}, M_{22})$, given by (16.32), (16.28), and (16.29). The relative individual condition numbers are then computed from $\gamma_k = c_k\|A_k\|_{\mathrm{F}}/\|X\|_{\mathrm{F}}$.

Relation (16.27) also gives

$$\xi \le \mathrm{est}_2(\delta) + O(\|\delta\|^2) := \|\left[M_1^0, M_2^0\right]\|_2\|\delta\|_2 + O(\|\delta\|^2), \ \delta \to 0,$$

and

$$\xi \le \mathrm{est}_3(\delta) + O(\|\delta\|_2) := \sqrt{\delta^\top M^0 \delta} + O(\|\delta\|^2), \ \delta \to 0,$$

where $[M_1^0, M_2^0] \in \mathbb{R}^{n^2 \times 2n^2}$ and $M^0 = [m_{kl}^0] \in \mathbb{R}_+^{2\times 2}$ is a symmetric matrix with elements $m_{kl}^0 = \left\|{M_k^0}^\top M_l^0\right\|_2$.

Since $c\delta \le \sqrt{\delta^\top M^0 \delta}$ (see [142]) we find the local perturbation estimate

$$\xi \le g(\delta) + O(\|\delta\|^2), \ \delta \to 0, \tag{16.34}$$

where

$$g(\delta) := \min\left\{\mathrm{est}_2(\delta), \mathrm{est}_3(\delta)\right\}. \tag{16.35}$$

The estimate (16.34), (16.35) allows to define the overall relative condition number as follows. Let $\delta_k = \varepsilon\|A_k\|_{\mathrm{F}}$, where $\varepsilon > 0$ (in floating point arithmetic the quantity ε may be taken as a multiple of the rounding unit). Then $g(\delta) = \varepsilon g(a^0)$, where $a^0 := [\|A_1\|_{\mathrm{F}}, \|A_2\|_{\mathrm{F}}]^\top$. Hence the relative perturbation in the solution can be estimated as $\|\delta X\|_{\mathrm{F}}/\|X\|_{\mathrm{F}} \le \gamma\varepsilon$, where $\gamma := g(a^0)/\|X\|_{\mathrm{F}}$ is the overall relative condition number of equation (E.3) at the particular solution X.

16.3.4 Nonlocal perturbation analysis

Suppose that the perturbed equation (16.26) has a solution with

$$\|\delta X\|_{\rm F} < 1/\|X^{-1}\|_2 = \sigma_{\min}(X).$$

Equation (16.26) may be written in the form

$$\mathcal{L}(\delta X) = \Phi_0(\delta A) + \Phi_1(\delta X, \delta A) + \Phi_2(\delta X, \delta A),$$

where

$$\begin{aligned}
\Phi_0(\delta A) &:= \delta A_1 + \sigma A_2^{\rm H} X^{-1}\delta A_2 + \sigma\delta A_2^{\rm H} X^{-1} A_2 + \sigma\delta A_2^{\rm H} X^{-1}\delta A_2, \\
\Phi_1(\delta X, \delta A) &:= -\sigma A_2^{\rm H} X^{-1}\delta X X^{-1}\delta A_2 - \sigma\delta A_2^{\rm H} X^{-1}\delta X X^{-1} A_2 - \\
&\quad\ \sigma\delta A_2^{\rm H} X^{-1}\delta X X^{-1}\delta A_2, \\
\Phi_2(\delta X, \delta A) &:= \sigma(A_2 + \delta A_2)^{\rm H} E (A_2 + \delta A_2), \\
E &:= (X^{-1}\delta X)^2 (I_n + X^{-1}\delta X)^{-1} X^{-1}.
\end{aligned}$$

The above relations are based on the identity

$$(X + \delta X)^{-1} = X^{-1} - X^{-1}\delta X X^{-1} + (X^{-1}\delta X)^2 (I_n + X^{-1}\delta X)^{-1} X^{-1}.$$

As a result we get the operator equation

$$\delta X = \Pi(\delta X, \delta A) := \Pi_0(\delta A) + \Pi_1(\delta X, \delta A) + \Pi_2(\delta X, \delta A), \tag{16.36}$$

where $\Pi_r = \mathcal{L}^{-1}(\Phi_r)$.

We shall show that under certain conditions on the F-norms δ_k of δA_k the operator $\Pi(\cdot, \delta A)$ maps a central ball $\mathcal{B}_\rho$ of diameter $\rho = f(\delta)$ into itself, where f is continuous and $f(0) = 0$. Hence according to the Schauder fixed point principle, see Appendix D, the operator equation (16.36) has a solution $\delta X \in \mathcal{B}_\rho$. Finally the estimate $\|\delta X\|_{\rm F} \le f(\delta)$ is the desired nonlocal perturbation estimate for δ belonging to a certain set $\Omega \subset \mathbb{R}_+^2$ containing the origin.

Suppose that $\xi \le \rho$, where $\rho < 1/\|X^{-1}\|_2 = \sigma_{\min}(X)$ is a positive quantity. Then, after some calculations, we obtain the inequality

$$\|F(\delta X, \delta A)\|_{\rm F} \le h(\rho, \delta) := a_0(\delta) + a_1(\delta)\rho + \frac{a_2(\delta)\rho^2}{1 - \mu\rho}.$$

Here $\mu := \|X^{-1}\|_2$,

$$\begin{aligned}
a_0(\delta) &:= g(\delta) + c_1\mu\delta_2^2, \\
a_1(\delta) &:= a_{11}\delta_2 + a_{12}\delta_2^2, \\
a_2(\delta) &:= a_{20} + a_{21}\delta_2 + a_{22}\delta_2^2
\end{aligned}$$

and

$$
\begin{aligned}
a_{11} &:= \mu\left(\left\|L^{-1}\left(I_n \otimes \left(A_2^{\mathrm{H}} X^{-1}\right)\right)\right\|_2 + \left\|L^{-1}\left(\left(X^{-1}A_2\right)^{\top} \otimes I_n\right) P_{n^2}\right\|_2\right), \\
a_{12} &:= c_1\mu^2, \\
a_{20} &:= \mu^3 \left\|L^{-1}\left(A_2^{\top} \otimes A_2^{\mathrm{H}}\right)\right\|_2, \\
a_{21} &:= \mu^3 \left\|L^{-1}\left(A_2^{\top} \otimes I_n\right) P_{n^2} + L^{-1}\left(I_n \otimes A_2^{\mathrm{H}}\right)\right\|_2, \\
a_{22} &:= c_1\mu^3.
\end{aligned}
$$

The function h is a Lyapunov majorant for the operator Π, see [85, 137]. The corresponding majorant equation $\rho = h(\rho, \delta)$ is equivalent (for $\rho < 1/\mu$) to the quadratic equation

$$(a_2(\delta) + \mu(1 - a_1(\delta)))\rho^2 - (1 - a_1(\delta) + \mu a_0(\delta))\rho + a_0(\delta) = 0.$$

Denote

$$d(\delta) := (1 - a_1(\delta) + \mu a_0(\delta))^2 - 4a_0(\delta)(a_2(\delta) + \mu(1 - a_1(\delta))).$$

Consider the domain

$$\Omega := \left\{\delta \in \mathbb{R}_+^2 \ : \ a_1 - \mu a_0 + 2\sqrt{a_0(a_2 + \mu(1 - a_1))} \leq 1\right\}. \tag{16.37}$$

If $\delta \in \Omega$ then the majorant equation $\rho = h(\rho, \delta)$ has a root

$$\rho(\delta) = f(\delta) := \frac{2a_0(\delta)}{1 - a_1 + \mu a_0 + \sqrt{d(\delta)}}. \tag{16.38}$$

Hence for $\delta \in \Omega$ the operator $\Pi(., \delta A)$ maps the set $\mathcal{B}_{f(\delta)}$ into itself, where

$$\mathcal{B}_r := \left\{x \in \mathbb{C}^{n^2} \ : \ \|x\|_2 \leq r\right\}$$

is the closed central ball of radius $r \geq 0$. Then according to Schauder fixed point principle there exists a solution $\delta X \in \mathcal{B}_{f(\delta)}$ of equation (16.36).

Thus we have the following result.

Theorem. *Let $\delta \in \Omega$, where Ω is given in (16.37). Then the nonlocal perturbation bound $\|\delta X\|_{\mathrm{F}} \leq f(\delta)$ is valid for equation (E.3), where $f(\delta)$ is determined by (16.38).*

As an example consider the complex fractional-affine matrix equation $X - A_1 - \sigma A_2^{\mathrm{H}} X^{-1} A_2 = 0$ with matrices

$$A_1 = \begin{bmatrix} 0.6192 + 0.3963i & -0.5293 - 0.3246i & -0.2048 - 0.8099i \\ -0.5293 - 0.3246i & -0.0546 + 1.2761i & -1.1566 - 0.3197i \\ -0.2048 - 0.8099i & -1.1566 - 0.3197i & 0.2078 + 0.1764i \end{bmatrix}, \ \sigma = +1,$$

$$A_1 = \begin{bmatrix} 1.3808+1.6037i & 0.5293+0.3246i & 0.2048+0.8099i \\ 0.5293+0.3246i & 2.0546+0.7239i & 1.1566+0.3197i \\ 0.2048+0.8099i & 1.1566+0.3197i & 1.7922+1.8236i \end{bmatrix}, \ \sigma = -1,$$

$$A_2 = \begin{bmatrix} 0.2190+0.0535i & 0.6793+0.0077i & 0.5194+0.4175i \\ 0.0470+0.5297i & 0.9347+0.3834i & 0.8310+0.6868i \\ 0.6789+0.6711i & 0.3835+0.0668i & 0.0346+0.5890i \end{bmatrix}$$

The perturbations in the data are taken as

$$\delta A_2 = \delta X = 10^{(-k)} \begin{bmatrix} 1+i & 1+i & 1+i \\ 1+i & 1+i & 1+i \\ 1+i & 1+i & 1+i \end{bmatrix},$$

$$\delta A_1 = 10^{(-k)} \begin{bmatrix} -0.0448+0.0168i & 0.1832+0.1004i & -0.0689+0.1286i \\ 0.1832+0.1004i & -0.1901-0.8913i & 0.6213-0.1385i \\ -0.0689+0.1286i & 0.6213-0.1385i & 0.1007+0.4331i \end{bmatrix},$$

$$\sigma = +1,$$

$$\delta A_1 = 10^{(-k)} \begin{bmatrix} 2.0448+1.9832i & 1.8168+1.8996i & 2.0689+1.8714i \\ 1.8168+1.8996i & 2.1901+2.8913i & 1.3787+2.1385i \\ 2.0689+1.8714i & 1.3787+2.1385i & 1.8993+1.5669i \end{bmatrix},$$

$$\sigma = -1$$

for $k = 10, 9, ..., 2$.

This problem was designed so as to have solutions $X = I_3$ and $X + \delta X = I_3 + \delta X$ of the unperturbed and perturbed equation respectively.

The perturbation $\|\delta X\|_F$ in the solution is estimated by the local bounds $\text{est}_2(\delta)$, $\text{est}_3(\delta)$ from Section **3** and the nonlocal bound (16.38), (16.37) from Section **4**.

The cases when the nonlocal estimate is not valid since the existence condition $\delta \in \Omega$ is violated, are denoted by asterisk.

The results obtained for different values of k are shown at Table 16.3, for the equation with $\sigma = 1$. When k decreases from 10 to 2 the nonlocal estimate is slightly more pessimistic than the local bounds est2(δ), est3(δ). We also see that for this particular example the bound est3(δ) is superior to est2(δ).

Table 16.3: Exact perturbation and perturbation bounds for $\sigma = 1$

k	$\|\delta X\|_F$	est2	est3	$\rho(\delta)$ (16.38)
10	4.24×10^{-10}	1.48×10^{-9}	1.43×10^{-9}	1.43×10^{-9}
9	4.24×10^{-9}	1.48×10^{-8}	1.43×10^{-8}	1.43×10^{-8}
8	4.24×10^{-8}	1.48×10^{-7}	1.43×10^{-7}	1.43×10^{-7}
7	4.24×10^{-7}	1.48×10^{-6}	1.43×10^{-6}	1.43×10^{-6}
6	4.24×10^{-6}	1.48×10^{-5}	1.43×10^{-5}	1.43×10^{-5}
5	4.24×10^{-5}	1.48×10^{-4}	1.43×10^{-4}	1.44×10^{-4}
4	4.24×10^{-4}	1.48×10^{-3}	1.43×10^{-3}	1.44×10^{-3}
3	4.24×10^{-3}	1.47×10^{-2}	1.43×10^{-2}	1.48×10^{-2}
2	4.24×10^{-2}	1.47×10^{-1}	1.43×10^{-1}	*

16.4 Notes and references

The local bounds of the type est(δ), presented in Section 16.2.5, had been proposed in [133].

There is a number of papers devoted to the perturbation analysis of discrete-time Riccati equations arising in linear systems theory [151, 153, 210, 214, 213, 130]. Until recently, however, the results for the complex case had not been clarified. Here the treatment in [213] for the standard Riccati equation should be complemented with the analysis from [147], see also [215].

Perturbation analysis of the periodic discrete-time Riccati equation is done in [163].

Backward errors for the standard discrete-time Riccati equations are analyzed in [211].

Perturbation analysis of the special symmetrix fractional-affine equation from Section 16.3 is given in [241, 123, 132, 216].

Appendix A

Elements of algebra and analysis

A.1 Introductory remarks

In this book we study perturbations in matrix equations, and, in a less extent, problems of existence and uniqueness of the solution to such equations. Hence, a basic knowledge of algebra and analysis is assumed. For convenience of the reader in this appendix we recall some facts from algebra (including linear algebra) and analysis that are used in the book. A good introduction to this subject is the classical textbook [29].

A.2 Sets and functions

A *set* X is defined by the characteristic property of its elements x, $X = \{x : s(x)\}$, where $s(x)$ is a statement about x. Thus, x is an *element* (or a *point*) of X, denoted as $x \in X$, if and only if the statement $s(x)$ holds. A set is also denoted by explicitly describing its elements, e.g., $X = \{x, y, \ldots\}$.

If x is not an element of X we write $x \notin X$. The set X is a *subset* of the set Y if $x \in X$ implies $x \in Y$. In the latter case we write $X \subset Y$. Two sets X and Y are *equal*, written as $X = Y$, if they consist of the same elements, or equivalently, if and only if $X \subset Y$ and $Y \subset X$. The *union* $X \cup Y$ of the sets X and Y is the set of all x with $x \in X$ or $x \in Y$. The *intersection* $X \cap Y$ is the set of all x with $x \in X$ and $x \in Y$.

The set, containing no elements, is referred to as *empty* set and is denoted by $\emptyset$. The empty set is a subset of any set. A set $\{x\}$ containing a single element x is called a *singleton*. An element of a set can itself be a set. Also, an object x must be distinguished from the singleton $\{x\}$ containing x as its single element.

Example A.1 The set $\{\emptyset\} \neq \emptyset$ is a singleton with element $\emptyset$. $\diamondsuit$

Two sets X and Y are *disjoint* if $X \cap Y = \emptyset$. The *complement* $Y \backslash X$ of the set X relative to the set Y is the set of all x, such that $x \in Y$ and $x \notin X$. Obviously $X \backslash X = \emptyset$ and, more generally, $Y \subset X$ implies $Y \backslash X = \emptyset$.

The set $X \times Y$ of all ordered pairs (x, y) with $x \in X$ and $y \in Y$ is called the *Cartesian product* of X and Y. We also write $X \times X = X^2$. The *Cartesian n-th degree power* X^n, $2 \leq n \in \mathbb{N}$, of the set X is defined inductively as $X^n := X \times X^{n-1}$, or as the set of ordered n-tuples $(x_1, x_2, \ldots, x_n)$ with $x_i \in X$. The set of all pairs (x, x) with $x \in X$ is the *diagonal of a set* of $X \times X$.

A subset R of $X \times Y$ is said to be a *relation*. The relation R is *functional* if for each $x \in X$ there is an unique $y \in Y$, such that $(x, y) \in R$. In this case we also say that the relation R defines a *function*, or a *mapping* $f : X \to Y$ with a *domain* X and *co-domain* Y. For every $x \in X$ the (unique) element $y \in Y$, such that $(x, y) \in R$, is said to be the *image* of x under f and is denoted as $y = f(x)$. We also say that x is the *argument*, and y is the *value* of the function f at the point x.

Let $f : X \to Y$ be a function and $A \subset X$, $B \subset Y$. The set $f(A) := \{f(x) : x \in A\}$ of the images of the elements of A under f is called the *image* of A under f, or simply the image of A if the underlying function f is preassumed. The set $f^{-1}(B) := \{x \in X : f(x) \in B\}$ is the *pre-image* of B. When $B = \{y\}$ is a singleton we write $f^{-1}(y)$ instead of $f^{-1}(\{y\})$. Similarly, if $f^{-1}(B)$ is the singleton $\{x\}$ we write $f^{-1}(B) = x$. If in particular the pre-image of $y \in Y$ is the singleton $\{x\}$ we write $f^{-1}(y) = x$.

The function $f : X \to Y$ is *onto*, or a *surjection*, if $Y = f(X)$, i.e. if each $y \in Y$ is the image of some $x \in X$ under f. The function f is an *injection* if $x_1 \neq x_2$ implies $f(x_1) \neq f(x_2)$, or equivalently, if the pre-image of each $y \in Y$ contains at most one element. The function f is a *bijection*, or one-to-one function, if it is simultaneously a surjection and an injection. In the latter case there exists an *inverse function* $f^{-1} : Y \to X$, which maps each $y \in Y$ into its (unique) pre-image $x = f^{-1}(y)$.

When dealing with objects such as systems of vectors, spectra of matrices, etc., it is convenient to consider *collections*, or sets with repeated elements, such as $\Sigma = \{\alpha, \alpha, \beta\}$. From set-theoretical point of view a collection is indistinguishable from the set, obtained by deleting the repeated elements. For example, as a set $\{\alpha, \alpha, \beta\}$ is the same as $\{\alpha, \beta\}$. Note that a finite collection with n elements is different from the corresponding vector (or ordered n-tuple) having the same elements in a certain order.

We now define some operations with sets (or with collections) such as summation and multiplication, which are useful in the description of the spectra of linear matrix operators. These operations are different from the standard set operations such as union, intersection, complement, etc.

Let A be a commutative algebra over the field $\mathbb{F}$, i.e. (i) A is a linear space over $\mathbb{F}$, and (ii) in $A \times A$ a multiplication $(x, y) \mapsto xy = yx \in A$ is defined for each $x, y \in A$, which obeys the distributive law $(x + y)z = xz + yz$.

Let X, Y be subsets of A. We define the *sum* and the *product* of X and Y as $X + Y := \{x + y : x \in X,\ y \in Y\} \subset A$ and $XY := \{xy : x \in X,\ y \in Y\} \subset A$. For $n \in \mathbb{Z}$ we also define the product nX by $nX := \{nx : x \in X\} \subset A$ and, if $n \in \mathbb{N}$, the power $X^{(n)} := \{x^n : x \in X\} \subset A$. If all elements of X are invertible then we may define $X^{(n)}$ for negative integers n as well.

It is easy to verify that $nX \subset X + \cdots + X$ (n summands), $X^{(n)} \subset X \cdots X$ (n factors) as well as $(X + Y)Z \subset (XZ) + (YZ)$ and $(m + n)X \subset mX + nX$ for $n, m \in \mathbb{N}$.

Finally we define the *difference* of the sets X and Y as $X - Y := X + (-1)Y = \{x - y : x \in X,\ y \in Y\} \subset A$.

A.3 Algebraic systems

In this section we recall some basic facts about algebraic systems.

Let Γ be a nonvoid set. A *unary operation* on Γ is a function $\Gamma \to \Gamma$. An *n-ary operation*, or simply an *operation* on Γ, where $n \in \mathbb{N}$, is a function $\Gamma^n \to \Gamma$. An *algebraic system* is a set Γ together with one or more operations on it. Binary operations $\Gamma \times \Gamma \to \Gamma$ are of special interest when studying algebraic systems. Among them are various types of summation and multiplication.

A *group* is a set Γ of elements $\alpha, \beta, \gamma, \ldots$, together with a binary operation $\circ : \Gamma \times \Gamma \to \Gamma$ (called a *group operation* or a *composition law*) with the following properties.

- *Associative law:* $\alpha \circ (\beta \circ \gamma) = (\alpha \circ \beta) \circ \gamma$ for all α, β, γ.
- *Identity law:* There exists a *neutral element*, or an *identity* $\varepsilon \in \Gamma$, such that $\alpha \circ \varepsilon = \varepsilon \circ \alpha = \alpha$ for all α.
- *Inverse law:* For every α there exists an *inverse*, denoted as α^{-1}, satisfying $\alpha \circ \alpha^{-1} = \alpha^{-1} \circ \alpha = \varepsilon$.

It is easy to show that in any group the equations $\xi \circ \alpha = \beta$ and $\alpha \circ \eta = \beta$ in ξ and η, respectively, have unique solutions $\xi = \beta \circ \alpha^{-1}$ and $\eta = \alpha^{-1} \circ \beta$. Hence, each of the equalities $\alpha \circ \beta = \alpha \circ \gamma$ and $\beta \circ \alpha = \gamma \circ \alpha$ implies $\beta = \gamma$. As a consequence we see that there is exactly one identity ε and for every α the inverse α^{-1} is unique.

Sometimes the group operation $\circ$ is called *multiplication*. Then we simply write $\alpha \circ \beta = \alpha\beta$ and denote the identity as $\varepsilon = 1$ or $\varepsilon = 1_\Gamma$. In this case we have a *multiplicative group*. When the group operation is called *addition* we write $\alpha \circ \beta = \alpha + \beta$ and refer to Γ as an *additive group*. Here we denote the identity as $\varepsilon = 0$ or $\varepsilon = 0_\Gamma$ and the inverse of α as $-\alpha$.

A group $(\Gamma, \circ)$ is *commutative*, or *Abelian*, if the group operation is commutative: $\alpha \circ \beta = \beta \circ \alpha$ for all α, β.

The following sets are commutative groups:

- The sets $\mathbb{Q}^*$, $\mathbb{R}^*$ and $\mathbb{C}^*$ of nonzero rational, real and complex numbers, respectively, are multiplicative groups.
- The sets $\mathbb{Q}_+^*$ and $\mathbb{R}_+^*$ of positive rational and real numbers are multiplicative groups.
- The sets $\mathbb{Z}$, $\mathbb{Q}$, $\mathbb{R}$ and $\mathbb{C}$ of integer, rational, real and complex numbers, respectively, are additive groups.
- The set of complex numbers $\exp(\imath\varphi)$, where $\varphi \in \mathbb{R}$, is a multiplicative group.

A *field* Φ is an algebraic system with two binary operations (called *field operations*), namely *addition* $(\alpha, \beta) \mapsto \alpha + \beta$ and *multiplication* $(\alpha, \beta) \mapsto \alpha\beta$, such that

- Under addition Φ is a commutative group with identity, called *zero* and denoted 0 or 0_Φ.
- Under multiplication the nonzero elements of Φ form a commutative group with identity, called *unit* and denoted 1 or 1_Φ.
- The distributive law $\alpha(\beta + \gamma) = \alpha\beta + \alpha\gamma$ is valid for all $\alpha, \beta, \gamma \in \Phi$.

Let $(\Gamma, \circ)$ be a group with identity $\varepsilon \in \Gamma$ and $\Delta \subset \Gamma$. If $(\Delta, \circ)$ is again a group, it is called a *subgroup* of $(\Gamma, \circ)$. When the group operation is not mentioned explicitly, we say that $\Delta \subset \Gamma$ is a subgroup of the group Γ. Obviously $\{\varepsilon\}$ and Γ are the smallest and the largest subgroups of Γ. They are called *trivial* subgroups. A subgroup is *proper* if it is not trivial.

If Δ_1 and Δ_2 are subgroups of Γ then their intersection $\Delta_1 \cap \Delta_2$ is again a subgroup of Γ. It is the largest subgroup of Γ, contained in both Δ_1 and Δ_2. Dually, the smallest subgroup of Γ, containing Δ_1 and Δ_2, consists of all products of powers of elements of Δ_1 and Δ_2. It is called the *join* of Δ_1 and Δ_2.

Let two multiplicative groups Γ and Δ be given. The function $h : \Gamma \to \Delta$ is a *homomorphism* of Γ to Δ if $h(\alpha\beta) = h(\alpha)h(\beta)$. Under the homomorphism h the identity 1_Γ of Γ goes to the identity 1_Δ of Δ, i.e., $h(1_\Gamma) = 1_\Delta$. The set of all $\alpha \in \Gamma$, such that $h(\alpha) = 1_\Delta$, is a subgroup of Γ, called the *kernel* of the homomorphism h, and denoted as $\mathrm{Ker}(h)$. Thus, $\mathrm{Ker}(h) := \{\alpha \in \Gamma : h(\alpha) = 1_\Delta\} = h^{-1}(1_\Delta)$ is the pre-image of 1_Δ under h. If the homomorphism $h : \Gamma \to \Delta$ is a bijection, it is called *isomorphism* between the groups Γ and Δ. An isomorphism $\Gamma \to \Gamma$ is called *automorphism*. For instance the function $\alpha \mapsto \alpha^{-1}$ is an automorphism on Γ. Note that a bijection $\Gamma \to \Gamma$ is not necessarily an automorphism. For instance, if $\alpha \neq 1_\Gamma$ is fixed, the bijection $\beta \mapsto \alpha\beta$ is not an automorphism since $1_\Gamma \mapsto \alpha \neq 1_\Gamma$.

A.4 Linear algebra

We first consider the fundamental concept of a linear space.

Let V be a set and $\mathbb{E}$ be a field (such as $\mathbb{C}$, $\mathbb{R}$, $\mathbb{Q}$, etc.). The elements of V are called *vectors* and are denoted by $x, y, \ldots, z$ and 0_V. The elements of $\mathbb{E}$ will be referred to as *scalars* (note that the case $V = \mathbb{E}$ is not excluded) and denoted by $\alpha, \beta, \ldots, \gamma$. The unit and zero element of $\mathbb{E}$ are denoted as 1 and 0. The pair $(V, \mathbb{E})$ is said to be a *linear* (or *vector*) *space* if two algebraic operations – *multiplication* of a scalar and a vector $(\alpha, x) \mapsto \alpha x = x\alpha \in V$ and *summation* of two vectors $(x, y) \mapsto x + y = y + x \in V$ are defined with the following additional properties, valid for all $x, y, z \in V$ and $\alpha, \beta \in \mathbb{E}$:

1. $(x + y) + z = x + (y + z) = x + y + z$;
2. $(\alpha + \beta)(x + y) = \alpha x + \alpha y + \beta x + \beta y$;
3. $1x = x$ and $\alpha(\beta x) = (\alpha\beta)x = \alpha\beta x$;
4. there exists a *zero vector* $0_V \in V$, such that $x + 0_V = x$ and $\alpha 0_V = 0x = 0_V$.

The fact that $(V, \mathbb{E})$ is a linear space is expressed by saying that V is a linear space over $\mathbb{E}$. If V is a linear space over $\mathbb{E}$ it is also a linear space over any subfield $\mathbb{E}'$ of $\mathbb{E}$. However, the properties of the linear spaces $(V, \mathbb{E})$ and $(V, \mathbb{E}')$ may be quite different.

Examples of linear spaces over a field $\mathbb{E}$ are as follows.

- The vector space $\mathbb{E}^n$ and the matrix space $\mathbb{E}^{m,n}$.
- The set of all polynomials in one or more indeterminates of a degree, not exceeding a given number, and with coefficients from $\mathbb{E}$.
- The set of functions $V \to W$, where W is a linear space over $\mathbb{E}$.

We consider column $n \times 1$ and row $1 \times n$ vectors x with real or complex elements $(x)_i = x_i$ and denote them as $x = [x_i]$. Consider for example the set $\mathbb{F}^n$ of column n-vectors with elements from the field $\mathbb{F} = \mathbb{R}$ or $\mathbb{F} = \mathbb{C}$. The set $\mathbb{F}^n$ is a *linear space* over $\mathbb{E}$ in the following sense. The (element-wise) *multiplication* $\mathbb{F} \times \mathbb{F}^n \to \mathbb{F}^n$ of a scalar β and a vector $x = [x_i]$ is defined via $(\beta, x) \mapsto \beta x = [\beta x_i]$ and the (element-wise) *summation* $\mathbb{F}^n \times \mathbb{F}^n \to \mathbb{F}^n$ of two vectors $x = [x_i]$ and $y = [y_i]$ is given by $(x, y) \mapsto x + y = [x_i + y_i]$. The vector with zero elements is called the *zero vector* and is denoted by 0, or $0_{n\times 1}$.

To recall the concepts of linear dependence, rank and dimension which are fundamental in the theory of linear spaces, let $S := \{x, y, \ldots, z, \ldots\}$ be a system of one or more vectors from a vector space V over $\mathbb{F}$, and let $\Sigma := \{\alpha, \beta, \ldots \gamma, \ldots\} \subset \mathbb{F}$ be a collection of scalars, which is in one-to-one correspondence with S, i.e., for every $s \in S$, there is a $\sigma \in \Sigma$ and vice versa (note that the case when the set

S and hence, Σ is infinite, is not excluded). The collection Σ is called *trivial* if $\alpha = \beta = \cdots = \gamma = \cdots = 0$.

A vector $u := \alpha x + \beta y + \cdots + \gamma z + \cdots$ is a *linear combination* of the vectors from S. Thus, given the system S, every collection Σ of scalars defines a linear combination of vectors from S. The linear combination u is zero if Σ is trivial. However, the vector u may be zero even if Σ is not trivial, i.e., if some of the scalars are not zero as in the case $u = x + y$, where $S = \{x, y\}$ and $y = -x$.

A system S is said to be *linearly independent* if $u = 0$ implies that Σ is trivial. Alternatively, a system S is *linearly dependent* if $u = 0$ for some nontrivial collection Σ of scalars. If a system S is linearly independent (or dependent), we say that its vectors $x, y, \ldots, z$ are linearly independent (or dependent).

If a system S contains one vector x, i.e., $S = \{x\}$, then it is linearly independent if and only if $x \neq 0$. Equivalently, a system $\{x\}$ is linearly dependent if and only if $x = 0$. When a system S consists of two or more vectors it is linearly dependent if and only if one of the vectors may be expressed as a linear combination of the other vectors. A linearly independent system S cannot contain the zero vector. Indeed, if for example $0 = x \in S$ then the nontrivial linear combination $1 \cdot 0 + 0 \cdot y + \cdots + 0 \cdot z + \cdots +$ (with a collection of scalars $\{1, 0, \ldots\}$) is zero and hence, the system is S linearly dependent.

A finite system $S = \{x, y, \ldots, z\}$ is of *rank* r if it contains r linearly independent vectors and each subsystem of S with more than r vectors is linearly dependent. We also say that the rank of S is the (maximum) number of its linearly independent vectors. Thus, a system S is of zero rank if and only if $x = y = \cdots = z = 0$.

A linear space V over $\mathbb{E}$ has an important integer characteristic, called *dimension*, which may be defined as follows. A linear space, containing only the zero vector, is of dimension *zero*. Let now n be a positive integer. The linear space V is *n-dimensional* if there is a linearly independent system $S \subset V$, containing n vectors, and any system with more than n vectors from V is linearly dependent. If, for any n, there exist a linearly independent system with n vectors, the linear space is *infinite-dimensional.* The dimension of V is denoted by $\dim(V)$.

A linearly independent system $S = \{x, y, \ldots, z\}$ is a *basis* for the linear space V if every vector $v \in V$ may be represented as a linear combination $v = \alpha x + \beta y + \cdots + \gamma y$ of vectors from S. In this case the representation, i.e., the choice of scalars, is unique. The basis itself is, of course, not unique. If $\dim(V) < \infty$ and S is a basis for V, then $\dim(V) = \operatorname{rank}(S)$.

A set $X \subset V$ is said to be a *(linear) subspace* of V if for every $\alpha, \beta \in \mathbb{E}$ and $x, y \in X$ we have $\alpha x + \beta y \in X$, i.e., if any linear combination of vectors from X belongs to X. Given a system of vectors $S \subset V$, the set of all linear combinations of vectors from S is a subspace, called the *span* of S and denoted as $\operatorname{span}(S)$. A basis for the subspace X is any linearly independent system S, such that $\operatorname{span}(S) = X$. The *dimension* $\dim(X)$ of the subspace X of a finite dimensional space V is the rank of its basis.

We consider $m \times n$ matrices $A = [a_{ij}]$ with real or complex elements $(A)_{ij} = a_{ij}$, i.e., $A \in \mathbb{F}^{m\times n}$, where $\mathbb{F} = \mathbb{R}$ or $\mathbb{F} = \mathbb{C}$. Matrices with elements from smaller fields such as $\mathbb{Q}$ are not considered, since the solution of a nonlinear matrix equation with rational coefficients is generically not a matrix over $\mathbb{Q}$, as in $x^2 = 2$. At the same time matrix equations with real coefficients may have complex solutions as in the simplest case $x^2 = -1$.

The standard operations with matrices are the element-wise multiplication with a scalar $\mathbb{F} \times \mathbb{F}^{m\times n} \to \mathbb{F}^{m\times n}$, defined by $(\beta, A) \mapsto \beta A = A\beta = [\beta a_{ij}]$, the element-wise *summation* $\mathbb{F}^{m\times n} \times \mathbb{F}^{m\times n} \to \mathbb{F}^{m\times n}$, defined by $(A, B) \mapsto A + B$, $(A+B)_{ij} = a_{ij} + b_{ij}$ and the row by column *multiplication* $\mathbb{F}^{m\times n} \times \mathbb{F}^{n\times p} \to \mathbb{F}^{m\times p}$, given by $(A, B) \mapsto AB$, where $(AB)_{ij} = \sum_{k=1}^{n} a_{ik}b_{kj}$. Other types of matrix multiplications are described in Appendix C.

The matrix space $\mathbb{F}^{m\times n}$ is a linear space over $\mathbb{F}$ of dimension mn. A basis for $\mathbb{F}^{m\times n}$ is the set of mn matrices $E_{ij}(m, n)$. Here $E_{ij}(m, n)$ is an $m \times n$ matrix with a single nonzero element, equal to 1, in position (i, j).

The set $\mathbb{R}^{m\times n}$ will always be considered as a linear space over $\mathbb{R}$ and then its (real) dimension is mn. However, the set $\mathbb{C}^{m\times n}$ may be considered as a linear space over $\mathbb{C}$ with a complex dimension mn, or as a linear space over $\mathbb{R}$ and then it is of real dimension $2mn$. The terms 'real' and 'complex' here are interpreted as follows. A space V is of real dimension l if one needs l real scalars to determine a vector (or a point) from V. Similarly, V is of complex dimension l if a vector from V is determined in general by l complex scalars.

The number of linearly independent columns of a matrix A is equal to the number of its linearly independent rows. This number is called the *rank* of A and is denoted by $\text{rank}(A)$. Thus, if A is $m \times n$ then $\text{rank}(A) \le \min\{m, n\}$. The matrix A is of *full rank* if $\text{rank}(A) = \min\{m, n\}$. In turn, a full rank $m \times n$ matrix A is either of *full row rank* if $\text{rank}(A) = m$, or of *full column rank* if $\text{rank}(A) = n$. Therefore to say that the matrix A is of full row rank simply means that it is of full rank and the number of its rows is less than or equal to the number of its columns.

If $A \in \mathbb{F}^{m\times n}$, the span of the columns of A is said to be the *range* (or *image*) of A and is denoted by $\text{Rg}(A)$. It is the image of $\mathbb{F}^n$ under the linear mapping $x \mapsto Ax$, namely $\text{Rg}(A) := \{Ax : x \in \mathbb{F}^n\} \subset \mathbb{F}^m$. The set of solutions to the equation $Ax = 0$ is the *kernel* of A and is denoted by $\text{Ker}(A) := \{x : Ax = 0\} \subset \mathbb{F}^n$. It is easy to see that $\text{Rg}(A)$ is a subspace of $\mathbb{F}^m$ and $\text{Ker}(A)$ is a subspace of $\mathbb{F}^n$. Moreover, $\dim(\text{Rg}(A)) = r$ and $\dim(\text{Ker}(A)) = m - r$, where $r = \text{rank}(A)$.

A square matrix A is *invertible* if there exists another matrix B of the same size such that $AB = I$, where I is the unit matrix. In this case we also have $BA = I$, the matrix B is referred to as the *inverse* of A and is denoted as A^{-1}.

If A is a rectangular matrix of full rank then we may define its left and right inverses as follows. A matrix B is a *right* (respectively *left*) *inverse* of A if $AB = I$ (respectively if $BA = I$). Thus, a matrix is simultaneously right and left invertible if and only if it is square and invertible. A square matrix is invertible if and only if it is of full rank. In this case we also say that the matrix is *nonsingular*.

If a matrix A is not square, it has a right (respectively left) inverse if and only if it is of full row (respectively column) rank. In this case the corresponding inverse is not unique. For a matrix A of full row rank over $\mathbb{F}$ a right inverse is $A^{\mathrm{H}}(AA^{\mathrm{H}})^{-1}$. If A is of full column rank a left inverse is $(A^{\mathrm{H}}A)^{-1}A^{\mathrm{H}}$. Here, A^{H} denotes the conjugate transpose of A and analogously $A^{\top}$ denotes the transpose of A.

Let A be an $n \times n$ real or complex matrix. A pair (λ, x), where λ is a number and $x \neq 0$ is an n-vector, is said to be an *eigenpair* of A, if $Ax = \lambda x$. The number λ is an *eigenvalue* and the vector x is an *eigenvector* of the matrix A. The eigenvalues of A are uniquely determined. We have $(\lambda I_n - A)x = 0$ and since $x \neq 0$, then λ satisfies the *characteristic equation* $\chi_A(\lambda) := \det(\lambda I_n - A) = 0$ of the matrix A. At the same time the eigenvectors are determined within a nonzero scalar factor. Indeed, if x is an eigenvector of A, corresponding to the eigenvalue λ, then for every nonzero number α the vector αx is also an eigenvector of A, corresponding to the same eigenvalue. It is usually assumed that the eigenvectors are normed as $\|x\|_2 = 1$.

There are various types of matrices according to their form, pattern of specified elements (for example zero and/or unit elements) and other properties.

A matrix $A = [a_{ij}]$ (not necessarily square) is:

- *diagonal*, if $a_{ij} = 0$ for $i \neq j$;
- *upper triangular*, if $a_{ij} = 0$ for $i > j$;
- *strictly upper triangular*, if $a_{ij} = 0$ for $i \geq j$;
- *lower triangular* if $a_{ij} = 0$ for $i < j$;
- *strictly lower triangular*, if $a_{ij} = 0$ for $i \leq j$.

A square matrix A is:

- *orthogonal*, if $A^{\top}A = I$. In this case we also have $AA^{\top} = I$;
- *unitary*, if $A^{\mathrm{H}}A = I$. In this case we also have $AA^{\mathrm{H}} = I$;
- *normal* if $A^{\mathrm{H}}A = AA^{\mathrm{H}}$;
- *symmetric* if $A^{\top} = A$;
- *skew-symmetric* if $A^{\top} = -A$;
- *Hermitian* if $A^{\mathrm{H}} = A$;
- *skew-Hermitian* if $A^{\mathrm{H}} = -A$;

- *positive definite* if $x^{\mathrm{H}}Ax > 0$ for all nonzero vectors $x \in \mathbb{F}^n$;
- *nonnegative definite* if $x^{\mathrm{H}}Ax \geq 0$ for all nonzero vectors $x \in \mathbb{F}^n$.

A.5 Normed spaces

The concept of a length of a vector is naturally generalized for abstract objects in the following way. Let x be a real or complex n-th vector with elements x_i. Then its *Euclidean length*, or *2-norm*, is defined as $\|x\| = \|x\|_2 = \sqrt{|x_1|^2 + \cdots + |x_n|^2}$. Thus, the length is a nonnegative function with three important properties.

First, if $x \neq 0$ then $\|x\| > 0$. Second, if we multiply the vector x by the scalar α, then the length of the new vector αx is $|\alpha|$ times the length of x, i.e., $\|\alpha x\| = |\alpha|\,\|x\|$. This may be interpreted as *homogenity* of the length. And third, if x and y are two n-vectors, then they, together with their difference $x - y$, form a triangle and the length of the third vector $x - y$ does not exceed the sum of the lengths of the other two, i.e., $\|x - y\| \leq \|x\| + \|y\|$. Replacing y with $-y$ and using the second property we have the more symmetric relation $\|x + y\| \leq \|x\| + \|y\|$, called the *triangle inequality*.

Given a linear space, we may introduce the concepts of a norm, which is similar to the concept of length, considered above.

Let V be a linear space over $\mathbb{E}$ with a zero element 0_V. A function $\|\cdot\| : V \to \mathbb{R}$ is said to be a *norm*, if it satisfies the following conditions:

1. $\|x\| \neq 0$ if $x \neq 0_V$ (*nontriviality* of the norm);
2. $\|\alpha x\| = |\alpha|\,\|x\|$ (*homogenity* of the norm);
3. $\|x + y\| \leq \|x\| + \|y\|$ (the *triangle inequality* for norms).

Two important properties of norms can be deduced as follows. Setting $\alpha = 0$ in Condition 2 we see that $\|0_V\| = 0$. Furthermore, setting $y = -x$ in Condition 3 and using Condition 2 and the identity $\|0_V\| = 0$ we obtain $\|0_V\| = 0 \leq \|x\| + \|-x\| = 2\|x\|$, i.e., $\|x\| \geq 0$. Thus, the norm is a nonnegative homogeneous function, satisfying the triangle inequality.

The triple $(V, \|\cdot\|, \mathbb{E})$ is said to be a *normed space.* We also say that $(V, \|\cdot\|)$ is a normed space over $\mathbb{E}$, or even more briefly that V is a normed space. If we have several normed spaces $V, W, \ldots$ the corresponding norms are denoted as $\|\cdot\|_V$, $\|\cdot\|_W$, etc.

Let V be a linear space over $\mathbb{F} = \mathbb{C}$ or $\mathbb{F} = \mathbb{R}$. The function $V \times V \to \mathbb{F}$ is called a *scalar product* if it is: not identically zero, semi-linear in its first argument and linear in its second argument. Hence, if $(x, y) \in \mathbb{F}$ is the scalar product of $x, y \in \mathcal{V}$, and $\lambda \in \mathbb{F}$, then $(\lambda x, y) = \overline{\lambda}(x, y)$, $(x, \lambda y) = \lambda(x, y)$ and $\overline{(x, y)} = (y, x)$.

Example A.2 For $\mathbb{F}$ equal to $\mathbb{C}$ or $\mathbb{R}$ a scalar product in $\mathbb{F}^n$ is $(x, y) = x^{\mathrm{H}}Ay$, where $A = A^{\mathrm{H}} \in \mathbb{F}^{n\times n}$ is a positive definite matrix. For $A = I$ we have the *standard* scalar product $(x, y) = x^{\mathrm{H}}y$. $\diamondsuit$

Having a scalar product (x, y) in a linear space V over $\mathbb{C}$ or $\mathbb{R}$, the function $x \mapsto \sqrt{(x, x)}$ is a norm in V.

The norm $\|x - y\|$ of the difference of two vectors x and y is the *distance* between x and y. With the notions of norm and distance it is convenient to use the geometric language and in particular to call the elements of V 'points'.

A *sequence* in a set V is a function $x : \mathbb{N} \to V$. Setting $x_i = x(i) \in V$ for $i \in \mathbb{N}$ we denote the sequence as $\{x_i\}_1^\infty$ or briefly $\{x_i\}$. Thus, the sequence is a numbered infinite collection. With certain abuse of notation we also write $\{x_i\} \subset V$ (instead of the rigorous $x(\mathbb{N}) \subset V$).

The sequence $\{x_i\}$ of points x_i from a normed space V *converges* to a point $a \in V$ if for every $\varepsilon > 0$ there exists $n = n(\varepsilon) \in \mathbb{N}$ such that $\|x_i - a\| < \varepsilon$ for all $i \geq n$. The sequence X is a *Cauchy sequence* if for every $\varepsilon > 0$ there exists $n = n(\varepsilon) \in \mathbb{N}$ such that $\|x_i - x_j\| < \varepsilon$ for all $i, j \geq n$.

The normed space V is said to be a *Banach space* if every Cauchy sequence $X \subset V$ converges to a point $a \in V$. The finite dimensional spaces over $\mathbb{F} = \mathbb{R}$ or $\mathbb{F} = \mathbb{C}$, are Banach spaces. According to the Bolzano-Weierstrass theorem [96] every bounded sequence $\{x_i\}_{i=1}^\infty$ in a finite dimensional space over $\mathbb{F}$ has a convergent subsequence $\{x_{i_k}\}_{k=1}^\infty$.

Let $x \in V$ and $\rho \geq 0$. The sets $B_\rho(x) := \{y \in V : \|y - x\| < \rho\}$, $\overline{B}_\rho(x) := \{y \in V : \|y - x\| \leq \rho\}$ and $S_\rho(x) := \{y \in V : \|y - x\| = \rho\}$ are called *open ball*, *closed ball* and *sphere*, respectively, centered at the point x and with radius ρ (for $\rho = 0$ we have $B_0(x) = \emptyset$ and $\overline{B}_0(x) = S_0(x) = \{x\}$).

The set $X \subset V$ is *open* if for every $x \in X$ there exists a nonempty open ball $B_\rho(x) \subset X$. The set X is *closed* if its complement $V\backslash X$ is open. Any (open) set, containing a particular point $x \in V$, is said to be a *(open) neighborhood* of x.

A set may be open, closed, neither open nor closed or even open and closed simultaneously.

Example A.3 The empty set $\emptyset$ and the whole space V are open as well as closed. The open ball $B_\rho(x)$ is an open set, while the closed ball $\overline{B}_\rho(x)$ and the sphere $S_\rho(x)$ are closed sets. The set $B_1(x)\backslash\{x\}$ of all $y \neq x$ with $\|y - x\| \leq 1$ is neither open nor closed.

A more subtle example is the set of all vectors $x = [x_1, \ldots, x_n]^\top$ with entries $x_i \in \mathbb{Q}$, satisfying $\|x\| \leq 1$. This set is neither open nor closed in $\mathbb{R}^n$. $\diamondsuit$

Let $X \subset V$. A point $x \in X$ is a *boundary point* for the set X if every $B_\rho(x)$ contains points from X as well as points from $V\backslash X$. Note that a boundary point of X may not belong to X. The set of boundary points of X is called the *boundary* of X and is denoted by ∂X. The union $\overline{X} := X \cup \partial X$ of the set X and its boundary

∂X is the *closure* of X. A set X is closed if and only if $X = \overline{X}$. The set of all $x \in X$, for which X is an open neighborhood, is said to be the *interior* of X and is denoted by X^o. Thus, X is open if and only if $X = X^o$.

Example A.4 The sphere $S_\rho(x)$ is the boundary of the open $B_\rho(x)$ as well as of the closed ball $\overline{B}_\rho(x)$. The set $\overline{B}_\rho(x) = B_\rho(x) \cup S_\rho(x)$ is the closure of $B^o_\rho(x)$. Also, $B_\rho(x)$ is the interior of $\overline{B}_\rho(x)$ provided $\rho > 0$. $\diamondsuit$

A set $X \subset V$ is *bounded* if it is contained in a ball of finite radius.

A family of open sets $\{O_i\}_{i \in \mathcal{I}}$ is an *open cover* of X if their union $\cup_{i \in \mathcal{I}} O_i$ is equal to X. A set X is *compact* if for every open cover there is a finite sub-cover. In a finite dimensional space over $\mathbb{F}$ a set is compact if and only if it is closed and bounded.

A subset X of a linear space V is *convex* if for every $x, y \in X$ we have $tx + (1-t)x \in X$ for $0 \le t \le 1$.

Example A.5 If $\rho > 0$, then the balls $B_\rho(x)$ and $\overline{B}_\rho(x)$ are convex, while the sphere $S_\rho(x)$ is not convex. $\diamondsuit$

For $x \in V$ and $X \subset V$ the quantity $\operatorname{dist}(x, X) := \inf\{\|x - y\| : y \in X\}$ is the *distance* between the point x and the set X. The quantity $\operatorname{diam}(X) := \sup\{\|x - y\| : x, y \in X\}$ is the *diameter* of the set X.

A.6 Matrix functions

In this section we consider the problems of continuity and differentiability of *matrix functions*, i.e. of matrix-valued functions of matrix arguments. First we discuss the corresponding problems for functions in normed spaces.

Let $f : D \to W$ be a function with a domain $D \subset V$, where V and W are normed spaces over the field $\mathbb{E} \subset \mathbb{F}$ with norms $\|\cdot\|$, where $\mathbb{F}$ stands for $\mathbb{R}$ or $\mathbb{C}$ (we use the same notation for norms $\|\cdot\|_V$ and $\|\cdot\|_W$ in V and W). Typically we assume that $V = \mathbb{F}^n$ and $W = \mathbb{F}^m$.

If $D \subset \mathbb{R}$ is an interval and $W = \mathbb{R}^m$, then the function f (or its image $f(D) \subset \mathbb{R}^m$) is interpreted as a *curve*. If $D \subset \mathbb{R}^{m-1}$ and $W = \mathbb{R}^m$ then f (or its image $f(D)$) defines a *surface* in $\mathbb{R}^m$ (under this definition, in $\mathbb{R}^2$ curves are surfaces and vice versa).

When $W = \mathbb{F}^m$, to determine a function $f : D \to W$ means to determine its components $f_i : D \to \mathbb{F}$, $i = 1, \ldots, m$. In this case we write $f = [f_1, \ldots, f_m]^\top$ or simply $f = (f_1, \ldots, f_m)$.

A function f is *continuous at the point* $x_0 \in D$ if for all $\varepsilon > 0$ there exists $\delta = \delta(\varepsilon, x_0) > 0$, such that $\|f(x) - f(x_0)\| < \varepsilon$ for any $x \in D$ with $\|x - x_0\| < \delta$. The function f is *continuous on the set* $E \subset D$ if it is continuous at all points of

E. A function $f = (f_1, \ldots, f_m) : D \to \mathbb{F}^m$ is continuous at $x_0 \in D$ if and only if every component $f_i : D \to \mathbb{F}$ of f is continuous at x_0.

The function $f : D \to W$ is *bounded* on the set $E \subset$ if the image $f(E)$ of E under f is a bounded set. The function f is *locally bounded* on D if it is bounded on each bounded subset $E \subset D$.

If $D \subset \mathbb{F}^n$ is compact (i.e., closed and bounded in this particular case) and the function $f : D \to \mathbb{R}$ is continuous on D, then f is bounded on D and reaches its minimum and maximum values on D.

A function $f : D \to W$ is *uniformly continuous* on the set $E \subset D$ if for every $\varepsilon > 0$ there exists $\delta = \delta(\varepsilon) > 0$ such that $\|f(x) - f(x_0)\| < \varepsilon$ for all $x, x_0 \in E$ with $\|x - x_0\| < \delta$. Note that uniform continuity is not connected with a particular point from E, but with the whole set E.

A difference between the (usual) continuity and uniform continuity of a function on a set is as follows. Recall from the above definitions that the function $f : D \to W$ is continuous on $E \subset D$ if for every $\varepsilon > 0$ and $x_0 \in E$ there exists $\delta = \delta(\varepsilon, x_0) > 0$ such that $\|f(x) - f(x_0)\| < \varepsilon$ for all $x \in E$ with $\|x - x_0\| < \delta$. Thus, for the usual continuity the quantity $\delta = \delta(\varepsilon, x_0)$ depends on ε and x_0, while in the uniform continuity it depends only on ε.

Uniform continuity on E implies continuity on E, but the opposite may not be true. Let the function $f : D \to W$ be continuous on D and $\delta(\varepsilon, x_0)$ be the quantity in the corresponding (ε, δ)-definition. If $\delta^0(\varepsilon) := \inf\{\delta(\varepsilon, x_0) : x_0 \in D\} > 0$ then the function is uniformly continuous. Indeed, in this case we may take δ to be δ^0 in the definition of uniform continuity. If a function $f : D \to W$ is continuous then it is also uniformly continuous on every compact subset E of D.

Example A.6 Consider the scalar real function $x \mapsto x^2$, defined on the interval $[0, a) \subset \mathbb{R}$, where $a > 0$. We have $\left|x^2 - x_0^2\right| = |x - x_0||x + x_0|$. If $|x - x_0| < \delta$ then $|x + x_0| < 2|x_0| + \delta$. In this case the inequalities $|x - x_0||x + x_0| < \delta(2|x_0| + \delta) \leq \varepsilon$ yield $\delta = \delta(\varepsilon, x_0) := \varepsilon / \left(|x_0| + \sqrt{x_0^2 + \varepsilon}\right)$.

If $a < \infty$, then we have $\delta(\varepsilon, x_0) \geq \delta^0(\varepsilon) := \varepsilon / \left(a + \sqrt{a^2 + \varepsilon}\right) > 0$ and the function $x \mapsto x^2$ is uniformly continuous on $[0, a)$. If the interval J is not bounded $(a = \infty)$, then the infimum of $\delta(\varepsilon, x_0)$ in $x_0 \in [0, \infty)$ is zero. Therefore the function $x \mapsto x^2$ is continuous but not uniformly continuous on $[0, \infty)$. $\diamondsuit$

If the function $f : D \to W$ is continuous at the point $x_0 \in D$ and the sequence $\{x_i\} \subset D$ converges to x_0 then the sequence $\{f(x_i)\} \subset W$ converges to $f(x_0)$. But f may not be continuous at some point x_0, or even may not be defined at x_0, and still the sequence $\{f(x_i)\}$ may converge to some point $y_0 \in W$. Of course, in this case x_0 must either belong to D or be 'close' to D in the following sense.

The point x_0 is an *accumulation point* for the set D if there is a sequence from D, which converges to x_0. Let x_0 be an accumulation point for the set D. The function $f : D \to W$ has a *limit* y_0 at the point x_0, denoted as $\lim_{x \to x_0} f(x) = y_0$, if the sequence $\{f(x_i)\}$ converges to y_0 provided the sequence $\{x_i\} \subset D$ converges to x_0.

Consider a sequence of functions $\{f_i\}$, mapping the set D into W. For a given $x \in D$ we have the sequence of points $\{f_i(x)\} \subset W$. Suppose that the limit $f(x) := \lim_{i\to\infty} f_i(x)$ exists for every $x \in D$. Then we say that the function sequence $\{f_i\}$ *converges point-wise* on D to the function f. If the functions f_i are continuous then the limit function f may or may not be continuous. To ensure that a limit of continuous functions is itself a continuous function we need the stronger concept of uniform convergence of a function sequence.

A function sequence $\{f_i\}$ with $f_i : D \to W$ *converges uniformly* to the function $f : D \to W$ if for every $\varepsilon > 0$ there exists $N = N(\varepsilon) \geq 1$ such that for all $i \geq N$ and all $x \in D$ we have $\|f_i(x) - f(x)\| < \varepsilon$. If a sequence of continuous functions $\{f_i\}$ with a common domain D converges uniformly to the function f, then the limit function f is continuous on D.

A function $f : V \to W$ is called *linear* if, for all scalars $\alpha, \beta \in \mathbb{E}$ and vectors $x, y \in V$, it is fulfilled that $f(\alpha x + \beta y) = \alpha f(x) + \beta f(y)$. The function $f : V \to W$ is *affine* if $f(x) = b + l(x)$, where $b \in W$ and the function $l : V \to W$ is linear. The affine function, defined via $f(x) = y_0 - l(x_0) + l(x)$, takes a prescribed value y_0 at the point x_0. Linear and affine functions are often referred to as linear and affine *operators*.

Linear and affine operators in vector and matrix spaces may be defined as follows. A linear operator $f : \mathbb{F}^n \to \mathbb{F}^m$ is defined via $f(x) = Ax$, and an affine operator – via $f(x) = Ax + b$, where $A \in \mathbb{F}^{m\times n}$ and $b \in \mathbb{F}^m$. A linear operator $L : \mathbb{F}^{m\times n} \to \mathbb{F}^{p\times q}$ may be defined by $L(X) = \sum_{k=1}^{r} A_i X B_i$, $X \in \mathbb{F}^{m\times n}$, where A_i and B_i are given $p \times m$ and $n \times q$ matrices, respectively. An affine operator F has the form $F(X) = B + L(X)$, where $B \in \mathbb{F}^{p\times q}$ and L is a linear operator.

A function $f : V \to W$ is called *homogeneous* (*absolutely homogeneous*) of order $k \in \mathbb{N}$ if $f(\alpha x) = \alpha^k f(x)$ (if $f(\alpha x) = |\alpha|^k f(x)$).

Let V and W be linear spaces over $\mathbb{C}$. A function $f : V \to W$ is *semi-homogeneous* of order $k \in \mathbb{N}$ if $f(\alpha x) = \overline{\alpha}^k f(x)$. A function f is *semi-linear* if it is additive and semi-homogeneous of first order, i.e., $f(x + y) = f(x) + f(y)$ and $f(\alpha x) = \overline{\alpha} f(x)$.

A function $f : V_1 \times V_2 \to W$, where V_1 and V_2 are linear spaces, is *bilinear* if for fixed $x_i \in V_i$ the functions $f(\cdot, x_2) : V_1 \to W$ and $f(x_1, \cdot) : V_2 \to W$ are linear. Similarly, the function $f : V_1 \times \cdots \times V_n \to W$ of n arguments is *multi-linear* if it is linear in each of its arguments.

Let a function $f : D \to W$ be defined on the open set $D \subset V$ and let x be a fixed point from D. A function f is said to be *Fréchet differentiable* (or simply *differentiable*) at the point x if there exists a linear operator $l : V \to W$ such that $f(x + h) = f(x) + l(h) + \omega(h)$ for all $h \in V$ with $x + h \in D$, where the function $\omega : V \to W$ satisfies $\lim_{h\to 0} \|\omega(h)\|/\|h\| = 0$. The linear operator $l(\cdot) : V \to W$ depends on both the function f and the point x. It is called the *Fréchet derivative* of f at x and is denoted as $f'(x)(\cdot)$, $f_x(x)(\cdot)$ or $J(x)(\cdot)$. When it exists, the Fréchet derivative is unique. Below we describe the Fréchet derivatives of some matrix-valued functions of matrix arguments.

If $V = W = \mathbb{R}$ then the Fréchet derivative $f'(x) \in \mathbb{R}$ is the standard derivative of the real-valued function f of a real argument at the point $x \in D$. If $V = W = \mathbb{C}$ then the Fréchet derivative $f'(z) \in \mathbb{C}$ is the standard derivative of the complex-valued function of a complex argument at the point $z \in D$. If $V = \mathbb{F}^n$ and $W = \mathbb{F}^m$ then the Fréchet derivative of $f = [f_1, \ldots, f_m]^\top$ at $x \in D$ is the $m \times n$ matrix (known also as the *Jacobi* matrix of f at x)

$$f'(x) = J(x) := \left[\frac{\partial f_i}{\partial x_j}(x)\right].$$

If $f : V_1 \times \cdots \times V_n \to W$ is a function of n (matrix) arguments $x_1, \ldots, x_n$, then we define the *partial Fréchet derivative* $l_i(\cdot) = f_{x_i}(x)(\cdot) : V_i \to W$ of f in the argument x_i at the point $x = (x_1, \ldots, x_n)$ via $f(x + h) = f(x) + l_i(h_i) + \omega(h_i)$, where $h = (h_1, \ldots, h_n)$ and $h_k = 0$ for $k \neq i$.

The existence and uniqueness of the solution as well as the perturbation analysis problems for nonlinear equations are often treated on the basis of the *implicit function theorem*. The implicit function theorem gives conditions which guarantee that the solution of a nonlinear equation, depending on a parameter, exists and continuously depends on this parameter. Usually the existence of the solution is claimed locally, in an open neighborhood of a fixed solution of the equation.

Let the continuous function $f(\cdot,\cdot) : A \times X \to Y$ be given, where A, X and Y are open subsets of the finite dimensional normed spaces $\mathcal{X}$, $\mathcal{A}$ and $\mathcal{Y}$, respectively, with norms $\|\cdot\|$, where $\mathcal{X}$ and $\mathcal{Y}$ are isomorphic.

Let the point (a_0, x_0) satisfy the equation $f(a_0, x_0) = 0$. We are interested in conditions for solvability of the equation

$$f(a, x) = 0 \tag{A.1}$$

in a neighborhood of (a_0, x_0) in the form $x = \varphi(a)$, where φ is a continuous function, satisfying $\varphi(a_0) = x_0$. Setting $\delta a := a - a_0$ and $\delta x := x - x_0$ we get $\delta x = \varphi(a_0 + \delta a) - \varphi(a_0)$.

Theorem A.7 *[173] Suppose that the partial Fréchet derivative $f_x(a_0, x_0)(\cdot) : \mathcal{X} \to \mathcal{Y}$ of f in x at the point (a_0, x_0) exists and is invertible.*

Then there is an open set $D \subset A$ and a continuous function $\varphi : D \to X$, such that $x_0 = \varphi(a_0)$ and $f(a, \varphi(a)) = 0$, $a \in D$.

If $f = [f_1, \ldots, f_n]^\top : \mathbb{F}^n \times \mathbb{F}^m \to \mathbb{F}^n$, and $x = [x_1, \ldots, x_n]^\top$ then the conditions of the implicit function theorem reduce to the existence of the nonsingular Jacobi matrix

$$\left[\frac{\partial f}{\partial x}(a_0, x_0)\right] = \left[\frac{\partial f_i}{\partial x_j}(a_0, x_0)\right]_{i,j=1}^{n}.$$

When the partial Fréchet derivative $f_a(a_0, x_0)$ of f in a at (a_0, x_0) also exists, we have $x = x_0 - f_x^{-1}(a_0, x_0) \circ f_a(a_0, x_0)(a - a_0) + \omega(a - a_0)$, or

$$\delta x = -f_x^{-1}(a_0, x_0) \circ f_a(a_0, x_0)(\delta a) + \omega(\delta a),$$

where $\lim_{z \to 0} \|\omega(z)\|/\|z\| = 0$.

The above concepts apply to matrix valued functions of matrix arguments as follows. Let $F : D \to \mathbb{F}^{p \times q}$ be a matrix-valued function of a matrix argument, defined in an open neighborhood $D \subset \mathbb{F}^{m \times n}$ of the matrix $X \in \mathbb{F}^{m \times n}$. The function F is said to be *differentiable* at the point X if there exists a linear operator $\mathcal{L} : \mathbb{F}^{m \times n} \to \mathbb{F}^{p \times q}$ such that

$$F(X + H) = F(X) + \mathcal{L}(H) + \alpha(H), \tag{A.2}$$

where the matrix-valued function $\alpha : \mathbb{F}^{m \times n} \to \mathbb{F}^{p \times q}$ satisfies

$$\lim_{H \to 0} \frac{\|\alpha(H)\|}{\|H\|} = 0. \tag{A.3}$$

The linear operator $\mathcal{L}(\cdot)$ in (A.2) in general depends on X, i.e., $\mathcal{L}(\cdot) = \mathcal{L}(\cdot, X)$. It is called the *Fréchet derivative* of F at the point X and is denoted by $\mathcal{L}(\cdot) = F'(X)(\cdot)$. If it exists, the Fréchet derivative is unique. The value $F'(X)(H)$ of the Fréchet derivative is the (best) linear approximation to the increment $F(X + H) - F(X)$.

If the function F is differentiable for all $X \in D$ it is *differentiable on the set* D.

Let now $F : D_1 \times \cdots \times D_r \to \mathbb{F}^{p \times q}$ be a function of the multi-matrix argument $X = (X_1, \ldots, X_r)$, where $D_i \subset \mathbb{F}^{m_i \times n_i}$ are open neighborhoods of some points $X_{i0} \in \mathbb{F}^{m_i \times n_i}$. The function F is said to be *differentiable* in X_i at the point $X_0 = (X_{10}, \ldots, X_{r0})$ if there exists a linear operator $\mathcal{L}_i(\cdot) : \mathbb{F}^{m_i \times n_i} \to \mathbb{F}^{p \times q}$ such that

$$F(X + H) = F(X) + \mathcal{L}_i(H_i) + \alpha_i(H_i),$$

where $H = (H_1, \ldots, H_r)$, $H_k = 0$ for $k \neq i$, and α_i satisfies (A.3). The linear operator $\mathcal{L}_i(\cdot)$ is said to be the *partial Fréchet derivative* of F in X_i at the point X and is denoted by $\mathcal{L}_i(\cdot) = F_{X_i}(X)(\cdot)$, or briefly $F_{X_i}(\cdot)$ if the point X at which it is calculated is clear from the context.

The Fréchet derivative of matrix-valued functions has many properties of the standard derivative of scalar functions of a scalar argument as shown below.

If $F = F_1 + F_2$, where $F_1, F_2 : D \to \mathbb{F}^{p \times q}$ and $D \subset \mathbb{F}^{m \times n}$ then

$$F'(X) = F_1'(X) + F_2'(X),$$

or, more generally,

$$(F_1(X) + \cdots + F_k(X))' = F_1'(X) + \cdots + F_k'(X).$$

Here and in the rest of this section we suppose that the corresponding Fréchet derivatives exist.

If $F = G \circ H : \mathbb{F}^{m\times n} \to \mathbb{F}^{p\times q}$ is a composition of the differentiable matrix-valued functions $G : \mathbb{F}^{r\times s} \to \mathbb{F}^{p\times q}$ and $H : \mathbb{F}^{m\times n} \to \mathbb{F}^{r\times s}$, then F is also differentiable, and $F' = G' \circ H'$, or equivalently,

$$F'(X) = G'(H(X))H'(X).$$

This is known as the *chain rule* and it may be extended for a function $F = F_1 \circ F_2 \circ \cdots \circ F_k$ as

$$F'(X) = F_1' \circ F_2' \circ \cdots \circ F_k'(X).$$

If $F = F_1F_2$ is a product of two functions $F_1 : D \to \mathbb{F}^{p\times s}$ and $F_2 : D \to \mathbb{F}^{s\times q}$ then

$$F'(X) = F_1'(X)F_2(X) + F_1(X)F_2'(X).$$

This is the *Leibnitz rule*, which can be extended to $k > 2$ factors as follows. If $F(X) = F_1(X)F_2(X)\cdots F_k(X)$ then

$$\begin{aligned} F'(X) &= F_1'(X)F_2(X)\cdots F_k(X) + F_1(X)F_2'(X)\cdots F_k(X) \\ &+ \cdots + F_1(X)F_2(X)\cdots F_k'(X). \end{aligned}$$

Another example of differentiation of composite functions is when F may be represented as $F(X) = F_1(X)F_2^{-1}(X)F_3(X)$, where $F_1 : \mathbb{F}^{m\times n} \to \mathbb{F}^{p\times s}$, $F_2 : \mathbb{F}^{m\times n} \to \mathbb{F}^{s\times s}$ and $F_3 : \mathbb{F}^{m\times n} \to \mathbb{F}^{s\times q}$. In this case

$$\begin{aligned} F'(X) &= F_1'(X)F_2^{-1}(X)F_3(X) + F_1(X)F_2^{-1}(X)F_3'(X) \\ &\quad - F_1(X)F_2^{-1}(X)F_2'(X)F_2^{-1}(X)F_3(X). \end{aligned}$$

Example A.8 Let $F(X) = AX + XB + XCX + DX^{-1}E$, where $A, B, C, D, E \in \mathbb{F}^{n\times n}$ and $X \in \mathbb{F}^{n\times n}$ is nonsingular matrix. Then $F'(X)(H) = (A + XC)H + H(B + CX) - DX^{-1}HX^{-1}E$. $\diamondsuit$

A.7 Transformation groups

Let S be a set and Γ be a group of automorphisms $S \to S$. The elements of Γ are called *transformations* and Γ itself is called a *transformation group* on the set S. The group Γ defines an *equivalence relation* $\equiv$ on S according to the rule $x \equiv y$ if there is a transformation $\gamma \in \Gamma$ such that $y = \gamma(x)$.

The set of all elements $y \in S$, equivalent to a given $x \in S$, is said to be the *orbit* of x and is denoted as $\Gamma(x)$, or as $[x] := \{y \in S : y \equiv x\} = \{\gamma(x) : \gamma \in \Gamma\}$. We note that for every $y \in S$ either both x and y belong to one orbit (i.e., $[x] = [y]$), or $[x] \cap [y] = \emptyset$. Thus, the set S is divided into disjoint orbits. The set $S/\equiv$ of all orbits is called the *orbit space* (or the *factor-space*) of S relative to the action of Γ. The mapping $x \mapsto [x]$, which assigns to every member of S its orbit from $S/\equiv$, is called the *canonical projection* and is denoted by $\pi : S \to S/\equiv$.

A function $f : S \to T$, where T is a given set, is said to be an *invariant* relative to Γ if $x \equiv y$ implies $f(x) = f(y)$. This means that f is constant on the orbits, induced by Γ. It is usually assumed that an invariant is surjective, i.e., that $f(S) = T$, since this can easily be achieved. If in addition $f(x) = f(y)$ implies $x \equiv y$, then the function f is a *complete invariant* for the action of Γ on the set S.

If $\Gamma_0 \subset \Gamma$ is a subgroup of Γ, then there exist complete invariants $f : S \to T$ and $f_0 : S \to T_0$ for Γ and Γ_0 respectively, such that f is a 'part' of f_0 in the sense that f_0 may be represented as $f_0 = (f, g)$, where $g : S \to T_0$ is another invariant for Γ_0.

A subset $C \subset S$ is called a *canonical set* for Γ if it contains exactly one member x_c of each orbit $[x]$. The element $x_c \in C$ is the *canonical form* of x relative to Γ. In this case the mapping $x \mapsto x_c$ of S onto C is a complete invariant for Γ. When S is a set of objects with internal structure (such as general matrices), the canonical set usually is a set of objects of simplified structure (e.g., triangular matrices). The construction of canonical sets is an important task in the analysis of the action of transformation groups, and of matrix transformation groups in particular.

A.8 Notes and references

Elements of algebra and analysis can be found in classical textbooks such as [29, 96].

Appendix B

Unitary and orthogonal decompositions

B.1 Introductory remarks

This appendix is an introduction to unitary and orthogonal decompositions (or factorizations) of a matrix. First we consider elementary unitary matrices. Then we describe the unitary-triangular, or QR decomposition, as well as some related matrix decompositions. We also present the Schur decomposition of a square matrix, and the polar and singular value decomposition of an arbitrary matrix.

Transformations with unitary and orthogonal matrices (we identify transformations with the corresponding matrices when the type of action of the transformation group is clear from the context) are useful in practical computations, since the absolute values of their elements do not exceed 1 and hence, the elements and the norms of the transformed matrices are not changed much. This is important, since most matrix computations are performed in finite precision arithmetic, where the rounding errors are usually proportional to the magnitude of the computed quantities.

Recall that a matrix $U \in \mathbb{F}^{n\times n}$ is *unitary* if $U^{\mathrm{H}}U = I_n$, and *orthogonal* if $U^{\top}U = I_n$. A complex unitary or a real orthogonal matrix U is also called *orthonormed*, since its columns u_i satisfy the conditions $u_i^H u_j = \delta_{ij}$, where δ_{ij} is the Kronecker delta.

The sets of $n \times n$ unitary and orthogonal matrices over $\mathbb{F}$ (where $\mathbb{F} = \mathbb{R}$ or $\mathbb{F} = \mathbb{C}$) are denoted by $\mathcal{U}(n)$ and $\mathcal{O}(n, \mathbb{F})$, respectively. They are multiplicative groups under the standard matrix multiplication.

The spectral, or 2-norm $\|A\|_2 = \max\{\|Ax\|_2 : \|x\|_2 = 1\}$ and the Frobenius, or F-norm $\|A\|_{\mathrm{F}} = \sqrt{\mathrm{tr}(A^{\mathrm{H}}A)}$ play an important role in connection with unitary transformations. The reason is that these norms are *unitary invariant.* in the sense that they preserve the norm of the transformed matrix: $\|S\|_2 = \|A\|_2$ and $\|S\|_F = \|A\|_F$. Also, we have the useful inequality $\|AB\|_F \le \|A\|_2 \|B\|_F$ which is easily generalized to

$$\|A_1 A_2 \cdots A_r\|_F \le \|A_k\|_F \prod_{i \ne k} \|A_i\|_2 \tag{B.1}$$

for each $k \in \{1, \ldots, r\}$.

If $U \in \mathcal{U}(n)$ or $U \in \mathcal{O}(n, \mathbb{R})$ then $\|U\|_2 = 1$ and $\|U\|_F = \sqrt{n}$. Thus, $\mathcal{U}(n) \subset \mathbb{C}^{n \times n}$ and $\mathcal{O}(n, \mathbb{R}) \subset \mathbb{R}^{n \times n}$ are closed balls of radius $\sqrt{n}$ relative to the distance, induced by the scalar products $(U, V) = \mathrm{tr}(U^H V)$ and $(U, V) = \mathrm{tr}(U^\top V)$ (or by the Frobenius norm).

If a square (complex) matrix U is represented as $U = U_0 + \imath U_1$, where U_0, U_1 are real, then it is unitary if and only if the matrix $U^{\mathbb{R}} := \begin{bmatrix} U_0 & -U_1 \\ U_1 & U_0 \end{bmatrix}$ is orthogonal.

In many applications a general matrix A is decomposed as a product, involving unitary or orthogonal matrices, namely $A = USV^H$ or $A = USV^\top$, where the matrix S has the size of A, while U and V are unitary or orthogonal matrices. The matrices U and V may not be independent (for example V may be equal to U), or one of them may be the identity matrix or a permutation matrix. The matrix S typically has a simple condensed form, e.g., triangular or diagonal. It reflects the *invariant structure* of A under unitary transformations $A \mapsto S = U^H AV$.

B.2 Elementary unitary matrices

A general unitary matrix may be decomposed into a product of "elementary" unitary matrices. There are several types of matrices, considered as elementary. Among them are the plane (or Givens) rotations and the elementary (or Householder) reflections.

An *elementary complex plane rotation* (*Givens rotation*) is a unitary matrix, which differs from the identity matrix in at most four positions, occupied by the elements of a 2×2 unitary matrix and has determinant 1. More precisely, a *rotation* in the (p, q)-plane, $p < q$, is an $n \times n$ matrix R_{pq}, whose (i, k) elements r_{ik} are determined as follows. The 2×2 matrix

$$\begin{bmatrix} r_{pp} & r_{pq} \\ r_{qp} & r_{qq} \end{bmatrix} \tag{B.2}$$

is unitary and r_{ik} is the Kronecker delta if $\{i, k\} \cap \{p, q\} = \emptyset$.

An *elementary real plane rotation* is defined similarly. It is a real orthogonal matrix with the structure of R_{pq}, where the matrix (B.2) is orthogonal.

Another type of elementary unitary matrices are the elementary reflections. Let $u \in \mathbb{C}^n$ be a nonzero vector. The matrix

$$H(u) := I_n - \frac{2uu^{\mathrm{H}}}{u^{\mathrm{H}}u} \in \mathbb{C}^{n\times n}$$

is said to be a *elementary complex* (or *Householder*) *reflection*. It follows from that $H(u) = H(\alpha u)$ for each nonzero scalar α. The matrix $H(u)$ is both Hermitian and unitary. *Elementary real reflections* are defined similarly as

$$H(v) := I_n - \frac{2vv^{\top}}{v^{\top}v} \in \mathbb{R}^{n\times n},\ 0 \neq v \in \mathbb{R}^n.$$

The matrix $H(v)$ is both symmetric and orthogonal.

The multiplication of a vector $x \in \mathbb{C}^n$ by a reflection $H(u)$ is reduced to the calculation of a single scalar product $u^{\mathrm{H}}x$, multiplication of a vector by a scalar and substraction of two vectors according to

$$H(u)x = x - \left(\frac{2u^{\mathrm{H}}x}{u^{\mathrm{H}}u}\right)u.$$

In particular we have

$$H(u)u = u - \left(\frac{2u^{\mathrm{H}}u}{u^{\mathrm{H}}u}\right)u = u - 2u = -u$$

and $\det(H(u)) = -1$.

A multiplication with $H(u)$ reflects any vector relative to $\mathrm{Ker}(u^{\mathrm{H}})$. Based on this we have an elegant solution to the following problem. Given two different vectors $x, y \in \mathbb{C}^n$ of equal 2-norm, find a unitary matrix U, which transforms x into y, i.e. $y = Ux$. It follows from the reflection property of $H(u)$ that a solution of this problem is $U = H(x - y)$, i.e.,

$$H(x-y)x = y,\ \|x\|_2 = \|y\|_2 > 0.$$

It is often necessary to transform a nonzero vector x into a form with only one nonzero element in the k-th position. Suppose that the vector $x \in \mathbb{C}^n$ is not proportional to the k-th column e_k of the identity matrix I_n. Let $\alpha \in \mathbb{C}$ and $|\alpha| = 1$. Then the required transformation is

$$H(x-y)x = y,\ y := \alpha\|x\|_2 e_k \neq x. \tag{B.3}$$

In the real case we have $\alpha = \pm 1$ and

$$H(x-y)x = y,\ y := \pm\|x\|_2 e_k. \tag{B.4}$$

The choice of α in (B.3), respectively of the sign in (B.4), is done from numerical considerations in order to avoid possible cancellations in substracting close quantities. If the argument of x_k is φ_k, i.e., $x_k = \rho_k \exp(\imath\varphi_k)$, then we choose the argument of α as $\varphi_k + \pi$ which gives $\alpha = -\exp(\imath\varphi_k)$. In this way the k-th element of the vector $x - y$ becomes $(\rho_k + \|x\|_2)\exp(\imath\varphi_k)$.

In the real case if x_k is nonnegative (respectively negative) we choose $y = \|x\|_2 e_k$ (respectively $y = -\|x\|_2 e_k$).

Since the matrix $H(x \mp \|x\|_2 e_k)$ is both Hermitian and unitary we have

$$x = H(x-y)y = \alpha\|x\|_2 H(x-y)e_k = \alpha\|x\|_2 h_k(x-y),$$

where $h_k(x-y)$ is the k-th column of $H(x-y)$.

Now we may solve the following problem. Given a unit vector $x \in \mathbb{C}^n$ find a $n \times (n-1)$ matrix V such that the matrix $U := [x, V]$ is unitary. If x is colinear to some column e_k of I_n, then V contains the other columns of I_n. Suppose now that x is not colinear to a column of I_n. Let $h_1, \ldots, h_n$ be the columns of the reflection $H(x \mp e_1)$ which transforms x into e_1. Then a solution is $V = [h_2, \ldots, h_n]$. Indeed, in this case $x = \pm h_1$.

B.3 QR decomposition

Using a finite number of orthogonal or unitary transformations it is possible to construct, the unitary-triangular, or QR decomposition of a general rectangular matrix. First we define the echelon form of a matrix.

Let $A = [a_{ij}]$ be an $m \times n$ matrix of rank $r \geq 1$. Denote by $k_1, \ldots, k_r$ the numbers of the first r linearly independent columns of A. Let $s \in \{1, \ldots, r\}$ be a given integer. We say that A is in *row s-echelon form* if $a_{ij} = 0$ for $i = 1, \ldots, s$ and $j < k_i$ as well as for $l = 1, \ldots, s$ and $i > k_l$, $j = k_l$. The matrix A is in *row echelon form* if it is in row r-echelon form (for completeness we assume that the zero matrix is in row echelon form and that every matrix is in row 0-echelon form).

Thus, the row echelon form is a matrix A with $a_{i,k_i} \neq 0$ for $i = 1, \ldots, r$ and zero elements before and below each element a_{i,k_i}. If A is in row s-echelon form then $a_{i,k_i} \neq 0$ for $i = 1, \ldots, s$. Also, if A is in row echelon form then $a_{ij} = 0$ for $i > r$. It is obvious also that if A is in row s-echelon form with $s \geq 1$ it is also in row l-echelon form for $l = 1, \ldots, s-1$.

The row echelon form A is an upper triangular matrix, and even an upper trapezoidal matrix if $k_r > r$, e.g. if the first r rows of A are not linearly independent.

An important observation is that if a matrix A of rank r is in row k-echelon form, then $A = \begin{bmatrix} A_k & \times \\ 0 & A_{k+1} \end{bmatrix}$, where the $k \times r_k$ matrix A_k is in row echelon form and the $(m-k) \times (n-r_k)$ matrix A_{k+1} is of rank $r-k$. the corresponding matrix.

Given a general $m \times n$ matrix A of rank r, we may construct a unitary matrix $Q \in \mathcal{U}(m)$, such that

$$A = QR, \tag{B.5}$$

where the $m \times n$ matrix R is in row echelon form. Note that if $r < m$ then the last $m - r$ rows of R are zero and we have

$$A = QR = [Q_1, Q_2] \begin{bmatrix} R_1 \\ 0 \end{bmatrix} = Q_1 R_1, \tag{B.6}$$

where R_1 is an $r \times n$ full row rank upper triangular matrix in row echelon form and Q_1 is the matrix, formed by the first r columns of Q. The factorizations $A = QR$ and $A = Q_1 R_1$ are referred to as the *QR decomposition* and the *condensed* (or *skinny*) *QR decomposition* of A.

Sometimes the QR decomposition includes a right multiplier, $A = QR\Pi$, where Π is the $n \times n$ permutation matrix, chosen so that the first r columns of $A\Pi$ are linearly independent. This is called *QR decomposition with column pivoting.*

There is also a triangular-unitary, or *LQ decomposition* $A = LP^{\mathrm{H}}$, where L is in column echelon form and $P \in \mathcal{U}(n)$. If $r < n$ then the last $n - r$ columns of L are zero and

$$A = LP^{\mathrm{H}} = [L_1, 0] \begin{bmatrix} P_1^{\mathrm{H}} \\ P_2^{\mathrm{H}} \end{bmatrix} = L_1 P_1^{\mathrm{H}}.$$

Here L_1 is of full column rank and P_1 is the matrix, formed by the first r columns of P.

If the matrix A is real, then all matrices in the QR and LQ decompositions may be chosen real with Q and P being orthogonal.

If some additional assumptions on R and L are imposed, then these matrices will be canonical forms for the actions $A \mapsto R = Q^{\mathrm{H}} A$ and $A \mapsto AP$ of $\mathcal{U}(m)$ and $\mathcal{U}(n)$ on $\mathbb{C}^{m \times n}$. To achieve this, one may impose the requirement that the pivots in R and L are real and positive.

Let $l_1, \ldots, l_r$ be the numbers of the first r linearly independent columns of A and hence, of L. The canonical forms R and L contain at most $r(n+1) - \sum_{i=1}^{r} k_i$ and $r(m+1) - \sum_{i=1}^{r} l_i$ nonzero elements (among them r positive), respectively, which constitute the algebraic invariant of A relative to the left and right multiplicative actions of $\mathcal{U}(m)$ and $\mathcal{U}(n)$. Generically $k_i = l_i = i$ and we have $r(2n - r + 1)/2$ and $r(2m - r + 1)/2$ scalar algebraic invariants, respectively. At the same time the integer r-tuples $(k_1, \ldots, k_r)$ and $(l_1, \ldots, l_r)$ constitute the arithmetic invariant for the above actions. The arithmetic and the algebraic invariants form a complete set of invariants for the multiplicative action of the corresponding unitary group.

Using the QR decomposition it is easy to solve the following problem. Given an $m \times n$ matrix X with $n < m$ orthonormal columns, find a $m \times (m - n)$ matrix Y such that the matrix $[X, Y]$ is unitary. If $X = QR$ is a QR decomposition of X, then Y is the matrix, formed by the last $m - n$ columns of Q.

If the rank r of A is less than $\min\{m, n\}$ then the canonical forms R and L may be further compressed, resulting in the *QCP (or URV)-decomposition*, described below. Let $R_1 = [C_1, 0]P^{\mathrm{H}}$ be the LQ decomposition of the matrix R_1 in (B.6). Then we have the *QCP decomposition*

$$A = QCP^{\mathrm{H}} = Q\begin{bmatrix} C_1 & 0_{r\times(n-r)} \\ 0_{(m-r)\times r} & 0_{(m-r)\times(n-r)} \end{bmatrix} P^{\mathrm{H}}, \tag{B.7}$$

where the $r \times r$ matrix C_1 is nonsingular. The QCP decomposition may also be written in the condensed form $A = Q_1 C_1 P_1^{\mathrm{H}}$, where Q_1 and P_1 are the matrices, formed by the first r columns of Q and P, respectively.

In contrast to the singular value decomposition, described below, the QCP decomposition is achieved in a finite number of steps and may be easily updated. The decomposition (B.7) allows also to derive easily the polar decomposition and the singular value decomposition, considered below.

B.4 Schur decomposition

One of the most useful results in applied linear algebra is the following theorem of Schur. It allows to obtain the spectrum of a general square matrix using only unitary (or orthogonal) transformations.

Theorem B.1 [83] *Let $A \in \mathbb{F}^{n\times n}$. Then there exists $U \in \mathcal{U}(n)$, such that*

$$A = UTU^{\mathrm{H}}, \tag{B.8}$$

where

$$T = U^{\mathrm{H}}AU = \begin{bmatrix} t_{11} & t_{12} & \dots & t_{1n} \\ 0 & t_{22} & \dots & t_{2n} \\ \vdots & \vdots & \ddots & \vdots \\ 0 & 0 & \dots & t_{nn} \end{bmatrix}. \tag{B.9}$$

The decomposition (B.8), (B.9) is called the *Schur decomposition* of the matrix A.

The diagonal elements t_{ii} of T are the eigenvalues $\lambda_i = \lambda_i(A)$ of A. The upper triangular matrix T is said to be the *Schur form* of A. The columns of the unitary matrix U form the *Schur basis* of $\mathbb{F}^n$ relative to A (or, briefly, the Schur basis for A). The pair (T, U) is referred to as the *Schur system* of the matrix A.

In this statement of the problem the Schur system of a matrix, and the Schur form, in particular, is not unique. That is why we do not call *this* Schur form canonical. For canonical forms of square matrices relative to the unitary similarity action see [197]. Note that it is possible to achieve any ordering of the eigenvalues of A on the diagonal of T.

If the matrix A is real and has only real eigenvalues then the matrix U may be chosen real and orthogonal, i.e., $U \in \mathcal{O}(n, \mathbb{R})$. If, however, A is real but has at least one pair of complex conjugate eigenvalues, then U cannot be real with T being upper triangular and hence, complex.

Let for instance the real matrix A have n_1 real and $2n_2$ complex eigenvalues ($n_1 + 2n_2 = n$). Set $p = n_1 + n_2$. In this case there exist $W \in \mathcal{O}(n, \mathbb{R})$ and a real upper triangular $p \times p$ block Schur form

$$T^0 = W^\top AW = \begin{bmatrix} T_{11} & T_{12} & \dots & T_{1p} \\ 0 & T_{22} & \dots & T_{2p} \\ \vdots & \vdots & \ddots & \vdots \\ 0 & 0 & \dots & T_{pp} \end{bmatrix}$$

of A, where the blocks T_{ii} are 1×1 and equal to λ_i when they correspond to real eigenvalues λ_i, or 2×2 when they correspond to pairs of complex conjugate eigenvalues $\alpha_i \pm \imath\beta_i$ of A.

A diagonal 2×2 block, corresponding to the eigenvalues $\alpha \pm \imath\beta$, may be further reduced to $\begin{bmatrix} \alpha & \delta \\ \gamma & \alpha \end{bmatrix}$ with $\gamma\delta = -\beta^2$, or $\begin{bmatrix} \gamma & -\beta \\ \beta & \delta \end{bmatrix}$ with $\gamma + \delta = 2\alpha$.

In contrast to the QR decomposition, the Schur decomposition in general cannot be constructed by a finite number of algebraic operations (arithmetic operations plus taking roots). This is a principal limitation following from the famous Abel-Ruffini-Galois theorem [29], which states that the roots of a general algebraic equation of degree ≥ 5 cannot be expressed by its coefficients in a finite number of algebraic operations. Now, if a general $n \times n$ matrix with $n \geq 5$ could be transformed into Schur form by a finite algebraic algorithm, then this would be true for the *companion matrix*

$$C_p := \begin{bmatrix} 0 & 1 & \dots & 0 & 0 \\ 0 & 0 & \dots & 0 & 0 \\ \vdots & \vdots & \ddots & \vdots & \vdots \\ 0 & 0 & \dots & 0 & 1 \\ -a_n & -a_{n-1} & \dots & -a_2 & -a_1 \end{bmatrix}$$

of a general polynomial $p(\lambda) = \lambda^n + a_1\lambda^{n-1} + \cdots + a_n$ of n-th degree. But the eigenvalues $\lambda_i(C_p)$ of the matrix C_p are the roots of the polynomial p, which cannot be computed by a finite algebraic procedure. Hence, an algorithm for the computation of the Schur decomposition of a general matrix must be iterative. An example of such an algorithm is the famous QR algorithm of Francis and Kublanovskaya [83].

If the matrix A is normal, i.e. $A^{\mathrm{H}}A = AA^{\mathrm{H}}$, then its Schur form T from (B.9) is diagonal. Indeed, it follows from the identities $A^{\mathrm{H}}A = UT^{\mathrm{H}}TU^{\mathrm{H}}$ and $AA^{\mathrm{H}} = UTT^{\mathrm{H}}U^{\mathrm{H}}$ that A is normal if and only if its Schur form T is normal, i.e., $T^{\mathrm{H}}T = TT^{\mathrm{H}}$. If we set $t = [t_{12}, \dots, t_{1n}] \in \mathbb{F}_{n-1}$ then a direct computation shows that the $(1,1)$ element of $T^{\mathrm{H}}T$ is $|t_{11}|^2$, while the $(1,1)$ element of TT^{H} is

$|t_{11}|^2 + \|t\|_2^2$. Hence, $t = 0$ and T must be of the form $T = \mathrm{diag}(t_{11}, T_2)$, where T_2 is $(n-1) \times (n-1)$ upper triangular matrix. But now the normality of T is equivalent to normality of T_2. Hence $T_2 = \mathrm{diag}(t_{22}, T_3)$ and $T = \mathrm{diag}(t_{11}, t_{22}, T_3)$. After a total of $n - 1$ such steps we come to the conclusion that T is a diagonal matrix.

The Schur decomposition allows to compute analytic functions of normal matrices as follows (analytic functions of general matrices may be computed using the Jordan decomposition).

Let $f : D \to \mathbb{C}$ be an analytic function,

$$f(z) = \sum_{k=0}^{\infty} a_k (z-a)^k, \ z \in D,$$

in the domain $D \subset \mathbb{C}$, defined on the spectrum of the normal $n \times n$ matrix A (i.e., $\mathrm{spect}(A) \subset D$) with Schur decomposition $A = UTU^{\mathrm{H}}$. In view of the normality of A we have $T = \mathrm{diag}(t_{11}, \ldots, t_{nn})$. Now we may define the matrix-valued function f as follows (we use the same letter f for the scalar-valued function and for its matrix-valued counterpart)

$$f(A) = \sum_{k=0}^{\infty} a_k (A - aI_n)^k, \ \mathrm{spect}(A) \subset D.$$

The expression $f(A)$ is correctly defined if $\|A - aI_n\|$ is smaller than the distance from the point $a \in \mathbb{C}$ to the boundary of D. Since $A^k = UT^kU^{\mathrm{H}}$ we may compute $f(A)$ from $f(A) := Uf(T)U^{\mathrm{H}}$, where $f(T) := \mathrm{diag}(f(t_{11}), \ldots, f(t_{nn}))$. In particular, if A is Hermitian and nonnegative definite, then T is real diagonal with nonnegative diagonal elements and we may compute the *nonnegative definite square root* of A as $A^{1/2} := U \mathrm{diag}\left(\sqrt{t_{11}}, \ldots, \sqrt{t_{nn}}\right) U^{\mathrm{H}}$.

B.5 Polar decomposition

A direct consequence of the QR and Schur decompositions is the so called *polar decomposition* of a square matrix A. Suppose first that A is nonsingular. Then the matrices $A^{\mathrm{H}}A$ and AA^{H} are Hermitian positive definite and normal in particular. Consider the matrix $U_l := A(A^{\mathrm{H}}A)^{-1/2}$. We have

$$U_l U_l^{\mathrm{H}} = A(A^{\mathrm{H}}A)^{-1/2}(A^{\mathrm{H}}A)^{-1/2}A^{\mathrm{H}} = A(A^{\mathrm{H}}A)^{-1}A^{\mathrm{H}} = AA^{-1}A^{-\mathrm{H}}A^{\mathrm{H}} = I_n$$

and hence, U_l is unitary. Now we have the identity

$$A = U_l (A^{\mathrm{H}}A)^{1/2}, \tag{B.10}$$

where the matrix $(A^{\mathrm{H}}A)^{1/2}$ is Hermitian positive definite.

The decomposition (B.10) is called a *polar decomposition* of A. There is also a similar polar decomposition, defined as

$$A = (AA^{\mathrm{H}})^{1/2}U_r, \tag{B.11}$$

where the matrix $(AA^{\mathrm{H}})^{1/2}$ is Hermitian positive definite and the matrix $U_r := (AA^{\mathrm{H}})^{-1/2}A$ is unitary.

The decompositions (B.10) and (B.11) are generalizations of the polar decomposition $z = \sqrt{z\bar{z}}\exp(\imath\varphi) = |z|\exp(\imath\varphi)$ of the complex number $z \neq 0$ and this is the origin of their name.

Decompositions of type (B.10) and (B.11) are valid also when A is a singular $n \times n$ matrix of rank $r < n$. Here the Hermitian factors are nonnegative definite and the unitary factors are defined in a different way. Indeed, in this case the we obtain

$$A = Q \begin{bmatrix} C_1 & 0 \\ 0 & 0 \end{bmatrix} P^{\mathrm{H}} = Q_1 C_1 P_1^{\mathrm{H}},$$

where $P = [P_1, P_2]$, $Q = [Q_1, Q_2] \in \mathcal{U}(n)$, the matrices Q_1 and P_1 are $n \times r$ and the $r \times r$ matrix C_1 is nonsingular. A straightforward calculation shows the existence of the polar decompositions

$$A = V_l (A^{\mathrm{H}}A)^{1/2}, \ V_l := \left[Q_1 C_1 \left(C_1^{\mathrm{H}} C_1\right)^{-1/2}, Q_2\right] P^{\mathrm{H}} \in \mathcal{U}(n), \tag{B.12}$$

and

$$A = (AA^{\mathrm{H}})^{1/2} V_r, \ V_r := Q \left[P_1 C_1^{\mathrm{H}} \left(C_1 C_1^{\mathrm{H}}\right)^{-1/2}, P_2\right]^{\mathrm{H}} \in \mathcal{U}(n). \tag{B.13}$$

Note that here

$$(A^{\mathrm{H}}A)^{1/2} = P \begin{bmatrix} \left(C_1^{\mathrm{H}} C_1\right)^{1/2} & 0_{r \times (n-r)} \\ 0_{(n-r) \times r} & 0_{(n-r) \times (n-r)} \end{bmatrix} P^{\mathrm{H}}$$

and

$$(AA^{\mathrm{H}})^{1/2} = Q \begin{bmatrix} \left(C_1 C_1^{\mathrm{H}}\right)^{1/2} & 0_{r \times (n-r)} \\ 0_{(n-r) \times r} & 0_{(n-r) \times (n-r)} \end{bmatrix} Q^{\mathrm{H}}.$$

A polar decomposition of a rectangular matrix may also be defined as the product of a Hermitian matrix and a matrix with orthonormal columns or rows. For instance, if A is rectangular with full column rank then relation (B.10) is valid, where U_l is a rectangular matrix with orthonormed columns. Similarly, if A is rectangular of full row rank then (B.11) is valid with U_r being a rectangular matrix with orthonormed rows.

For real matrices we have similar results, replacing "Hermitian" by "symmetric" and "unitary" by "orthogonal".

B.6 Singular value decomposition

As a direct consequence of the QR decomposition, the Schur decomposition and the polar decomposition we may deduce the singular value decomposition of a general matrix. This decomposition is widely used in matrix theory and its applications.

A *singular value decomposition* (or briefly *SVD*) of the $m \times n$ matrix A is the product $A = USV^{\mathrm{H}}$ if A is complex, or $A = USV^{\top}$ if A is real, where the matrices U, V are unitary in the complex case and orthogonal in the real case. The matrix S in both cases is real diagonal with nonnegative diagonal elements ordered in descending magnitude. It is a canonical form of A under the above action of $\mathcal{U}(m) \times \mathcal{U}(n)$ or $\mathcal{O}(m, \mathbb{R}) \times \mathcal{O}(n, \mathbb{R})$. The description of S depends on the integers m, n and $r = \operatorname{rank}(A)$. In the trivial case $A = 0_{m\times n}$ we have $S = 0_{m\times n}$. If $r \geq 1$ we have four possibilities: $r = m = n$, $r = n < m$, $r = m < n$ and $r < \min\{m, n\}$.

1. The case $r = m = n$. Consider the polar decomposition $A = U_l(A^{\mathrm{H}}A)^{1/2}$ of A, where $U_l = A(A^{\mathrm{H}}A)^{-1/2}$. Since the matrix $(A^{\mathrm{H}}A)^{1/2}$ is Hermitian positive definite, its Schur decomposition is $(A^{\mathrm{H}}A)^{1/2} = V\Sigma V^{\mathrm{H}}$, where $V \in \mathcal{U}(n)$ and $\Sigma = \operatorname{diag}(\sigma_1, \ldots, \sigma_r)$, $\sigma_1 \geq \cdots \geq \sigma_r > 0$. Hence, we have the singular value decomposition $A = USV^{\mathrm{H}}$, $S = \Sigma$, where $U := U_lV \in \mathcal{U}(r)$ and the matrix $U_l = A(A^{\mathrm{H}}A)^{-1/2} \in \mathcal{U}(r)$ is the left unitary factor in the polar decomposition $A = U_l(A^{\mathrm{H}}A)^{1/2}$ of A. Thus, we have proved the existence of the SVD for nonsingular matrices.

2. The case $r = n < m$. Here the QR decomposition of A is $A = QR = Q_1R_1$, where $Q = [Q_1, Q_2] \in \mathcal{U}(m)$, the matrix Q_1 is $m \times r$ and the $r \times r$ upper triangular matrix R_1 is nonsingular. If $R_1 = U_1\Sigma V^{\mathrm{H}}$ is the SVD of R_1, where $U_1, V \in \mathcal{U}(r)$, then the SVD of A is $A = U \begin{bmatrix} \Sigma \\ 0_{m-r} \end{bmatrix} V^{\mathrm{H}}$, where $U := [Q_1U_1, Q_2]$.

3. The case $r = m < n$. The LQ decomposition of A here is $A = LP^{\mathrm{H}} = L_1P_1^{\mathrm{H}}$, where $P = [P_1, P_2] \in \mathcal{U}(n)$, the matrix P_1 is $n \times r$ and the $r \times r$ lower triangular matrix L_1 is nonsingular. If $L_1 = U\Sigma V_1^{\mathrm{H}}$ is the SVD of L_1, where $U, V_1 \in \mathcal{U}(r)$, then the SVD of A is $A = U[\Sigma, 0_{n-r}]V^{\mathrm{H}}$, where $V := [P_1V_1, P_2]$.

4. The case $r < \min\{m, n\}$. Consider the compressed QCP decomposition $A = QCP^{\mathrm{H}} = Q_1C_1P_1^{\mathrm{H}}$ from (B.7), where the matrices Q_1 and P_1 are formed by the first r columns of Q and P respectively. Let $A_1 = U_1\Sigma V_1^{\mathrm{H}}$ be the SVD of the nonsingular $r \times r$ matrix A_1. Then the SVD of A is

$$A = USV^{\mathrm{H}} = U \begin{bmatrix} \Sigma & 0_{r\times(n-r)} \\ 0_{(m-r)\times r} & 0_{(m-r)\times(n-r)} \end{bmatrix} V^{\mathrm{H}}, \qquad \text{(B.14)}$$

where $U := [Q_1U_1, U_2]$, $P = [P_1V_1, P_2]$.

The singular value decomposition of a general matrix A may always be written in the form (B.14), where any of the zero matrices $0_{p\times q}$ is considered void if $p = 0$ or $q = 0$.

The numbers $\sigma_i = \sigma_i(A) \geq 0$ are called the *singular values* of the matrix A.

Their number is $\min\{m,n\}$. The first r of them, where $r = \operatorname{rank}(A)$, are the positive eigenvalues of $(A^{\mathrm{H}}A)^{1/2}$ or $(AA^{\mathrm{H}})^{1/2}$.

Since

$$A^{\mathrm{H}}A = V\operatorname{diag}(\Sigma^2, 0_{(n-r)\times(n-r)})V^{\mathrm{H}}$$

and

$$AA^{\mathrm{H}} = U\operatorname{diag}(\Sigma^2, 0_{(m-r)\times(m-r)})U^{\mathrm{H}},$$

we see that U and V are the matrices of eigenvectors of $A^{\mathrm{H}}A$ and AA^{H}, respectively. The columns of U and V are also referred to as the *left* and *right singular vectors* of A. Using the left and right singular vectors of A, the SVD of A may be written as

$$A = U_1\Sigma V_1^{\mathrm{H}} = \sum_{i=1}^{r} \sigma_i u_i v_i^{\mathrm{H}},$$

where $r = \operatorname{rank}(A)$ and the matrices U, V are partitioned as $U = [U_1, U_2]$, $V = [V_1, V_2]$ with U_1 being $m \times r$ and V_1 being $n \times r$. Hence, we get the following orthonormed bases for the subspaces $\operatorname{Rg}(A)$ and $\operatorname{Ker}(A)$:

$$\operatorname{Rg}(A) = \operatorname{Rg}(U_1) = \operatorname{Ker}\left(U_2^{\mathrm{H}}\right), \ \operatorname{Ker}(A) = \operatorname{Ker}\left(V_1^{\mathrm{H}}\right) = \operatorname{Rg}(V_2).$$

The SVD is also used in the determination of the so called *pseudo-inverse* of an arbitrary matrix.

Consider the SVD (B.14), where $0 \le r \le \min\{m,n\}$, i.e., the case $A = 0_{m\times n}$ is not excluded. As usual, the matrices $0_{p\times 0}$ and $0_{0\times q}$ are considered void.

The $n \times m$ matrix

$$A^{\dagger} := VS^{\dagger}U^{\mathrm{H}} := V \begin{bmatrix} \Sigma^{-1} & 0_{r\times(m-r)} \\ 0_{(n-r)\times r} & 0_{(n-r)\times(m-r)} \end{bmatrix} U^{\mathrm{H}} \tag{B.15}$$

is called a *pseudo-inverse* of the $m \times n$ matrix A. Since there are other pseudo-inverses as well, this particular one is also referred to as the *Moore-Penrose pseudo-inverse*. The pseudo-inverse $A^{\dagger}$ exists for any matrix A. In particular, $0_{m\times n}^{\dagger} = 0_{n\times m}$ and $(A^{\dagger})^{\dagger} = A$.

All solutions of the *least squares problem*

$$\min\{\|Ax - b\|_2 : x \in \mathbb{F}^n\}$$

are given by

$$x = A^{\dagger}b + \left(I_n - A^{\dagger}A\right)c,$$

where the vector $c \in \mathbb{F}^n$ is arbitrary. Under the additional requirement $\|x\|_2 \to \min$ the solution $x^0 = A^{\dagger}b$ is unique.

B.7 Notes and references

Unitary and orthogonal matrix decompositions are considered in most books on linear algebra and matrix theory, see [70, 71, 83, 107, 228] and [157, 36, 54, 224].

Appendix C

Kronecker product of matrices

C.1 Introductory remarks

In this appendix we present some basic facts about the Kronecker product of matrices which is a very useful tool in analyzing and solving matrix equations.

C.2 Definitions and properties

Let the matrices $A = [a_{ij}] \in \mathbb{F}^{m\times n}$ and $B \in \mathbb{F}^{p\times q}$ be given.

Definition C.1 The matrix

$$A \otimes B := [a_{ij}B] = \begin{bmatrix} a_{11}B & a_{12}B & \dots & a_{1n}B \\ a_{21}B & a_{22}B & \dots & a_{2n}B \\ \vdots & \vdots & \ddots & \vdots \\ a_{m1}B & a_{m2}B & \dots & a_{mn}B \end{bmatrix} \in \mathbb{F}^{mp\times nq}$$

is called the *Kronecker product* (or the *tensor product*) of the matrices A and B.

The Kronecker product $A \otimes B$ is an $m \times n$ block matrix, whose (i,j)-block is the $p \times q$ matrix $a_{ij}B$. In the above representation A in turn may be a block matrix, i.e., Definition C.1 is valid with a_{ij} being arbitrary $m_i \times n_j$ matrices. Note that no restrictions on the sizes of A and B are imposed for the matrix $A \otimes B$ to exist.

Usually the standard matrix product, the Kronecker product and the standard matrix summation are considered as algebraic operations of decreasing priority, e.g., the expression $E = ((AB) \otimes C) + D$ is written without brackets as $E = AB \otimes$

$C+D$. However, to avoid misunderstandings, we shall consider all multiplications as operations of equal priority, resulting here in the expression $E=(AB)\otimes C+D$.

Some important applications of the Kronecker product to the theory of matrix equations and other areas are based on the *column-wise vector representation* of the product AXB, namely (we assume that the standard matrix products are correctly defined)

$$\mathrm{vec}(AXB)=(B^\top\otimes A)\mathrm{vec}(X) \tag{C.1}$$

and in particular

$$\mathrm{vec}(AX)=(I_n\otimes A)\mathrm{vec}(X),\ \ \mathrm{vec}(XB)=\left(B^\top\otimes I_m\right)\mathrm{vec}(X).$$

As a direct corollary we have

$$\|AX\|_{\mathrm{F}}=\|\mathrm{vec}(AX)\|_2=\|(I_n\otimes A)\mathrm{vec}(X)\|_2\leq\|I_n\otimes A\|_2\|\mathrm{vec}(X)\|_2=\|A\|_2\|X\|_{\mathrm{F}}.$$

Here we have used the fact that $I_n\otimes A=\mathrm{diag}(A,\ldots,A)$ and hence, $\|I_n\otimes A\|_2=\|A\|_2$. Similarly, $\|XB\|_{\mathrm{F}}=\|B^\top X^\top\|_{\mathrm{F}}\leq\|B\|_2\|X\|_{\mathrm{F}}$ and

$$\|AXB\|_{\mathrm{F}}\leq\|A\|_2\|XB\|_{\mathrm{F}}\leq\|A\|_2\|C\|_2\|X\|_{\mathrm{F}}.$$

The generalization of this result to the product of any number of matrices now follows by inspection.

If the matrix X is $m\times n$ then

$$\mathrm{vec}(X^\top)=P_{m,n}\mathrm{vec}(X),$$

where

$$P_{m,n}=\sum_{i=1}^{m}\sum_{j=1}^{n}E_{ij}(m,n)\otimes E_{ji}(m,n)\in\mathbb{R}^{mn\times mn}$$

and $E_{ij}(m,n)\in\mathbb{R}^{m\times n}$ is defined in Appendix 10.17. Here $P_{m,n}$ is a permutation matrix (its columns are a permutation of the columns of the identity matrix I_{mn}), called the *vec-permutation matrix*. It has the property

$$P_{m,n}=P_{n,m}^\top=P_{n,m}^{-1}. \tag{C.2}$$

The matrix $P_{n,n}$ is denoted also as P_{n^2}.

The matrices $A\otimes B$ and $B\otimes A$ have equal sizes $mp\times nq$. This is in contrast to the standard matrix product, where one (or both) of the products AB and BA may not be defined, or AB and BA may be defined but have different sizes.

The Kronecker product is in general not commutative, i.e. $A\otimes B\neq B\otimes A$. In addition, $A\neq I\otimes A$ and $A\neq A\otimes I$ unless $I=1$ (the scalar unit). However, the Kronecker product is associative, and distributive relative to the standard summation:

$$\begin{aligned}(A\otimes B)\otimes C &= A\otimes(B\otimes C)=A\otimes B\otimes C, \qquad \text{(C.3)}\\ (A+B)\otimes C &= A\otimes C+B\otimes C,\ C\otimes(A+B)=C\otimes A+C\otimes B.\end{aligned}$$

A basic relation between the standard matrix product and the Kronecker product is

$$(A \otimes B)(C \otimes D) = (AC) \otimes (BD). \tag{C.4}$$

Furthermore we have

$$(A \otimes B)^\top = A^\top \otimes B^\top, \ \overline{A \otimes B} = \overline{A} \otimes \overline{B}, \ (A \otimes B)^{\mathrm{H}} = A^{\mathrm{H}} \otimes B^{\mathrm{H}}. \tag{C.5}$$

If the matrices A and B are square and nonsingular then their Kronecker product $A \otimes B$ is also square and nonsingular, and

$$(A \otimes B)^{-1} = A^{-1} \otimes B^{-1}. \tag{C.6}$$

Example C.2 Let the matrices A and B be nonsingular and A be a 2×2 matrix. Then

$$(A \otimes B)^{-1} = \frac{1}{\det(A)} \begin{bmatrix} a_{22}B^{-1} & -a_{12}B^{-1} \\ -a_{21}B^{-1} & a_{11}B^{-1} \end{bmatrix}.$$

◊

The transposition and inversion of a Kronecker product do not invert the order of factors, in contrast to the standard matrix product, where $(AB)^\top = B^\top A^\top$ and $(AB)^{-1} = B^{-1}A^{-1}$. To invert the order of multiplication in a Kronecker product, one may use the formula [107]

$$(A \otimes B)P_{n,q} = P_{m,p}(B \otimes A), \tag{C.7}$$

or, equivalently, $A \otimes B = P_{m,p}B \otimes AP_{q,n}$.

Using (C.7) we may derive an expression, similar to (C.1), for the *row-wise vector representation* of the product AXB. Denote by

$$\mathrm{row}(X) = [\xi_1, \xi_2, \ldots, \xi_m] \in \mathbb{F}^{1\times mn}$$

the *row-wise vectorization* of the $m \times n$ matrix X with rows $\xi_1, \ldots, \xi_m \in \mathbb{F}^{1\times n}$. We have $\mathrm{row}(X) = \mathrm{vec}^\top(X^\top) = \mathrm{vec}^\top(X)P_{n,m}$. Representing both sides of the relation $Y = AXB$ as row vectors we obtain the row-wise counterpart of (C.1)

$$\mathrm{row}(AXB) = \mathrm{row}(X)(A^\top \otimes B). \tag{C.8}$$

The singular values of the matrix $A \otimes B$ are the products $\sigma_i(A)\sigma_k(B)$, where $\sigma_i(A)$ and $\sigma_k(B)$ are the singular values of A and B. Indeed, let $A = U_A S_A V_A^{\mathrm{H}}$ and $B = U_B S_B V_B^{\mathrm{H}}$ be the SVD of A and B, respectively. Then $A \otimes B = (U_A \otimes U_B)(S_A \otimes S_B)\left(V_A^{\mathrm{H}} \otimes V_B^{\mathrm{H}}\right)$ is the SVD of $A \otimes B$ (up to reordering of the diagonal of $S_A \otimes S_B$). Since the matrices S_A and S_B are diagonal with diagonal elements $\sigma_i(A)$ and $\sigma_k(B)$, respectively, the diagonal elements of $S_A \otimes S_B$ are all possible products $\sigma_i(A)\sigma_k(B)$.

Similarly, if $A \in \mathbb{F}^{m\times m}$ and $B \in \mathbb{F}^{n\times n}$, then the eigenvalues of their Kronecker product $A\otimes B$ are all possible products $\lambda_i(A)\lambda_k(B)$, where $\lambda_i(A)$ and $\lambda_k(B)$ are the eigenvalues of A and B. Indeed, let $A = W_A T_A W_A^{\mathrm{H}}$ be the Schur decomposition of A. Then $A \otimes B = (W_A \otimes I_n)(T_A \otimes B)\left(W_A^{\mathrm{H}} \otimes I_n\right)$ and the eigenvalues of $A \otimes B$ are those of $T_A \otimes B$. But $T_A \otimes B$ is an upper triangular block matrix with diagonal blocks $\lambda_i(A)B$. Each of these blocks has spectrum $\lambda_i(A)\mathrm{spect}(B)$. The union of these spectra is the collection $\mathrm{spect}(A)\otimes\mathrm{spect}(B)$ of all possible products $\lambda_i(A)\lambda_k(B)$ (for operations with collections see Section A.2).

If $A \in \mathbb{C}^{m\times m}$, $B \in \mathbb{C}^{n\times n}$, $C \in \mathbb{C}^{n\times m}$, $x \in \mathbb{C}^l$ and F is an analytic function, then we have

$$\begin{aligned}
\det(B \otimes A) &= (\det(B))^m(\det(A))^n, \ \mathrm{tr}(B \otimes A) = \mathrm{tr}(B)\mathrm{tr}(A),\\
\exp(B \otimes A) &= \exp(B)\exp(A),\\
F(I_n \otimes A) &= I_n \otimes F(A), \ F(A \otimes I_n) = F(A) \otimes I_n,\\
C \otimes x &= (I_n \otimes x)C, \ C \otimes x^\top = C\left(I_m \otimes x^\top\right).
\end{aligned}$$

Let $A \in \mathbb{C}^{m\times m}$ and $B \in \mathbb{C}^{n\times n}$. The *Kronecker sum* of the matrices B and A is the matrix

$$B \oplus A := B \otimes I_m + I_n \otimes A \in \mathbb{C}^{mn\times mn}.$$

Note that the Kronecker summation is not commutative.

The eigenvalues of $B \oplus A$ are all possible sums $\lambda_i(A) + \lambda_k(B)$. Indeed, let $B = W_B T_B W_B^{\mathrm{H}}$ be the Schur decomposition of B. Then the matrix $B \oplus A$ is similar to

$$\left(U_B^{\mathrm{H}} \otimes I_m\right)(B \oplus A)(U_B \otimes I_m) = T_B \otimes I_m + I_n \otimes A = T_B \oplus A.$$

The matrix $T_B \oplus A$ is $n \times n$ upper block triangular with $m \times m$ diagonal blocks $A + \lambda_k(B)I_m$. The spectrum of such a block is $\mathrm{spect}(A) + \{\lambda_k(B)\}$. The whole spectrum of $T_B \oplus A$ and hence, of $B \oplus A$ is the union of the spectra of the diagonal blocks, which is exactly the set $\mathrm{spect}(A) + \mathrm{spect}(B)$.

Thus we have the problem of finding a simple expression for the spectrum of the matrix $M := A\otimes B + C\otimes D$, where A, C are $m\times m$ and B, D are $n\times n$. This is only possible if some special structure of the involved matrices is preassumed. Suppose for instance that the matrices A and C have a joint Schur basis U, i.e., that there is a unitary matrix U, such that the matrices $T_A := U^{\mathrm{H}}AU$ and $T_C := U^{\mathrm{H}}CU$ are upper triangular. Then the matrix M is similar to $\widetilde{M} = \left(U^{\mathrm{H}} \otimes I_n\right)M(U \otimes I_n) = T_A \otimes B + T_C \otimes D$. The matrix $\widetilde{M}$ is $m \times m$ upper block triangular with $n \times n$ diagonal blocks $\lambda_i(A)B + \lambda_k(C)D$. Thus, the spectrum of $\widetilde{M}$ and hence, of M is the sum of collections

$$\mathrm{spect}(M) = \sum_{i,k=1}^{m} \mathrm{spect}(\lambda_i(A)B + \lambda_k(C)D)).$$

The above results are directly applicable to the analysis of spectra of linear matrix operators.

Example C.3 Let $A \in \mathbb{C}^{m\times m}$ and $B \in \mathbb{C}^{n\times n}$. Consider the operators $\mathcal{L}_c$ and $\mathcal{L}_d$ in the continuous- and discrete-time Sylvester equations $\mathcal{L}_c(X) := AX + XB = C$ and $\mathcal{L}_d(X) := AXB - X = C$, respectively, where X is an $m \times n$ unknown matrix. The spectrum of a linear operator is the spectrum of its matrix. Hence,

$$\mathrm{spect}(\mathcal{L}_c) = \mathrm{spect}\left(I_n \otimes A + B^\top \otimes I_m\right) = \mathrm{spect}(A) + \mathrm{spect}(B)$$

and

$$\mathrm{spect}(\mathcal{L}_d) = \mathrm{spect}\left(B^\top \otimes A - I_{mn}\right) = \mathrm{spect}(A)\mathrm{spect}(B) - \{1\}.$$

$\diamondsuit$

C.3 Notes and references

More detailed information about the Kronecker product and sum of matrices and their applications may be found in [19, 230, 84], see also [107, 157, 231].

Appendix D

Fixed point principles

D.1 Introductory remarks

Among the most powerful tools to study the problems of existence and uniqueness of the solutions of various classes of equations, including equations arising in perturbation analysis of matrix problems, are the *topological fixed point principles* named after *Banach* and *Schauder* (the Schauder principle in its finite dimensional version is known also as the *Brauer* principle). In this section we briefly state these principles for operators in finite dimensional spaces.

D.2 Banach principle

Consider a finite dimensional space $\mathcal{X}$ endowed with the norm $\|\cdot\|$ and let $\Pi : \mathcal{B}_\alpha \to \mathcal{X}$ be a (nonlinear) operator, defined in the ball $\mathcal{B}_\alpha := \{x \in \mathcal{X} : \|x\| \leq \alpha\}$ for some $\alpha > 0$. We are interested in the existence and uniqueness of solutions to the operator equation

$$x = \Pi(x). \tag{D.1}$$

The solutions of (D.1) are called *fixed points* of the operator Π.

Definition D.1 *The operator* Π *is said to be a* contraction *(or a* contractive operator*) if there exists a nonnegative constant* $l < 1$ *such that* Π *satisfies the* Lipschitz condition

$$\|\Pi(x) - \Pi(y)\| \leq l\|x - y\|, \ \ x, y \in \mathcal{B}_\alpha.$$

The quantity $l = l(\alpha)$ *is the* Lipschitz constant *of* Π.

The main result for contractions is formulated in Theorem D.2 below.

Theorem D.2 *(Banach principle). Let the inequality*

$$\alpha_0 := \frac{\|\Pi(0)\|}{1-l} \leq \alpha$$

be fulfilled for the operator $\Pi : \mathcal{B}_\alpha \to \mathcal{X}$ *with Lipschitz constant* $l < 1$.

Then the following assertions hold.

1. *The contractive operator* Π *has a unique fixed point* $\xi \in \mathcal{B}_\alpha$ *and* $\|\xi\| \leq \alpha_0$.

2. *The unique solution* ξ *of equation (D.1) may be obtained as the limit point of the iterative process*

$$x_{k+1} = \Pi(x_k), \ k = 0, 1, \ldots, \tag{D.2}$$

where the point $x_0 \in \mathcal{B}_{\alpha_0}$ *is arbitrary.*

3. *The rate of convergence of the iteration (D.2) to the solution* ξ *is determined by*

$$\|\xi - x_k\| \leq \frac{\theta l^k}{1-l}, \tag{D.3}$$

where $\theta := \|x_1 - x_0\| = \|\Pi(x_0) - x_0\|$.

Proof. First we show that the operator Π maps the set $\mathcal{B}_{\alpha_0}$ into itself which yields that the sequence $\{x_k\}$ is well defined. Indeed, for $x \in \mathcal{B}_{\alpha_0}$ we have

$$\begin{aligned} \|\Pi(x)\| &= \|\Pi(x) - \Pi(0) + \Pi(0)\| \leq \|\Pi(x) - \Pi(0)\| + \|\Pi(0)\| \\ &\leq l\|u\| + \|\Pi(0)\| \leq l\alpha_0 + \|\Pi(0)\| = \alpha_0. \end{aligned}$$

Furthermore, we have

$$\begin{aligned} \|x_{k+1} - x_k\| &= \|\Pi(x_k) - \Pi(x_{k-1})\| \leq l\|x_k - x_{k-1}\| \leq l^2\|x_{k-1} - x_{k-2}\| \\ \ldots &\leq l^k\|x_1 - x_0\| = l^k\theta. \end{aligned}$$

Using the last inequality, we may estimate the quantity $\|x_{k+m} - x_k\|$ for $m > 2$ subtracting and adding some terms in order to get differences of the type $x_{i+1} - x_i$, whose norms have been already estimated. We have $x_{k+m} - x_k = \sum_{i=k}^{k+m-1}(x_{i+1} - x_i)$ and

$$\begin{aligned} \|x_{k+m} - x_k\| &= \left\|\sum_{i=k}^{k+m-1}(x_{i+1} - x_i)\right\| \leq \sum_{i=k}^{k+m-1}\|x_{i+1} - x_i\| \\ &\leq \theta\sum_{i=k}^{k+m-1} l^i = \theta l^k \sum_{i=0}^{m-1} l^i < \frac{\theta l^k}{1-l}. \end{aligned} \tag{D.4}$$

Since $l < 1$ we have $\lim_{k\to\infty}\|x_{k+m} - x_k\| = 0$ and hence $\{x_k\}$ is a Cauchy sequence. Thus, it is convergent to some element $\xi \in \mathcal{B}_{\alpha_0}$, namely $\lim_{k\to\infty} x_k = \xi$. Passing to the limit $k \to \infty$ on both sides of (D.2) we see that ξ is a fixed point of Π, i.e., $\xi = \Pi(\xi)$.

In turn, letting $m \to \infty$ in (D.4) we have $x_{k+m} \to \xi$ and hence the estimate (D.3) for the rate of convergence holds.

To demonstrate that the solution $\xi \in \mathcal{B}_{\alpha_0}$ is unique in $\mathcal{B}_\alpha$, suppose that there is another solution $\eta \in \mathcal{B}_\alpha$, different from ξ. Then

$$0 < \|\xi - \eta\| = \|\Pi(\xi) - \Pi(\eta)\| \leq l\|\xi - \eta\| < \|\xi - \eta\|.$$

This is a contradiction, which proves that the solution is unique. □

Example D.3 Suppose that we have a scalar equation $f(x) = 0$, where the function f is defined on the interval $[-\alpha, \alpha]$ and satisfies there the two-sided Lipschitz condition $m|x - y| \leq |f(x) - f(y)| \leq M|x - y|$, where $0 < m \leq M < \infty$. Then we can rewrite the equation in an equivalent operator form $x = \Pi(x) := x - Kf(x)$, $K := 2/(M + m)$. The operator Π is a contraction with Lipschitz constant $l := (M - m)/(M + m) < 1$. Hence, the equation has a unique root $\xi \in [-\alpha, \alpha]$, such that $|\xi| \leq \alpha_0$, provided that $\alpha_0 := |f(0)|/(1 - l) = |f(0)|/(mK) \leq \alpha$. ◇

D.3 Generalized Banach principle

Consider now the case when the space $\mathcal{X}$ is endowed with the generalized norm $|\cdot| : \mathcal{X} \to \mathbb{R}_+^s$, $s > 1$. Suppose that the operator $\Pi : \mathcal{B}_\rho \to \mathcal{X}$ satisfies the *generalized Lipschitz condition*

$$|\Pi(x) - \Pi(y)| \preceq L|x - y|; \ \ x, y \in \mathcal{B}_\rho.$$

Here $B_\rho := \{x \in \mathcal{X} : |x| \preceq \rho\} \subset \mathcal{X}$ is the generalized ball centered at the origin and of generalized radius $\rho \in \mathbb{R}_+^s$, while $L = L(\rho) \in \mathbb{R}_+^{s \times s}$ is the *Lipschitz matrix* of the mapping Π.

Definition D.4 *The operator* Π *is said to be a* generalized contraction *on the set* $\mathcal{B}_\rho$ *if the matrix* L *is* convergent, *i.e. if its* spectral radius rad(L) *is less than 1.*

We recall that according to the Perron-Frobenius theorem [26] the spectral radius of a nonnegative matrix is equal to its largest nonnegative eigenvalue. For a nonnegative convergent matrix L the matrix $I_s - L$ is invertible and the matrix $(I - L)^{-1}$ is well defined and nonnegative.

Example D.5 Let $\Pi = [\Pi_1, \ldots, \Pi_q]^\top : D \to \mathbb{F}^q$, where $D \subset \mathbb{F}^q$. If the Jacobi matrix

$$\Pi'(x) = [J_{ij}(x)] := \left[\frac{\partial \Pi_i(x)}{\partial x_j}\right] \in \mathbb{F}^{q \times q}, \ x = [x_1, \ldots, x_q]^\top$$

exists, then the Lipschitz matrix may be taken as $L = [l_{ij}]$, where l_{ij} is the supremum of $|J_{ij}(x)|$ in x over the domain D of Π. ◇

For generalized contractions we have similar results as for usual contractions according to the following theorem.

Theorem D.6 *(Generalized Banach principle). Let the inequality*

$$\rho_0 := (I - L)^{-1}|\Pi(0)| \preceq \rho$$

be fulfilled.

Then the following statements hold.

1. *The contractive operator Π has a unique fixed point $\xi \in \mathcal{B}_\rho$ and $|\xi| \preceq \rho_0$.*

2. *The unique solution ξ of the operator equation (D.1) may be obtained as the limit point of the iterative process*

$$x_{k+1} = \Pi(x_k), \; k = 0, 1, \ldots, \tag{D.5}$$

where $x_0 \in \mathcal{B}_{\rho_0}$ is arbitrary.

3. *The rate of convergence of the approximations x_k to the solution ξ is determined by*

$$|\xi - x_k| \preceq L^k (I - L)^{-1} \beta, \tag{D.6}$$

where $\beta := |x_1 - x_0| = |\Pi(x_0) - x_0| \in \mathbb{R}^s_+$.

Proof. We have

$$\begin{aligned} |\Pi(x)| &= |\Pi(x) - \Pi(0) + \Pi(0)| \preceq |\Pi(x) - \Pi(0)| + |\Pi(0)| \\ &\preceq L|x| + |\Pi(0)| \preceq L\rho_0 + |\Pi(0)| = \rho_0 \end{aligned}$$

and the operator Π maps the set $\mathcal{B}_{\rho_0}$ into itself. Hence, the sequence $\{u_k\}_0^\infty$ is well defined via (D.5).

Furthermore

$$\begin{aligned} |x_{k+1} - x_k| &= |\Pi(x_k) - \Pi(x_{k-1})| \preceq L|x_k - x_{k-1}| \\ &\preceq L^2 |x_{k-1} - x_{k-2}| \preceq \cdots \preceq L^k |x_1 - x_0| := L^k \beta \end{aligned}$$

and

$$\begin{aligned} |x_{k+m} - x_k| &= \left| \sum_{i=k}^{k+m-1} (x_{i+1} - x_i) \right| \preceq \sum_{i=k}^{k+m-1} |x_{i+1} - x_i| \\ &\preceq \left(\sum_{i=k}^{k+m-1} L^i \right) \beta = L^k \left(\sum_{i=0}^{m-1} L^i \right) \beta \preceq L^k (I - L)^{-1} \beta. \end{aligned} \tag{D.7}$$

Since $\lim_{k\to\infty} L^k = 0$ then the sequence $\{x_k\}$ is convergent to an element $\xi \in \mathcal{B}_{\rho_0}$. Passing to the limit $k \to \infty$ in (D.5), we see that $\xi = \Pi(\xi)$, i.e., the operator equation (D.1) has ξ as a solution.

Setting $m \to \infty$ in (D.7) we get the rate of convergence (D.6) for the generalized norms of the differences $\xi - x_k$.

To show that the operator Π has no other fixed points in $\mathcal{B}_\rho$ except $\xi \in \mathcal{B}_{\rho_0}$, let $\eta \in \mathcal{B}_\rho$ be any solution of the equation $x = \Pi(x)$. Then

$$|\xi - \eta| = |\Pi(\xi) - \Pi(\eta)| \preceq L|\xi - \eta| \preceq L^2|\xi - \eta| \preceq \cdots$$

and $|\xi-\eta| \preceq L^k|\xi-\eta|$ for all $k \in \mathbb{N}$. Taking the limit $k \to \infty$ in view of $\mathrm{rad}(L) < 1$, we obtain $|\xi - \eta| = 0$, i.e. $\eta = \xi$. □

A norm-wise estimate for the rate of convergence in case of generalized contractions may be obtained as follows. For each positive $\varepsilon < 1 - \mathrm{rad}(L)$ there exists [107] a norm $\|\cdot\| : \mathbb{R}^s \to \mathbb{R}_+$ such that $\|L\| = \mathrm{rad}(L) + \varepsilon < 1$ for the corresponding operator norm $\|L\|$ of $L \in \mathbb{R}^{s\times s}$. Hence, if $\|x\| \le \|\,|x|\,\|$ for the norms $\|\cdot\| : \mathcal{X} \to \mathbb{R}_+$ and $|\cdot| : \mathcal{X} \to \mathbb{R}^s_+$, we have also a norm-wise estimate for the rate of convergence:

$$\|\xi - x_k\| \le \|L^k(I - L)^{-1}\beta\| \le \|L\|^k\|(I - L)^{-1}\beta\|.$$

Example D.7 Consider the equation $f(x) = 0$, where $f : \mathbb{F}^q \to \mathbb{F}^q$. Suppose that f satisfies the generalized Lipschitz condition $|f(x) - f(y)| \preceq M|x - y|$, $M = [M_{ij}] \in \mathbb{R}^{q\times q}_+$, as well as the lower growth bounds $m_i|h| \preceq |f(x+he_i) - f(x)|$, $m_i > 0$, where e_i is the i-th column of the unit matrix I_q. Then we may rewrite the equation in an equivalent operator form $x = \Pi(x) := x - Kf(x)$, where $K := \mathrm{diag}(K_1, \ldots, K_q)$, $K_i := 2/(M_{ii} + m_i)$. As in Example D.3, the operator Π satisfies the generalized Lipschitz condition $|\Pi(x) - \Pi(y)| \preceq L|x - y|$, where the elements l_{ij} of the matrix L are determined from

$$l_{ij} := \begin{cases} (M_{ii} - m_i)/(M_{ii} + m_i) & \text{if} \quad i = j \\ 2M_{ij}/(M_{ii} + m_i) & \text{if} \quad i \neq j. \end{cases}$$

Hence, the equation will have a unique solution if $\mathrm{rad}(L) < 1$. If $m := \min\{m_i\}$ and $\mu := \max\{M_{ij} : i \neq j\}$ then the inequality $\mathrm{rad}(L) < 1$ will be fulfilled provided that $\mu < (q - 1)m$. ◊

The use of generalized contractions is very useful in many applications, including some important problems in perturbation analysis. In particular, there are problems for which it is easier to show that the equivalent operator is a generalized contraction rather than a contraction. This will be the case when the operator is a generalized contraction with a Lipschitz matrix L such that $\|L\| \ge 1$ for some of the commonly used matrix norms.

Example D.8 Let $L = \begin{bmatrix} \lambda_1 & l \\ 0 & \lambda_2 \end{bmatrix}$, where $\lambda_1, \lambda_2 \in [0, 1)$ and $l > 0$. Any Hölder p-norm of L satisfies $\|L\|_p \geq l$ and may become arbitrary large for large l. At the same time the spectral radius $\mathrm{rad}(L) = \max\{\lambda_1, \lambda_2\}$ does not depend on l and is always less than 1. ◊

The main application of generalized contractions in perturbation analysis is in the derivation of component-wise perturbation bounds which are usually more informative in comparison with the norm-wise bounds.

D.4 Schauder principle

Consider now the implementation of another powerful topological fixed point principle, the so called *Schauder* (or *Brauer*) principle, which gives sufficient conditions for existence of solutions to the operator equation $x = \Pi(x)$ in $\mathcal{X}$.

If $\Pi : S \to \mathcal{X}$ is a map and $T \subset S$, we denote by $\Pi(T)$ the set of all $\Pi(x)$ when x varies over T.

Theorem D.9 *(Schauder principle). Let the operator $\Pi : B \to \mathcal{X}$ be continuous and $\Pi(B) \subset B$, where $B \subset \mathcal{X}$ is a convex compact.*

Then the operator equation $x = \Pi(x)$ has a solution $\xi \in B$.

Proof. For a complete proof see [173, 117, 34]. However, it is instructive to give the proof in the scalar real case $\mathcal{X} = \mathbb{R}$. Here the nontrivial convex compact sets are the closed intervals with different end points, say $B = [0, 1]$. Let $\Pi : B \to B$ be a continuous function. If $\Pi(0) = 0$ or $\Pi(1) = 1$ then Π has a fixed point $\xi = 0$ or $\xi = 1$ and there is nothing to prove. Therefore assume that $\Pi(0) > 0$ and $\Pi(1) < 1$, and consider the function $\psi : B \to B$, defined from $\psi(x) = x - \Pi(x)$. According to the last two inequalities we have $\psi(0) = -\Pi(0) < 0$ and $\psi(1) = 1 - \Pi(1) > 0$. By a continuity argument, there exists a point $\xi \in (0, 1)$ such that $\psi(\xi) = 0$ which is equivalent to $\xi = \Pi(\xi)$. □

Since in most applications the operator $\Pi : \mathcal{X} \to \mathcal{X}$ is continuous, to apply the Schauder principle one must construct a suitable convex compact set $B \subset \mathcal{X}$, and then to show that $\Pi(B) \subset B$.

We see that the price of the substantial reduction in the requirements, imposed on the equivalent operator Π in the Schauder principle (no Lipschitz conditions, only continuity), is that we claim only existence but not uniqueness of the solution. Thus, the Schauder principle is applicable to problems with nonunique solutions.

Conditions for an operator to be a contraction or a generalized contraction (in order to use the Banach principle), or to map a certain compact convex set into itself (so that to apply the Schauder principle), may be formulated using the technique of Lyapunov majorants [85, 135, 127], see Section 5.

D.5 Notes and references

There are many standard text books and review articles that discuss fixed point principles and their applications, e.g., [117, 173, 34, 55].

Appendix E

Sylvester operators

E.1 Introductory remarks

In this appendix Sylvester operators in real and complex matrix spaces are studied, which include as particular cases the operators arising in the theory of linear time-invariant systems. Let $\mathcal{M} : \mathbb{F}^{m\times n} \to \mathbb{F}^{p\times q}$ be a linear operator, where $\mathbb{F} = \mathbb{R}$ or $\mathbb{F} = \mathbb{C}$. The operator $\mathcal{M}$ is *elementary* if there exist matrices $A \in \mathbb{F}^{p\times m}$ and $B \in \mathbb{F}^{q\times n}$, such that $\mathcal{M}(X) = AXB$. Every $\mathcal{M}$ can be represented as a sum of minimum number of elementary operators, called the *Sylvester index* of $\mathcal{M}$. An expression for the Sylvester index of a general linear operator $\mathcal{M}$ is given. For this purpose a special permutation operator $\mathcal{V}_{p,m} : \mathbb{F}^{pq\times mn} \to \mathbb{F}^{pm\times nq}$ is considered, such that the image $\mathcal{V}_{p,m}(B^\top \otimes A)$ of the matrix $B^\top \otimes A$ of the nonzero elementary operator $\mathcal{M}$ is equal to the rank 1 matrix $\mathrm{vec}(A)\mathrm{row}(B)$. The application of $\mathcal{V}_{p,m}$ reduces a sum of Kronecker products of matrices to the standard product of two matrices.

E.2 Basic concepts

Denote by $\mathbf{Lin}(p, m, n, q, \mathbb{F})$ the linear space of linear matrix operators $\mathcal{M} : \mathbb{F}^{m\times n} \to \mathbb{F}^{p\times q}$, i.e., $\mathcal{M}(X) \in \mathbb{F}^{p\times q}$, $X \in \mathbb{F}^{m\times n}$. In what follows a linear operator will often depend on a collection of $2r$ matrices

$$C := (A_1, B_1, \ldots, A_r, B_r) \in \Sigma_r := (\mathbb{F}^{p\times m} \times \mathbb{F}^{n\times q})^r, \tag{E.1}$$

where $A_k \in \mathbb{F}^{p\times m}$, $B_k \in \mathbb{F}^{n\times q}$. To emphasize this dependence we write $\mathcal{E}_r(C) \in \mathbf{Lin}(p, m, n, q, \mathbb{F})$ for the operator itself and $\mathcal{E}_r(C)(X) \in \mathbb{F}^{p\times q}$ for its matrix value at a given X. Thus, we have a family of operators $\{\mathcal{E}_r(C)\}_{C\in\Sigma_r}$ and $\mathcal{E}_r$ may be considered as a mapping

$$\mathcal{E}_r(\cdot)(\cdot) : \Sigma_r \times \mathbb{F}^{m\times n} \to \mathbb{F}^{p\times q}, \tag{E.2}$$

quadratic in its first argument $C \in \Sigma_r$ and linear in its second argument $X \in \mathbb{F}^{m\times n}$.

An operator $\mathcal{M} \in \mathbf{Lin}(p,m,n,q,\mathbb{F})$ may be determined as follows. Let pq vectors $m_{i,j} \in \mathbb{F}^{mn}$, $i = 1,\dots,p$, $j = 1,\dots,q$, be given. Define the linear functionals $\mu_{i,j} : \mathbb{F}^{m\times n} \to \mathbb{F}$ from $\mu_{i,j}(X) := m_{i,j}^\top \mathrm{vec}(X) \in \mathbb{F}$, $X \in \mathbb{F}^{m\times n}$. Then the operator $\mathcal{M} : \mathbb{F}^{m\times n} \to \mathbb{F}^{p\times q}$, given by $\mathcal{M}(X) = [\mu_{i,j}(X)]_{i,j=1}^{p,q}$, is a linear matrix operator. The matrix $M := \mathrm{Mat}(\mathcal{M}) \in \mathbb{F}^{pq\times mn}$ of $\mathcal{M}$, is defined via the equality $\mathrm{vec}(\mathcal{L}(X)) = M\mathrm{vec}(X)$ and hence

$$M = [m_{1,1}, m_{2,1} \dots, m_{p,1}, m_{1,2}, m_{2,2}, \dots, m_{p,2}, \dots, m_{1,q}, m_{2,q} \dots, m_{p,q}]^\top.$$

Definition E.1 *The operator* $\mathcal{E}_1(A,B) \in \mathbf{Lin}(p,m,n,q,\mathbb{F})$, *such that* $\mathcal{E}_1(A,B)(X) := AXB$, $X \in \mathbb{F}^{m\times n}$, *where* $A \in \mathbb{F}^{p\times m}$ *and* $B \in \mathbb{F}^{n\times q}$, *is called an* elementary Sylvester operator *with a pair of generating matrices* (A,B).

The *zero operator* $0_{p,m,n,q} \in \mathbf{Lin}(p,m,n,q,\mathbb{F})$ and the *identity operator* $1_{m,n} \in \mathbf{Lin}(m,n,\mathbb{F})$ are elementary Sylvester operators $\mathcal{E}_1(A,B)$ with pairs of generating matrices $(A, 0_{n\times q})$ (or $(0_{p\times m}, B)$) and (I_m, I_n), respectively, where $A \in \mathbb{F}^{p\times m}$ (or $B \in \mathbb{F}^{n\times q}$) is arbitrary. A pair (A,B), corresponding to the zero operator (with at least one of its components A or B being zero), is said to be a *trivial pair*.

Let a matrix $2r$-tuple as defined in (E.1) be given. Consider a nonzero operator $\mathcal{E}_r(C) \in \mathbf{Lin}(p,m,n,q,\mathbb{F})$, which is represented as a sum of r nonzero elementary Sylvester operators $\mathcal{E}_1(A_k,B_k)$, i.e.,

$$\mathcal{E}_r(C)(X) := \sum_{k=1}^{r} \mathcal{E}_1(A_k,B_k)(X) = \sum_{k=1}^{r} A_k X B_k\,, \ X \in \mathbb{F}^{m\times n}. \tag{E.3}$$

Operators of the form (E.3) are called *Sylvester operators*.

Every $\mathcal{M} \in \mathbf{Lin}(p,m,n,q,\mathbb{F})$ may be represented in the form (E.3), i.e., $\mathcal{M} = \mathcal{E}_r(C)$ for some r and C. Applying the vec operation to the expression for $\mathcal{E}_r(C)(X)$ we get

$$\mathrm{vec}(\mathcal{E}_r(C)(X)) = E_r(C)\mathrm{vec}(X), \tag{E.4}$$

where

$$E_r = E_r(C) := \mathrm{Mat}(\mathcal{E}_r(C)) = \sum_{k=1}^{r} B_k^\top \otimes A_k \in \mathbb{F}^{pq\times mn} \tag{E.5}$$

is the matrix of the operator $\mathcal{E}_r(C)$.

Using the vec operator and its inverse, $\mathrm{vec}_{p,q}^{-1} : \mathbb{F}^{pq} \to \mathbb{F}^{p\times q}$, any operator $\mathcal{M} \in \mathbf{Lin}(p,m,n,q,\mathbb{F})$ and its matrix representation $M \in \mathbb{F}^{pq\times mn}$ are related via the relations $\mathrm{vec}(\mathcal{M}(X)) = M\mathrm{vec}(X)$ and $\mathcal{M}(X) = \mathrm{vec}_{p,q}^{-1}(M\mathrm{vec}(X))$, $X \in \mathbb{F}^{m\times n}$. There exist different collections $C \in \Sigma_r$, such that $\mathcal{M}$ has a representation of type (E.3), i.e., $\mathcal{M} = \mathcal{E}_r(C)$ for some collection C, which satisfies the bilinear matrix equation

$$\sum_{k=1}^{r} B_k^\top \otimes A_k = M. \tag{E.6}$$

Definition E.2 *The minimum number* $\ell \in \mathbb{N}$*, such that the nonzero operator* $\mathcal{M} \in \mathbf{Lin}(p, m, n, q, \mathbb{F})$ *may be represented as a sum of* ℓ *elementary Sylvester operators, is said to be the* Sylvester index *of* $\mathcal{M}$ *and is denoted by* $\mathrm{ind}_{p,m,n,q}(\mathcal{M})$*. The zero operator is of Sylvester index 1. Any representation of* $\mathcal{M}$ *as a sum of minimum number of elementary operators is called a* condensed representation. *We also abbreviate* $\mathrm{ind}_{m,n} := \mathrm{ind}_{m,m,n,n}$ *and* $\mathrm{ind}_n := \mathrm{ind}_{n,n,n,n}$.

For $\mathcal{M} \in \mathbf{Lin}(p, m, n, q, \mathbb{F})$ we have

$$\|\mathcal{M}(X)\|_{\mathrm{F}} = \|\mathrm{vec}(\mathcal{M}(X))\|_2 \leq \|M\|_2 \|\mathrm{vec}(X)\|_2 = \|M\|_2 \|X\|_{\mathrm{F}}$$

with equality holding if $\mathrm{vec}(X)$ is a right singular vector of the matrix M, corresponding to its maximum singular value $\|M\|_2$. Hence, we may define a norm in $\mathbf{Lin}(p, m, n, q, \mathbb{F})$ as follows.

Definition E.3 *The* (Frobenius) norm *of* $\mathcal{M} \in \mathbf{Lin}(p, m, n, q, \mathbb{F})$ *is*

$$\|\mathcal{M}\|_{\mathrm{F}} := \max\{\|\mathcal{M}(X)\|_{\mathrm{F}} : \|X\|_{\mathrm{F}} = 1\} = \|M\|_2.$$

Other norms as $\|\mathcal{M}\|_{\alpha,\beta} := \max\{\|\mathcal{M}(X)\|_\alpha : \|X\|_\beta = 1\}$; $\alpha, \beta \geq 1$, where $\|\cdot\|_\alpha$ is a Hölder norm, may also be used. Here convenient expressions for $\|\cdot\|_{\alpha,\beta}$ are known only for $\alpha = \beta = 2$ when $\mathcal{M}$ is the standard continuous-time $X \mapsto A^*XE + E^*XA$ or discrete-time $X \mapsto A^*XA - E^*XE$ Lyapunov operator of (generically) Sylvester index 2, see e.g., [95, 68].

E.3 Representations

Consider the problem of representing a general linear matrix operator $\mathcal{M}$ with associated matrix M in the form (E.3). The dimension (real or complex) of $\mathbf{Lin}(p, m, n, q, \mathbb{F}) \simeq \mathbb{F}^{pq \times mn} \simeq \mathbb{F}^{pmnq}$ is $pmnq$. In particular, for every matrix $M \in \mathbb{F}^{pq \times mn}$ there exists $C \in \Sigma_r$ with $r = \mathrm{ind}_{p,m,n,q}(\mathcal{M})$, such that the associated matrix $E_r(C)$ of the operator $\mathcal{E}_r(C) \in \mathbf{Lin}(p, m, n, q, \mathbb{F})$ is equal to M, i.e., $E_r(C) = M$.

Relation (E.6) may be considered also as an equation for both $r \in \mathbb{N}$ and $C \in \Sigma_r$. A particular solution is obtained as follows. Partition the matrix $M \in \mathbb{F}^{pq \times mn}$ into nq blocks of size $p \times m$ as

$$M = [M_{i,j}],\ M_{i,j} \in \mathbb{F}^{p \times m};\ i = 1, \ldots, q,\ j = 1, \ldots, n. \tag{E.7}$$

Then M may be written as $M = \sum_{i,j=1}^{q,n} E_{i,j}(q,n) \otimes M_{i,j}$. Therefore, in view of (E.6), a possible solution for C is $A_k = M_{i,j}$, $B_k = E_{j,i}(n,q)$, $k = k(i,j) := i + (j-1)q$, in which the number of nontrivial pairs (A_k, B_k) is the number of nonzero blocks $M_{i,j}$ of M, which is at most nq. Thus the resulting operator $\mathcal{E}_r(C)$ and hence $\mathcal{M}$ are of Sylvester index at most nq. A similar argument shows that this index is at most pm.

Next we calculate the Sylvester index of a linear operator and construct a representation of type (E.3). For this purpose we introduce a special linear matrix operator $\mathcal{V}_{p,m}$, defined on matrix spaces $\mathbb{F}^{p_0 \times m_0}$ when $p|p_0$ and $m|m_0$.

Let $p, m, p_0, m_0 \in \mathbb{N}$ be given integers such that $p|p_0$ and $m|m_0$. Then each matrix $Z \in \mathbb{F}^{p_0 \times m_0}$ may be partitioned into nq blocks $Z_{i,j}$ of size $p \times m$, where $q := p_0|p$, $n := m_0|m$:

$$Z = \begin{bmatrix} Z_{1,1} & Z_{1,2} & \cdots & Z_{1,n} \\ Z_{2,1} & Z_{2,2} & \cdots & Z_{2,n} \\ \vdots & \vdots & \ddots & \vdots \\ Z_{q,1} & Z_{q,2} & \cdots & Z_{q,n} \end{bmatrix}, Z_{i,j} \in \mathbb{F}^{p \times m}. \tag{E.8}$$

Definition E.4 Set $z_{i,j} := \operatorname{vec}(Z_{i,j})$. The linear operator

$$\mathcal{V}_{p,m} : \mathbb{F}^{pq \times mn} \to \mathbb{F}^{pm \times nq} \tag{E.9}$$

is defined by

$$\mathcal{V}_{p,m}[Z] := [z_{1,1}, z_{2,1}, \ldots, z_{q,1}, z_{1,2}, z_{2,2}, \ldots, z_{q,2}, \ldots, z_{1,n}, z_{2,n}, \ldots, z_{q,n}]. \tag{E.10}$$

The properties of the operator $\mathcal{V}_{p,m}$ are described in the next two propositions.

Proposition E.5 *The operator $\mathcal{V}_{p,m}$ is a permutation operator, for which the following relations hold*

$$\mathcal{V}_{p,q} \circ \mathcal{V}_{p,m} = \mathcal{V}_{p,q} \circ \mathcal{V}_{p,n} = \mathcal{V}_{q,p} \circ \mathcal{V}_{q,m} = \mathcal{V}_{q,p} \circ \mathcal{V}_{q,n} = 1_{p,m,n,q}. \tag{E.11}$$

Proof. The proof follows by inspection. □

Proposition E.6 *Let $M \in \mathbb{F}^{pq \times mn}$, $A \in \mathbb{F}^{p \times m}$ and $B = [b_{i,j}] \in \mathbb{F}^{n \times q}$. Then*

$$\begin{aligned} \mathcal{V}_{1,1}(M) &= (\operatorname{vec}(M))^\top,\ \mathcal{V}_{1,mn}(M) = M^\top,\ \mathcal{V}_{p,m}(M) = (\mathcal{V}_{q,n}(\Pi_{n,q} M \Pi_{n,m}))^\top, \\ \mathcal{V}_{pq,1}(M) &= M,\ \mathcal{V}_{pq,mn}(M) = \operatorname{vec}(M),\ \mathcal{V}_{p,m}(\Pi_{m,p}) = \Pi_{m,p}, \end{aligned}$$

and

$$\mathcal{V}_{p,m}(B^\top \otimes A) = \operatorname{row}(B) \otimes \operatorname{vec}(A) = \operatorname{vec}(A)\operatorname{row}(B). \tag{E.12}$$

Proof. Relations (E.12) follow from the definition of $\mathcal{V}_{p,m}$. To prove (E.12) we note that

$$B^\top \otimes A = \sum_{i,j=1}^{n,q} b_{i,j} E_{i,j}(q,n) \otimes A = \begin{bmatrix} b_{1,1}A & b_{2,1}A & \ldots & b_{n,1}A \\ b_{1,2}A & b_{2,2}A & \ldots & b_{n,2}A \\ \vdots & \vdots & \ddots & \vdots \\ b_{1,q}A & b_{2,q}A & \ldots & b_{n,q}A \end{bmatrix}$$

and hence

$$\begin{aligned}\mathcal{V}_{p,m}(B^\top \otimes A) &= [b_{1,1}\mathrm{vec}(A), b_{1,2}\mathrm{vec}(A), \ldots, b_{1,q}\mathrm{vec}(A),\\ &\quad b_{2,1}\mathrm{vec}(A), b_{2,2}\mathrm{vec}(A), \ldots, b_{2,q}\mathrm{vec}(A), \ldots,\\ &\quad b_{n,1}\mathrm{vec}(A), b_{n,2}\mathrm{vec}(A), \ldots, b_{n,q}\mathrm{vec}(A)]\\ &= \mathrm{row}(B) \otimes \mathrm{vec}(A) = \mathrm{vec}(A)\mathrm{row}(B)\end{aligned}$$

as claimed. □

The operator $\mathcal{V}_{p,m}$ allows the reduction of a sum of Kronecker products of matrices into a product of two matrices. Thus one may solve efficiently equation (E.6).

Suppose that $M \in \mathbb{F}^{pq\times mn}$ is the matrix representation of the operator $\mathcal{M} \in \mathbf{Lin}(p,m,n,q,\mathbb{F})$, partitioned as in (E.7), and set $M^\# := \Pi_{p,q}M\Pi_{n,m} = [M^\#_{k,l}]$, where $M^\#_{k,l} \in \mathbb{F}^{q\times n}$, $k = 1,\ldots,p$, $l = 1,\ldots,m$.

Using the operator $\mathcal{V}_{\bullet,\bullet}$ define the matrices

$$\begin{aligned}\mathbf{M} &:= \mathcal{V}_{p,m}(M) = [\mathrm{vec}(M_{1,1}),\ldots,\mathrm{vec}(M_{q,1}),\ldots,\mathrm{vec}(M_{1,n}),\ldots,\mathrm{vec}(M_{q,n})]\\ &\in \mathbb{F}^{pm\times qn}\end{aligned}$$

and

$$\begin{aligned}\mathbf{M}^\# &:= \mathcal{V}_{q,n}(M^\#) = \left[\mathrm{vec}(M^\#_{1,1}),\ldots,\mathrm{vec}(M^\#_{p,1}),\ldots,\mathrm{vec}(M^\#_{1,m}),\ldots,\mathrm{vec}(M^\#_{p,m})\right]\\ &\in \mathbb{F}^{qn\times pm}.\end{aligned}$$

Now we can determine the Sylvester index of an arbitrary operator $\mathcal{M} \in \mathbf{Lin}(p,m,n,q,\mathbb{F})$ and construct a matrix collection $C \in \Sigma_r$ such that $\mathcal{M} = \mathcal{E}_r(C)$.

Proposition E.7 *Let $M \in \mathbb{F}^{pq\times mn}$ be the matrix representation of the operator $\mathcal{M} \in \mathbf{Lin}(p,m,n,q,\mathbb{F})$. Then*

$$\mathrm{ind}_{p,m,n,q}(\mathcal{M}) = \mathrm{ind}_{p,m,n,q}(M) = \mathrm{ind}_{q,n,m,p}(M) = \max\{1,\rho(M)\},$$

where $\rho(M) := \mathrm{rank}(\mathbf{M}) = \mathrm{rank}(\mathbf{M}^\#)$.

Proof. It follows from Proposition E.6 that for given $r \in \mathbb{N}$ equation (E.6) for $C = (A_1,B_1,\ldots,A_r,B_r)$ may be written as a bilinear equation

$$\mathbf{AB} = \mathbf{M} \tag{E.13}$$

in the unknown matrices

$$\mathbf{A} := [\mathrm{vec}(A_1),\mathrm{vec}(A_2),\ldots,\mathrm{vec}(A_r)] \in \mathbb{F}^{pm\times r}, \tag{E.14}$$

$$\mathbf{B} := [\mathrm{vec}(B_1),\mathrm{vec}(B_2),\ldots,\mathrm{vec}(B_r)]^\top\Pi_{q,n} = \begin{bmatrix}\mathrm{row}(B_1)\\ \mathrm{row}(B_2)\\ \vdots\\ \mathrm{row}(B_r)\end{bmatrix} \in \mathcal{F}^{r\times nq}.$$

Equation (E.13), (E.14) is fundamental for determining the indices as well as for the construction of the linear matrix operator $\mathcal{M}$ as a sum of elementary operators, provided the matrix M of $\mathcal{M}$ is given.

Let $\Theta_r(M) \subset \mathbb{F}^{pm\times r} \times \mathbb{F}^{r\times nq}$ be the set of solutions of (E.13). We shall show that $\Theta_r(M) \neq \emptyset$ if and only if $r \geq \rho(M)$ and hence equation (E.13) is solvable for $r = \rho(M)$. The proof is constructive and provides explicit expressions for $\Theta_{\rho(M)}(M)$.

In the trivial case $\mathcal{M} = 0_{p,m,n,q}$ we have $r = 1$ by definition and the solution of (E.13) may be taken as $(\mathbf{A}, 0_{1\times nq})$ or $(0_{pm\times 1}, \mathbf{B})$ with $\max\{pm, nq\}$ free parameters. Hence,

$$\Theta_1[0_{p,m,n,q}] = (\mathbb{F}^{pm} \times \{0_{1\times nq}\}) \bigcup (\{0_{pm\times 1}\} \times \mathbb{F}^{1\times nq}).$$

Consider the general case $\mathcal{M} \neq 0_{p,m,n,q}$. It follows from (E.13) that

$$\rho(M) \leq \min\{\mathrm{rank}(\mathbf{A}), \mathrm{rank}(\mathbf{B})\} \leq r.$$

We prove that if $r = \rho(M)$, then (E.13) is explicitly solved. Consider the three possible cases.

1. If $r = \rho(M) = pm \leq nq$ then the solution set is

$$\Theta_r(M) = \left\{(P, P^{-1}\mathbf{M}) : P \in \mathcal{GL}(pm, \mathbb{F})\right\}.$$

2. If $r = \rho(M) = nq < mn$ then the solution set is

$$\Theta_r(M) = \left\{(\mathbf{M}P^{-1}, P) : P \in \mathcal{GL}(nq, \mathbb{F})\right\}.$$

3. If $r = \rho(M) < \min\{pm, nq\}$ then $\mathbf{M}$ may be decomposed as

$$\mathbf{M} = U\mathrm{diag}\left(I_r, 0_{(pm-r)\times(nq-r)}\right) V^{-1},$$

where $U \in \mathcal{GL}(pm, \mathbb{F})$, $V \in \mathcal{GL}(nq, \mathbb{F})$. Thus, the solution set may be represented as

$$\Theta_r(M) = \left\{\left(U^{-1}\begin{bmatrix} P \\ 0_{(pm-r)\times r}\end{bmatrix}, \left[P^{-1}, 0_{r\times(nq-r)}\right] V\right) : P \in \mathcal{GL}(r, \mathbb{F})\right\}.$$

Similar arguments hold for the transposed operator with a matrix $M^{\#}$, showing that $\mathrm{ind}_{p,m,n,q}(M) = \mathrm{ind}_{q,n,m,p}(M)$. Note finally that $\mathbf{M}^{\#} = \mathbf{M}^{\top}$, see Proposition E.6. □

We see from the proof of Proposition E.7 that in the nontrivial case $\mathcal{M} \neq 0_{p,m,n,q}$ the set of all collections C in the condensed representation of $\mathcal{M}$ is isomorphic to $\mathcal{GL}(r, \mathbb{F})$, where r is the Sylvester index of $\mathcal{M}$. Hence, it is an open algebraic variety (of real or complex dimension r^2) in the corresponding Zariski topology.

When $\mathcal{M} \neq 0_{p,m,n,q}$ the solution set $\Theta_r(M)$ of (E.13) with $r = \nu(M)$ is parametrized by the r^2 free elements of the matrix $P \in \mathcal{GL}(r, \mathbb{F})$. Note that the matrix equation (E.13) is equivalent to $pmnq$ scalar quadratic equations in $r(pm+nq)$ scalar unknowns (the elements of $\mathbf{A}$ and $\mathbf{B}$). Hence, we may expect that in general the solution set $\Theta_r(M)$ is a k-parameter family, where $k := r(pm+nq)-pmnq$ is the number of unknowns minus the number of equations. Since $r^2-k = (pm-r)(nq-r)$ and generically $\nu(M) = \min\{pm, nq\}$, we may indeed expect that $k = r^2$.

Example E.8 Consider the *transposition operator* $\mathcal{T}_{m,n} \in \mathbf{Lin}(n, m, n, m, \mathbb{F})$, acting as $\mathcal{T}_{m,n}(X) = X^\top$. The matrix representation of $\mathcal{T}_{m,n}$ is $\Pi_{m,n}$. Since $\mathcal{V}_{n,m}(\Pi_{m,n}) = \Pi_{m,n}$ (see Proposition E.6) and $\mathrm{rank}(\Pi_{m,n}) = mn$, we see that $\mathrm{ind}_{n,m,n,m}(\mathcal{T}_{m,n}) = mn$. In particular we have [107]

$$X^\top = \sum_{i,j=1}^{n,m} E_{i,j}(n,m) X E_{i,j}(n,m).$$

Consider the case when $mn = pq$ and the operator $\mathcal{M} \in \mathbf{Lin}(p, m, n, q, \mathbb{F})$ is invertible, i.e., its associated matrix $M \in \mathbb{F}^{mn \times mn}$ is nonsingular. For some classes of invertible operators it may be shown that

$$\mathrm{ind}_{p,m,n,q}(M) = \mathrm{ind}_{m,p,q,n}(M^{-1}). \tag{E.15}$$

It is interesting to determine whether (E.15) holds for all invertible operators $\mathcal{M} \in \mathbf{Lin}(p, m, n, q, \mathbb{F})$.

E.4 Notes and references

Linear matrix equations and linear matrix operators have been studied since the pioneering work of Sylvester and Kronecker [152, 215, 214], see also [196, 193, 229] and [8]. Now there are hundreds of papers, surveys and many books, e.g., [12, 10, 69, 106, 107, 205, 228] devoted to the analysis, existence, uniqueness and representation of the solution and also to the numerical algorithms and software to solve various types of linear matrix equations. Most of the existing results, however, are connected with particular classes of such matrix equations.

The problem of representing a general linear matrix operator has only recently been studied in [125].

Appendix F

Lyapunov operators

F.1 Introductory remarks

In this appendix Lyapunov operators in real and complex matrix spaces are studied, which include as particular cases the operators arising in the theory of linear time-invariant systems.

A linear operator $\mathcal{L} : \mathbb{F}^{n\times n} \to \mathbb{F}^{n\times n}$ is a *Lyapunov operator* if

$$(\mathcal{L}(X))^* = \mathcal{L}(X^*),$$

where the star denotes transposition in the real case and complex conjugate transposition in the complex case. Characterizations and parametrizations of the sets of real and complex Lyapunov operators are given and their dimensions are determined. Relevant *Lyapunov indices* for Lyapunov operators are introduced and calculated. Similar results are given also for several classes of Lyapunov-like linear and pseudo-linear operators. The concept of *Lyapunov singular values* of a Lyapunov operator is introduced and the application of these values to the sensitivity and a posteriori error analysis of Lyapunov equations is discussed.

Despite of the existence of a large amount of literature on Lyapunov equations and operators some general properties of finite-dimensional Lyapunov operators, however, have not been studied to a sufficient extent. In particular, the notion of the minimal singular value of a Lyapunov operator is sometimes misused. Introducing the new concept of Lyapunov singular values of a Lyapunov operator, some well-known estimates in the sensitivity theory of matrix equations may be substantially improved.

In this appendix we denote by $\Omega(n,\mathbb{F}) \subset \mathbb{F}^{2n\times 2n}$ the set of all matrices $L \in \mathbb{F}^{2n\times 2n}$ such that $LP_{n^2} = P_{n^2}\overline{L}$. We use the notation from Appendices E and 10.17.

F.2 Real operators

An important class of linear operators are the Lyapunov operators which are automorphisms in $\mathbb{F}^{n\times n}$. In this section we consider the class of real Lyapunov operators in $\mathbf{Lin}(n,\mathbb{R})$.

Definition F.1 *An operator* $\mathcal{L}\in\mathbf{Lin}(n,\mathbb{R})$ *is called a* real Lyapunov operator *if*

$$(\mathcal{L}(X))^\top=\mathcal{L}(X^\top),\ X\in\mathbb{R}^{n\times n}.$$

We denote by $\mathbf{Lyap}(n,\mathbb{R})\subset\mathbf{Lin}(n,\mathbb{R})$ *the set of real Lyapunov operators.*

It follows from Definition F.1 that $X=X^\top\Rightarrow\mathcal{L}(X)=(\mathcal{L}(X))^\top$ and $X=-X^\top\Rightarrow\mathcal{L}(X)=-(\mathcal{L}(X))^\top$ provided $\mathcal{L}\in\mathbf{Lyap}(n,\mathbb{R})$. Hence, the subspaces $\mathbf{Her}(n,\mathbb{R})$ of symmetric and $\mathbf{SHer}(n,\mathbb{R})$ of skew-symmetric real matrices are invariant subspaces for operators from $\mathbf{Lyap}(n,\mathbb{R})$ (see also [40], where the particular case $\mathcal{L}(X)=A^{\mathrm{H}}X+XA$ has been considered).

Below we need the operator $\mathcal{V}_n:=\mathcal{V}_{n,n}:\mathbb{F}^{2n\times 2n}\to\mathbb{F}^{2n\times 2n}$, defined by (E.9), (E.10) for $p=m=n=q$, which in the given case is an involutary permutation, $\mathcal{V}_n^2=1_{n,n}$.

The set $\mathbf{Lyap}(n,\mathbb{R})$ itself is a linear subspace of $\mathbf{Lin}(n,\mathbb{R})$, which may be characterized in the next proposition.

Proposition F.2 *The following four statements are equivalent:*

(i) $\mathcal{L}\in\mathbf{Lyap}(n,\mathbb{R})$.

(ii) *There exists* $\mathcal{M}\in\mathbf{Lin}(n,\mathbb{R})$, *such that*

$$\mathcal{L}(X)=\mathcal{M}(X)+(\mathcal{M}(X^\top))^\top,\ X\in\mathbb{F}^{n\times n},$$

i.e., $\mathcal{L}(X)=\sum_{k=1}^{r}\left(A_kXB_k+B_k^\top XA_k^\top\right)$, *or equivalently*

$$L:=\mathrm{Mat}(\mathcal{L})=\sum_{k=1}^{r}\left(B_k^\top\otimes A_k+A_k\otimes B_k^\top\right),$$

where $A_k,B_k\in\mathbb{R}^{n\times n}$ *are given matrices.*

(iii) $L\in\Omega(n,\mathbb{R})$, *where* $\Omega(n,\mathbb{R})$ *is the subspace of real* $n^2\times n^2$ *matrices* L, *satisfying the equation* $P_{n^2}L=LP_{n^2}$.

(iv) *The matrix* $\mathbf{L}:=\mathcal{V}_n(L)$ *is symmetric.*

Proof. The equivalence between (i) and (ii) follows from the definitions. To prove (iii) we perform the vec operation on both sides of the characteristic equation

$(\mathcal{L}(X))^\top = \mathcal{L}(X^\top)$ of the Lyapunov operator $\mathcal{L}$ with associated matrix L, which gives

$$\begin{aligned} \mathrm{vec}((\mathcal{L}(X))^\top) &= \mathrm{vec}(\mathcal{L}(X^\top)), \\ P_{n^2}\mathrm{vec}(\mathcal{L}(X)) &= L\mathrm{vec}(X^\top), \\ P_{n^2}L\mathrm{vec}(X) &= LP_{n^2}\mathrm{vec}(X) \end{aligned}$$

for all $X \in \mathbb{R}^{n\times n}$ and hence, $P_{n^2}L = LP_{n^2}$.

To prove (iv) note that the relation $\sum_{k=1}^r \left(B_k^\top \otimes A_k + A_k \otimes B_k^\top\right) = L$ from (ii) is an equation for the matrices $A_1, B_1, \ldots, A_r, B_r$, similar to (E.6). After some calculations we get the following counterpart of the bilinear equation (E.13), (E.14):

$$\mathbf{AB} + (\mathbf{AB})^\top = \mathbf{L} \tag{F.1}$$

and hence, the matrix $\mathbf{L}$ is symmetric. □

Representations of $\mathcal{L} \in \mathbf{Lyap}(n, \mathbb{R})$ as in Proposition F.2(ii) usually arise in the theory of continuous-time standard and descriptor dynamical systems. They involve $2r$ terms and cannot be condensed in the sense of Definition F.1. In particular, the representation of the Lyapunov operator $X \mapsto DXD^\top$ (of Sylvester index 1) in the form (ii) requires two terms, e.g. $r = 1$ and $A_1 = D$, $B_1 = D^\top/2$.

As in the case of a general Sylvester operator $\mathcal{M} \in \mathbf{Lin}(n, \mathbb{F})$, the real Lyapunov operator $\mathcal{L} \in \mathbf{Lyap}(n, \mathbb{R})$ may be represented in a condensed form as a sum of $\mathrm{ind}_n(\mathcal{L})$ elementary linear operators (not necessarily Lyapunov) but in this case the formal symmetry in Proposition F.2(ii) may be lost. To preserve this symmetry, characterizing Lyapunov operators, we introduce the following two symmetric representations that hold for every nonzero operator $\mathcal{L} \in \mathbf{Lyap}(n, \mathbb{R})$.

The *continuous-time representation* is of the form

$$\mathcal{L}(X) = \sum_{k=1}^{\ell_c} \left(A_k X B_k + B_k^\top X A_k^\top\right), \quad X \in \mathbb{R}^{n\times n}, \tag{F.2}$$

while the *discrete-time representation* is

$$\mathcal{L}(X) = \sum_{j=1}^{\ell_d} \varepsilon_j D_j X D_j^\top, \quad X \in \mathbb{R}^{n\times n}, \tag{F.3}$$

where $\varepsilon_j = \pm 1$ and $D_j, A_k, B_k \in \mathbb{R}^{n\times n}$. Obviously $2\ell_c \geq \mathrm{ind}_n(\mathcal{L})$ and $\ell_d \geq \mathrm{ind}_n(\mathcal{L})$.

Mixed representations as $\mathcal{L}(X) = DXD^\top + A^\top X + XA$ may be reduced to some of the above two types (F.2) or (F.3).

Definition F.3 *The representations (F.2) and (F.3) of* $\mathcal{L} \in \mathbf{Lyap}(n, \mathbb{R})$ *are said to be* cl-condensed *and* dl-condensed *respectively, if there are no representations of* $\mathcal{L}$ *of the corresponding types with less terms. The numbers* $\mathrm{clind}_n(\mathcal{L}) := 2\ell_c$

and $\mathrm{dlind}_n(\mathcal{L}) := \ell_d$ *in the cl-condensed representation and in the dl-condensed representation are called the* continuous-time Lyapunov index *and the* discrete-time Lyapunov index *of* $\mathcal{L}$.

Example F.4 Let $\lambda_1, \lambda_2 \in \mathcal{R}$, with $\lambda_1 > 0$. Then the operator $\mathcal{L} \in \mathbf{Lin}(2, \mathbb{R})$, defined via

$$\mathcal{L}(X) := \begin{bmatrix} \lambda_1^2 x_{11} & \lambda_1\lambda_2 x_{12} \\ \lambda_1\lambda_2 x_{21} & \lambda_2^2 x_{22} \end{bmatrix}, \quad X = \begin{bmatrix} x_{11} & x_{12} \\ x_{21} & x_{22} \end{bmatrix},$$

has both its Lyapunov indices equal to 2. It admits the cl-condensed representation $\mathcal{L}(X) = AX + XA$ and the dl-condensed representation $\mathcal{L}(X) = D_1XD_1 - D_2XD_2$, where $A := \mathrm{diag}(\lambda_1, \lambda_2)$ and

$$D_1 := \mathrm{diag}\left(\sqrt{2\lambda_1}, \frac{\lambda_1 + \lambda_2}{\sqrt{2\lambda_1}}\right), \quad D_2 := \mathrm{diag}\left(0, \frac{\lambda_1 - \lambda_2}{\sqrt{2\lambda_1}}\right).$$

Explicit expressions for the Lyapunov indices of Lyapunov operators are given below. Obviously $\mathrm{ind}_n(\mathcal{L}) \geq \mathrm{clind}_n(\mathcal{L}), \mathrm{dlind}_n(\mathcal{L})$. In fact we will show that the Sylvester index of a Lyapunov operator is equal to its discrete-time Lyapunov index.

Proposition F.5 *The continuous-time and the discrete-time Lyapunov indices of the nonzero operator* $\mathcal{L} \in \mathbf{Lyap}(n, \mathbb{R})$ *are determined by*

$$\begin{aligned} \mathrm{clind}_n(\mathcal{L}) &= 2\max\{\nu_+(\mathbf{L}), \nu_-(\mathbf{L})\}, \\ \mathrm{dlind}_n(\mathcal{L}) &= \nu_+(\mathbf{L}) + \nu_-(\mathbf{L}) = \mathrm{rank}[\mathbf{L}]. \end{aligned}$$

In particular, the Sylvester and the discrete-time Lyapunov index of an operator $\mathcal{L} \in \mathbf{Lyap}(n, \mathbb{R})$ coincide, i.e.

$$\mathrm{ind}_n(\mathcal{L}) = \mathrm{dlind}_n(\mathcal{L}) \geq \mathrm{clind}_n(\mathcal{L}).$$

Proof. Consider first the continuous-time case and set $\mathbf{C} = \mathbf{AB}$ in equation (F.1). Hence, the number $r := \mathrm{clind}_n(\mathcal{L})/2$ may be computed from

$$r = \min\left\{\mathrm{rank}(\mathbf{C}) : \mathbf{C} \in \mathbb{R}^{n^2 \times n^2}, \mathbf{C} + \mathbf{C}^\top = \mathbf{L}\right\}.$$

Denoting $\alpha := \nu_+(\mathbf{L})$, $\beta := \nu_-(\mathbf{L})$ and $\gamma := \alpha + \beta$ we will show that $r = \alpha$. Indeed, there exists $P \in \mathcal{GL}(n^2, \mathbb{R})$ such that the matrix $\mathbf{L}$ is factorized as $\mathbf{L} = P\Delta_{\mathbf{L}}P^\top$, where $\Delta_{\mathbf{L}} := \mathrm{diag}(2I_\alpha, -2I_\beta, 0_{n^2-\gamma})$. Setting $\mathbf{C} = PYP^\top$ we obtain that r is the minimum of the ranks of the matrices Y, such that $Y + Y^\top = \Delta_{\mathbf{L}}$. The general form of Y is

$$Y = \begin{bmatrix} I_\alpha + Y_{11} & -Y_{21}^\top & -Y_{31}^\top \\ Y_{21} & -I_\beta + Y_{22} & -Y_{32}^\top \\ Y_{31} & Y_{32} & Y_{33} \end{bmatrix}, \tag{F.4}$$

where the matrices $Y_{11} \in \mathbf{SHer}(\alpha,\mathbb{R})$, $Y_{21} \in \mathbb{R}^{\beta\times\alpha}$, $Y_{31} \in \mathbb{R}^{(n^2-\gamma)\times\alpha}$, $Y_{22} \in \mathbf{SHer}(\beta,\mathbb{R})$, $Y_{32} \in \mathbb{R}^{(n^2-\gamma)\times\beta}$, $Y_{33} \in \mathbf{SHer}(n^2-\gamma,\mathbb{R})$ are arbitrary. Suppose w.l.o.g. that $\alpha \geq \beta$. The eigenvalues $\lambda(I_\alpha + Y_{11})$ of the matrix $I_\alpha + Y_{11}$ are equal to $1 + \lambda(Y_{11})$. In turn, Y_{11} has its eigenvalues on the imaginary axis, i.e., the eigenvalues of $I_\alpha + Y_{11}$ have real part 1. Hence, the diagonal block $I_\alpha + Y_{11}$ of Y in (F.4) is nonsingular and $\operatorname{rank}(Y) \geq \operatorname{rank}[I_\alpha + Y_{11}] = \alpha$. Moreover, for certain Y the equality $\operatorname{rank}[Y] = \alpha$ is achieved. To see this, take Y_{11}, Y_{31}, Y_{22} and Y_{33} as zero matrices, and let $Y_{21} := [I_\beta, 0_{\beta\times(\alpha-\beta)}]$. Then $Y_{21}Y_{21}^\top = I_\beta$ and hence,

$$\begin{bmatrix} I_\alpha & 0 & 0 \\ -Y_{21} & I_\beta & 0 \\ 0 & 0 & I_{n^2-\gamma} \end{bmatrix} Y = \begin{bmatrix} I_\alpha & -Y_{21}^\top & 0 \\ 0 & 0 & 0 \\ 0 & 0 & 0 \end{bmatrix},$$

which yields $\operatorname{rank}(Y) = \alpha$. Therefore we may find matrices $\mathbf{A}$, $\mathbf{B}$, satisfying $\mathbf{AB} + (\mathbf{AB})^\top = \mathbf{L}$ with $r = \alpha$. Since the continuous-time representation of a Lyapunov operator has $2r$ terms, we have proved the first part of the proposition.

Consider now the discrete-time case. Denote $\mathrm{D} := [\operatorname{vec}(D_1), \ldots, \operatorname{vec}(D_r)] \in \mathbb{R}^{n^2\times r}$ and let $E \in \mathcal{GL}(r,\mathbb{R})$ be a diagonal matrix with elements $\varepsilon_j = \pm 1$ on the diagonal. Then the equation $\sum_{k=1}^r \varepsilon_k (D_k \otimes D_k) = L$ for the matrices $D_1, \ldots, D_k$ becomes $\mathrm{D}E\mathrm{D}^\top = \mathbf{L}$. We have $r \geq \gamma = \operatorname{rank}(\mathbf{L})$. Consider again the factorization $\mathbf{L} = P\Delta_{\mathbf{L}}P^\top$. Partitioning the matrix P as $P = [P_1, P_2]$ with $P_1 \in \mathbb{R}^{n^2\times\gamma}$, we may choose $r = \gamma$ and $\mathrm{D} = P_1$, $E = \operatorname{diag}(I_\alpha, -I_\beta)$. □

According to parts (i) and (iii) of Proposition F.2, a matrix $L \in \mathbb{R}^{n^2\times n^2}$ is the matrix representation of a Lyapunov operator if and only if it has the symmetry property $P_{n^2}L = LP_{n^2}$, or, equivalently, $L = P_{n^2}LP_{n^2}$. This leads to the following proposition.

Proposition F.6 *The subspace $\Omega(n,\mathbb{R}) \subset \mathcal{R}^{n^2\times n^2}$ of matrix representations of real Lyapunov operators is isomorphic to the subspace*

$$\operatorname{Ker}(I_{n^2} \otimes P_{n^2} - P_{n^2} \otimes I_{n^2}) = \operatorname{Ker}(P_{n^2} \otimes P_{n^2} - I_{n^4}) \subset \mathbb{R}^{n^4}. \tag{F.5}$$

Proof. Multiplying the last equation on the left with P_{n^2} and taking into consideration that $P_{n^2}^2 = I_{n^2}$, we also get $L = P_{n^2}LP_{n^2}$. The characterization of $\Omega(n,\mathbb{R})$ by the subspace (F.5) is obtained taking the vec operation on both sides of the equalities $P_{n^2}L - LP_{n^2} = 0_{n^2\times n^2}$ and $P_{n^2}LP_{n^2} - L = 0_{n^2\times n^2}$, namely $(I_{n^2} \otimes P_{n^2} - P_{n^2} \otimes I_{n^2})\operatorname{vec}(L) = 0_{n^4\times 1}$. □

Next we will give two explicit parametrizations of the set $\Omega(n,\mathbb{R})$, which in particular yield the dimension of the space of real Lyapunov operators. For this purpose we need the Jordan form J_n of P_{n^2}. The matrix P_{n^2} has two eigenvalues: $\lambda_1 = 1$ with multiplicity $n_1 := n(n+1)/2$ and $\lambda_2 = -1$ with multiplicity $n_2 := n(n-1)/2$. Thus, the Jordan form of P_{n^2} is

$$J_n = \Theta_n^\top P_{n^2}\Theta_n = \operatorname{diag}(I_{n_1}, -I_{n_2}), \tag{F.6}$$

where the orthogonal matrix $\Theta_n \in \mathcal{O}(n^2, \mathbb{R})$ may be obtained as follows. The permutation $L \mapsto P_{n^2} L$ leaves n rows of L at their positions $(k-1)n+k$, $k = 1, \ldots, n$, and interchanges the positions of the rows in the remaining $n(n-1)/2$ pairs of rows. Hence, there is a permutation matrix Θ_n', such that

$$(\Theta_n')^\top P_{n^2} \Theta_n' = \operatorname{diag}\left(I_n, \begin{bmatrix} 0 & 1 \\ 1 & 0 \end{bmatrix}, \ldots, \begin{bmatrix} 0 & 1 \\ 1 & 0 \end{bmatrix}\right).$$

Let

$$\Theta_n'' := \operatorname{diag}\left(I_n, \begin{bmatrix} \omega & -\omega \\ \omega & \omega \end{bmatrix}, \ldots, \begin{bmatrix} \omega & -\omega \\ \omega & \omega \end{bmatrix}\right), \ \omega := 1/\sqrt{2}.$$

Then

$$(\Theta_n' \Theta_n'')^\top P_{n^2} \Theta_n' \Theta_n'' = \operatorname{diag}(I_n, 1, -1, \ldots, 1, -1).$$

Let $\Theta_2''' = I_4$ and, if $n > 2$, let Θ_n''' be the permutation matrix, corresponding to the permutation $n + 2l \leftrightarrow n^2 + 1 - 2l$, $l = 1, \ldots, (n-1)(n-2)/2$, leaving the other elements of $\{1, \ldots, n^2\}$ unchanged. Then

$$\Theta_n = \Theta_n' \Theta_n'' \Theta_n'''. \tag{F.7}$$

Example F.7 For $n = 2$ the transformation of Π_2 into J_2 is done via

$$\Theta_2 = \begin{bmatrix} 1 & 0 & 0 & 0 \\ 0 & 0 & \omega & -\omega \\ 0 & 0 & \omega & \omega \\ 0 & 1 & 0 & 0 \end{bmatrix}, \ J_2 = \Theta_2^\top \Pi_2 \Theta_2 = \operatorname{diag}(1, 1, 1, -1).$$

Proposition F.8 *The subspace $\Omega(n, \mathbb{R})$ is parametrized as*

$$\begin{aligned} \Omega(n, \mathbb{R}) &= \mathcal{V}_n^{-1}(\mathbf{Her}(n^2, \mathbb{R})) \\ &= \left\{ \Theta_n \begin{bmatrix} L_{11} & 0 \\ 0 & L_{22} \end{bmatrix} \Theta_n^\top : L_{ii} \in \mathbb{R}^{n_i \times n_i} \right\}. \end{aligned}$$

In particular the (real) dimension of $\mathbf{Lyap}(n, \mathbb{R})$ *and* $\Omega(n, \mathbb{R})$ *is* $n_1^2 + n_2^2 = n^2(n^2 + 1)/2$.

Proof. The first parametrization of $\Omega(n, \mathbb{R})$ follows immediately from Proposition F.2(iv) and we see that the dimension of $\Omega(n, \mathbb{R})$ is that of $\mathbf{Her}(n^2, \mathbb{R})$, i.e., $n^2(n^2 + 1)/2$.

Consider the second parametrization. The matrix equation $P_{n^2} L = L P_{n^2}$ for the matrix L is equivalent to

$$J_n \widehat{L} = \widehat{L} J_n, \ \widehat{L} := \Theta_n^\top L \Theta_n. \tag{F.8}$$

The general solution of equation (F.8) is of the form $\widehat{L} = \operatorname{diag}(L_{11}, L_{22})$, where the matrices $L_{ii} \in \mathcal{R}^s n_i \times n_i$ are arbitrary, which completes the proof. □

Example F.9 For $n = 2$ and $n = 3$ the sets $\Omega(2,\mathbb{R})$ and $\Omega(3,\mathbb{R})$ are 10- and 45-dimensional real spaces with patterns Λ_2 and Λ_3 of the free parameters as follows:

$$\Lambda_2 = \begin{bmatrix} \underline{1} & 2 & 2 & \underline{3} \\ 4 & 5 & 6 & 7 \\ 4 & 6 & 5 & 7 \\ \underline{8} & 9 & 9 & \underline{10} \end{bmatrix}, \quad \Lambda_3 = \begin{bmatrix} \underline{1} & 2 & 3 & 2 & \underline{4} & 5 & 3 & 5 & \underline{6} \\ 7 & 8 & 9 & 10 & 11 & 12 & 13 & 14 & 15 \\ 16 & 17 & 18 & 19 & 20 & 21 & 22 & 23 & 24 \\ 7 & 10 & 13 & 8 & 11 & 14 & 9 & 12 & 15 \\ \underline{25} & 26 & 27 & 26 & \underline{28} & 29 & 27 & 29 & \underline{30} \\ 31 & 32 & 33 & 34 & 35 & 36 & 37 & 38 & 39 \\ 16 & 19 & 22 & 17 & 20 & 23 & 18 & 21 & 24 \\ 31 & 34 & 37 & 32 & 35 & 38 & 33 & 36 & 39 \\ \underline{40} & 41 & 42 & 41 & \underline{43} & 44 & 42 & 44 & \underline{45} \end{bmatrix}.$$

In both cases the underlined elements are in the positions, corresponding to the zero scalar identities for the elements of L in the matrix equation $P_{n^2}L = LP_{n^2}$.

If $\mathcal{M} \in \mathbf{Lin}(n,\mathbb{R})$ is a general Sylvester operator, then according to Definition E.3 we have $\|\mathcal{M}\|_{\mathrm{F}} := \sigma_{\max}(\mathcal{M}) := \sigma_1(\mathrm{Mat}(\mathcal{M})) = \max\{\|\mathcal{M}(X)\|_{\mathrm{F}} : \|X\|_{\mathrm{F}} = 1\}$. Similarly

$$\sigma_{\min}(\mathcal{M}) := \sigma_{n^2}(\mathrm{Mat}(\mathcal{M})) = \min\{\|\mathcal{M}(X)\|_{\mathrm{F}} : \|X\|_{\mathrm{F}} = 1\}$$

and if $\mathcal{M} \in \mathbf{Lin}(n,\mathbb{R})$ is invertible, then $\|\mathcal{M}^{-1}\|_{\mathrm{F}} = 1/\sigma_{\min}(\mathcal{M})$.

For Lyapunov operators $\mathcal{L} \in \mathbf{Lyap}(n,\mathbb{R})$, however, in addition to the standard maximum and minimum singular values $\sigma_{\max}(\mathcal{L})$ and $\sigma_{\min}(\mathcal{L})$, we may also define the maximum and minimum *Lyapunov singular values*

$$\|\mathcal{L}\|_{\widetilde{\mathrm{F}}} := \widetilde{\sigma_{\max}}(\mathcal{L}) := \max\{\|\mathcal{L}(X)\|_{\mathrm{F}} : \|X\|_{\mathrm{F}} = 1,\ X = X^\top\}$$

and

$$\widetilde{\sigma_{\min}}(\mathcal{L}) := \min\{\|\mathcal{L}(X)\|_{\mathrm{F}} : \|X\|_{\mathrm{F}} = 1,\ X = X^\top\}.$$

If $\mathcal{L}$ is invertible, then $\|\mathcal{L}^{-1}\|_{\widetilde{\mathrm{F}}} = 1/\widetilde{\sigma_{\min}}(\mathcal{L})$. Obviously

$$\sigma_{\min}(\mathcal{L}) \le \widetilde{\sigma_{\min}}(\mathcal{L}) \le \widetilde{\sigma_{\max}}(\mathcal{L}) \le \sigma_{\max}(\mathcal{L}).$$

Each of these inequalities may be strict, i.e., the inequalities $\sigma_{\min}(\mathcal{L}) < \widetilde{\sigma_{\min}}(\mathcal{L})$ and $\widetilde{\sigma_{\max}}(\mathcal{L}) < \sigma_{\max}(\mathcal{L})$ are possible. Moreover, as we show below, the differences $\widetilde{\sigma_{\min}}(\mathcal{L}) - \sigma_{\min}(\mathcal{L})$ and $\sigma_{\max}(\mathcal{L}) - \widetilde{\sigma_{\max}}(\mathcal{L})$ may be arbitrarily large, see Example F.12.

Let $A \in \mathbb{R}^{n\times n}$ and $a := \mathrm{vec}(A) \in \mathcal{R}^{n^2}$. Using the notation $\mathrm{vec}_n^{-\top}(a) = (\mathrm{vec}_n^{-1}(a))^\top$ define the set $Z(n) := \left\{a \in \mathbb{R}^{n^2} : \mathrm{vec}_n^{-1}(a) = \mathrm{vec}_n^{-\top}(a)\right\}$, corresponding to the symmetric matrices $A = \mathrm{vec}_n^{-1}(a)$, which is an $n(n+1)/2$-dimensional subspace of $\mathbb{R}^{n^2}$. We will show that $Z(n) = \mathrm{Rg}(I_{n^2} + P_{n^2}) = \mathrm{Rg}(P_n)$, where

$P_n = [P_{n,ij}] \in \mathbb{R}^{n^2 \times n(n+1)/2}$, $i,j = 1,\ldots,n$, is an upper block-triangular matrix. The blocks $P_{n,ij} \in \mathbb{R}^{n\times j}$ are defined by

$$P_{n,ij} := \begin{cases} 0_{n\times j} & \text{if} \quad i > j, \\ \begin{bmatrix} I_i \\ 0_{(n-i)\times i} \end{bmatrix} & \text{if} \quad i = j, \\ E_{ji}(n,j) & \text{if} \quad i < j. \end{cases}$$

If L is the matrix representation of $\mathcal{L} \in \mathbf{Lyap}(n,\mathbb{R})$, then we can rewrite the expression for $\widetilde{\sigma_{\max}}(\mathcal{L})$ in the equivalent form

$$\begin{aligned} \widetilde{\sigma_{\max}}(\mathcal{L}) &= \max\left\{ \frac{\|La\|_2}{\|a\|_2} : 0 \neq a \in Z(n) \right\} = \max\left\{ \frac{\|LP_nb\|_2}{\|P_nb\|_2} : 0 \neq b \in \mathbb{R}^{n(n+1)/2} \right\} \\ &= \|LQ_n\|_2 = \sigma_{\max}(LQ_n), \end{aligned}$$

where

$$Q_n := P_n(P_n^\top P_n)^{-1} = [Q_{n,ij}] \in \mathbb{R}^{n^2\times n(n+1)/2};\ i,j = 1,\ldots,n, \qquad \text{(F.9)}$$

is an upper block-triangular projector ($Q_n^\top Q_n = I_{n(n+1)/2}$). The blocks $Q_{n,ij} \in \mathbb{R}^{n\times j}$ are given by $Q_{n,ij} = 0$ if $i > j$, $Q_{n,11} = [1,0,\ldots,0]^\top \in \mathbb{R}^n$, $Q_{n,kk} = [\operatorname{diag}(\omega I_{k-1},1),0]^\top$ and $Q_{n,ij} = \omega E_{ji}(n,j)$ if $i < j$, where $\omega := 1/\sqrt{2}$.

The matrices P_n and Q_n have the same sign-patterns, the only difference being that the nonzero elements of P_n are equal to 1, while the nonzero elements of Q_n are equal to 1 or ω.

Example F.10 The matrices Q_2, Q_3, Q_4 are

$$Q_2 = \left[\begin{array}{c|cc} 1 & 0 & 0 \\ 0 & \omega & 0 \\ \hline 0 & \omega & 0 \\ 0 & 0 & 1 \end{array}\right], Q_3 = \left[\begin{array}{c|cc|ccc} 1 & 0 & 0 & 0 & 0 & 0 \\ 0 & \omega & 0 & 0 & 0 & 0 \\ 0 & 0 & 0 & \omega & 0 & 0 \\ \hline 0 & \omega & 0 & 0 & 0 & 0 \\ 0 & 0 & 1 & 0 & 0 & 0 \\ 0 & 0 & 0 & 0 & \omega & 0 \\ \hline 0 & 0 & 0 & \omega & 0 & 0 \\ 0 & 0 & 0 & 0 & \omega & 0 \\ 0 & 0 & 0 & 0 & 0 & 1 \end{array}\right],$$

$$
Q_4 = \left[\begin{array}{c|cc|ccc|cccc}
1 & 0 & 0 & 0 & 0 & 0 & 0 & 0 & 0 & 0 \\
0 & \omega & 0 & 0 & 0 & 0 & 0 & 0 & 0 & 0 \\
0 & 0 & 0 & \omega & 0 & 0 & 0 & 0 & 0 & 0 \\
0 & 0 & 0 & 0 & 0 & 0 & \omega & 0 & 0 & 0 \\
\hline
0 & \omega & 0 & 0 & 0 & 0 & 0 & 0 & 0 & 0 \\
0 & 0 & 1 & 0 & 0 & 0 & 0 & 0 & 0 & 0 \\
0 & 0 & 0 & 0 & \omega & 0 & 0 & 0 & 0 & 0 \\
0 & 0 & 0 & 0 & 0 & 0 & 0 & \omega & 0 & 0 \\
\hline
0 & 0 & 0 & \omega & 0 & 0 & 0 & 0 & 0 & 0 \\
0 & 0 & 0 & 0 & \omega & 0 & 0 & 0 & 0 & 0 \\
0 & 0 & 0 & 0 & 0 & 1 & 0 & 0 & 0 & 0 \\
0 & 0 & 0 & 0 & 0 & 0 & 0 & 0 & \omega & 0 \\
\hline
0 & 0 & 0 & 0 & 0 & 0 & \omega & 0 & 0 & 0 \\
0 & 0 & 0 & 0 & 0 & 0 & 0 & \omega & 0 & 0 \\
0 & 0 & 0 & 0 & 0 & 0 & 0 & 0 & \omega & 0 \\
0 & 0 & 0 & 0 & 0 & 0 & 0 & 0 & 0 & 1
\end{array}\right].
$$

Similarly, we have for the minimum Lyapunov singular value

$$
\widetilde{\sigma_{\min}}(\mathcal{L}) = \sigma_{\min}(LQ_n)
$$

.

Definition F.11 *The singular values of the matrix LQ_n are called* Lyapunov singular values *of the Lyapunov operator $\mathcal{L}$ with associated matrix L. The set of Lyapunov singular values of $\mathcal{L}$ is denoted as $\widetilde{\sigma}(\mathcal{L}) := \sigma(LQ_n)$.*

To compare the standard and Lyapunov maximum and minimum singular values, consider the following example.

Example F.12 Let operators $\mathcal{L}_1, \mathcal{L}_2 \in \mathbf{Lyap}(n, \mathbb{R})$ be determined by

$$
\begin{aligned}
\mathcal{L}_1(X) &:= E_{11}XE_{22} + E_{22}XE_{11} - E_{12}XE_{12} - E_{21}XE_{21}, \\
\mathcal{L}_2(X) &:= X + \beta\mathcal{L}_1(X), \ X \in \mathbb{R}^{2\times 2},
\end{aligned}
$$

where $E_{ij} := E_{ij}(2,2)$ and $\beta > -1/2$. Setting $L_i := \mathrm{Mat}(\mathcal{L}_i)$ we have

$$
L_1 = \begin{bmatrix} 0 & 0 & 0 & 0 \\ 0 & 1 & -1 & 0 \\ 0 & -1 & 1 & 0 \\ 0 & 0 & 0 & 0 \end{bmatrix}, \quad L_2 = \begin{bmatrix} 1 & 0 & 0 & 0 \\ 0 & 1+\beta & -\beta & 0 \\ 0 & -\beta & 1+\beta & 0 \\ 0 & 0 & 0 & 1 \end{bmatrix}.
$$

Since $\sigma_{\max}(\mathcal{L}_1) = 2$ and $L_1Q_2 = 0_{4\times 3}$, the maximum singular value $\sigma_{\max}(\beta\mathcal{L}_1) = 2|\beta|$ of the operator $\beta\mathcal{L}_1$ may be arbitrarily larger than its maximum Lyapunov singular value $\widetilde{\sigma_{\max}}(\beta\mathcal{L}_1) = 0$. Furthermore, we have $\sigma(\mathcal{L}_2) = \{2\beta+1, 1, 1, 1\}$ and, since $L_2Q_2 = Q_2$, we obtain $\widetilde{\sigma}(\mathcal{L}_2) = \{1,1,1\}$. Then for large β the maximum singular value $\sigma_{\max}(\mathcal{L}_2) = 2\beta + 1$ of $\mathcal{L}_2$ is arbitrarily larger than its maximum Lyapunov singular value $\widetilde{\sigma_{\max}}(\mathcal{L}_2) = 1$. Finally, let $\beta = -1/2 + \varepsilon/2$, where $\varepsilon > 0$ is a small parameter. Then the minimum singular value $\sigma_{\min}(\mathcal{L}_2) = \varepsilon$ of $\mathcal{L}_2$ may be arbitrarily smaller than its minimum Lyapunov singular value, which is equal to 1.

The relationship between the sets of standard and Lyapunov singular values of a Lyapunov operator is revealed by the following assertion.

Proposition F.13 *If $\mathcal{L} \in \mathbf{Lyap}(n,\mathbb{R})$ then $\widetilde{\sigma}(\mathcal{L}) \subset \sigma(\mathcal{L})$.*

Proof. The set $\mathbf{Her}(n,\mathbb{R})$ is an invariant subspace of the operator $\mathcal{L} \in \mathbf{Lyap}(n,\mathbb{R})$. The orthogonal complement of that invariant subspace, the set $\mathbf{SHer}(n,\mathbb{R})$, is also an invariant subspace of $\mathcal{L}$. It follows that $\widetilde{\sigma}(\mathcal{L}) \subset \sigma(\mathcal{L})$. □

From an application viewpoint it is important to define the class of Lyapunov operators $\mathcal{L}$ with Sylvester index $\mathrm{ind}_n(\mathcal{L}) \leq 2$ such that

$$\sigma_{\min}(\mathcal{L}) = \widetilde{\sigma_{\min}}(\mathcal{L}) \text{ and } \sigma_{\max}(\mathcal{L}) = \widetilde{\sigma_{\max}}(\mathcal{L}). \tag{F.10}$$

As Example F.12 shows, for $\mathrm{ind}_n(\mathcal{L}) \geq 4$ it is possible that $\sigma_{\min}(\mathcal{L}) < \widetilde{\sigma_{\min}}(\mathcal{L})$ and/or $\sigma_{\max}(\mathcal{L}) > \widetilde{\sigma_{\max}}(\mathcal{L})$. The results in [40] can be extended to show that for $n = 3$ and $\mathrm{ind}_3(\mathcal{L}) = 2$ relation (F.10) is not valid in general.

If (F.10) holds, then for Lyapunov operators that are most used in practice, e.g. for the descriptor continuous- and discrete-time operators $\mathcal{L}_c$ and $\mathcal{L}_d$, given by $\mathcal{L}_c(X) = A^\top XE + E^\top XA$ and $\mathcal{L}_d(X) = A^\top XA - E^\top XE$, it is justified to use the minimum and maximum standard singular values, since they are equal to the corresponding Lyapunov singular values. For general Lyapunov operators, however, one should use the Lyapunov singular values, since they produce tighter bounds.

Note finally that the converse of Proposition F.13 is not true, i.e., the inclusion $\widetilde{\sigma}(\mathcal{M}) \subset \sigma(\mathcal{M})$ for some $\mathcal{M} \in \mathbf{Lin}(n,\mathbb{R})$ does not imply $\mathcal{M} \in \mathbf{Lyap}(n,\mathbb{R})$, as is demonstrated in the following example.

Example F.14 Let

$$M = \begin{bmatrix} 1 & 0 & 0 & 0 \\ 0 & 1 & 1 & 0 \\ 0 & 2 & 2 & 0 \\ 0 & 0 & 0 & 1 \end{bmatrix}.$$

Then $\widetilde{\sigma}(\mathcal{M}) = \sigma(MQ_2) = \{\sqrt{10},1,1\} \subset \sigma(M) = \{\sqrt{10},1,1,0\}$, but $M \notin \Omega(2,\mathbb{R})$ and hence, the corresponding $\mathcal{M}$ is not a Lyapunov operator.

F.3 Complex operators

The results for real Lyapunov operators have their counterparts for complex Lyapunov operators, defined next.

An operator $\mathcal{M} \in \mathbf{Lin}(n, \mathbb{C})$ may always be represented in the form (E.3), where $A_k, B_k \in \mathbb{C}^{n\times n}$. Definition E.3 is directly applicable to such operators and Proposition E.7 holds as well. Definition F.1 is modified as follows.

Definition F.15 *The complex operator $\mathcal{L} \in \mathbf{Lin}(n, \mathbb{C})$ is said to be a* Lyapunov operator *if*

$$(\mathcal{L}(X))^{\mathrm{H}} = \mathcal{L}(X^{\mathrm{H}}),\ X \in \mathbb{C}^{n\times n}.$$

The set of complex Lyapunov operators is denoted by $\mathbf{Lyap}(n, \mathbb{C})$.

It follows from Definition F.15 that $X = X^{\mathrm{H}} \Rightarrow \mathcal{L}(X) = (\mathcal{L}(X))^{\mathrm{H}}$ and $X = -X^{\mathrm{H}} \Rightarrow \mathcal{L}(X) = -(\mathcal{L}(X))^{\mathrm{H}}$ provided $\mathcal{L} \in \mathbf{Lyap}(n, \mathbb{C})$. Therefore $\mathbf{Her}(n, \mathbb{C})$ and $\mathbf{SHer}(n, \mathbb{C})$ are invariant sets for complex Lyapunov operators.

In the complex case, due to the nonlinearity of the complex conjugation, the set $\mathbf{Lyap}(n, \mathbb{C}) \subset \mathbf{Lin}(n, \mathbb{C})$ of Lyapunov operators is not a subspace of $\mathbf{Lin}(n, \mathbb{C})$ and the set $\Omega(n, C) \subset \mathbb{C}^{2n\times 2n}$ is not a subspace of $\mathbb{C}^{2n\times 2n}$ (these sets may become subspaces if we consider linear spaces of complex matrices with $\mathbb{R}$ as a field of scalars or if we pass to the representation $\mathbb{C}^{2n\times 2n} \simeq \mathbb{R}^{2n^2\times 2n^2}$).

We have the following analogue of Proposition F.2 in the complex case.

Proposition F.16 *The following four statements are equivalent:*

(i) $\mathcal{L} \in \mathbf{Lyap}(n, \mathbb{C})$.

(ii) *There exists $\mathcal{M} \in \mathbf{Lin}(n, \mathbb{C})$, such that*

$$\mathcal{L}(X) = \mathcal{M}(X) + (\mathcal{M}(X^{\mathrm{H}}))^{\mathrm{H}},\ X \in \mathbb{C}^{n\times n},$$

i.e., $\mathcal{L}(X) = \sum_{k=1}^{r} \left(A_k X B_k + B_k^{\mathrm{H}} X A_k^{\mathrm{H}}\right)$ *and*

$$L := \mathrm{Mat}(\mathcal{L}) = \sum_{k=1}^{r} \left(B_k^{\top} \otimes A_k + \overline{A}_k \otimes B_k^{\mathrm{H}}\right),$$

where $A_k, B_k \in \mathbb{C}^{n\times n}$ are given matrices.

(iii) $L \in \Omega(n, \mathbb{C})$, *where $\Omega(n, \mathbb{C})$ is the set of complex $n^2\times n^2$ matrices L, satisfying the equation $P_{n^2} L = \overline{L} P_{n^2}$.*

(iv) *The matrix $\mathbf{L} := \mathcal{V}_n(L)$ is Hermitian.*

Proof. The proof is similar to this of Proposition F.2. In particular we have the equation

$$\mathbf{AB} + (\mathbf{AB})^{\mathrm{H}} = \mathbf{L},$$

showing that $\mathbf{L}$ is Hermitian. □

If we represent $L \in \mathbb{C}^{n^2 \times n^2}$ as $L = S + \imath T$, where $S, T \in \mathbb{R}^{n^2 \times n^2}$, then Proposition F.16(iii) yields

$$P_{n^2} S = S P_{n^2}, \; P_{n^2} T = -T P_{n^2}. \tag{F.11}$$

Hence, we come to the following analogues of Propositions F.6 and F.8.

Proposition F.17 *The set $\Omega(n, \mathcal{C}) \subset \mathbb{C}^{2n \times 2n}$ of matrix representations of complex Lyapunov operators is isomorphic to the subspace*

$$\begin{aligned} & \operatorname{Ker}\left[\operatorname{diag}\left(I_{n^2} \otimes P_{n^2} - P_{n^2} \otimes I_{n^2}, I_{n^2} \otimes P_{n^2} + P_{n^2} \otimes I_{n^2}\right)\right] \\ = \; & \operatorname{Ker}\left[\operatorname{diag}\left(P_{n^2} \otimes P_{n^2} - I_{n^4}, P_{n^2} \otimes P_{n^2} + I_{n^4}\right)\right] \subset \mathbb{R}^{2n^4}. \end{aligned}$$

Proof. The proof follows directly from (F.11). □

Using the Jordan form (F.6) of P_{n^2} and the matrix Θ_n from (F.7) we can parametrize the set $\Omega(n, \mathbb{C})$ and determine its real dimension according to the following proposition.

Proposition F.18 *The set $\Omega(n, \mathbb{C})$ is parametrized as*

$$\Omega(n, \mathbb{C}) = \mathcal{V}_n^{-1}(\mathbf{Her}(n^2, \mathbb{C})) = \left\{ \Theta_n \begin{bmatrix} L_{11} & \imath L_{12} \\ \imath L_{21} & L_{22} \end{bmatrix} \Theta_n^{\top} : L_{ij} \in \mathbb{R}^{n_i \times n_j} \right\}.$$

In particular the real dimension of $\mathbf{Lyap}(n, \mathbb{C})$ *and* $\Omega(n, \mathbb{C})$ *is* n^4.

Proof. The first representation follows from Proposition F.16(iv). The second is based on equations (F.11) for the matrices S and T. Using the Jordan form J_n of P_{n^2} we obtain the equivalent equations

$$J_n \widehat{S} = \widehat{S} J_n, \; \widehat{S} := \Theta_n^{\top} S \Theta_n; \; J_n \widehat{T} = -\widehat{T} J_n, \; \widehat{T} := \Theta_n^{\top} T \Theta_n. \tag{F.12}$$

The general solution of (F.12) is of the form

$$\widehat{S} = \begin{bmatrix} L_{11} & 0 \\ 0 & L_{22} \end{bmatrix}, \; \widehat{T} = \begin{bmatrix} 0 & L_{12} \\ L_{21} & 0 \end{bmatrix},$$

where the matrices $L_{ij} \in \mathbb{R}^{n_i \times n_j}$ are arbitrary. □

Similarly to the real case, a complex Lyapunov operator $\mathcal{L} \in \mathbf{Lyap}(n, \mathbb{C})$ admits also the Hermitian representation

$$\mathcal{L}(X) = \sum_{j=1}^{\ell_d} \varepsilon_j D_j X D_j^{\mathrm{H}},$$

where $\varepsilon_j = \pm 1$ and $D_j \in \mathbb{C}^{n\times n}$. Accordingly, the concepts from Definition F.3 are easily extended to the case of complex Lyapunov operators. In particular we see that Proposition F.5 holds also in the complex case.

The maximum and minimum Lyapunov singular values of the operator $\mathcal{L} \in \mathbf{Lyap}(n, \mathbb{C})$ are defined as

$$\begin{aligned} \widetilde{\sigma}_{\max}(\mathcal{L}) &:= \max\{\|\mathcal{L}(X)\|_{\mathrm{F}} : \|X\|_{\mathrm{F}} = 1,\ X = X^{\mathrm{H}}\}, \\ \widetilde{\sigma}_{\min}(\mathcal{L}) &:= \min\{\|\mathcal{L}(X)\|_{\mathrm{F}} : \|X\|_{\mathrm{F}} = 1,\ X = X^{\mathrm{H}}\}, \end{aligned} \tag{F.13}$$

respectively.

The Lyapunov singular values of a complex Lyapunov operator $\mathcal{L} \in \mathbf{Lyap}(n, \mathbb{C})$ with matrix

$$L = S + \imath T;\ S \in \mathbf{Her}(n, \mathbb{R}),\ T \in \mathbf{SHer}(n, \mathbb{R}),$$

are defined as follows. Let

$$L^{\mathbb{R}} := \begin{bmatrix} S & -T \\ T & S \end{bmatrix} \in \mathbb{R}^{2n\times 2n}$$

be the real version of L. Let $X = Y + \imath Z$, where $Y, Z \in \mathbb{R}^{n\times n}$. Then the restriction $X = X^{\mathrm{H}}$ in (F.13) means that Y is symmetric and Z is skew-symmetric, i.e., $y := \mathrm{vec}(Y) \in \mathrm{Rg}(I_{n^2} - P_{n^2})$ and $z := \mathrm{vec}(Z) \in \mathrm{Rg}(I_{n^2} + P_{n^2})$. Furthermore, we have

$$\|\mathcal{L}(X)\|_{\mathrm{F}} = \left\| L^{\mathbb{R}} \begin{bmatrix} y \\ z \end{bmatrix} \right\|_2 .$$

Therefore, as in the real case, we obtain

$$\widetilde{\sigma}_{\max}(\mathcal{L}) = \|\Psi_n(L)\|_2 = \sigma_1(\Psi_n(L)),\ \widetilde{\sigma}_{\min}(\mathcal{L}) = \sigma_{n^2}(\Psi_n(L)),$$

where the matrix $\Psi_n(L)$ is defined by

$$\Psi_n(L) := L^{\mathbb{R}}\mathrm{diag}(Q_n, R_n) = \begin{bmatrix} SQ_n & -TR_n \\ TQ_n & SR_n \end{bmatrix} \in \mathbb{R}^{2n^2\times n^2}.$$

Here the matrix $R_n \in \mathbb{R}^{n^2\times n(n-1)/2}$ is obtained from Q_n (see (F.9) and Example F.10) by deleting the columns containing 1's (which are numbered as $k(k+1)/2$, $k = 1, \ldots, n$) and by changing the sign of each second element ω in each column of the reduced matrix. Formally this procedure is described as follows. Let $\Delta_n = [\Delta_n]_{i,j} := [\delta_{i(i+1)/2,j}] \in \mathbb{R}^{n(n+1)/2\times n(n-1)/2}$, where $\delta_{i,j}$ is the Kronecker delta, and

$$\mathcal{J} := \{(kn + l, k(k-1)/2 + l) : k = 1, \ldots, n-1,\ l = 1, \ldots, k\}.$$

Then the elements $[R_n]_{i,j}$ of the matrix R_n are given by

$$[R_n]_{i,j} = \begin{cases} [Q_n\Delta_n]_{i,j} & \text{if } (i,j) \notin \mathcal{J}, \\ -[Q_n\Delta_n]_{i,j} & \text{if } (i,j) \in \mathcal{J}. \end{cases}$$

Example F.19 The matrices R_2, R_3, R_4 are

$$
R_2 = \left[\begin{array}{c} 0 \\ \omega \\ \hline -\omega \\ 0 \end{array}\right], \; R_3 = \left[\begin{array}{c|cc} 0 & 0 & 0 \\ \omega & 0 & 0 \\ 0 & -\omega & 0 \\ \hline -\omega & 0 & 0 \\ 0 & 0 & 0 \\ 0 & 0 & \omega \\ \hline 0 & -\omega & 0 \\ 0 & 0 & -\omega \\ 0 & 0 & 0 \end{array}\right],
$$

$$
R_4 = \left[\begin{array}{c|cc|ccc} 0 & 0 & 0 & 0 & 0 & 0 \\ \omega & 0 & 0 & 0 & 0 & 0 \\ 0 & \omega & 0 & 0 & 0 & 0 \\ 0 & 0 & 0 & \omega & 0 & 0 \\ \hline -\omega & 0 & 0 & 0 & 0 & 0 \\ 0 & 0 & 0 & 0 & 0 & 0 \\ 0 & 0 & \omega & 0 & 0 & 0 \\ 0 & 0 & 0 & 0 & \omega & 0 \\ \hline 0 & -\omega & 0 & 0 & 0 & 0 \\ 0 & 0 & -\omega & 0 & 0 & 0 \\ 0 & 0 & 0 & 0 & 0 & 0 \\ 0 & 0 & 0 & 0 & 0 & \omega \\ \hline 0 & 0 & 0 & -\omega & 0 & 0 \\ 0 & 0 & 0 & 0 & -\omega & 0 \\ 0 & 0 & 0 & 0 & 0 & -\omega \\ 0 & 0 & 0 & 0 & 0 & 0 \end{array}\right].
$$

Definition F.20 *The* Lyapunov singular values *of the complex Lyapunov operator* $\mathcal{L} \in \mathbf{Lyap}(n, \mathbb{C})$ *with associated matrix* $L \in \Omega(n, \mathbb{C})$ *are the singular values of the matrix* $\Psi_n(L)$, *namely* $\widetilde{\sigma}(\mathcal{L}) := \sigma(\Psi_n(L))$.

A similar statement $\widetilde{\sigma}(\mathcal{L}) \subset \sigma(\mathcal{L})$ as in the real case can be stated for complex Lyapunov operators $\mathcal{L}$.

Consider now some problems concerning the inversion of Lyapunov operators. The operator $\mathcal{L} \in \mathbf{Lyap}(n, \mathbb{F})$ is invertible if and only if its matrix L is nonsingular, and in this case we have $\mathrm{Mat}(\mathcal{L}^{-1}) = L^{-1}$. In addition, the inverse of a Lyapunov operator is again a Lyapunov $\mathcal{L}$ operator since for $L \in \mathcal{GL}(n^2, \mathbb{F})$ the equations $P_{n^2} L = L P_{n^2}$ and $P_{n^2} L^{-1} = L^{-1} P_{n^2}$ are equivalent. Conditions for invertibility of general real and complex Lyapunov operators are given in [132].

The continuous-time Lyapunov indices of the operator and its inverse may differ, see Example F.22.

Consider the continuous-time and discrete-time Lyapunov operators from **Lyap**$(2,\mathbb{R})$.

Example F.21 Given the matrix $A \in \mathbb{R}^{2\times 2}$, the continuous-time operator $\mathcal{L}_{A,c} \in$ **Lyap**$(2,\mathbb{R})$ is defined by $\mathcal{L}_{A,c}(X) := \mathcal{E}_2(A^\top, I_2, I_2, A) = A^\top X + XA$, $X \in \mathbb{R}^{2\times 2}$. It is invertible if and only if $\mathrm{tr}(A) \neq 0$ and $\det(A) \neq 0$. Also, it is of index 1 if and only if A is a scalar multiple of I_2, and of index 2 otherwise.

The discrete-time operator $\mathcal{L}_{A,d} \in$ **Lyap**$(2,\mathbb{R})$, defined by

$$\mathcal{L}_{A,d}(X) = \mathcal{E}_2(A^\top, A, I_n, -I_n) = A^\top XA - X,\ X \in \mathbb{R}^{2\times 2},$$

is invertible if and only if $\det(A) \neq 1$ and $\mathrm{tr}(A) - \det(A) \neq 1$. It is of index 1 if and only if A is a scalar multiple of I_2, and of index 2 otherwise.

Example F.22 Let $A = \begin{bmatrix} \lambda & 1 \\ 0 & \lambda \end{bmatrix}$, where $0 \neq \lambda \in \mathbb{R}$. Then

$$\begin{aligned} \mathcal{L}_{A,c}(X) &= \begin{bmatrix} 2\lambda x_{11} & x_{11} + 2\lambda x_{12} \\ x_{11} + 2\lambda x_{21} & x_{21} + x_{12} + 2\lambda x_{22} \end{bmatrix},\ X = [x_{ij}] \in \mathbb{R}^{2\times 2}, \\ \mathcal{L}_{A,c}^{-1}(Y) &= l \begin{bmatrix} y_{11} & y_{12} - l y_{11} \\ y_{21} - l y_{11} & 2l^2 y_{11} - l y_{21} - l y_{12} + y_{22} \end{bmatrix},\ Y = [y_{ij}] \in \mathbb{R}^{2\times 2} \end{aligned}$$

where $l := 1/(2\lambda)$. Hence, the matrix $L_{A,c}$ of $\mathcal{L}_{A,c}$ is

$$L_{A,c} = \begin{bmatrix} 2\lambda & 0 & 0 & 0 \\ 1 & 2\lambda & 0 & 0 \\ 1 & 0 & 2\lambda & 0 \\ 0 & 1 & 1 & 2\lambda \end{bmatrix}.$$

The matrix

$$\mathcal{V}_2(L_{A,c}) = \begin{bmatrix} 2\lambda & 1 & 0 & 2\lambda \\ 1 & 0 & 0 & 1 \\ 0 & 0 & 2\lambda & 0 \\ 2\lambda & 1 & 0 & 2\lambda \end{bmatrix}$$

has two eigenvalues $2\lambda \pm \sqrt{4\lambda^2 + 2}$ of opposite sign and two zero eigenvalues. Hence, $\mathrm{clind}_2(\mathcal{L}_{A,c}) = \mathrm{dlind}_2(\mathcal{L}_{A,c}) = 2$. The matrix $\mathcal{V}_2(L_{A,c}^{-1}) \in$ **Her**$(4,\mathbb{R})$ has two eigenvalues of the same sign and two zero eigenvalues. Therefore $\mathrm{dlind}_2(\mathcal{L}_{A,c}^{-1}) = 2$, $\mathrm{clind}_2(\mathcal{L}_{A,c}^{-1}) = 4$. If for example $\lambda > 0$, then we have the following discrete-time representation $\mathcal{L}_{A,c}^{-1}(Y) = D_1 Y D_1^\top + D_2 Y D_2^\top$, where $D_1 := \sqrt{l} \begin{bmatrix} 1 & 0 \\ -l & 1 \end{bmatrix}$, $D_2 := l\sqrt{l} \begin{bmatrix} 0 & 0 \\ 1 & 0 \end{bmatrix}$.

So far we have made an analysis of general Lyapunov operators. In the following section, we discuss the application of these results to the sensitivity and a posteriori error analysis of Lyapunov equations.

F.4 Sensitivity and error analysis

Consider the Hermitian Lyapunov equation

$$\mathcal{L}(X) = Q, \ Q^{\mathrm{H}} = Q \neq 0, \tag{F.14}$$

with an invertible Lyapunov operator $\mathcal{L}$. The minimum symmetric singular value $\widetilde{\sigma_{\min}}(\mathcal{L})$ of $\mathcal{L}$ is a relevant measure for the sensitivity of the Lyapunov equation (F.14) relative to perturbations in the coefficient matrices of $\mathcal{L}$ and Hermitian perturbations $\Delta Q = \Delta Q^{\mathrm{H}}$ in the matrix Q.

Denote by $X_0 = X_0^{\mathrm{H}} = \mathcal{L}^{-1}(Q)$ the solution of (F.14) and let $X = X_0 + \delta X$ be the solution to the perturbed Lyapunov equation $\mathcal{L}(X) = Q + \delta Q$. We have $\delta X = \mathcal{L}^{-1}(\delta Q)$ and hence,

$$\|\delta X\|_{\mathrm{F}} \leq \|\mathcal{L}^{-1}\|_{\widetilde{\mathrm{F}}} \|\delta Q\|_{\mathrm{F}} = \frac{1}{\widetilde{\sigma_{\min}}(\mathcal{L})} \|\delta Q\|_{\mathrm{F}}.$$

In terms of relative perturbations it holds that

$$\varepsilon_X \leq \widetilde{\kappa} \varepsilon_Q, \ \widetilde{\kappa} := \frac{1}{\widetilde{\sigma_{\min}}(\mathcal{L})} \frac{\|Q\|_{\mathrm{F}}}{\|X_0\|_{\mathrm{F}}},$$

where $e_Z := \|\delta Z\|_{\mathrm{F}} / \|Z\|_{\mathrm{F}}$ and $\widetilde{\kappa}$ is the *relative condition number* of the Lyapunov equation (F.14) with respect to Hermitian perturbations in Q. Note that usually $Q = D^{\mathrm{H}} D$ and when the matrix D is perturbed, then the perturbation $\delta Q = \delta D^{\mathrm{H}} D + D^{\mathrm{H}} \delta D + \delta D^{\mathrm{H}} \delta D$ in Q is Hermitian.

Most of the perturbation bounds in the literature [95, 68] are based on $\sigma_{\min}(\mathcal{L})$ instead on $\widetilde{\sigma_{\min}}(\mathcal{L})$, e.g. the condition number is taken as

$$\kappa := \|Q\|_{\mathrm{F}} / (\|X_0\|_{\mathrm{F}} \sigma_{\min}(\mathcal{L})).$$

Since $\kappa \geq \widetilde{\kappa}$ may be much larger than $\widetilde{\kappa}$, it is clear that in case of Hermitian perturbations one should use the relevant sensitivity estimates, based on symmetric singular values instead on standard singular values of Lyapunov operators. At the same time sensitivity estimates, based on the standard singular values, should be used in case of non-Hermitian perturbations.

Consider now the a posteriori error analysis of equation (F.14). Suppose that $\widehat{X}$ is an approximate solution of equation (F.14). For example this may be the solution, produced by a numerical method in finite precision arithmetics. Then it is important to have a sharp computable bound on the actual relative error $\delta_{\widehat{X}} := \|\widehat{X} - X_0\|_{\mathrm{F}} / \|X_0\|_{\mathrm{F}}$. Such a tight bound may be derived using the symmetric singular values of $\mathcal{L}$ and in particular the symmetric relative condition number of $\mathcal{L}$, defined below.

Denote by $\widehat{Q} := \mathcal{L}(\widehat{X})$ the residual, corresponding to the approximate solution $\widehat{X}$. We have $\mathcal{L}(\widehat{X} - X_0) = \widehat{Q} - Q$, which gives $\widehat{X} - X_0 = \mathcal{L}^{-1}(\widehat{Q} - Q)$ and

$$\|\widehat{X} - X_0\|_{\mathrm{F}} \leq \frac{\|\widehat{Q} - Q\|_{\mathrm{F}}}{\widetilde{\sigma}_{\min}(\mathcal{L})}. \tag{F.15}$$

Since $\|Q\|_{\mathrm{F}} \leq \widetilde{\sigma}_{\max}(\mathcal{L})\|X_0\|_{\mathrm{F}}$, it holds that

$$\frac{1}{\|X_0\|_{\mathrm{F}}} \leq \frac{\widetilde{\sigma}_{\max}(\mathcal{L})}{\|Q\|_{\mathrm{F}}}. \tag{F.16}$$

Combining (F.15) and (F.16) we get the desired estimate

$$\delta_{\widehat{X}} \leq \widetilde{\mathrm{cond}_2}(\mathcal{L}) \frac{\|\widehat{Q} - Q\|_{\mathrm{F}}}{\|Q\|_{\mathrm{F}}},$$

where $\widetilde{\mathrm{cond}_2}(\mathcal{L}) := \frac{\widetilde{\sigma}_{\max}(\mathcal{L})}{\widetilde{\sigma}_{\min}(\mathcal{L})}$ is the *symmetric relative condition number* of $\mathcal{L}$ with respect to inversion. This condition number may be used also for a posteriori error analysis of approximate solutions to symmetric matrix Riccati equations.

F.5 Notes and references

Since the fundamental work of Lyapunov on stability of motion, Lyapunov matrix equations have been widely used in stability theory of differential equations [236], in the theory of linear-quadratic optimization and filtering [181], in the perturbation analysis of linear and nonlinear matrix equations [68, 95, 120, 151] and other fields of pure and applied mathematics. This has motivated a continuous interest to both the theory and numerical treatment of Lyapunov operators and equations [40, 41, 79, 203, 204, 216, 69] and also recently in the context of the analysis and numerical simulation of descriptor systems via generalized Lyapunov equations [167].

The results presented in this chapter are entirely based on the papers [126, 125].

Appendix G

Lyapunov-like operators

G.1 Introductory remarks

In this appendix we consider six more classes of Lyapunov operators and present their parametrizations and dimensions in particular. The proofs are similar to these from Appendix F and are omitted.

G.2 Skew-Lyapunov operators

Real skew-Lyapunov operators $\mathcal{L}$ from $\mathbf{Lin}(n, \mathbb{R})$ are defined via

$$(\mathcal{L}(X))^\top = -\mathcal{L}(X^\top), \ X \in \mathbb{R}^{n\times n},$$

and may be represented as

$$\mathcal{L}(X) = \sum_{k=1}^{r} \left(A_k X B_k - B_k^\top X A_k^\top \right); \ A_k, B_k \in \mathbb{R}^{n\times n}. \tag{G.1}$$

The matrix $L \in \mathbb{R}^{n^2 \times n^2}$ of a skew-Lyapunov operator satisfies $P_{n^2} L = -L P_{n^2}$ and has the form

$$L = \Theta_n \begin{bmatrix} 0 & L_{12} \\ L_{21} & 0 \end{bmatrix} \Theta_n^\top,$$

where the matrices $L_{ij} \in \mathbb{R}^{n_i \times n_j}$ are arbitrary. Hence the space of real skew-Lyapunov operators is of dimension $2n_1 n_2 = n^2(n^2-1)/2$. Since $\mathbf{AB} - (\mathbf{AB})^\top = \mathbf{L}$ then the matrix $\mathbf{L} := \mathcal{V}_n(L)$ of a real skew-Lyapunov operator $\mathcal{L}$ is skew-symmetric.

Complex skew-Lyapunov operators $\mathcal{L}$ from $\mathbf{Lin}(n, \mathbb{C})$ are defined by the relation

$$(\mathcal{L}(X))^{\mathrm{H}} = -\mathcal{L}(X^{\mathrm{H}}), \ X \in \mathbb{C}^{n\times n},$$

and may be represented as

$$\mathcal{L}(X) = \sum_{k=1}^{r} \left(A_k X B_k - B_k^{\mathrm{H}} X A_k^{\mathrm{H}} \right); \; A_k, B_k \in \mathbb{C}^{n\times n}. \tag{G.2}$$

The matrix $L = S + \jmath T \in \mathbb{C}^{n^2\times n^2}$ with $S, T \in \mathbb{R}^{n^2\times n^2}$ of a complex skew-Lyapunov operator $\mathcal{L}$ satisfies the equation $P_{n^2} L = -\overline{L} P_{n^2}$ and hence, $P_{n^2} S = -S P_{n^2}$, $P_{n^2} T = T P_{n^2}$. Thus

$$L = \Theta_n \begin{bmatrix} \jmath L_{11} & L_{12} \\ L_{21} & \jmath L_{22} \end{bmatrix} \Theta_n^{\top},$$

where the matrices $L_{ij} \in \mathbb{R}^{n_i \times n_j}$ are arbitrary. Hence, the space of complex skew-Lyapunov operators is of real dimension n^4. The matrix $\mathbf{L} := \mathcal{V}_n(L)$ for a complex skew-Lyapunov operator $\mathcal{L}$ is skew-Hermitian since $\mathbf{AB} - (\mathbf{AB})^{\mathrm{H}} = \mathbf{L}$.

The *skew-Lyapunov index* of a skew-Lyapunov operator is defined as the minimum number of terms in the representations (G.1) or (G.2) and may be determined as follows.

Consider the equation $\mathbf{C} - \mathbf{C}^* = \mathbf{L}$ in $\mathbf{C} := \mathbf{AB}$ for a skew-Lyapunov operator. The matrix $\mathbf{L}$ is congruent to the matrix

$$\operatorname{diag}\left(\begin{bmatrix} 0 & -1 \\ 1 & 0 \end{bmatrix}, \ldots, \begin{bmatrix} 0 & -1 \\ 1 & 0 \end{bmatrix}, 0_{(n^2-2r)\times(n^2-2r)} \right)$$

with r blocks of size 2×2 on the diagonal in the real case, and to the matrix

$$\operatorname{diag}\left(\imath I_\alpha, -\imath I_\beta, 0_{(n^2-\gamma)\times(n^2-\gamma)} \right)$$

in the complex case, where $\gamma := \alpha + \beta = \operatorname{rank}(\mathbf{L})$. Hence, the minimum achievable rank of $\mathbf{C}$ is the rank of $\mathbf{L}$. Thus, the skew-Lyapunov index of the skew-Lyapunov operator $\mathcal{L} \in \mathbf{Lin}(n, \mathbb{F})$ with matrix L is equal to the rank of the matrix $\mathbf{L} := \mathcal{V}_n(L)$.

G.3 Associated Lyapunov operators

Associated Lyapunov and Riccati equations have been considered in [148] in the real case and in [140] in the complex case. Below we present the parametrizations of associated Lyapunov operators.

Real associated Lyapunov operators $\mathcal{L}$ from $\mathbf{Lin}(n, \mathbb{R})$ are defined by

$$(\mathcal{L}(X))^{\top} = \mathcal{L}(X), \; X \in \mathbb{R}^{n\times n},$$

and are given by

$$\mathcal{L}(X) = \sum_{k=1}^{r} \left(A_k X B_k + B_k^{\top} X^{\top} A_k^{\top} \right); \; A_k, B_k \in \mathbb{R}^{n\times n}.$$

The matrix

$$L = \sum_{k=1}^{r} \left(B_k^\top \otimes A_k + A_k \otimes B_k^\top P_{n^2} \right) \in \mathbb{R}^{n^2 \times n^2}$$

of the associated Lyapunov operator $\mathcal{L}$ satisfies $P_{n^2} L = L$ and has the form

$$L = \Theta_n \begin{bmatrix} L_1 \\ 0_{n_2 \times n^2} \end{bmatrix} \Theta_n^\top,$$

where the matrix $L_1 \in \mathcal{R}^s n_1 \times n^2$ is arbitrary. Hence the space of real associated Lyapunov operators is of dimension $n^3(n+1)/2$. It may be shown that $\mathbf{L}_{ij} = \mathbf{L}_{ij}^\top$, where $\mathbf{L}_{ij}$ are the $n \times n$ blocks in the partition $\mathbf{L} = [\mathbf{L}_{ij}]$ of the matrix $\mathbf{L} := \mathcal{V}_n(L)$.

Complex associated Lyapunov operators $\mathbb{C}^{n\times n} \to \mathbb{C}^{n\times n}$ are defined by

$$(\mathcal{L}(X))^{\mathrm{H}} = \mathcal{L}(X), \;\; X \in \mathbb{C}^{n\times n},$$

and may be represented as

$$\mathcal{L}(X) = \sum_{k=1}^{r} \left(A_k X B_k + B_k^{\mathrm{H}} X^{\mathrm{H}} A_k^{\mathrm{H}} \right); \;\; A_k, B_k \in \mathbb{C}^{n\times n}.$$

Complex associated Lyapunov operators are not linear, but pseudo-linear operators, see Chapter 13 and [140]. For pseudo-linear operators $\mathcal{L} : \mathbb{C}^{n\times n} \to \mathbb{C}^{n\times n}$ we have $\mathcal{L}(X) = \mathcal{M}_1(X) + \mathcal{M}_2(X^{\mathrm{H}})$, $\mathcal{M}_i \in \mathbf{Lin}(n, \mathbb{C})$, and $\mathrm{vec}(\mathcal{L}(X)) = M_1 \mathrm{vec}(X) + M_2 P_{n^2} \mathrm{vec}(\overline{X})$, $M_i := \mathrm{Mat}(\mathcal{M}_i)$. Thus, the set of these pseudo-linear operators is of complex dimension $2n^4$.

For a complex associated Luapunov operator it is fulfilled $M_2 = \overline{M}_1$, i.e., $\mathrm{vec}(\mathcal{L}(X)) = \Lambda \mathrm{vec}(X) + \overline{\Lambda} \mathrm{vec}(\overline{X})$, $\Lambda \in \mathbb{C}^{n\times n}$. Hence, the set of complex associated Lyapunov operators is of complex dimension n^4.

The values of an associated Lyapunov operator are symmetric matrices in the real case and Hermitian matrices in the complex case. Hence, these operators are not surjective if considered as mappings $\mathbb{F}^{n\times n} \to \mathbb{F}^{n\times n}$ and thus, one should consider associated Lyapunov operators as mappings $\mathbb{F}^{n\times n} \to \mathbf{Her}(n, \mathbb{F})$.

G.4 Associated skew-Lyapunov operators

Real associated skew-Lyapunov operators $\mathcal{L}$ from $\mathbf{Lin}(n, \mathbb{R})$ are defined by

$$(\mathcal{L}(X))^\top = -\mathcal{L}(X), \;\; X \in \mathbb{R}^{n\times n},$$

and may be represented as

$$\mathcal{L}(X) = \sum_{k=1}^{r} \left(A_k X B_k - B_k^\top X^\top A_k^\top \right); \;\; A_k, B_k \in \mathbb{R}^{n\times n}.$$

The matrix

$$L = \sum_{k=1}^{r} \left(B_k^{\top} \otimes A_k - \left(A_k \otimes B_k^{\top} P_{n^2} \right) \right) \in \mathbb{R}^{n^2 \times n^2}$$

of an associated skew-Lyapunov operator satisfies $P_{n^2} L = -L$ and has the form

$$L = \Theta_n \begin{bmatrix} 0_{n_1 \times n^2} \\ L_2 \end{bmatrix} \Theta_n^{\top},$$

where the matrix $L_2 \in \mathbb{R}^{n_2 \times n^2}$ is arbitrary. Hence the space of real associated skew-Lyapunov operators is of dimension $n_2 n^2 = n^3(n-1)/2$.

It is easily proved that $\mathbf{L}_{ij} = -\mathbf{L}_{ij}^{\top}$, where $\mathbf{L}_{ij}$ are the $n \times n$ blocks in the partition of the matrix $\mathbf{L} := \mathcal{V}_n(L)$.

Complex associated skew-Lyapunov operators $\mathbb{C}^{n \times n} \to \mathbb{C}^{n \times n}$ are determined by

$$(\mathcal{L}(X))^{\mathrm{H}} = -\mathcal{L}(X), \ X \in \mathbb{C}^{n \times n},$$

and may be represented as

$$\mathcal{L}(X) = \sum_{k=1}^{r} \left(A_k X B_k - B_k^{\mathrm{H}} X^{\mathrm{H}} A_k^{\mathrm{H}} \right); \ A_k, B_k \in \mathbb{C}^{n \times n}.$$

These operators are pseudo-linear and satisfy $\mathrm{vec}(\mathcal{L}(X)) = \Lambda \mathrm{vec}(X) - \overline{\Lambda} \mathrm{vec}(\overline{X})$, $\Lambda \in \mathbb{C}^{n \times n}$. Thus the set of complex associated skew-Lyapunov operators is of complex dimension n^4.

The values of associated skew-Lyapunov operator are skew-symmetric matrices (in the real case) or skew-Hermitian matrices (in the complex case) and these operators are not surjective if considered as mappings $\mathbb{F}^{n \times n} \to \mathbb{F}^{n \times n}$. Hence, one may consider associated skew-Lyapunov operators as mappings $\mathbb{C}^{n \times n} \to \mathbf{SHer}(n, \mathbb{F})$.

G.5 Notes and references

Lyapunov-like and other related (bilinear and Riccati) operators have been considered in [125].

Appendix H

Notation

We usually use upper case for matrices, lower case for vectors and lower case Greek for scalars.

In what follows we list the common notation which is used throughout the text.

H.1 Sets and spaces

$\imath = \sqrt{-1}$ – the imaginary unit;

$\mathbb{R}$ and $\mathbb{C}$ – the fields of real and complex numbers;

$\mathbb{R}_+ = [0, \infty)$ – the set of nonnegative real numbers;

$\sup\{\mathcal{M}\}$ – the supremum (least upper bound) of the set $\mathcal{M} \subset \mathbb{R}$, i.e., the least real number such that $\alpha \in \mathcal{M}$ implies $\alpha \leq \sup\{\mathcal{M}\}$. The supremum of the empty set is assumed to be $-\infty$;

$\inf\{\mathcal{M}\}$ – the infimum (largest lower bound) of $\mathcal{M}$, i.e., the greatest number such that $\alpha \in \mathcal{M}$ implies $\alpha \geq \inf\{\mathcal{M}\}$. The infimum of the empty set is assumed to be ∞;

$\mathbb{F}$ – a replacement of either $\mathbb{R}$ or $\mathbb{C}$;

$\mathbb{C}_-$ – the open left complex half-plane;

$\mathbb{D}_1$ – the open unit complex disc, centered at the origin;

$\mathbb{F}^{m\times n}$ – the space of $m \times n$ matrices $A = [a_{ij}]$ with elements $a_{ij} \in \mathbb{F}$. The elements of A are also denoted as $(A)_{ij}$. The pair (m, n) is the size of $A \in \mathbb{F}^{m\times n}$;

$\mathbb{F}^n = \mathbb{F}^{n\times 1}$ – the space of column n-vectors $x = [x_i]$ with elements $x_i \in \mathbb{F}$;

$2^{\mathcal{X}}$ – the set of subsets of the set $\mathcal{X}$.

H.2 Matrices

I_n (or I) – the $n \times n$ identity matrix;

$0_{m\times n}$ (or 0) – the $m \times n$ zero matrix, or a zero matrix, whose size is clear from the context. If $m = 0$ or $n = 0$ the matrix $0_{m\times n}$ is void;

$E_{ij}(m,n) \in \mathbb{R}^{m\times n}$ – a matrix with a single nonzero element, equal to 1, in position (i,j); $E_{ij}(n) = E_{ij}(n,n)$;

$A^\top = [a_{ji}] \in \mathbb{F}^{n\times m}$ – the transpose of $A = [a_{ij}] \in \mathbb{F}^{m\times n}$;

$\overline{A} = [\overline{a}_{ij}] \in \mathbb{F}^{m\times n}$ – the complex conjugate of $A \in \mathbb{F}^{m\times n}$;

$A^{\mathrm{H}} = \overline{A}^\top \in \mathbb{F}^{n\times m}$ – the complex conjugate transpose of $A \in \mathbb{F}^{m\times n}$;

$\mathrm{Re}(A) \in \mathbb{R}^{m\times n}$ and $\mathrm{Im}(A) \in \mathbb{R}^{m\times n}$ - the real and imaginary parts of $A \in \mathbb{C}^{m\times n}$, i.e., $A = \mathrm{Re}(A) + \imath\mathrm{Im}(A)$;

$\mathrm{diag}(a_1,\ldots,a_n)$ – a (block) diagonal matrix with diagonal (block) elements a_i, where a_i are scalars or matrices;

$\mathrm{rank}(A)$ – the rank of A, equal to the number of its linearly independent columns or rows;

$\det(A)$ and $\mathrm{tr}(A)$ – the determinant and trace of A;

$\lambda_1(A),\ldots,\lambda_n(A) \in \mathbb{C}$ – the eigenvalues of $A \in \mathbb{F}^{n\times n}$, counted according to their algebraic multiplicity;

$\mathrm{spect}(A) = \{\lambda_1(A),\ldots,\lambda_n(A)\} \subset \mathbb{C}$ – the spectrum of $A \in \mathbb{F}^{n\times n}$. We note that $\mathrm{spect}(A)$ is a collection, i.e., a set with (possibly) repeated elements;

$\mathrm{rad}(A) = \max\{|z| : z \in \mathrm{spect}(A)\}$ – the spectral radius of A;

$\sigma_1(A) \geq \cdots \geq \sigma_k(A) \geq 0$ – the singular values of $A \in \mathbb{F}^{m\times n}$, where $k = \min\{m,n\}$. The positive singular values of A (whose number is $r = \mathrm{rank}(A)$) are the positive square roots of the positive eigenvalues of $A^{\mathrm{H}}A$ or AA^{H}. We also denote $\sigma_{\max}(A) = \sigma_1(A)$ and $\sigma_{\min}(A) = \sigma_r(A)$;

$\mathrm{Ker}(A) = \{x \in \mathbb{F}^n : Ax = 0\} \subset \mathbb{F}^n$ and $\mathrm{Rg}(A) = \{Ax : x \in \mathbb{F}^n\} \subset \mathbb{F}^m$ – the kernel and range (or image) of $A \in \mathbb{F}^{m\times n}$;

$A \otimes B = [a_{ij}B] \in \mathbb{F}^{mk\times nl}$ – the Kronecker product of $A = [a_{ij}] \in \mathbb{F}^{m\times n}$ and $B \in \mathbb{F}^{k\times l}$;

$\mathcal{GL}(n,\mathbb{F}) \subset \mathbb{F}^{n\times n}$ – the group of nonsingular $n \times n$ matrices over $\mathbb{F}$;

$\mathcal{U}(n) \subset \mathcal{GL}(n,\mathbb{C})$ – the group of unitary matrices $U \in \mathbb{C}^{n\times n}$ ($U^{\mathrm{H}}U = I_n$);

$\mathcal{O}(n,\mathbb{F}) \subset \mathbb{F}^{n\times n}$ – the group of orthogonal matrices $U \in \mathbb{F}^{n\times n}$ ($U^\top U = I_n$);

$\mathbf{Her}(n,\mathbb{F}) \subset \mathbb{F}^{n\times n}$ – the set of Hermitian matrices, satisfying $A^{\mathrm{H}} = A$. In the real case $\mathbf{Her}(n,\mathbb{R}) \subset \mathbb{R}^{n\times n}$ is the set of symmetric matrices;

$\nu_+(A)$, $\nu_-(A)$ and $\nu_0(A)$ – the number of positive, negative and zero eigenvalues of the matrix $A \in \mathbf{Her}(n,\mathbb{F})$;

$\mathbf{SHer}(n, \mathbb{F}) \subset \mathbb{F}^{n\times n}$ – the set of skew-Hermitian matrices, satisfying $A^{\mathrm{H}} = -A$. In the real case $\mathbf{SHer}(n, \mathbb{R}) \subset \mathbb{R}^{n\times n}$ is the set of skew-symmetric matrices;

We write $A \preceq B$ if $a_{ij} \leq b_{ij}$ for all i, j, where $A = [a_{ij}]$ and $B = [b_{ij}]$ are real matrices of equal size.

H.3 Matrix operators

For $A = [a_{ij}] \in \mathbb{F}^{m\times n}$ we denote by

$$\mathrm{vec}(A) = [a_{11}, \ldots, a_{m1}, a_{12}, \ldots, a_{m2}, \ldots, a_{1n}, \ldots, a_{mn}]^\top \in \mathbb{F}^{mn}$$

the column-wise vector representation of A. If A is represented by its columns $a_i \in \mathbb{F}^m$ as $A = [a_1, \ldots, a_n]$ then $\mathrm{vec}(A) = \left[a_1^\top, \ldots, a_n^\top\right]^\top$. When the size (m, n) of A is essential, we also write $\mathrm{vec}(A)$ as $\mathrm{vec}_{m,n}(A)$ ($\mathrm{vec}_n = \mathrm{vec}_{n,n}$) . We also consider $\mathrm{vec}_{m,n}$ as a linear operator $\mathbb{F}^{m\times n} \to \mathbb{F}^{mn}$;

$\mathrm{vec}_{m,n}^{-1} : \mathbb{F}^{mn} \to \mathbb{F}^{m\times n}$ – the inverse of $\mathrm{vec}_{m,n} : \mathbb{F}^{m\times n} \to \mathbb{F}^{mn}$, $\mathrm{vec}_n^{-1} = \mathrm{vec}_{n,n}^{-1}$;

$P_{m,n} \in \mathbb{R}^{mn\times mn}$ – the vec-permutation matrix such that

$$\mathrm{vec}(M^\top) = P_{m,n}\mathrm{vec}(M)$$

for $M \in \mathbb{F}^{m\times n}$, $P_{n^2} = P_{n,n}$;

$\mathrm{row}_{m,n}(A) = (\mathrm{vec}(A^\top))^\top = [\alpha_1, \ldots, \alpha_m] = (\mathrm{vec}(A))^\top P_{n,m} \in \mathbb{F}^{1\times mn}$ – the row-wise vector representation of the matrix $A = [\alpha_1^\top, \ldots, \alpha_m^\top]^\top \in \mathbb{F}^{m\times n}$, where $\alpha_i \in \mathbb{F}^{1\times n}$ are the rows of A. We also use the shorter notation $\mathrm{row}(A)$;

The matrix representation (or briefly the matrix) of a linear matrix operator $\mathcal{L}$ is denoted by $\mathrm{Mat}(\mathcal{L})$. If $Y = \mathcal{L}(X)$, where X and Y are matrices, then $\mathrm{vec}(Y) = \mathrm{Mat}(\mathcal{L})\mathrm{vec}(X)$.

$\mathbf{Lin}(p, m, n, q, \mathbb{F})$ – the space of linear matrix (Sylvester) operators $\mathcal{M} : \mathbb{F}^{m\times n} \to \mathbb{F}^{p\times q}$. We abbreviate $\mathbf{Lin}(m, n, \mathbb{F}) = \mathbf{Lin}(m, m, n, n, \mathbb{F})$ and

$$\mathbf{Lin}(n, \mathbb{F}) = \mathbf{Lin}(n, n, n, n, \mathbb{F}).$$

An operator $\mathcal{M} \in \mathbf{Lin}(p, m, n, q, \mathbb{F})$ may be represented as

$$\mathcal{M}(X) = \sum_{i=1}^{r} A_i X B_i, \ X \in \mathbb{F}^{m\times n},$$

where $A_i \in \mathbb{F}^{p\times m}$ and $B_i \in \mathbb{F}^{n\times q}$ are given matrices. In this case the matrix of $\mathcal{M}$ is

$$\mathrm{Mat}(\mathcal{M}) = \sum_{i=1}^{r} B_i^\top \otimes A_i\,;$$

$0_{p,m,n,q}$ and $1_{m,n}$ – the zero operator in $\mathbf{Lin}(p, m, n, q, \mathbb{F})$ and the identity operator in $\mathbf{Lin}(m, n, \mathbb{F})$, respectively;

$\mathbf{Lyap}(n, \mathbb{F}) \subset \mathbf{Lin}(n, \mathbb{F})$ – the set of Lyapunov operators $\mathcal{L}$, defined by $(\mathcal{L}(X))^{\mathrm{H}} = \mathcal{L}(X^{\mathrm{H}})$.

The singular values $\sigma_i(\mathcal{M})$ of an operator $\mathcal{M} \in \mathbf{Lin}(p, m, n, q, \mathbb{F})$ are the singular values $\sigma_i(M)$ of its matrix representation M, i.e., $\sigma_i(\mathcal{M}) = \sigma_i(M)$. For operators $\mathcal{L} \in \mathbf{Lyap}(n, \mathbb{F})$ we also define Lyapunov singular values $\widetilde{\sigma}_i(\mathcal{L})$.

H.4 Norms

$\|\cdot\|$ – a norm in $\mathbb{F}^n$ or $\mathbb{F}^{m\times n}$;

$\|x\|_p$ – a Hölder p-norm of $x \in \mathbb{F}^n$,

$$\|x\|_p = \left(\sum_{i=1}^{n} |x_i|^p\right)^{1/p}, \ p \geq 1.$$

In particular we have

$$\|x\|_1 = \sum_{i=1}^{n} |x_i|, \ \|x\|_2 = \left(\sum_{i=1}^{n} |x_i|^2\right)^{1/2}, \ \|x\|_\infty = \max\left\{|x_i| : 1 \leq i \leq n\right\};$$

$\|A\|_{p,q}$ – an induced norm of the matrix $A \in \mathbb{F}^{m\times n}$:

$$\|A\|_{p,q} = \max\left\{\|Ax\|_q : x \in \mathbb{F}^n, \ \|x\|_p = 1\right\}, \ p, q \geq 1.$$

We set $\|A\|_p = \|A\|_{p,p}$;

$\|A\|_{\mathrm{F}}$ – the Frobenius norm of $A = [a_{ij}] \in \mathbb{F}^{m\times n}$,

$$\|A\|_{\mathrm{F}} = \left(\mathrm{tr}(A^{\mathrm{H}}A)\right)^{1/2} = \left(\sum_{i,j=1}^{m,n} |a_{ij}|^2\right)^{1/2};$$

$|A| = [|a_{ij}|] \in \mathbb{R}_+^{m\times n}$ – the matrix absolute value of $A = [a_{ij}] \in \mathbb{F}^{m\times n}$;

$\|a\| = [\|a_1\|, \ldots, \|a_r\|]^\top \in \mathbb{R}_+^r$ – a generalized norm of the r-tuple

$$a = (a_1, \ldots, a_r),$$

where a_i are vectors or matrices. When all a_i are scalars, then the generalized norm agrees with the vector absolute value $|a|$;

For a linear or additive matrix operator $\mathcal{M} : \mathbb{F}^{m\times n} \to \mathbb{F}^{p\times q}$ we denote

$$\|\mathcal{M}\|_p = \max\{\|\mathcal{M}(X)\|_p : \|X\|_p = 1\},$$

where $p \in [1, \infty)$ of $p = \mathrm{F}$;

$\mathcal{B}_\rho(a) = \{x \in \mathbb{F}^n : \|x - a\| \leq \rho\} \subset \mathbb{F}^n$ – a closed ball, centered at $a \in \mathbb{F}^n$ and of radius $\rho \geq 0$. The same notation is used for a generalized ball $\mathcal{B}_\rho(a) = \{x \in \mathbb{F}^n : |x - a| \preceq \rho\} \subset \mathbb{F}^n$, centered at a and of generalized radius $\rho \in \mathbb{R}_+^n$.

H.5 Perturbation analysis

$\mathcal{A}, \mathcal{X}, \mathcal{Y}$ – (subsets of) linear finite dimensional spaces, isomorphic to $\mathbb{F}^p$, $\mathbb{F}^q$, $\mathbb{F}^r$, respectively. An example is $\mathcal{X} = \mathbb{F}^q$ or $\mathcal{X} = \mathbb{F}^{m \times n} \simeq \mathbb{F}^{mn}$. Typically $\mathcal{A}$ is the space of data and $\mathcal{X}$ is the space of results of a given problem;

$\| \cdot \|_{\mathcal{X}}$ – a norm in $\mathcal{X}$;

$\| \cdot \|_{\mathcal{X},\mathcal{Y}}$ – a norm in the space of linear operators $\mathcal{X} \to \mathcal{Y}$;

$C(\mathcal{X}, \mathcal{Y})$ – the space of continuous functions $\mathcal{X} \to \mathcal{Y}$;

Φ (or φ) : $\mathcal{A} \to \mathcal{X}$ – a function, defining a problem with data A (or a) $\in \mathcal{A}$ and result $X = \Phi(A) \in \mathcal{X}$ or $x = \varphi(a) \in \mathcal{X}$. The data A is usually a collection $(A_1, \dots, A_r)$ of matrices A_i;

$F : \mathcal{A} \times \mathcal{X} \to \mathcal{Y}$ – a continuous function, defining a computational problem with data A and result X via the equation $F(A, X) = 0$;

$\delta A \in \mathcal{A}$ and $\delta X \in \mathcal{X}$ – perturbations in $A \in \mathcal{A}$ and $X \in \mathcal{X}$, such that $X + \delta X$ is a solution of a perturbed problem with data $A + \delta A$;

$\delta_X = \|\delta X\|$ – an absolute norm perturbation in $X \in \mathcal{X}$;

$\varepsilon_X = \|\delta X\| / \|X\|$ – a relative norm perturbation in $X \neq 0$;

Ψ – a perturbation operator, defined via

$$\Psi(A, E) = \Phi(A + E) - \Phi(A).$$

With this notation we have $\delta X = \Psi(A, \delta A)$;

$\|\delta X\| \leq f(\|\delta A\|)$ or $\|\delta X\| \leq f(|\!|\!|\delta A|\!|\!|)$ – a perturbation bound for a problem $X = \Phi(A)$. Obtaining such bounds is the goal of perturbation analysis;

$F_X(A_0, X_0) : \mathcal{X} \to \mathcal{Y}$ – the partial Fréchet derivative of the mapping $F : \mathcal{A} \times \mathcal{X} \to \mathcal{Y}$ in X, evaluated at the point (A_0, X_0). Thus $F_X(A_0, X_0)$ is a linear operator from $\mathcal{X}$ to $\mathcal{Y}$. The partial Fréchet derivative of F in A at (A_0, X_0) is denoted $F_A(A_0, X_0)$ and it is a linear operator from $\mathcal{A}$ to $\mathcal{Y}$. Often the partial Fréchet derivatives are abbreviated as F_X and F_A;

$F_X(A_0, X_0)(Y) \in \mathcal{Y}$ and $F_A(A_0, X_0)(Z) \in \mathcal{Y}$ – the images of $Y \in \mathcal{X}$ and $Z \in \mathcal{A}$ under the linear mappings $F_X(A_0, X_0)$ and $F_A(A_0, X_0)$.

$\Pi : \mathcal{X} \to \mathcal{X}$ or $\Pi : \mathcal{X} \times \mathcal{A} \to \mathcal{X}$ – a (nonlinear) operator which is locally equivalent to the perturbation analysis problem. Usually the perturbation problem is rewritten as an operator equation $\delta X = \Pi(\delta X, \delta A)$. It is further shown that Π has a fixed point in a set $\mathcal{B} \subset \mathcal{X}$ of diameter $f(\|\delta A\|) = O(\|\delta A\|)$, $\delta A \to 0$. As a result a perturbation bound $\|\delta X\| \leq f(\|\delta A\|)$ follows.

H.6 Other notation

We denote by eps the roundoff unit of the finite precision arithmetic (the floating point computing environment in particular). For many computer platforms eps is of order 10^{-16}, see also [174].

The notation $m|n$ (m divides n) means that $m, n, n/m \in \mathbb{N}$.

The symbol $:=$ stands for "equal by definition" and $\square$ marks the end of proofs.

Bibliography

[1] E. Anderson, Z. Bai, C.H. Bischof, J.M. Demmel, J.J. Dongarra, J.J. Du Croz, A. Greenbaum, S.J. Hammarling, A. McKenney, S. Ostrouchov, and D.C. Sorensen. *LAPACK Users' Guide.* SIAM, Philadelphia, third edition, 1999.

[2] V. Angelova, M. Konstantinov, D. Gu, and P. Petkov. Perturbation bounds for general coupled matrix Riccati equations. *J. Math. Game Theory Alg.*, 2003. To appear.

[3] R. Aripirala and V.L. Syrmos. Sensitivity analysis of stable generalized Lyapunov equations. In *Proc. of the 32nd IEEE Conf. on Decision and Control*, pages 3144–3149, San Antonio, 1993.

[4] R. Aripirala and V.L. Syrmos. Sensitivity analysis and computable bounds for the generalized algebraic Riccati equation. In *Proc. of the 1994 Amer. Control Conf.*, pages 2680–2684, Baltimore, 1994.

[5] T. Başar and P. Bernhard. *$\mathcal{H}_\infty$-Optimal Control and Related Minimax Design Problems: A Dynamic Game Approach.* Systems and Control: Foundations and Applications. Birkhäuser, Boston, 1991.

[6] Y. Bar–Ness and G. Langholz. The solution of the matrix equation $XC - BX = D$ as an eigenvalue problem. *Int. J. Syst. Sci.*, 8:385–392, 1977.

[7] V. Bargmann, D. Montgomery, and M. J. Von Neumann. Solutions of linear systems of high order. Technical Report NORD-9596, Navy B. Ord., 1946. Also in J. V. Neumann. *Collected Works* (A. H. Taut, Ed.), Pergamon Press, London, 1963, vol. 5, pp. 421-478.

[8] J.B. Barlow, M.M. Monahemi, and D.P. O'Leary. Constrained matrix Sylvester equations. *SIAM J. Matrix Anal. Appl.*, 13:1–9, 1992.

[9] S. Barnett. *Matrices in Control Theory.* Van Nostrand Reinhold Co., London, 1971.

[10] S. Barnett. *Introduction to Mathematical Control Theory.* Oxford Univ. Press, Oxford, 1975.

[11] S. Barnett and C. Storey. Some applications of the Liapunov matrix equation. *J. Inst. Math. Appl.*, 4:33–42, 1968.

[12] S. Barnett and C. Storey. *Matrix Methods in Stability Theory.* Nelson, London, 1970.

[13] A.Y. Barraud. A numerical algorithm to solve $A^TXA - X = Q$. *IEEE Trans. Automat. Control*, AC-22:883–885, 1977.

[14] A.Y. Barraud. Comments on "The numerical solution of $A^TQ + QA = -C$". *IEEE Trans. Automat. Control*, AC-24:671–672, 1979.

[15] R.H. Bartels and G.W. Stewart. Algorithm 432: Solution of the matrix equation $AX + XB = C$. *Comm. ACM*, 15:820–826, 1972.

[16] H. Baumgärtel. *Analytic Perturbation Theory for Matrices and Operators.* Birkhäser Verlag, Basel, 1985.

[17] A.N. Beavers and E.D. Denman. A new solution method for the Lyapunov matrix equation. *SIAM J. Appl. Math.*, 29:416–421, 1975.

[18] P.R. Bélanger and T.P. McGillivray. Computational experience with the solution of the matrix Lyapunov equation. *IEEE Trans. Automat. Control*, AC-21:799–800, 1976.

[19] R. Bellman. Kronecker products and the second method of Lyapunov. *Mathematische Nach.*, 20:17–19, 1959.

[20] R. Bellman. *Introduction to Matrix Analysis.* McGraw-Hill, New York, second edition, 1970. Reprinted by SIAM Publications, 1997.

[21] P. Benner. *Contributions to the Numerical Solution of Algebraic Riccati Equations and Related Eigenvalue Problems.* Logos-Verlag, Berlin, 1997.

[22] P. Benner. Computational methods for linear-quadratic optimization. *Rend. Circ. Mat. Palermo, ser. II*, pages 21–56, 1999.

[23] P. Benner, R. Byers, V. Mehrmann, and H. Xu. Numerical methods for linear quadratic and H_∞ control problems. In G. Picci and D.S. Gillian, editors, *Dynamical Systems, Control, Coding, Computer Vision*, volume 25 of *Progress in Systems and Control Theory*, pages 203–222, Basel, 1999. Birkhäuser Verlag.

[24] P. Benner, R. Byers, V. Mehrmann, and H. Xu. Numerical solution of linear-quadratic control problems for descriptor systems. In *Proc. IEEE Conf. on Computer Aided Control Systems Design*, Hawaii, 1999. CD Rom.

[25] C.S. Berger. A numerical solution of the matrix equation $P = \phi P \phi^t + S$. *IEEE Trans. Automat. Control*, AC-16:381–382, 1971.

[26] A. Berman and R.J. Plemmons. *Nonnegative Matrices in the Mathematical Sciences*. SIAM, Philadelphia, 1994.

[27] P. Bernstein and W. Haddad. LQC control with H_∞ performance bound: A Riccati equation approach. *IEEE Trans. Automat. Control*, 34:293–305, 1989.

[28] R. Bhatia. Matrix factorizations and their perturbations. *Linear Algebra Appl.*, 197-198:245–276, 1994.

[29] G. Birkhoff and S. Mac Lane. *Modern Algebra*. Mackmillan Publ., New York, 1977.

[30] Å. Björck. Component-wise perturbation analysis and error bounds for linear least squares solutions. *BIT*, 31:238–244, 1991.

[31] C. Blendinger, V. Mehrmann, A. Steinbrecher, and R. Unger. Numerical simulation of train traffic in large networks via time-optimal control. Preprint 722, Institut für Mathematik, TU Berlin, 2001.

[32] D. Boley and B. Datta. Numerical methods for linear control systems. In C. Byrness et al., editor, *Systems and Control in 21 Century*, volume 22 of *Progress in Systems and Control Theory*, pages 51–74. Birkhäuser, 1997.

[33] T.L. Boullion and G.D. Poole. A characterization of the general solution of the matrix equation $AX + XB = C$. *Industrial Math.*, 20:91–95, 1970.

[34] G.E. Bredon. *Topology and Geometry*. Springer-Verlag, New York, 1993.

[35] R.S. Bucy. Structural stability for the Riccati equation. *SIAM J. Cont. Optim.*, 13:749–753, 1975.

[36] W. Bunse and A. Bunse-Gerstner. *Numerische Lineare Algebra*. Teubner, Stuttgart, 1985. In German.

[37] A. Bunse-Gerstner, R. Byers, and V. Mehrmann. Numerical methods for algebraic Riccati equations. In S. Bittanti, editor, *Proc. Workshop on the Riccati Equation in Control, Systems, and Signals; Como, Italy*, pages 107–116. Pitagora Editrice, Bologna, Italy, 1989.

[38] R. Byers. A LINPACK-style condition estimator for the equation $AX - XB^T = C$. *IEEE Trans. Automat. Control*, AC-29:926–928, 1984.

[39] R. Byers. Numerical condition of the algebraic Riccati equation. *Contemp. Math.*, 47:35–49, 1985.

[40] R. Byers and S. Nash. On the singular "vectors" of the Lyapunov operator. *SIAM J. Algebraic Discrete Methods*, 8:59–66, 1987.

[41] D. Carlson and R. Loewy. On ranges of Lyapunov transformations. *Linear Algebra Appl.*, 8:237–248, 1974.

[42] F. Chaitin-Chatelin and V. Frayssé. Elements of a condition theory for the computational analysis of algorithms. In *Iterative Methods in Linear Algebra*, pages 15–25, Amsterdam, 1991. North Holland.

[43] F. Chaitin-Chatelin and V. Frayssé. Qualitative computing. Technical report, CERFACS, Orsay, France, 1991.

[44] F. Chaitin-Chatelin and V. Frayssé. *Lectures on Finite Precision Computations*. SIAM, Philadelphia, 1996.

[45] K.-W.E. Chu. Singular value and generalized singular value decompositions and solution of linear matrix equations. *Linear Algebra Appl.*, 88/89:83–98, 1987.

[46] K.-W.E. Chu. Symmetric solutions of linear matrix equations by matrix decompositions. *Linear Algebra Appl.*, 119:35–50, 1989.

[47] A. Czornik and A. Swiernak. On the sensitivity of the coupled discrete-time Lyapunov equation. *IEEE Trans. Automat. Control*, 46:659–664, 2001.

[48] B.N. Datta and Y. Saad. Arnoldi methods for large Sylvester-like matrix equations and an associated algorithm for partial pole assignment algorithm. *Linear Algebra Appl.*, 156:225–244, 1991.

[49] K. Datta. The matrix equation $XA - BX = R$ and its application. *Linear Algebra Appl.*, 109:91–105, 1988.

[50] E.J. Davison and F.T. Man. The numerical solution of $A'Q + QA = -C$. *IEEE Trans. Automat. Control*, AC-13:448–449, 1968.

[51] J.W. Demmel. On condition numbers and the distance to the nearest ill-posed problem. *Numer. Math.*, 51:251–289, 1987.

[52] J.W. Demmel. The probability that a numerical analysis problem is difficult. *Math. Comp.*, 50:449–480, 1988.

[53] J.W. Demmel. Nearest defective matrices and the geometry of ill-conditioning. In *Reliable Numerical Computation*, pages 35–55, Oxford, 1990. Clarendon Press.

[54] J.W. Demmel. *Applied Numerical Linear Algebra*. SIAM, Philadelphia, 1997.

[55] V. Demyanov. Fixed point theorems in nonsmooth analysis and its applications. *Numer. Funct. Anal. Appl.*, 16:53–109, 1995.

[56] T.E. Djaferis and S.K. Mitter. Exact solution to Lyapunov's equation using algebraic methods. In *Proc. 1976 IEEE Conf. on Decision and Control incl. 15th Symposium on Adaptive Processes*, pages 1194–1200, Clearwater, FL, 1976.

[57] J. Doyle, K. Zhou, K. Glover, and B. Bodenheimer. Mixed H_2 and H_∞ performance objectives: Optimal control. *IEEE Trans. Automat. Control*, 39:1375–1387, 1994.

[58] J.C. Engwerda, A.C.M. Ran, and A.L. Rijkeboer. Necessary and sufficient conditions for the existence of a positive definite solution of the matrix equation $X + A^*X^{-1}A = Q$. *Linear Algebra Appl.*, 186:255–275, 1993.

[59] M.A. Epton. Methods for the solution of $AXD - BXC = E$ and its application in the numerical solution of implicit ordinary differential equations. *BIT*, 20:341–345, 1980.

[60] M. Eslami. *Theory of Sensitivity in Dynamic Systems.* Springer-Verlag, Berlin, 1994.

[61] L.E. Faibusovich. Algebraic Riccati equation and symplectic algebra. *Internat. J. Control*, 43:781–792, 1986.

[62] K.V. Fernando and H. Nicholson. Solution of the Lyapunov equation for the state matrix. *Electron. Lett.*, 17:204–205, 1981.

[63] A. Ferrante and B.C. Levy. Hermitian solutions of the equation $X = Q + NX^{-1}N^*$. *Linear Algebra Appl.*, 247:359–373, 1996.

[64] G. Forsythe and C.B. Moler. *Computer Solution of Linear Algebraic Systems.* Prentice Hall, Englewood Cliffs, N.J., 1967.

[65] P. Gahinet. *Perturbational and Topological Aspects of Sensitivity in Control Theory.* PhD thesis, ECE Dept., Univ. of California, Santa Barbara, CA, 1989.

[66] P. Gahinet and A.J. Laub. Computable bounds for the sensitivity of the algebraic Riccati equation. *SIAM J. Cont. Optim.*, 28:1461–1480, 1990.

[67] P. Gahinet and A.J. Laub. Algebraic Riccati equations and the distance to the nearest uncontrollable pair. *SIAM J. Cont. Optim.*, 30:765–786, 1992.

[68] P.M. Gahinet, A.J. Laub, C.S. Kenney, and G.A. Hewer. Sensitivity of the stable discrete-time Lyapunov equation. *IEEE Trans. Automat. Control*, 35:1209–1217, 1990.

[69] Z. Gajić and M.T.J. Qureshi. *Lyapunov Matrix Equation in System Stability and Control.* Academic Press, San Diego, 1995.

[70] F.R. Gantmacher. *Theory of Matrices*, volume 1. Chelsea, New York, 1959.

[71] F.R. Gantmacher. *Theory of Matrices*, volume 2. Chelsea, New York, 1959.

[72] J.D. Gardiner, A.J. Laub, J.J. Amato, and C.B. Moler. Solution of the Sylvester matrix equation $AXB^T + CXD^T = E$. *ACM Trans. Math. Software*, 18:223–231, 1992.

[73] J.D. Gardiner, M.R. Wette, A.J. Laub, J.J. Amato, and C.B. Moler. Algorithm 705: A Fortran-77 software package for solving the Sylvester matrix equation $AXB^T + CXD^T = E$. *ACM Trans. Math. Software*, 18:232–238, 1992.

[74] W. Gautschi. On the condition of algebraic equations. *Numer. Math.*, 21:405–424, 1973.

[75] W. Gautschi. Questions of numerical condition related to polynomials. In C. De Boor and G.H. Golub, editors, *Recent Advances in Numerical Analysis*, pages 45–72. Academic Press, New York, 1978.

[76] A.J. Geurts. A contribution to the theory of condition. *Numer. Math.*, 39:85–96, 1982.

[77] A.R. Ghavimi and A.J. Laub. Backward error, sensitivity and refinement of computed solutions of algebraic Riccati equations. *Numer. Alg. Appl.*, 2:29–49, 1995.

[78] A.R. Ghavimi and A.J. Laub. Computation of approximate null vectors of Sylvester and Lyapunov operators. *IEEE Trans. Automat. Control*, 40:387–391, 1995.

[79] W. Givens. Elementary divisors and some properties of the Lyapunov mapping $X \to AX + XA^*$. Technical Report ANL-6456, Argonne Nat. Lab., 1961.

[80] I. Gohberg and I. Koltracht. Mixed, componentwise, and structured condition numbers. *SIAM J. Matrix Anal. Appl.*, 14:688–704, 1993.

[81] I. Gohberg, P. Lancaster, and L. Rodman. On Hermitian solutions of the symmetric algebraic Riccati equation. *SIAM J. Cont. Optim.*, 24:1323–1334, 1986.

[82] G.H. Golub, S. Nash, and C. Van Loan. A Hessenberg–Schur method for the problem $AX + XB = C$. *IEEE Trans. Automat. Control*, AC-24:909–913, 1979.

[83] G.H. Golub and C.F. Van Loan. *Matrix Computations.* The Johns Hopkins Univ. Press, Baltimore, third edition, 1996.

[84] A. Graham. *Kronecker Products and Matrix Calculus with Applications.* Wiley, New York, 1981.

[85] E.A. Grebenikov and Yu.A. Ryabov. *Constructive Methods for Analysis of Nonlinear Systems.* Nauka, Moscow, 1979. In Russian.

[86] R.P. Guidorzi. Transformation approaches in the solution of the matrix equation $A^T X + XB = P$. *IEEE Trans. Automat. Control*, AC-17:377–379, 1972.

[87] C. h. Chen. Perturbation analysis for solutions of algebraic Riccati equations. *J. Comput. Math.*, 6:336–347, 1988.

[88] P. Hagander. Numerical solution of $A^T S + SA + Q = 0$. *Inform. Sci.*, 4:35–50, 1972.

[89] W.W. Hager. Condition estimates. *SIAM J. Sci. Statist. Comput.*, 5:311–316, 1984.

[90] S. Hammarling. Numerical solution of the discrete-time, convergent, non-negative definite Lyapunov equation. *Syst. Contr. Lett.*, 17:137–139, 1991.

[91] S.J. Hammarling. Numerical solution of the stable, non-negative definite Lyapunov equation. *IMA J. Numer. Anal.*, 2:303–323, 1982.

[92] P.C. Hansen. *Rank-Deficient and Discrete Ill-Posed Problems: Numerical Aspects of Linear Inversion.* SIAM, Philadelphia, 1998.

[93] C. He. On the distance to uncontrollability and the distance to instability and their relation to some condition numbers in control. *Numer. Math.*, 76:463–477, 1997.

[94] H.V. Henderson and S.R. Searle. The vec-permutation matrix, the vec operator and Kronecker products: A review. *Lin. Multilin. Alg.*, 9:271–288, 1981.

[95] G. Hewer and C. Kenney. The sensitivity of the stable Lyapunov equation. *SIAM J. Cont. Optim.*, 26:321–344, 1988.

[96] E. Hewitt and K. Stromberg. *Real and Abstract Analysis.* Springer-Verlag, New York, 1965. Third Printing 1975.

[97] D.J. Higham and N.J. Higham. Backward error and condition of structured linear systems. *SIAM J. Matrix Anal. Appl.*, 13:162–175, 1992.

[98] D.J. Higham and N.J. Higham. Componentwise perturbation theory for linear systems with multiple right-hand sides. *Linear Algebra Appl.*, 174:111–129, 1992.

[99] N. J. Higham. Perturbation theory and backward error for $AX - XB = C$. *BIT*, 33:124–136, 1993.

[100] N.J. Higham. A survey of componentwise perturbation theory in numerical linear algebra. In W. Gautchi, editor, *Mathematics of Computation 1943-1993: A Half Century of Computational Mathematics*, volume 48 of *Proc. of Symposia in Applied Mathematics*, pages 49–77. Amer. Math. Soc., Providence, RI, USA, 1994.

[101] N.J. Higham. *Accuracy and Stability of Numerical Algorithms.* SIAM, Philadelphia, second edition, 2002.

[102] N.J. Higham, M. Konstantinov, V. Mehrmann, and P. Petkov. Sensitivity of computational control problems. *Control Systems Mag.*, 6, 2003. (To be published).

[103] A.S. Hodel. Recent applications of the Lyapunov equation in control theory. In R. Beauwens and P. de Groen, editors, *Iterative Methods in Linear Algebra*, pages 217–227. Elsevier (North-Holland), Amsterdam, 1992.

[104] A.S. Hodel and K.R. Poolla. Heuristic approaches to the solution of very large sparse Lyapunov and algebraic Riccati equations. In *Proc. 27th IEEE Conf. on Decision and Control*, pages 2217–2222, Austin, TX, 1988.

[105] A.S. Hodel and K.R. Poolla. Numerical solution of very large, sparse Lyapunov equations through approximate power iteration. In *Proc. 29th IEEE Conf. on Decision and Control*, pages 291–296, Honolulu, HI, 1990.

[106] R.A. Horn and C.R. Johnson. *Matrix Analysis.* Cambridge Univ. Press, Cambridge, 1985.

[107] R.A. Horn and C.R. Johnson. *Topics in Matrix Analysis.* Cambridge Univ. Press, Cambridge, 1991.

[108] V. Ionescu, C. Oară, and M. Weiss. *Generalized Riccati Theory and Robust Control: A Popov Function Approach.* John Wiley and Sons Inc., Chichester, UK, 1999.

[109] A. Jameson. Solution of the equation $AX + XB = C$ by inversion of a $M \times M$ or $N \times N$ matrix. *SIAM J. Appl. Math.*, 16:1020–1023, 1968.

[110] E.A. Jonckheere. New bound on the sensitivity of the solution of the Lyapunov equation. *Linear Algebra Appl.*, 60:57–64, 1984.

[111] W. Anderson Jr., T. Morley, and G. Trapp. Positive solutions to $X = A - BX^{-1}B^*$. *Linear Algebra Appl.*, 134:53–62, 1990.

[112] B. Kågström. A perturbation analysis of the generalized Sylvester equation. *SIAM J. Matrix Anal. Appl.*, 15:1045–1060, 1994.

[113] B. Kågström and P. Poromaa. Distributed and shared memory block algorithms for the triangular Sylvester equation with sep^{-1} estimators. *SIAM J. Matrix Anal. Appl.*, 13:90–101, 1992.

[114] B. Kågström and L. Westin. Generalized Schur methods with condition estimators for solving the generalized Sylvester equation. *IEEE Trans. Automat. Control*, 34:745–751, 1989.

[115] W. Kahan. Numerical linear algebra. *Canadian Math. Bull.*, 9:757–801, 1966.

[116] W. Kahan. A survey of error analysis. In *Proc. IFIP Congress*, pages 1241–1239, Amsterdam, 1971.

[117] L. V. Kantorovich and G. P. Akilov. *Functional Analysis in Normed Spaces.* Pergamon, New York, 1964.

[118] V. Kapila and W. Haddad. A multivariable extension of the Tsypkin criterion using a Lyapunov-function approach. *IEEE Trans. Automat. Control*, 41:149–159, 1996.

[119] T. Kato. *Perturbation Theory for Linear Operators.* Springer-Verlag, Berlin, second edition, 1980. Reprint 1995.

[120] C. Kenney and G. Hewer. The sensitivity of the algebraic and differential Riccati equations. *SIAM J. Cont. Optim.*, 28:50–69, 1990.

[121] C. Kenney and A.J. Laub. Condition estimates for matrix functions. *SIAM J. Matrix Anal. Appl.*, 10:191–209, 1989.

[122] A. Kielbasinski and H. Schwetlick. *Numerische Lineare Algebra.* Verlag Harri Deutsch, Berlin, 1988.

[123] M. Konstantinov and V. Angelova. Sensitivity analysis of the differential matrix Riccati equation based on the associated linear differential system. *Adv. Comp. Math.*, 7:295–301, 1997.

[124] M. Konstantinov, V. Angelova, P. Petkov, D. Gu, and V. Tsachouridis. Perturbation analysis of coupled matrix Riccati equations. *Linear Algebra Appl.*, 359:197–218, 2002.

[125] M. Konstantinov, V. Mehrmann, and P. Petkov. On properties of general Sylvester and Lyapunov operators. *Linear Algebra Appl.*, 312:35–71, 2000.

[126] M. Konstantinov, V. Mehrmann, P. Petkov, and D. Gu. Structural properties and parametrizations of Lyapunov and Lyapunov–like operators. Technical Report 99–6, Dept. of Engineering, Leicester Univ., Leicester, UK, 1999.

[127] M. Konstantinov, V. Mehrmann, P. Petkov, and D.W. Gu. A general framework for the perturbation theory of matrix equations. Technical Report Prep. 760, Inst. Math., TU-Berlin, 2002.

[128] M. Konstantinov, S. Patarinski, P. Petkov, and N. Christov. Mutual observation in linear systems and synthesis under incomplete information. *Math. Balkanica*, 6:88–98, 1976. in Russian.

[129] M. Konstantinov, P. Petkov, and V. Angelova. Sensitivity of general discrete algebraic Riccati equations. In *Proc. 28 Spring Conf. of Union of Bulgar. Mathematicians*, pages 128–136, Montana, Bulgaria, 1999.

[130] M. Konstantinov, P. Petkov, V. Angelova, and D. Gu. Perturbation analysis of fractional affine matrix equations. Technical Report 98–12, Dept. of Engineering, Leicester Univ., Leicester, UK, 1998.

[131] M. Konstantinov, P. Petkov, and D.W. Gu. Improved perturbation bounds for general quadratic matrix equations. *Numer. Func. Anal. Optim.*, 20:717–736, 1999.

[132] M. Konstantinov, P. Petkov, D.W. Gu, and V. Mehrmann. Sensitivity of general Lyapunov equations. Technical Report 98–15, Dept. of Engineering, Leicester Univ., Leicester, UK, 1998.

[133] M. Konstantinov, P. Petkov, D.W. Gu, and I. Postlethwaite. Numerical issues in linear control. part i. Technical Report 93–65, Dept. of Engineering, Leicester Univ., Leicester, UK, 1993.

[134] M. Konstantinov, P. Petkov, D.W. Gu, and I. Postlethwaite. Perturbation techniques for linear control problems. Technical Report 95-7, Dept. of Engineering, Leicester Univ., Leicester, UK, 1995.

[135] M. Konstantinov, P. Petkov, D.W. Gu, and I. Postlethwaite. Perturbation analysis in finite dimensional spaces. Technical Report 96–18, Dept. of Engineering, Leicester Univ., Leicester, UK, 1996.

[136] M. Konstantinov, P. Petkov, A. Linnemann, J. Kawelke, D.W. Gu, and I. Postlethwaite. Sensitivity of system norms. *Int. J. Control*, 72:84–95, 1998.

[137] M. Konstantinov, P. Petkov, V. Mehrmann, and D. Gu. Additive matrix operators. In *Proc. 30 Spring Conf. of Union of Bulgar. Mathematicians*, pages 169–175, Borovetz, Bulgaria, 2001.

[138] M. Konstantinov, I. Popchev, and V. Angelova. Conditioning and sensitivity of the difference matrix Riccati equation. In *Proc. Amer. Control Conf. ACC'95*, volume 1, pages 466–467, Seattle, 1995.

[139] M. Konstantinov, I. Popchev, and V. Angelova. On the sensitivity estimation of the matrix differential Riccati equation. In *Proc. Third Europ. Control Conf. ECC'95*, volume 3, pages 2084–2086, Rome, 1995.

[140] M. Konstantinov, M. Stanislavova, and P. Petkov. Perturbation bounds and characterisation of the solution of the associated algebraic Riccati equation. *Linear Algebra Appl.*, 285:7–31, 1998.

[141] M. M. Konstantinov, P. Hr. Petkov, D.-W. Gu, and I. Postlethwaite. Nonlocal sensitivity analysis of the eigensystem of a matrix with distinct eigenvalues. *Numer. Funct. Anal. and Optim.*, 18, 1997.

[142] M.M. Konstantinov, N.D. Christov, and P.Hr. Petkov. Perturbation analysis of linear control problems. In *Prepr. 10 IFAC Congress*, volume 9, pages 16–21, Munich, 1987.

[143] M.M. Konstantinov and G.B. Pelova. Sensitivity of the solutions to differential matrix Riccati equations. *IEEE Trans. Automat. Control*, 36:213–215, 1991.

[144] M.M. Konstantinov, P.H. Petkov, and N.D. Christov. Conditioning of the continuous-time H_∞ optimisation problem. In *Proc. Third Europ. Control Conf. ECC'95*, volume 1, pages 613–618, Rome, 1995.

[145] M.M. Konstantinov and P.Hr. Petkov. A note on "Perturbation theory for alegbraic Riccati equations". *SIAM J. Matrix Anal. Appl.*, 21:327, 1999.

[146] M.M. Konstantinov and P.Hr. Petkov. Condition and error estimates in the solution of Lyapunov and Riccati equations. *Math. Balkanica (N.S.)*, 15:139–153, 2001.

[147] M.M. Konstantinov and P.Hr. Petkov. The method of splitting operators and Lyapunov majorants in perturbation linear algebra and control. *Numer. Func. Anal. Optim.*, 23:529–572, 2002.

[148] M.M. Konstantinov, P.Hr. Petkov, and N.D. Christov. The associated algebraic Riccati equation. In *Proc. Third Internat. Conf. on Systems Engr.*, pages 530–537, Wright State Univ., Dayton, 1984.

[149] M.M. Konstantinov, P.Hr. Petkov, and N.D. Christov. Perturbation analysis of the continuous and discrete matrix Riccati equations. In *Proc. 1986 Amer. Control Conf.*, pages 636–639, Seattle, 1986.

[150] M.M. Konstantinov, P.Hr. Petkov, and N.D. Christov. Perturbation analysis of matrix quadratic equations. *SIAM J. Sci. Statist. Comput.*, 11:1159–1163, 1990.

[151] M.M. Konstantinov, P.Hr. Petkov, and N.D. Christov. Perturbation analysis of the discrete Riccati equation. *Kybernetica* (Prague), 29:18–29, 1993.

[152] L. Kronecker. Algebraische Reduction der Schaaren bilineare Formen. *S.B. Akad. Berlin*, pages 1225–1237, 1880.

[153] V. Kučera. The matrix equation $AX + XB = C$. *SIAM J. Appl. Math.*, 26:15–25, 1974.

[154] V. Kučera. Algebraic Riccati equation: Symmetric and definite solutions. In S. Bittanti, editor, *Lecture Notes of the Workshop on "The Riccati Equation in Control, Systems and Signal"*, pages 73–75. Pitagora Editrice, Bologna, Italy, 1989.

[155] P. Lancaster. Explicit solutions of linear matrix equations. *SIAM Rev.*, 12:544–566, 1970.

[156] P. Lancaster and L. Rodman. *The Algebraic Riccati Equation.* Oxford Univ. Press, Oxford, 1995.

[157] P. Lancaster and M. Tismenetsky. *The Theory of Matrices.* Academic Press, Orlando, FL, second edition, 1985.

[158] J.P. LaSalle and S. Lefschetz. *Stability by Liapunov's Direct Method with Applications.* Academic Press, New York, 1961.

[159] A.J. Laub. Invariant subspace methods for the numerical solution of Riccati equations. In S. Bittanti, A.J. Laub, and J.C. Willems, editors, *The Riccati Equation*, pages 163–196. Springer-Verlag, Berlin, 1991.

[160] D.K. Lika and Yu.A. Ryabov. *Iterative Methods and Lyapunov Majorant Equations in Non-Linear Oscillation Theory.* Shtiinca, Kishinev, 1974. In Russian.

[161] W. Lin and J. Sun. Perturbation analysis of the periodic discrete-time algebraic Riccati equation. *SIAM J. Matrix Anal. Appl.*, 24:411–438, 2002.

[162] A.M. Lyapunov. *General Problem of Stability of Motion.* Gostehizdat, Moscow, 1950. In Russian.

[163] A.M. Lyapunov. A finite series solution of the matrix equation $AX - XB = C$. *SIAM J. Appl. Math.*, 14:490–495, 1966.

[164] E.-C. Ma. A finite series solution of the matrix equation $AX - XB = C$. *SIAM J. Appl. Math.*, 14:490–495, 1966.

[165] C.C. Mac Duffee. *The Theory of Matrices.* Chelsea, New York, 1946.

[166] R. Mathias. Condition estimation for matrix functions via the Schur decomposition. *SIAM J. Matrix Anal. Appl.*, 16:565–578, 1995.

[167] V. Mehrmann. *The Autonomous Linear Quadratic Control Problem, Theory and Numerical Solution.* Number 163 in Lecture Notes in Control and Information Sciences. Springer-Verlag, Heidelberg, 1991.

[168] V. Mehrmann and H. Xu. Numerical methods in control. *J. Comput. Appl. Math.*, 123:371–394, 2000.

[169] B. Meini. Efficient computation of the extreme solutions of $X + A^*X^{-1}A = Q$ and $X - A^*X^{-1}A = Q$. *Math. Comp.*, 71:1189–1204, 2001.

[170] B. Molinari. Algebraic solution of matrix linear equations in control theory. *Proc. IEE*, 116:1748–1754, 1969.

[171] M. Mrabti and A. Hmamed. Bounds for the solution of the Lyapunov matrix equation–A unified approach. *Syst. Contr. Lett.*, 18:73–81, 1992.

[172] J. Von Neumann and H. Goldstein. Numerical inverting of matrices of high order. *Bull. AMS*, 53:1021–1099, 1947.

[173] J. Ortega and W. Rheinboldt. *Iterative Solution of Nonlinear Equations in Several Variables*, volume 30 of *Classics in Applied Mathematics.* SIAM, Philadelphia, 2000.

[174] M. Overton. *Computing with IEEE Floating Point Arithmetic.* SIAM, Philadelphia, 2001.

[175] I.S. Pace and S. Barnett. Comparison of numerical methods for solving Lyapunov matrix equations. *Internat. J. Control*, 15:907–915, 1972.

[176] R.V. Patel and M. Toda. On norm bounds for algebraic Riccati and Lyapunov equations. *IEEE Trans. Automat. Control*, AC-23:87–88, 1978.

[177] S. Peng and C.E. de Souza. Bounds on the solution of the algebraic matrix Riccati equation under perturbations in the coefficients. *Syst. Contr. Lett.*, 15:175–181, 1990.

[178] S. Peng and C.E. de Souza. On bounds for perturbed discrete-time algebraic Riccati equations. In H. Kimura and S. Kodama, editors, *Mathematical Theory of Systems, Control, Networks and Signal Processing. Proc. of the International Symposium MTNS-91, Kobe, Japan, June 1991*, pages 9–14. Mita Press, Tokyo, 1992.

[179] P.H. Petkov, M.M. Konstantinov, and V. Mehrmann. DGRSVX and DMSRIC: Fortan 77 subroutines for solving continuous-time matrix algebraic Riccati equations with condition and accuracy estimates. Technical Report SFB393/98-16, Fak. für Mathematik, TU Chemnitz, Chemnitz, Germany, 1998.

[180] P.Hr. Petkov, N.D. Christov, and M.M. Konstantinov. A new approach to the perturbation analysis of linear control problems. In *Prepr. 11th IFAC Congress*, pages 311–316, Tallin, 1990.

[181] P.Hr. Petkov, N.D. Christov, and M.M. Konstantinov. *Computational Methods for Linear Control Systems.* Prentice-Hall, Hemel Hempstead, 1991.

[182] P.Hr. Petkov, N.D. Christov, and M.M. Konstantinov. Numerical issues in the solution of matrix Riccati equations. In *Proc. Second European Control Conf. ECC'93*, pages 2379–2384, Groningen, 1993.

[183] P.Hr. Petkov, M.M. Konstantinov, D.W. Gu, and I. Postlethwaite. Numerical issues in the solution of continuous-time matrix algebraic Riccati equations. Technical Report 96–13, Dept. of Engineering, Leicester Univ., Leicester, UK, 1996.

[184] A.C.M. Ran and L. Rodman. Perturbation analysis of algebraic matrix Riccati equations. In *Proc. 29th IEEE Conf. on Decision and Control*, pages 1855–1856, Honolulu, HI, 1990.

[185] A.C.M. Ran and L. Rodman. Stable hermitian solutions of discrete algebraic Riccati equations. *Math. Control Signals Systems*, 5:165–193, 1992.

[186] A.C.M. Ran and L. Rodman. Stable solutions of real algebraic matrix Riccati equations. *SIAM J. Cont. Optim.*, 30:63–81, 1992.

[187] W. Rheinboldt. On measures of ill-conditioning for non-linear equations. *Math. Comp.*, 30:104–111, 1976.

[188] J.R. Rice. A theory of condition. *SIAM J. Numer. Anal.*, 3:287–310, 1966.

[189] J.R. Rice. *Matrix Computation and Mathematical Software.* McGraw-Hill, New York, 1981.

[190] J.L Rigal and J. Gaches. On the compatibility of a given solution with the data of a linear system. *J. Assoc. Comput. Mach.*, 14:543–548, 1967.

[191] J. Rohn. New condition numbers for matrices and linear systems. *Computing*, 41:167–169, 1989.

[192] I.G. Rosen and C. Wang. A multilevel technique for the approximate solution of operator Lyapunov and algebraic Riccati equations. *SIAM J. Numer. Anal.*, 32:514–541, 1995.

[193] W.E. Roth. The equations $AX - YB = C$ and $AX - XB = C$ in matrices. *Proc. Amer. Math. Soc.*, 3:392–396, 1952.

[194] D. Rothschild and A. Jameson. Comparison of four numerical algorithms for solving the Lyapunov matrix equation. *Internat. J. Control*, 11:181–198, 1970.

[195] S.M. Rump. Estimation of the sensitivity of linear and nonlinear algebraic problems. *Linear Algebra Appl.*, 153:1–34, 1991.

[196] D.E. Rutherford. On the solution of the matrix equation $AX + XB = C$. *Nederl. Akad. Wetensch. Proc. Ser. A*, 35:53–59, 1932.

[197] H. Shapiro. A survey of canonical forms and invariants under similarity. *Linear Algebra Appl.*, 147:101–167, 1991.

[198] M.A. Shayman. Geometry of the algebraic Riccati equation, Part I. *SIAM J. Cont. Optim.*, 21:375–394, 1983.

[199] M.A. Shayman. Geometry of the algebraic Riccati equation, Part II. *SIAM J. Cont. Optim.*, 21:395–409, 1983.

[200] V. Sima. *Algorithms for Linear Quadratic Optimization.* Marcel Dekker Inc., New York, 1996.

[201] R.D Skeel. Iterative refinement implies numerical stability for Gaussian elimination. *Math. Comp.*, 35:817–832, 1980.

[202] G. Starke and W. Niethammer. SOR for $AX - XB = C$. *Linear Algebra Appl.*, 154–156:355–375, 1991.

[203] P. Stein. On the ranges of two functions of positive definite matrices. *J. Algebra*, 2:350–353, 1965.

[204] P. Stein and A. Pfeffer. On the ranges of two functions of positive definite matrices II. *ICC Bull.*, 6:81–86, 1967.

[205] G.W. Stewart. *Introduction to Matrix Computations.* Academic Press, New York, 1973.

[206] G.W. Stewart and J.-G. Sun. *Matrix Perturbation Theory*. Academic Press, New York, 1990.

[207] T. Stykel. Numerical solution and perturbation theory for generalized Lyapunov equations. *Lin. Alg. Appl.*, 349:155–185, 2002.

[208] J.-G. Sun. Sensitivity analysis of the discrete-time algebraic Riccati equation. Technical Report UMINF 96.08, Dept. of Computing Science, Univ. of Umeå, Umeå, Sweden, 1996.

[209] J.-G. Sun. Backward error for the discrete-time algebraic Riccati equation. *Linear Algebra Appl.*, 259:183–208, 1997.

[210] J.-G. Sun. Residual bounds of approximate solutions of the algebraic Riccati equation. *Numer. Math.*, 76:249–263, 1997.

[211] J.-G. Sun. Perturbation theory for algebraic Riccati equations. *SIAM J. Matrix Anal. Appl.*, 19:39–65, 1998.

[212] J.-G. Sun. Sensitivity analysis of the discrete-time algebraic Riccati equation. *Lin. Alg. Appl.*, 275–276:595–615, 1998.

[213] J.-G. Sun. Condition numbers of algebraic Riccati equations in Frobenius norm. *Lin. Alg. Appl.*, 350:237–261, 2002.

[214] J.J. Sylvester. Sur la solution du cas le plus général des équations linéaires en quantités binaires, c'est-à-dire en quaternions ou en matrices du second ordre. *C.R. Acad. Sci. Paris*, 99:117–118, 1884.

[215] J.J. Sylvester. Sur l'équation en matrices $px = xq$. *C.R. Acad. Sci. Paris*, 99:67–71, 1884.

[216] O. Taussky and H. Wielandt. On the matrix function $AX + X'A'$. *Arch. Rat. Mech. Anal.*, 9:93–96, 1962.

[217] A.J. Telford and J.B. Moore. On the existence of solutions to nonsymmetric algebraic Riccati equations. In S. Bittanti, editor, *Lecture Notes of the Workshop on "The Riccati Equation in Control, Systems and Signal"*, pages 83–86. Pitagora Editrice, Bologna, Italy, 1989.

[218] A.N. Tikhonov. Regularization of incorrectly posed problems. *Soviet. Math. Dokl.*, 4:1624–1627, 1963.

[219] A.N. Tikhonov. Solution of incorrectly formulated problems and the regularization method. *Soviet. Math. Dokl.*, 4:1036–1038, 1963.

[220] A.N. Tikhonov and V.Y. Arsenin. *Solutions of Ill–Posed Problems*. Wiley, New York, 1977.

[221] A.N. Tikhonov, A.V. Goncharov, V.V. Stepanov, and A.G. Yagola. *Numerical Methods for the Solution of Ill–Posed Problems.* Kluwer, Dordrecht, 1995.

[222] A. Trampus. A canonical basis for the matrix transformation $X \rightarrow AX + XB$. *J. Math. Anal. Appl.*, 14:242–252, 1966.

[223] A. Trampus. A canonical basis for the matrix transformation $X \rightarrow AXB$. *J. Math. Anal. Appl.*, 14:153–160, 1966.

[224] L.N. Trefethen and D. Bau III. *Numerical Linear Algebra.* SIAM, Philadelphia, 1997.

[225] C.-C. Tsui. On the solution to matrix equation $TA - FT = LC$ and its applications. *SIAM J. Matrix Anal. Appl.*, 14:33–44, 1993.

[226] A. Turing. Rounding-off errors in matrix processes. *Quart. J. Mech. Appl. Math.*, 1:287–308, 1948.

[227] E. Tyan and P. Bernstein. Anti-windup compensator synthesis for systems with saturation actuators. *Int. J. Robust. Nonlin. Control*, 5:321–337, 1995.

[228] D.S. Watkins. *Fundamentals of Matrix Computations.* John Wiley and Sons Inc., 1991.

[229] J.H.M. Wedderburn. Note on the linear matrix equation. *Proc. Edinburgh Math. Soc.*, 22:49–53, 1904.

[230] R.J. Weidner and R.J. Mulholland. Kronecker product representation for the solution of the general linear matrix equation. *IEEE Trans. Automat. Control*, AC-25:563–564, 1980.

[231] A. Weinmann. *Uncertain Models and Robust Control.* Springer-Verlag, Wien, 1991.

[232] J.H. Wilkinson. *Rounding Errors in Algebraic Processes.* Prentice-Hall, Englewood Cliffs, 1963.

[233] J.H. Wilkinson. *The Algebraic Eigenvalue Problem.* Oxford Univ. Press, Oxford, 1965.

[234] J.H. Wilkinson. Modern error analysis. *SIAM Rev.*, 13:548–568, 1971.

[235] H.K. Wimmer. Explicit solutions of the matrix equation $\sum A^i X D^i = C$. *SIAM J. Matrix Anal. Appl.*, 13:1123–1130, 1992.

[236] W.M. Wonham. *Linear Multivariable Control: A Geometric Approach.* Springer-Verlag, New York, third edition, 1985.

[237] S. Xu. Sensitivity analysis of the algebraic Riccati equations. *Numer. Math.*, 75:121–134, 1996.

[238] X. Zhan. Computing the external positive definite solutions of a matrix equation. *SIAM J. Comput.*, 17:1167–1174, 1996.

[239] K. Ziętak. The Chebyshev solution of the linear matrix equation $AX+YB = C$. *Numer. Math.*, 46:455–478, 1985.

[240] K. Ziętak. On a particular case of the inconsistent linear matrix equation $AX + YB = C$. *Linear Algebra Appl.*, 66:249–258, 1985.

[241] I.E. Ziedan. Explicit solution of the Lyapunov-matrix equation. *IEEE Trans. Automat. Control*, AC-17:379–381, 1972.

Index